Jeep Liberty & Dodge Nitro Automotive Repair Manual

by Len Taylor, Tim Imhoff and John H Haynes

Member of the Guild of Motoring Writers

Models covered:

Jeep Liberty - 2002 through 2012

Dodge Nitro - 2007 through 2011

Does not include information specific to diesel models

(50035-8Y11)

Haynes Group Limited
Haynes North America, Inc.

www.haynes.com

Acknowledgements

Wiring diagrams originated exclusively for Haynes North America, Inc. by Valley Forge Technical Information Services.

A book in the Haynes Automotive Repair Manual Series

ISBN-13: 978-1-62092-346-7
ISBN-10: 1-62092-346-7

Library of Congress Control Number 2018963714

While every attempt is made to ensure that the information in this man-ual is correct, no liability can be accepted by the authors or publishers for loss, damage or injury caused by any errors in, or omissions from, the information given.

Contents

Haynes mechanic and photographer with a 2003 Jeep Liberty

About this manual

Its purpose

The purpose of this manual is to help you get the best value from your vehicle. It can do so in several ways. It can help you decide what work must be done, even if you choose to have it done by a dealer service department or a repair shop; it provides information and procedures for routine maintenance and servicing; and it offers diagnostic and repair procedures to follow when trouble occurs.

We hope you use the manual to tackle the work yourself. For many simpler jobs, doing it yourself may be quicker than arranging an appointment to get the vehicle into a shop and making the trips to leave it and pick it up. More importantly, a lot of money can be saved by avoiding the expense the shop must pass on to you to cover its labor and overhead costs. An added benefit is the sense of satisfaction and accomplishment that you feel after doing the job yourself.

Using the manual

The manual is divided into Chapters. Each Chapter is divided into numbered Sections, which are headed in bold type between horizontal lines. Each Section consists of consecutively numbered paragraphs.

At the beginning of each numbered Section you will be referred to any illustrations which apply to the procedures in that Section. The reference numbers used in illustration captions pinpoint the pertinent Section and the Step within that Section. That is, illustration 3.2 means the illustration refers to Section 3 and Step (or paragraph) 2 within that Section.

Procedures, once described in the text, are not normally repeated. When it's necessary to refer to another Chapter, the reference will be given as Chapter and Section number. Cross references given without use of the word "Chapter" apply to Sections and/or paragraphs in the same Chapter. For example, "see Section 8" means in the same Chapter.

References to the left or right side of the vehicle assume you are sitting in the driver's seat, facing forward.

Even though we have prepared this manual with extreme care, neither the publisher nor the author can accept responsibility for any errors in, or omissions from, the information given.

NOTE

A **Note** provides information necessary to properly complete a procedure or information which will make the procedure easier to understand.

CAUTION

A **Caution** provides a special procedure or special steps which must be taken while completing the procedure where the Caution is found. Not heeding a Caution can result in damage to the assembly being worked on.

WARNING

A **Warning** provides a special procedure or special steps which must be taken while completing the procedure where the Warning is found. Not heeding a Warning can result in personal injury.

Introduction to the Jeep Liberty and Dodge Nitro

Engines available in these models are a 2.4L four-cylinder (through 2005), a 3.7L V6 or a 4.0L V6. All models are equipped with the On Board Diagnostic Second-Generation (OBD-II) computerized engine management system that controls virtually every aspect of engine operation. OBD-II monitors emissions and powertrain control system components for signs of degradation and engine operation for any malfunction that could affect emissions, turning on the CHECK ENGINE light if any faults are detected.

Chassis layout is conventional, with the engine mounted at the front and the power being transmitted through a five-speed or six-speed manual transmission or a four-speed or five-speed automatic transmission (depending on model year), and a driveshaft to the solid rear axle. On 4WD models, a transfer case transmits power to a front differential by way of a CV-jointed driveshaft, and then to the front wheels through the CV-jointed driveaxles.

Front suspension is independent with upper and lower control arms and coil springs. The rear suspension features upper arm(s) (a single A-shaped design on 2007 and earlier models, and two straight control arms on 2008 and later models), trailing lower arms, coil springs and, on 2008 and later models, a track bar. All models use front and rear stabilizer bars, and gas-filled shock absorbers.

The power-assisted rack-and-pinion steering is mounted on the front suspension crossmember.

The brakes are power-assisted front disc and rear disc or drum with an Anti-Lock Brake System (ABS) available as an option.

Vehicle identification numbers

Modifications are a continuing and unpublicized process in vehicle manufacturing. Since spare parts lists and manuals are compiled on a numerical basis, the individual vehicle numbers are necessary to correctly identify the component required.

Vehicle Identification Number (VIN)

This very important identification number is stamped on a plate attached to the A-pillar on the driver's side of the vehicle (see illustration). The VIN also appears on the Vehicle Certificate of Title and Registration. It contains information such as where and when the vehicle was manu-

factured, the model year and the body style.

VIN engine and model year codes

Two particularly important pieces of information found in the VIN are the engine code and the model year code. Counting from the left, the engine code letter designation is the 8th digit and the model year code letter designation is the 10th digit.

On the models covered by this manual the engine codes are:

1	2.4L four-cylinder engine
K	3.7L V6 engine
X	4.0L V6 engine

On the models covered by this manual the model year codes are:

2	2002
3	2003
4	2004
5	2005
6	2006
7	2007
8	2008
9	2009
A	2010
B	2011
C	2012

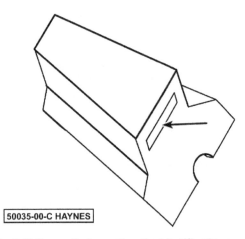

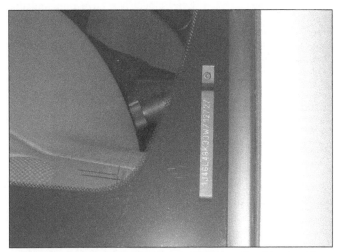

The VIN is visible through the windshield on the driver's side

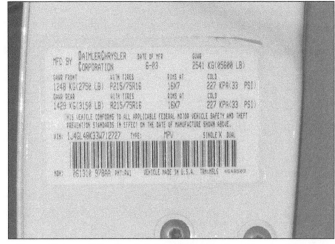

The vehicle certification label is affixed to the driver's side door

50035-00-C HAYNES

On the 2.4L four-cylinder engine, the identification number is located on the left rear of the engine block

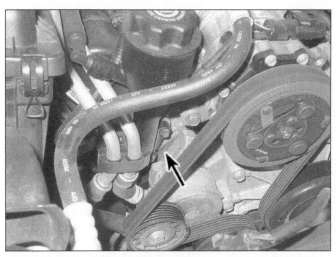

On the 3.7L V6 engine, the identification number is located on a machined pad at the right front corner of the engine

On the 42RLE automatic transmission, the identification label is affixed to the upper left side of the transmission bellhousing; it is only partially visible from the engine compartment (arrow)

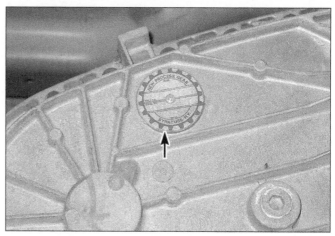

The transfer case identification tag is located on the rear side of the case

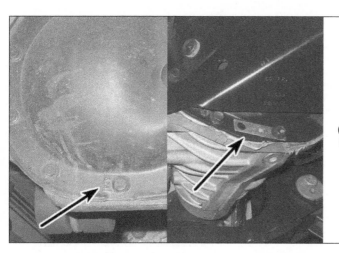

The identification tags on the differentials are attached to the differential covers (rear differential is at left, front differential is at right)

Automatic transmission identification number

The 45RFE transmission ID number is on a label affixed to the left side of the case, just above the oil pan. The 42RLE automatic transmission ID number is on a label affixed to the upper left side of the bellhousing **(see illustration)**. On 545RFE five-speed automatic transmissions, the tag is just above the fluid pan on the left side of the case.

Transfer case identification number

The transfer case ID tag is located on the rear side of the case **(see illustration)**.

Differential identification number

On both front and rear differentials, the ID tags are attached to the differential covers **(see illustration)**.

Vehicle Emissions Control Information label

This label is found in the engine compartment. See Chapter 6 for more information on this label.

Vehicle Certification Label

The Vehicle Certification Label is attached to the rear of the driver's side door **(see illustration)**. Information on this label includes the name of the manufacturer, the month and year of production, Gross Vehicle Weight Rating (GVWR), VIN, vehicle type, type of rear wheels, bar code, time of final assembly, paint and trim codes and the country of origin. This label is especially useful for matching the color and type of paint for repair work.

Engine identification number

On 2.4L four-cylinder engines, the serial number is located on the left rear side of the engine block. On the 3.7L V6, the serial number is located on the right front side of the engine block **(see illustrations)**. On the 4.0L V6, the serial number is located on the rear of the engine block just below the left cylinder head.

Manual transmission identification number

On the 2.4L model, the NV1500 manual transmission ID tag is located below the shifter, on the left side of the case. The tag on the right side is the part number. On the 3.7L V6 model, the NV3550 manual transmission ID tag is located on the top of the transmission case, in front of the shift tower.

Recall information

Vehicle recalls are carried out by the manufacturer in the rare event of a possible safety-related defect. The vehicle's registered owner is contacted at the address on file at the Department of Motor Vehicles and given the details of the recall. Remedial work is carried out free of charge at a dealer service department.

If you are the new owner of a used vehicle which was subject to a recall and you want to be sure that the work has been carried out, it's best to contact a dealer service department and ask about your individual vehicle - you'll need to furnish them your Vehicle Identification Number (VIN).

The table below is based on information provided by the National Highway Traffic Safety Administration (NHTSA), the body which oversees vehicle recalls in the United States. The recall database is updated constantly. For the latest information on vehicle recalls, check the NHTSA website at www.nhtsa.gov, www.safercar.gov or call the NHTSA hotline at 1-888-327-4236.

Recall date	Recall campaign number	Model(s) affected	Concern
October 04, 2001	01V311000	2002	On some models, the driver's side knee blocker trim panel assembly can disengage, fall, and startle the operator. This could increase the risk of a crash.
December 05, 2001	01V373000	2002	On some models, in the event of a severe frontal offset collision, sharp edges on the power steering pressure hose bracket could cut the front impact sensor wiring insulation. This could delay or prevent the deployment of the airbag system, increasing the risk of injury to the front seat occupants.
November 10, 2003	03V460000	2002, 2003	On some models, the lower control arm ball-joints can experience a loss of lubrication that can lead to corrosion, accelerated wear, and possible separation from the steering knuckle. A balljoint that has separated from the knuckle could result in loss of vehicle directional control and could result in a crash.
February 03, 2004	04V048000	2004	On some models, due to a software programming error, the tire, wheel, and inflation pressure information was inadvertently omitted from the vehicle certification label. The vehicle owner would not be aware of the tire, wheel, and inflation pressure information for their vehicle.

Recall date	Recall campaign number	Model(s) affected	Concern
March 04, 2004	04V111000	2004	On some models, certain Remote Keyless Entry (RKE) input may cause the Body Control Module (BCM) software to actuate the door lock motors continuously. This can cause the lock motor bearings to overheat and seize. If this occurs, the door lock system will become inoperative.
July 08, 2004	04V337000	2004	On some models equipped with 3.7L engines, two valve cover studs may interfere with the fuel injector and alternator wiring harnesses. Damage to the alternator wiring harness could result in an underhood fire.
October 06, 2005	05V460000	2006 – Liberty 4X4	On some models equipped with 42RLE automatic transmissions, the cup plug that retains the park pawl anchor shaft may be improperly installed. If the shaft moves out of position, the vehicle may not be able to achieve Park position. If this occurs and the parking brake is not applied, the vehicle may roll away and cause a crash.
August 03, 2006	06V288000	2002, 2003, 2004, 2005, 2006	On some models, the constant-tension front lower balljoints may experience contamination. If the vehicle is operated for an extended period with this condition, the balljoint may eventually experience wear and looseness. When the driver continues to operate the vehicle with this condition, the contamination can lead to corrosion, accelerated wear and possibly separation from the steering knuckle, causing a loss of vehicle steering control and increasing the risk of a crash.
September 7, 2006	06V341000	2007	On some models equipped with automatic transmissions, the software programmed into the Powertrain Control Module can cause a momentary lock up of the drive wheels if the vehicle is traveling over 40 mph and the operator shifts from drive to neutral and back to drive. If the drive wheels locked up. loss of vehicle control could occur increasing the risk of a crash.
December 26, 2006	06V493000	2007	On some models, the Antilock Brake System (ABS) control module software may cause the rear brakes to lock up during certain braking conditions. This could result in a loss of vehicle control and cause a crash.
March 09, 2007	07V089000	2006, 2007	On some models equipped with Valeo Heating, Ventilation and Air Conditioning (HVAC), the blower motor may overheat. This could cause an interior fire.
July 3, 2007	07V291000	2007	On certain vehicles, the Totally Integrated Power Module (TIPM) was programmed with software that may allow the engine to stall under certain operating conditions, which could cause a crash.

Recall date	Recall campaign number	Model(s) affected	Concern
September 14, 2007	07V434000	2007	On certain vehicles, the software programmed into the ABS electronic control module may allow a momentary delay in braking when coasting up a hill during certain conditions, which could cause a crash.
September 09, 2008	08V455000	2008	On some models, the windshield and/or fixed rear quarter glass may have been assembled using incorrect urethane adhesive. This could affect glass retention. If the glass separates while driving, it could strike another vehicle or injure a pedestrian.
October 08, 2008	08V525000	2007, 2008	On some models equipped with manual transmissions, the parking brake lever clutch drum may distort and reduce the effectiveness of the parking brake system. This could allow the vehicle to move inadvertently and cause a crash.
April 09, 2009	09V119000	2002, 2003	On some models, the front suspension upper control arm balljoint may experience a loss of lubrication which could lead to corrosion, accelerated wear, and possible separation while driving. Failure of the balljoint could result in a loss of steering capability, increasing the risk of a crash
November 6, 2009	09V438000	2007	On certain vehicles, the windshield wiper system may experience a condition over time where high usage of the variable pause wiper function ("delay" mode) may eventually affect primary wiper function, rendering the wiper system inoperable. Inoperative wipers under inclement weather conditions could cause impaired visibility increasing the risk of a crash.
January 11, 2010	10V009000	2010	Some models may have been built with an improperly formed or missing brake booster input rod retaining clip. This could result in brake failure, which could cause a crash.
July 07, 2010	10V315000	2010	Some models may have been built with an improperly formed master cylinder-to-Hydraulic Control Unit (HCU) brake tube assembly end flare. This could lead to loss of brake fluid and reduced braking performance, increasing the risk of a crash
November 03, 2010	10V550000	2008	Some models are equipped with windshield wiper motors that contain excess adhesive that may allow for water intrusion, which could result in intermittent and eventually inoperative windshield wiper operation. A failing or inoperative windshield wiper system will reduce a driver's visibility, which could result in a crash.

Recall date	Recall campaign number	Model(s) affected	Concern
June 08, 2011	11V315000	2011	Some models may have been built with a missing or incorrectly installed steering column pivot rivet. A missing or incorrectly installed rivet could compromise the ability of the steering column to support the occupant loads in the event of a frontal crash, decreasing the effectiveness of the frontal impact safety system. As a result, the condition may increase the potential for injury in a frontal crash.
March 05, 2012	12V085000	2004, 2005, 2006, 2007	Some models may be equipped with rear lower control arms that can experience a fracture due to excessive corrosion from application of road salts during winter weather conditions. A fracture of the rear lower control arm may result in a loss of vehicle control and may lead to a crash
November 08, 2012	12V527000	2002, 2003, 2004	On some models, a component in the airbag control module may fail, causing the front airbags, side curtain airbags, and/or seatbelt pretensioners to deploy inadvertently while the vehicle is being operated. Inadvertent deployment of the airbags may increase the risk of injury and the possibility of a crash.
April 03, 2013	13V121000	2007, 2008	On some models, the transfer case heat shield may drop down and rub on the driveshaft, weakening the driveshaft and potentially causing it to break. If the driveshaft breaks, a section of it can strike the transfer case tunnel with enough force that the airbag computer will believe the vehicle is in a crash and thus deploy the airbags. Additionally, the driveshaft failure will result in a loss of motive power. Either condition may increase the risk of a crash.
June 19, 2013	13V252000	2002, 2003, 2004, 2005, 2006, 2007	On some models, the fuel tanks are at risk of failure and leakage in certain rear impacts. A fuel leak in the presence of an ignition source may result in a fire.
July 02, 2013	13V282000	2011, 2012	On some models, electrical over-stress of a resistor in the occupant restraint control module may lead to the non-deployment of the active head restraints during a rear impact collision. In the event of a crash necessitating the deployment of the active head restraints, their non-deployment could increase the risk of injury.
October 10, 2017	17V624000	2012	Some models equipped a component within the Occupant Restraint Controller (ORC) may fail and prevent the active headrests from deploying in the event of a rear-end crash. If the active headrests do not deploy in certain rear-end crashes, the front seat occupants have an increased risk of injury.

Buying parts

Replacement parts are available from many sources, which generally fall into one of two categories - authorized dealer parts departments and independent retail auto parts stores. Our advice concerning these parts is as follows:

Retail auto parts stores: Good auto parts stores will stock frequently needed components which wear out relatively fast, such as clutch components, exhaust systems, brake parts, tune-up parts, etc. These stores often supply new or reconditioned parts on an exchange basis, which can save a considerable amount of money. Discount auto parts stores are often very good places to buy materials and parts needed for general vehicle maintenance such as oil, grease, filters, spark plugs, belts, touch-up paint, bulbs, etc. They also usually sell tools and general accessories, have convenient hours, charge lower prices and can often be found not far from home.

Authorized dealer parts department: This is the best source for parts which are unique to the vehicle and not generally available elsewhere (such as major engine parts, transmission parts, trim pieces, etc.).

Warranty information: If the vehicle is still covered under warranty, be sure that any replacement parts purchased - regardless of the source - do not invalidate the warranty!

To be sure of obtaining the correct parts, have engine and chassis numbers available and, if possible, take the old parts along for positive identification.

Maintenance techniques, tools and working facilities

Maintenance techniques

There are a number of techniques involved in maintenance and repair that will be referred to throughout this manual. Application of these techniques will enable the home mechanic to be more efficient, better organized and capable of performing the various tasks properly, which will ensure that the repair job is thorough and complete.

Fasteners

Fasteners are nuts, bolts, studs and screws used to hold two or more parts together. There are a few things to keep in mind when working with fasteners. Almost all of them use a locking device of some type, either a lockwasher, locknut, locking tab or thread adhesive. All threaded fasteners should be clean and straight, with undamaged threads and undamaged corners on the hex head where the wrench fits. Develop the habit of replacing all damaged nuts and bolts with new ones. Special locknuts with nylon or fiber inserts can only be used once. If they are removed, they lose their locking ability and must be replaced with new ones.

Rusted nuts and bolts should be treated with a penetrating fluid to ease removal and prevent breakage. Some mechanics use turpentine in a spout-type oil can, which works quite well. After applying the rust penetrant, let it work for a few minutes before trying to loosen the nut or bolt. Badly rusted fasteners may have to be chiseled or sawed off or removed with a special nut breaker, available at tool stores.

If a bolt or stud breaks off in an assembly, it can be drilled and removed with a special tool commonly available for this purpose. Most automotive machine shops can perform this task, as well as other repair procedures, such as the repair of threaded holes that have been stripped out.

Flat washers and lockwashers, when removed from an assembly, should always be replaced exactly as removed. Replace any damaged washers with new ones. Never use a lockwasher on any soft metal surface (such as aluminum), thin sheet metal or plastic.

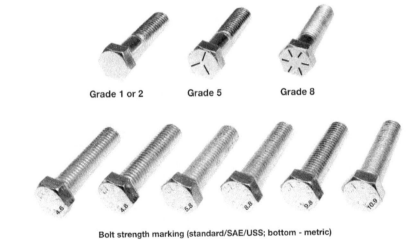

Grade 1 or 2   Grade 5   Grade 8

Bolt strength marking (standard/SAE/USS; bottom - metric)

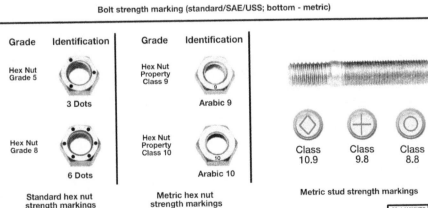

Grade	Identification	Grade	Identification
Hex Nut Grade 5	3 Dots	Hex Nut Property Class 9	Arabic 9
Hex Nut Grade 8	6 Dots	Hex Nut Property Class 10	Arabic 10

Standard hex nut strength markings

Metric hex nut strength markings

Class 10.9   Class 9.8   Class 8.8

Metric stud strength markings

00-1 HAYNES

Fastener sizes

For a number of reasons, automobile manufacturers are making wider and wider use of metric fasteners. Therefore, it is important to be able to tell the difference between standard (sometimes called U.S. or SAE) and metric hardware, since they cannot be interchanged.

All bolts, whether standard or metric, are sized according to diameter, thread pitch and length. For example, a standard 1/2 - 13 x 1 bolt is 1/2 inch in diameter, has 13 threads per inch and is 1 inch long. An M12 - 1.75 x 25 metric bolt is 12 mm in diameter, has a thread pitch of 1.75 mm (the distance between threads) and is 25 mm long. The two bolts are nearly identical, and easily confused, but they are not interchangeable.

In addition to the differences in diameter, thread pitch and length, metric and standard bolts can also be distinguished by examining the bolt heads. To begin with, the distance across the flats on a standard bolt head is measured in inches, while the same dimension on a metric bolt is sized in millimeters (the same is true for nuts). As a result, a standard wrench should not be used on a metric bolt and a metric wrench should not be used on a standard bolt. Also, most standard bolts have slashes radiating out from the center of the head to denote the grade or strength of the bolt, which is an indication of the amount of torque that can be applied to it. The greater the number of slashes, the greater the strength of the bolt. Grades 0 through 5 are commonly used on automobiles. Metric bolts have a property class (grade) number, rather than a slash, molded into their heads to indicate bolt strength. In this case, the higher the number, the stronger the bolt. Property class numbers 8.8, 9.8 and 10.9 are commonly used on automobiles.

Strength markings can also be used to distinguish standard hex nuts from metric hex nuts. Many standard nuts have dots stamped into one side, while metric nuts are marked with a number. The greater the number of dots, or the higher the number, the greater the strength of the nut.

Metric studs are also marked on their ends according to property class (grade). Larger studs are numbered (the same as metric bolts), while smaller studs carry a geometric code to denote grade.

It should be noted that many fasteners, especially Grades 0 through 2, have no distinguishing marks on them. When such is the case, the only way to determine whether it is standard or metric is to measure the thread pitch or compare it to a known fastener of the same size.

Standard fasteners are often referred to as SAE, as opposed to metric. However, it should be noted that SAE technically refers to a non-metric fine thread fastener only. Coarse thread non-metric fasteners are referred to as USS sizes.

Since fasteners of the same size (both standard and metric) may have different strength ratings, be sure to reinstall any bolts,

Metric thread sizes

	Ft-lbs	Nm
M-6	6 to 9	9 to 12
M-8	14 to 21	19 to 28
M-10	28 to 40	38 to 54
M-12	50 to 71	68 to 96
M-14	80 to 140	109 to 154

Pipe thread sizes

1/8	5 to 8	7 to 10
1/4	12 to 18	17 to 24
3/8	22 to 33	30 to 44
1/2	25 to 35	34 to 47

U.S. thread sizes

1/4 - 20	6 to 9	9 to 12
5/16 - 18	12 to 18	17 to 24
5/16 - 24	14 to 20	19 to 27
3/8 - 16	22 to 32	30 to 43
3/8 - 24	27 to 38	37 to 51
7/16 - 14	40 to 55	55 to 74
7/16 - 20	40 to 60	55 to 81
1/2 - 13	55 to 80	75 to 108

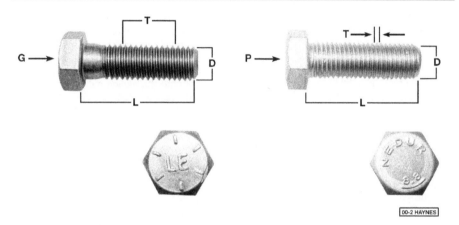

Standard (SAE and USS) bolt dimensions/grade marks

G   Grade marks (bolt strength)
L   Length (in inches)
T   Thread pitch (number of threads per inch)
D   Nominal diameter (in inches)

Metric bolt dimensions/grade marks

P   Property class (bolt strength)
L   Length (in millimeters)
T   Thread pitch (distance between threads in millimeters)
D   Diameter

studs or nuts removed from your vehicle in their original locations. Also, when replacing a fastener with a new one, make sure that the new one has a strength rating equal to or greater than the original.

Tightening sequences and procedures

Most threaded fasteners should be tightened to a specific torque value (torque is the twisting force applied to a threaded component such as a nut or bolt). Overtightening the fastener can weaken it and cause it to break, while undertightening can cause it to eventually come loose. Bolts, screws and studs, depending on the material they are made of and their thread diameters, have specific torque values, many of which are noted in the Specifications at the beginning of each Chapter. Be sure to follow the torque recommendations closely. For fasteners not assigned a specific torque, a general torque value chart is presented here as a guide. These torque values are for dry (unlubricated) fasteners threaded into steel or cast iron (not aluminum). As was previously mentioned, the size and grade of a fastener determine the amount of torque that can safely be applied to it. The figures listed here are approximate for Grade 2 and Grade 3 fasteners. Higher grades can tolerate higher torque values.

Fasteners laid out in a pattern, such as cylinder head bolts, oil pan bolts, differential cover bolts, etc., must be loosened or tightened in sequence to avoid warping the com-

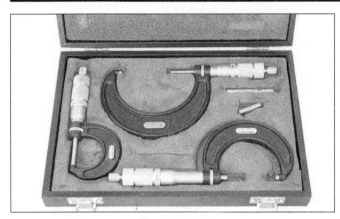

Micrometer set

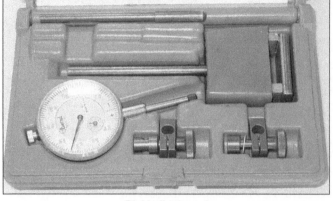

Dial indicator set

ponent. This sequence will normally be shown in the appropriate Chapter. If a specific pattern is not given, the following procedures can be used to prevent warping.

Initially, the bolts or nuts should be assembled finger-tight only. Next, they should be tightened one full turn each, in a criss-cross or diagonal pattern. After each one has been tightened one full turn, return to the first one and tighten them all one-half turn, following the same pattern. Finally, tighten each of them one-quarter turn at a time until each fastener has been tightened to the proper torque. To loosen and remove the fasteners, the procedure would be reversed.

Component disassembly

Component disassembly should be done with care and purpose to help ensure that the parts go back together properly. Always keep track of the sequence in which parts are removed. Make note of special characteristics or marks on parts that can be installed more than one way, such as a grooved thrust washer on a shaft. It is a good idea to lay the disassembled parts out on a clean surface in the order that they were removed. It may also be helpful to make sketches or take instant photos of components before removal.

When removing fasteners from a component, keep track of their locations. Sometimes threading a bolt back in a part, or putting the washers and nut back on a stud, can prevent mix-ups later. If nuts and bolts cannot be returned to their original locations, they should be kept in a compartmented box or a series of small boxes. A cupcake or muffin tin is ideal for this purpose, since each cavity can hold the bolts and nuts from a particular area (i.e. oil pan bolts, valve cover bolts, engine mount bolts, etc.). A pan of this type is especially helpful when working on assemblies with very small parts, such as the carburetor, alternator, valve train or interior dash and trim pieces. The cavities can be marked with paint or tape to identify the contents.

Whenever wiring looms, harnesses or connectors are separated, it is a good idea to identify the two halves with numbered pieces of masking tape so they can be easily reconnected.

Gasket sealing surfaces

Throughout any vehicle, gaskets are used to seal the mating surfaces between two parts and keep lubricants, fluids, vacuum or pressure contained in an assembly.

Many times these gaskets are coated with a liquid or paste-type gasket sealing compound before assembly. Age, heat and pressure can sometimes cause the two parts to stick together so tightly that they are very difficult to separate. Often, the assembly can be loosened by striking it with a soft-face hammer near the mating surfaces. A regular hammer can be used if a block of wood is placed between the hammer and the part. Do not hammer on cast parts or parts that could be easily damaged. With any particularly stubborn part, always recheck to make sure that every fastener has been removed.

Avoid using a screwdriver or bar to pry apart an assembly, as they can easily mar the gasket sealing surfaces of the parts, which must remain smooth. If prying is absolutely necessary, use an old broom handle, but keep in mind that extra clean up will be necessary if the wood splinters.

After the parts are separated, the old gasket must be carefully removed and the gasket surfaces cleaned. If you're working on cast iron or aluminum parts, stubborn gasket material can be soaked with rust penetrant or treated with a special chemical to soften it so it can be easily scraped off. **Caution:** *Never use gasket removal solutions or caustic chemicals on plastic or other composite components.* A scraper can be fashioned from a piece of copper tubing by flattening and sharpening one end. Copper is recommended because it is usually softer than the surfaces to be scraped, which reduces the chance of gouging the part. Some gaskets can be removed with a wire brush, but regardless of the method used, the mating surfaces must be left clean and smooth. If for some reason the gasket surface is gouged, then a gasket sealer thick enough to fill scratches will have to be used during reassembly of the components. For most applications, a non-drying (or semi-drying) gasket sealer should be used.

Hose removal tips

Warning: *If the vehicle is equipped with air conditioning, do not disconnect any of the A/C hoses without first having the system depressurized by a dealer service department or a service station.*

Hose removal precautions closely parallel gasket removal precautions. Avoid scratching or gouging the surface that the hose mates against or the connection may leak. This is especially true for radiator hoses. Because of various chemical reactions, the rubber in hoses can bond itself to the metal spigot that the hose fits over. To remove a hose, first loosen the hose clamps that secure it to the spigot. Then, with slip-joint pliers, grab the hose at the clamp and rotate it around the spigot. Work it back and forth until it is completely free, then pull it off. Silicone or other lubricants will ease removal if they can be applied between the hose and the outside of the spigot. Apply the same lubricant to the inside of the hose and the outside of the spigot to simplify installation.

As a last resort (and if the hose is to be replaced with a new one anyway), the rubber can be slit with a knife and the hose peeled from the spigot. If this must be done, be careful that the metal connection is not damaged.

If a hose clamp is broken or damaged, do not reuse it. Wire-type clamps usually weaken with age, so it is a good idea to replace them with screw-type clamps whenever a hose is removed.

Tools

A selection of good tools is a basic requirement for anyone who plans to maintain and repair his or her own vehicle. For the owner who has few tools, the initial investment might seem high, but when compared to the spiraling costs of professional auto maintenance and repair, it is a wise one.

To help the owner decide which tools are needed to perform the tasks detailed in this manual, the following tool lists are offered: *Maintenance and minor repair, Repair/overhaul* and *Special*.

The newcomer to practical mechanics should start off with the *maintenance and*

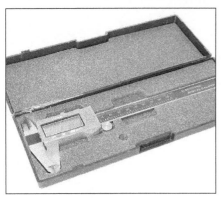

Dial caliper

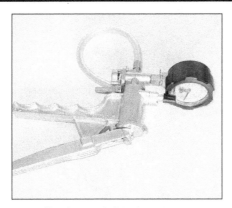

Hand-operated vacuum pump

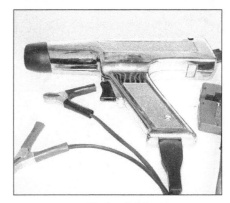

Timing light

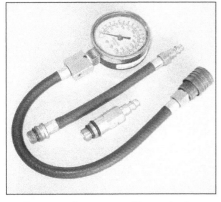

Compression gauge with spark plug
hole adapter

Damper/steering wheel puller

General purpose puller

Hydraulic lifter removal tool

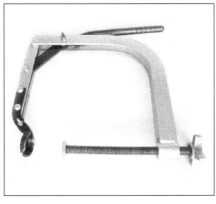

Valve spring compressor

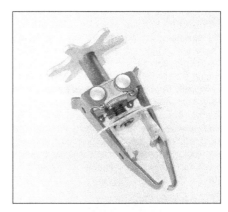

Valve spring compressor

Ridge reamer

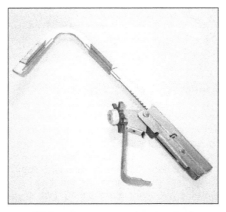

Piston ring groove cleaning tool

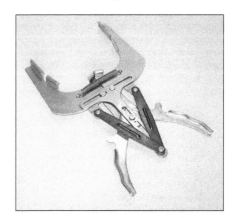

Ring removal/installation tool

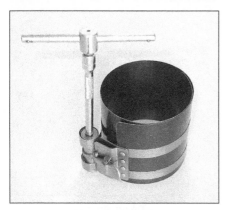

Ring compressor

Cylinder hone

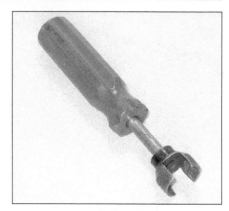

Brake hold-down spring tool

Torque angle gauge

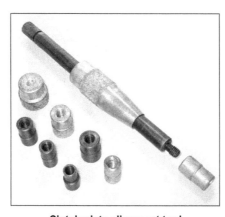

Clutch plate alignment tool

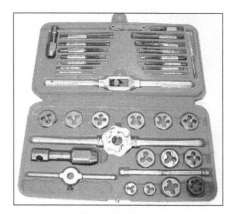

Tap and die set

minor repair tool kit, which is adequate for the simpler jobs performed on a vehicle. Then, as confidence and experience grow, the owner can tackle more difficult tasks, buying additional tools as they are needed. Eventually the basic kit will be expanded into the *repair and overhaul* tool set. Over a period of time, the experienced do-it-yourselfer will assemble a tool set complete enough for most repair and overhaul procedures and will add tools from the special category when it is felt that the expense is justified by the frequency of use.

Maintenance and minor repair tool kit

The tools in this list should be considered the minimum required for performance of routine maintenance, servicing and minor repair work. We recommend the purchase of combination wrenches (box-end and open-end combined in one wrench). While more expensive than open end wrenches, they offer the advantages of both types of wrench.

> *Combination wrench set (1/4-inch to 1 inch or 6 mm to 19 mm)*
> *Adjustable wrench, 8 inch*
> *Spark plug wrench with rubber insert*
> *Spark plug gap adjusting tool*
> *Feeler gauge set*
> *Brake bleeder wrench*
> *Standard screwdriver (5/16-inch x 6 inch)*

> *Phillips screwdriver (No. 2 x 6 inch)*
> *Combination pliers - 6 inch*
> *Hacksaw and assortment of blades*
> *Tire pressure gauge*
> *Grease gun*
> *Oil can*
> *Fine emery cloth*
> *Wire brush*
> *Battery post and cable cleaning tool*
> *Oil filter wrench*
> *Funnel (medium size)*
> *Safety goggles*
> *Jackstands (2)*
> *Drain pan*

Note: *If basic tune-ups are going to be part of routine maintenance, it will be necessary to purchase a good quality stroboscopic timing light and combination tachometer/dwell meter. Although they are included in the list of special tools, it is mentioned here because they are absolutely necessary for tuning most vehicles properly.*

Repair and overhaul tool set

These tools are essential for anyone who plans to perform major repairs and are in addition to those in the maintenance and minor repair tool kit. Included is a comprehensive set of sockets which, though expensive, are invaluable because of their versatility, especially when various extensions and drives are available. We recommend the 1/2-inch drive over the 3/8-inch drive. Although the larger drive is bulky and more expensive, it has the

capacity of accepting a very wide range of large sockets. Ideally, however, the mechanic should have a 3/8-inch drive set and a 1/2-inch drive set.

> *Socket set(s)*
> *Reversible ratchet*
> *Extension - 10 inch*
> *Universal joint*
> *Torque wrench (same size drive as sockets)*
> *Ball peen hammer - 8 ounce*
> *Soft-face hammer (plastic/rubber)*
> *Standard screwdriver (1/4-inch x 6 inch)*
> *Standard screwdriver (stubby - 5/16-inch)*
> *Phillips screwdriver (No. 3 x 8 inch)*
> *Phillips screwdriver (stubby - No. 2)*
> *Pliers - vise grip*
> *Pliers - lineman's*
> *Pliers - needle nose*
> *Pliers - snap-ring (internal and external)*
> *Cold chisel - 1/2-inch*
> *Scribe*
> *Scraper (made from flattened copper tubing)*
> *Centerpunch*
> *Pin punches (1/16, 1/8, 3/16-inch)*
> *Steel rule/straightedge - 12 inch*
> *Allen wrench set (1/8 to 3/8-inch or 4 mm to 10 mm)*
> *A selection of files*
> *Wire brush (large)*
> *Jackstands (second set)*
> *Jack (scissor or hydraulic type)*

Note: *Another tool which is often useful is an electric drill with a chuck capacity of 3/8-inch and a set of good quality drill bits.*

Special tools

The tools in this list include those which are not used regularly, are expensive to buy, or which need to be used in accordance with their manufacturer's instructions. Unless these tools will be used frequently, it is not very economical to purchase many of them. A consideration would be to split the cost and use between yourself and a friend or friends. In addition, most of these tools can be obtained from a tool rental shop on a temporary basis.

This list primarily contains only those tools and instruments widely available to the public, and not those special tools produced by the vehicle manufacturer for distribution to dealer service departments. Occasionally, references to the manufacturer's special tools are included in the text of this manual. Generally, an alternative method of doing the job without the special tool is offered. However, sometimes there is no alternative to their use. Where this is the case, and the tool cannot be purchased or borrowed, the work should be turned over to the dealer service department or an automotive repair shop.

Valve spring compressor
Piston ring groove cleaning tool
Piston ring compressor
Piston ring installation tool
Cylinder compression gauge
Cylinder ridge reamer
Cylinder surfacing hone
Cylinder bore gauge
Micrometers and/or dial calipers
Hydraulic lifter removal tool
Balljoint separator
Universal-type puller
Impact screwdriver
Dial indicator set
Stroboscopic timing light (inductive pick-up)
Hand operated vacuum/pressure pump
Tachometer/dwell meter
Universal electrical multimeter
Cable hoist
Brake spring removal and installation tools
Floor jack

Buying tools

For the do-it-yourselfer who is just starting to get involved in vehicle maintenance and repair, there are a number of options available when purchasing tools. If maintenance and minor repair is the extent of the work to be done, the purchase of individual tools is satisfactory. If, on the other hand, extensive work is planned, it would be a good idea to purchase a modest tool set from one of the large retail chain stores. A set can usually be bought at a substantial savings over the individual tool prices, and they often come with a tool box.

As additional tools are needed, add-on sets, individual tools and a larger tool box can be purchased to expand the tool selection. Building a tool set gradually allows the cost of the tools to be spread over a longer period of time and gives the mechanic the freedom to choose only those tools that will actually be used.

Tool stores will often be the only source of some of the special tools that are needed, but regardless of where tools are bought, try to avoid cheap ones, especially when buying screwdrivers and sockets, because they won't last very long. The expense involved in replacing cheap tools will eventually be greater than the initial cost of quality tools.

Care and maintenance of tools

Good tools are expensive, so it makes sense to treat them with respect. Keep them clean and in usable condition and store them properly when not in use. Always wipe off any dirt, grease or metal chips before putting them away. Never leave tools lying around in the work area. Upon completion of a job, always check closely under the hood for tools that may have been left there so they won't get lost during a test drive.

Some tools, such as screwdrivers, pliers, wrenches and sockets, can be hung on a panel mounted on the garage or workshop wall, while others should be kept in a tool box or tray. Measuring instruments, gauges, meters, etc. must be carefully stored where they cannot be damaged by weather or impact from other tools.

When tools are used with care and stored properly, they will last a very long time. Even with the best of care, though, tools will wear out if used frequently. When a tool is damaged or worn out, replace it. Subsequent jobs will be safer and more enjoyable if you do.

How to repair damaged threads

Sometimes, the internal threads of a nut or bolt hole can become stripped, usually from overtightening. Stripping threads is an all-too-common occurrence, especially when working with aluminum parts, because aluminum is so soft that it easily strips out.

Usually, external or internal threads are only partially stripped. After they've been cleaned up with a tap or die, they'll still work. Sometimes, however, threads are badly damaged. When this happens, you've got three choices:

1) *Drill and tap the hole to the next suitable oversize and install a larger diameter bolt, screw or stud.*
2) *Drill and tap the hole to accept a threaded plug, then drill and tap the plug to the original screw size. You can also buy a plug already threaded to the original size. Then you simply drill a hole to the specified size, then run the threaded*

plug into the hole with a bolt and jam nut. Once the plug is fully seated, remove the jam nut and bolt.
3) *The third method uses a patented thread repair kit like Heli-Coil or Slimsert. These easy-to-use kits are designed to repair damaged threads in straight-through holes and blind holes. Both are available as kits which can handle a variety of sizes and thread patterns. Drill the hole, then tap it with the special included tap. Install the Heli-Coil and the hole is back to its original diameter and thread pitch.*

Regardless of which method you use, be sure to proceed calmly and carefully. A little impatience or carelessness during one of these relatively simple procedures can ruin your whole day's work and cost you a bundle if you wreck an expensive part.

Working facilities

Not to be overlooked when discussing tools is the workshop. If anything more than routine maintenance is to be carried out, some sort of suitable work area is essential.

It is understood, and appreciated, that many home mechanics do not have a good workshop or garage available, and end up removing an engine or doing major repairs outside. It is recommended, however, that the overhaul or repair be completed under the cover of a roof.

A clean, flat workbench or table of comfortable working height is an absolute necessity. The workbench should be equipped with a vise that has a jaw opening of at least four inches.

As mentioned previously, some clean, dry storage space is also required for tools, as well as the lubricants, fluids, cleaning solvents, etc. which soon become necessary.

Sometimes waste oil and fluids, drained from the engine or cooling system during normal maintenance or repairs, present a disposal problem. To avoid pouring them on the ground or into a sewage system, pour the used fluids into large containers, seal them with caps and take them to an authorized disposal site or recycling center. Plastic jugs, such as old antifreeze containers, are ideal for this purpose.

Always keep a supply of old newspapers and clean rags available. Old towels are excellent for mopping up spills. Many mechanics use rolls of paper towels for most work because they are readily available and disposable. To help keep the area under the vehicle clean, a large cardboard box can be cut open and flattened to protect the garage or shop floor.

Whenever working over a painted surface, such as when leaning over a fender to service something under the hood, always cover it with an old blanket or bedspread to protect the finish. Vinyl covered pads, made especially for this purpose, are available at auto parts stores.

Booster battery (jump) starting

Observe the following precautions when using a booster battery to start a vehicle:

a) *Before connecting the booster battery, make sure the ignition switch is in the Off position.*
b) *Turn off the lights, heater and other electrical loads.*
c) *Your eyes should be shielded. Safety goggles are a good idea.*
d) *Make sure the booster battery is the same voltage as the dead one in the vehicle.*
e) *The two vehicles MUST NOT TOUCH each other.*
f) *Make sure the transmission is in Park.*
g) *If the booster battery is not a maintenance-free type, remove the vent caps and lay a cloth over the vent holes.*

Connect the red jumper cable to the positive (+) terminals of each battery.

Connect one end of the black cable to the negative (-) terminal of the booster battery. The other end of this cable should be connected to a good ground on the engine block **(see illustration)**. Make sure the cable will not come into contact with the fan, drivebelts or other moving parts of the engine.

Start the engine using the booster battery, then, with the engine running at idle speed, disconnect the jumper cables in the reverse order of connection.

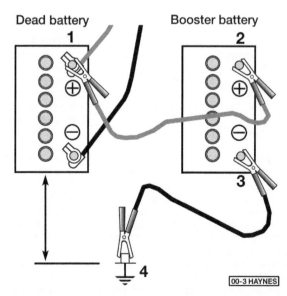

Make the booster battery cable connections in the numerical order shown (note that the negative cable of the booster battery is NOT attached to the negative terminal of the dead battery)

Jacking and towing

Jacking

Warning: *The jack supplied with the vehicle should only be used for changing a tire or placing jackstands under the frame. Never work under the vehicle or start the engine while this jack is being used as the only means of support.*

The vehicle should be on level ground. Place the shift lever in Park, if you have an automatic, or Reverse if you have a manual transmission. Block the wheel diagonally opposite the wheel being changed. Set the parking brake.

Remove the spare tire and jack from stowage. Loosen the wheel lug nuts about 1/4-to-1/2 turn each.

Place the jack under the vehicle in the indicated position **(see illustrations)**. Turn the jack handle clockwise until the tire clears the ground. Remove the lug nuts and pull the wheel off, then install the spare.

Install the lug nuts with the beveled edges facing in. Tighten them snugly. Don't attempt to tighten them completely until the vehicle is lowered or it could slip off the jack. Turn the jack handle counterclockwise to lower the vehicle. Remove the jack and tighten the lug nuts in a diagonal pattern.

Install the cover and be sure it's snapped into place all the way around.

Stow the tire, jack and wrench. Unblock the wheels.

Towing

Two-wheel drive models can be towed from the rear, with the rear wheels off the ground. Vehicles with four-wheel drive must only be towed with all four wheels off the ground.

Equipment specifically designed for towing should be used. It must be attached to the main structural members of the vehicle, not the bumpers or brackets.

Safety is a major consideration when towing and all applicable state and local laws must be obeyed. A safety chain must be used at all times.

The parking brake must be released and the transmission must be in Neutral. The steering must be unlocked (ignition switch in the Off position). Remember that power steering and power brakes won't work with the engine off.

Front jacking location

Rear jacking location

DECIMALS to MILLIMETERS

Decimal	mm	Decimal	mm
0.001	0.0254	0.500	12.7000
0.002	0.0508	0.510	12.9540
0.003	0.0762	0.520	13.2080
0.004	0.1016	0.530	13.4620
0.005	0.1270	0.540	13.7160
0.006	0.1524	0.550	13.9700
0.007	0.1778	0.560	14.2240
0.008	0.2032	0.570	14.4780
0.009	0.2286	0.580	14.7320
		0.590	14.9860
0.010	0.2540		
0.020	0.5080		
0.030	0.7620		
0.040	1.0160	0.600	15.2400
0.050	1.2700	0.610	15.4940
0.060	1.5240	0.620	15.7480
0.070	1.7780	0.630	16.0020
0.080	2.0320	0.640	16.2560
0.090	2.2860	0.650	16.5100
		0.660	16.7640
0.100	2.5400	0.670	17.0180
0.110	2.7940	0.680	17.2720
0.120	3.0480	0.690	17.5260
0.130	3.3020		
0.140	3.5560		
0.150	3.8100		
0.160	4.0640	0.700	17.7800
0.170	4.3180	0.710	18.0340
0.180	4.5720	0.720	18.2880
0.190	4.8260	0.730	18.5420
		0.740	18.7960
0.200	5.0800	0.750	19.0500
0.210	5.3340	0.760	19.3040
0.220	5.5880	0.770	19.5580
0.230	5.8420	0.780	19.8120
0.240	6.0960	0.790	20.0660
0.250	6.3500		
0.260	6.6040		
0.270	6.8580	0.800	20.3200
0.280	7.1120	0.810	20.5740
0.290	7.3660	0.820	21.8280
		0.830	21.0820
0.300	7.6200	0.840	21.3360
0.310	7.8740	0.850	21.5900
0.320	8.1280	0.860	21.8440
0.330	8.3820	0.870	22.0980
0.340	8.6360	0.880	22.3520
0.350	8.8900	0.890	22.6060
0.360	9.1440		
0.370	9.3980		
0.380	9.6520		
0.390	9.9060		
		0.900	22.8600
0.400	10.1600	0.910	23.1140
0.410	10.4140	0.920	23.3680
0.420	10.6680	0.930	23.6220
0.430	10.9220	0.940	23.8760
0.440	11.1760	0.950	24.1300
0.450	11.4300	0.960	24.3840
0.460	11.6840	0.970	24.6380
0.470	11.9380	0.980	24.8920
0.480	12.1920	0.990	25.1460
0.490	12.4460	1.000	25.4000

FRACTIONS to DECIMALS to MILLIMETERS

Fraction	Decimal	mm	Fraction	Decimal	mm
1/64	0.0156	0.3969	33/64	0.5156	13.0969
1/32	0.0312	0.7938	17/32	0.5312	13.4938
3/64	0.0469	1.1906	35/64	0.5469	13.8906
1/16	0.0625	1.5875	9/16	0.5625	14.2875
5/64	0.0781	1.9844	37/64	0.5781	14.6844
3/32	0.0938	2.3812	19/32	0.5938	15.0812
7/64	0.1094	2.7781	39/64	0.6094	15.4781
1/8	0.1250	3.1750	5/8	0.6250	15.8750
9/64	0.1406	3.5719	41/64	0.6406	16.2719
5/32	0.1562	3.9688	21/32	0.6562	16.6688
11/64	0.1719	4.3656	43/64	0.6719	17.0656
3/16	0.1875	4.7625	11/16	0.6875	17.4625
13/64	0.2031	5.1594	45/64	0.7031	17.8594
7/32	0.2188	5.5562	23/32	0.7188	18.2562
15/64	0.2344	5.9531	47/64	0.7344	18.6531
1/4	0.2500	6.3500	3/4	0.7500	19.0500
17/64	0.2656	6.7469	49/64	0.7656	19.4469
9/32	0.2812	7.1438	25/32	0.7812	19.8438
19/64	0.2969	7.5406	51/64	0.7969	20.2406
5/16	0.3125	7.9375	13/16	0.8125	20.6375
21/64	0.3281	8.3344	53/64	0.8281	21.0344
11/32	0.3438	8.7312	27/32	0.8438	21.4312
23/64	0.3594	9.1281	55/64	0.8594	21.8281
3/8	0.3750	9.5250	7/8	0.8750	22.2250
25/64	0.3906	9.9219	57/64	0.8906	22.6219
13/32	0.4062	10.3188	29/32	0.9062	23.0188
27/64	0.4219	10.7156	59/64	0.9219	23.4156
7/16	0.4375	11.1125	15/16	0.9375	23.8125
29/64	0.4531	11.5094	61/64	0.9531	24.2094
15/32	0.4688	11.9062	31/32	0.9688	24.6062
31/64	0.4844	12.3031	63/64	0.9844	25.0031
1/2	0.5000	12.7000	1	1.0000	25.4000

Conversion factors

Length (distance)

Inches (in)	X	25.4	= Millimeters (mm)	X	0.0394	= Inches (in)
Feet (ft)	X	0.305	= Meters (m)	X	3.281	= Feet (ft)
Miles	X	1.609	= Kilometers (km)	X	0.621	= Miles

Volume (capacity)

Cubic inches (cu in; in^3)	X	16.387	= Cubic centimeters (cc; cm^3)	X	0.061	= Cubic inches (cu in; in^3)
Imperial pints (Imp pt)	X	0.568	= Liters (l)	X	1.76	= Imperial pints (Imp pt)
Imperial quarts (Imp qt)	X	1.137	= Liters (l)	X	0.88	= Imperial quarts (Imp qt)
Imperial quarts (Imp qt)	X	1.201	= US quarts (US qt)	X	0.833	= Imperial quarts (Imp qt)
US quarts (US qt)	X	0.946	= Liters (l)	X	1.057	= US quarts (US qt)
Imperial gallons (Imp gal)	X	4.546	= Liters (l)	X	0.22	= Imperial gallons (Imp gal)
Imperial gallons (Imp gal)	X	1.201	= US gallons (US gal)	X	0.833	= Imperial gallons (Imp gal)
US gallons (US gal)	X	3.785	= Liters (l)	X	0.264	= US gallons (US gal)

Mass (weight)

Ounces (oz)	X	28.35	= Grams (g)	X	0.035	= Ounces (oz)
Pounds (lb)	X	0.454	= Kilograms (kg)	X	2.205	= Pounds (lb)

Force

Ounces-force (ozf; oz)	X	0.278	= Newtons (N)	X	3.6	= Ounces-force (ozf; oz)
Pounds-force (lbf; lb)	X	4.448	= Newtons (N)	X	0.225	= Pounds-force (lbf; lb)
Newtons (N)	X	0.1	= Kilograms-force (kgf; kg)	X	9.81	= Newtons (N)

Pressure

Pounds-force per square inch (psi; lbf/in^2; lb/in^2)	X	0.070	= Kilograms-force per square centimeter (kgf/cm^2; kg/cm^2)	X	14.223	= Pounds-force per square inch (psi; lbf/in^2; lb/in^2)
Pounds-force per square inch (psi; lbf/in^2; lb/in^2)	X	0.068	= Atmospheres (atm)	X	14.696	= Pounds-force per square inch (psi; lbf/in^2; lb/in^2)
Pounds-force per square inch (psi; lbf/in^2; lb/in^2)	X	0.069	= Bars	X	14.5	= Pounds-force per square inch (psi; lbf/in^2; lb/in^2)
Pounds-force per square inch (psi; lbf/in^2; lb/in^2)	X	6.895	= Kilopascals (kPa)	X	0.145	= Pounds-force per square inch (psi; lbf/in^2; lb/in^2)
Kilopascals (kPa)	X	0.01	= Kilograms-force per square centimeter (kgf/cm^2; kg/cm^2)	X	98.1	= Kilopascals (kPa)

Torque (moment of force)

Pounds-force inches (lbf in; lb in)	X	1.152	= Kilograms-force centimeter (kgf cm; kg cm)	X	0.868	= Pounds-force inches (lbf in; lb in)
Pounds-force inches (lbf in; lb in)	X	0.113	= Newton meters (Nm)	X	8.85	= Pounds-force inches (lbf in; lb in)
Pounds-force inches (lbf in; lb in)	X	0.083	= Pounds-force feet (lbf ft; lb ft)	X	12	= Pounds-force inches (lbf in; lb in)
Pounds-force feet (lbf ft; lb ft)	X	0.138	= Kilograms-force meters (kgf m; kg m)	X	7.233	= Pounds-force feet (lbf ft; lb ft)
Pounds-force feet (lbf ft; lb ft)	X	1.356	= Newton meters (Nm)	X	0.738	= Pounds-force feet (lbf ft; lb ft)
Newton meters (Nm)	X	0.102	= Kilograms-force meters (kgf m; kg m)	X	9.804	= Newton meters (Nm)

Vacuum

Inches mercury (in. Hg)	X	3.377	= Kilopascals (kPa)	X	0.2961	= Inches mercury
Inches mercury (in. Hg)	X	25.4	= Millimeters mercury (mm Hg)	X	0.0394	= Inches mercury

Power

Horsepower (hp)	X	745.7	= Watts (W)	X	0.0013	= Horsepower (hp)

Velocity (speed)

Miles per hour (miles/hr; mph)	X	1.609	= Kilometers per hour (km/hr; kph)	X	0.621	= Miles per hour (miles/hr; mph)

Fuel consumption*

Miles per gallon, Imperial (mpg)	X	0.354	= Kilometers per liter (km/l)	X	2.825	= Miles per gallon, Imperial (mpg)
Miles per gallon, US (mpg)	X	0.425	= Kilometers per liter (km/l)	X	2.352	= Miles per gallon, US (mpg)

Temperature

Degrees Fahrenheit = ($°C \times 1.8$) + 32   Degrees Celsius (Degrees Centigrade; °C) = ($°F - 32$) x 0.56

*It is common practice to convert from miles per gallon (mpg) to liters/100 kilometers (l/100km), where mpg (Imperial) x l/100 km = 282 and mpg (US) x l/100 km = 235

Safety first!

Regardless of how enthusiastic you may be about getting on with the job at hand, take the time to ensure that your safety is not jeopardized. A moment's lack of attention can result in an accident, as can failure to observe certain simple safety precautions. The possibility of an accident will always exist, and the following points should not be considered a comprehensive list of all dangers. Rather, they are intended to make you aware of the risks and to encourage a safety conscious approach to all work you carry out on your vehicle.

Essential DOs and DON'Ts

DON'T rely on a jack when working under the vehicle. Always use approved jackstands to support the weight of the vehicle and place them under the recommended lift or support points.

DON'T attempt to loosen extremely tight fasteners (i.e. wheel lug nuts) while the vehicle is on a jack - it may fall.

DON'T start the engine without first making sure that the transmission is in Neutral (or Park where applicable) and the parking brake is set.

DON'T remove the radiator cap from a hot cooling system - let it cool or cover it with a cloth and release the pressure gradually.

DON'T attempt to drain the engine oil until you are sure it has cooled to the point that it will not burn you.

DON'T touch any part of the engine or exhaust system until it has cooled sufficiently to avoid burns.

DON'T siphon toxic liquids such as gasoline, antifreeze and brake fluid by mouth, or allow them to remain on your skin.

DON'T inhale brake lining dust - it is potentially hazardous (see *Asbestos* below).

DON'T allow spilled oil or grease to remain on the floor - wipe it up before someone slips on it.

DON'T use loose fitting wrenches or other tools which may slip and cause injury.

DON'T push on wrenches when loosening or tightening nuts or bolts. Always try to pull the wrench toward you. If the situation calls for pushing the wrench away, push with an open hand to avoid scraped knuckles if the wrench should slip.

DON'T attempt to lift a heavy component alone - get someone to help you.

DON'T rush or take unsafe shortcuts to finish a job.

DON'T allow children or animals in or around the vehicle while you are working on it.

DO wear eye protection when using power tools such as a drill, sander, bench grinder, etc. and when working under a vehicle.

DO keep loose clothing and long hair well out of the way of moving parts.

DO make sure that any hoist used has a safe working load rating adequate for the job.

DO get someone to check on you periodically when working alone on a vehicle.

DO carry out work in a logical sequence and make sure that everything is correctly assembled and tightened.

DO keep chemicals and fluids tightly capped and out of the reach of children and pets.

DO remember that your vehicle's safety affects that of yourself and others. If in doubt on any point, get professional advice.

Steering, suspension and brakes

These systems are essential to driving safety, so make sure you have a qualified shop or individual check your work. Also, compressed suspension springs can cause injury if released suddenly - be sure to use a spring compressor.

Airbags

Airbags are explosive devices that can **CAUSE** injury if they deploy while you're working on the vehicle. Follow the manufacturer's instructions to disable the airbag whenever you're working in the vicinity of airbag components.

Asbestos

Certain friction, insulating, sealing, and other products - such as brake linings, brake bands, clutch linings, torque converters, gaskets, etc. - may contain asbestos or other hazardous friction material. Extreme care must be taken to avoid inhalation of dust from such products, since it is hazardous to health. If in doubt, assume that they do contain asbestos.

Fire

Remember at all times that gasoline is highly flammable. Never smoke or have any kind of open flame around when working on a vehicle. But the risk does not end there. A spark caused by an electrical short circuit, by two metal surfaces contacting each other, or even by static electricity built up in your body under certain conditions, can ignite gasoline vapors, which in a confined space are highly explosive. Do not, under any circumstances, use gasoline for cleaning parts. Use an approved safety solvent.

Always disconnect the battery ground (-) cable at the battery before working on any part of the fuel system or electrical system. Never risk spilling fuel on a hot engine or exhaust component. It is strongly recommended that a fire extinguisher suitable for use on fuel and electrical fires be kept handy in the garage or workshop at all times. Never try to extinguish a fuel or electrical fire with water.

Fumes

Certain fumes are highly toxic and can quickly cause unconsciousness and even death if inhaled to any extent. Gasoline vapor falls into this category, as do the vapors from some cleaning solvents. Any draining or pouring of such volatile fluids should be done in a well ventilated area.

When using cleaning fluids and solvents, read the instructions on the container carefully. Never use materials from unmarked containers.

Never run the engine in an enclosed space, such as a garage. Exhaust fumes contain carbon monoxide, which is extremely poisonous. If you need to run the engine, always do so in the open air, or at least have the rear of the vehicle outside the work area.

The battery

Never create a spark or allow a bare light bulb near a battery. They normally give off a certain amount of hydrogen gas, which is highly explosive.

Always disconnect the battery ground (-) cable at the battery before working on the fuel or electrical systems.

If possible, loosen the filler caps or cover when charging the battery from an external source (this does not apply to sealed or maintenance-free batteries). Do not charge at an excessive rate or the battery may burst.

Take care when adding water to a non maintenance-free battery and when carrying a battery. The electrolyte, even when diluted, is very corrosive and should not be allowed to contact clothing or skin.

Always wear eye protection when cleaning the battery to prevent the caustic deposits from entering your eyes.

Household current

When using an electric power tool, inspection light, etc., which operates on household current, always make sure that the tool is correctly connected to its plug and that, where necessary, it is properly grounded. Do not use such items in damp conditions and, again, do not create a spark or apply excessive heat in the vicinity of fuel or fuel vapor.

Secondary ignition system voltage

A severe electric shock can result from touching certain parts of the ignition system (such as the spark plug wires) when the engine is running or being cranked, particularly if components are damp or the insulation is defective. In the case of an electronic ignition system, the secondary system voltage is much higher and could prove fatal.

Hydrofluoric acid

This extremely corrosive acid is formed when certain types of synthetic rubber, found in some O-rings, oil seals, fuel hoses, etc. are exposed to temperatures above 750-degrees F (400-degrees C). The rubber changes into a charred or sticky substance containing the acid. *Once formed, the acid remains dangerous for years. If it gets onto the skin, it may be necessary to amputate the limb concerned.*

When dealing with a vehicle which has suffered a fire, or with components salvaged from such a vehicle, wear protective gloves and discard them after use.

Automotive chemicals and lubricants

A number of automotive chemicals and lubricants are available for use during vehicle maintenance and repair. They include a wide variety of products ranging from cleaning solvents and degreasers to lubricants and protective sprays for rubber, plastic and vinyl.

Cleaners

Carburetor cleaner and choke cleaner is a strong solvent for gum, varnish and carbon. Most carburetor cleaners leave a dry-type lubricant film which will not harden or gum up. Because of this film it is not recommended for use on electrical components.

Brake system cleaner is used to remove brake dust, grease and brake fluid from the brake system, where clean surfaces are absolutely necessary. It leaves no residue and often eliminates brake squeal caused by contaminants.

Electrical cleaner removes oxidation, corrosion and carbon deposits from electrical contacts, restoring full current flow. It can also be used to clean spark plugs, carburetor jets, voltage regulators and other parts where an oil-free surface is desired.

Demoisturants remove water and moisture from electrical components such as alternators, voltage regulators, electrical connectors and fuse blocks. They are non-conductive and non-corrosive.

Degreasers are heavy-duty solvents used to remove grease from the outside of the engine and from chassis components. They can be sprayed or brushed on and, depending on the type, are rinsed off either with water or solvent.

Lubricants

Motor oil is the lubricant formulated for use in engines. It normally contains a wide variety of additives to prevent corrosion and reduce foaming and wear. Motor oil comes in various weights (viscosity ratings) from 0 to 50. The recommended weight of the oil depends on the season, temperature and the demands on the engine. Light oil is used in cold climates and under light load conditions. Heavy oil is used in hot climates and where high loads are encountered. Multi-viscosity oils are designed to have characteristics of both light and heavy oils and are available in a number of weights from 0W-20 to 20W-50.

Gear oil is designed to be used in differentials, manual transmissions and other areas where high-temperature lubrication is required.

Chassis and wheel bearing grease is a heavy grease used where increased loads and friction are encountered, such as for wheel bearings, balljoints, tie-rod ends and universal joints.

High-temperature wheel bearing grease is designed to withstand the extreme temperatures encountered by wheel bearings in disc brake equipped vehicles. It usually contains molybdenum disulfide (moly), which is a dry-type lubricant.

White grease is a heavy grease for metal-to-metal applications where water is a problem. White grease stays soft under both low and high temperatures (usually from -100 to +190-degrees F), and will not wash off or dilute in the presence of water.

Assembly lube is a special extreme pressure lubricant, usually containing moly, used to lubricate high-load parts (such as main and rod bearings and cam lobes) for initial start-up of a new engine. The assembly lube lubricates the parts without being squeezed out or washed away until the engine oiling system begins to function.

Silicone lubricants are used to protect rubber, plastic, vinyl and nylon parts.

Graphite lubricants are used where oils cannot be used due to contamination problems, such as in locks. The dry graphite will lubricate metal parts while remaining uncontaminated by dirt, water, oil or acids. It is electrically conductive and will not foul electrical contacts in locks such as the ignition switch.

Moly penetrants loosen and lubricate frozen, rusted and corroded fasteners and prevent future rusting or freezing.

Heat-sink grease is a special electrically non-conductive grease that is used for mounting electronic ignition modules where it is essential that heat is transferred away from the module.

Sealants

RTV sealant is one of the most widely used gasket compounds. Made from silicone, RTV is air curing, it seals, bonds, waterproofs, fills surface irregularities, remains flexible, doesn't shrink, is relatively easy to remove, and is used as a supplementary sealer with almost all low and medium temperature gaskets.

Anaerobic sealant is much like RTV in that it can be used either to seal gaskets or to form gaskets by itself. It remains flexible, is solvent resistant and fills surface imperfections. The difference between an anaerobic sealant and an RTV-type sealant is in the curing. RTV cures when exposed to air, while an anaerobic sealant cures only in the absence of air. This means that an anaerobic sealant cures only after the assembly of parts, sealing them together.

Thread and pipe sealant is used for sealing hydraulic and pneumatic fittings and vacuum lines. It is usually made from a Teflon compound, and comes in a spray, a paint-on liquid and as a wrap-around tape.

Chemicals

Anti-seize compound prevents seizing, galling, cold welding, rust and corrosion in fasteners. High-temperature anti-seize, usually made with copper and graphite lubricants, is used for exhaust system and exhaust manifold bolts.

Anaerobic locking compounds are used to keep fasteners from vibrating or working loose and cure only after installation, in the absence of air. Medium strength locking compound is used for small nuts, bolts and screws that may be removed later. High-strength locking compound is for large nuts, bolts and studs which aren't removed on a regular basis.

Oil additives range from viscosity index improvers to chemical treatments that claim to reduce internal engine friction. It should be noted that most oil manufacturers caution against using additives with their oils.

Gas additives perform several functions, depending on their chemical makeup. They usually contain solvents that help dissolve gum and varnish that build up on carburetor, fuel injection and intake parts. They also serve to break down carbon deposits that form on the inside surfaces of the combustion chambers. Some additives contain upper cylinder lubricants for valves and piston rings, and others contain chemicals to remove condensation from the gas tank.

Miscellaneous

Brake fluid is specially formulated hydraulic fluid that can withstand the heat and pressure encountered in brake systems. Care must be taken so this fluid does not come in contact with painted surfaces or plastics. An opened container should always be resealed to prevent contamination by water or dirt.

Weatherstrip adhesive is used to bond weatherstripping around doors, windows and trunk lids. It is sometimes used to attach trim pieces.

Undercoating is a petroleum-based, tar-like substance that is designed to protect metal surfaces on the underside of the vehicle from corrosion. It also acts as a sound-deadening agent by insulating the bottom of the vehicle.

Waxes and polishes are used to help protect painted and plated surfaces from the weather. Different types of paint may require the use of different types of wax and polish. Some polishes utilize a chemical or abrasive cleaner to help remove the top layer of oxidized (dull) paint on older vehicles. In recent years many non-wax polishes that contain a wide variety of chemicals such as polymers and silicones have been introduced. These non-wax polishes are usually easier to apply and last longer than conventional waxes and polishes.

Troubleshooting

Contents

Engine

1 Engine will not rotate when attempting to start

1   Battery terminal connections loose or corroded. Check the cable terminals at the battery; tighten cable clamp and/or clean off corrosion as necessary (see Chapter 1).
2   Battery discharged or faulty. If the cable ends are clean and tight on the battery posts, turn the key to the On position and switch on the headlights or windshield wipers. If they won't run, the battery is discharged.
3   Automatic transmission not engaged in Park (P) or Neutral (N).
4   Broken, loose or disconnected wires in the starting circuit.
5   Starter motor pinion jammed in driveplate ring gear. Remove starter (Chapter 5) and inspect pinion and driveplate (Chapter 2).
6   Starter solenoid faulty (Chapter 5).
7   Starter motor faulty (Chapter 5).
8   Clutch pedal neutral start switch faulty.
9   Ignition switch faulty (Chapter 12).
10   Engine seized. Try to turn the crankshaft with a large socket and breaker bar on the pulley bolt.
11   Starter relay faulty (Chapter 5).

2 Engine rotates but will not start

1   Fuel tank empty.
2   Battery discharged (engine rotates slowly).
3   Battery terminal connections loose or corroded.
4   Fuel not reaching fuel injectors. Check for clogged fuel filter or lines and defective fuel pump. Also make sure the tank vent lines aren't clogged (Chapter 4).
5   Low cylinder compression. Check as described in Chapter 2.
6   Water in fuel. Drain tank and fill with new fuel.
7   Defective ignition coil(s) (Chapter 5).
8   Dirty or clogged fuel injector(s) (Chapter 4).
9   Wet or damaged ignition components (Chapters 1 and 5).
10   Worn, faulty or incorrectly gapped spark plugs (Chapter 1).
11   Broken, loose or disconnected wires in the starting circuit (see previous Section).
12   Broken, loose or disconnected wires at the ignition coil or faulty coil (Chapter 5).
13   Timing chain failure or wear affecting valve timing (Chapter 2).
14   Fuel injection or engine control systems failure (Chapters 4 and 6).
15   Defective MAF sensor (Chapter 6)

3 Starter motor operates without turning engine

1   Starter pinion sticking. Remove the starter (Chapter 5) and inspect.
2   Starter pinion or driveplate teeth worn or broken. Remove the inspection cover and inspect.

4 Engine hard to start when cold

1   Battery discharged or low. Check as described in Chapter 1.
2   Fuel not reaching the fuel injectors. Check the fuel filter, lines and fuel pump (Chapters 1 and 4).
3   Defective spark plugs (Chapter 1).
4   Defective engine coolant temperature sensor (Chapter 6).
5   Fuel injection or engine control systems malfunction (Chapters 4 and 6).

5 Engine hard to start when hot

1   Air filter dirty (Chapter 1).
2   Bad engine ground connection.
3   Fuel injection or engine control systems malfunction (Chapters 4 and 6).

6 Starter motor noisy or engages roughly

1   Pinion or driveplate teeth worn or broken. Remove the inspection cover on the left side of the engine and inspect.
2   Starter motor mounting bolts loose or missing.

7 Engine starts but stops immediately

1   Loose or damaged wire harness connections at coil or alternator.
2   Intake manifold vacuum leaks. Make sure all mounting bolts/nuts are tight and all vacuum hoses connected to the manifold are attached properly and in good condition.
3   Insufficient fuel pressure (see Chapter 4).
4   Fuel injection or engine control systems malfunction (Chapters 4 and 6).

8 Engine lopes while idling or idles erratically

1   Vacuum leaks. Check mounting bolts at the intake manifold for tightness. Make sure that all vacuum hoses are connected and in good condition. Use a stethoscope or a length of fuel hose held against your ear to listen for vacuum leaks while the engine is running. A hissing sound will be heard. A soapy water solution will also detect leaks. Check the intake manifold gasket surfaces.
2   Leaking EGR valve or plugged PCV valve (see Chapters 1 and 6).

3   Air filter clogged (Chapter 1).
4   Fuel pump not delivering sufficient fuel (Chapter 4).
5   Leaking head gasket. Perform a cylinder compression check (Chapter 2).
6   Timing belt/chain(s) worn (Chapter 2).
7   Camshaft lobes worn (Chapter 2).
8   Valves burned or otherwise leaking (Chapter 2).
9   Ignition system not operating properly (Chapters 1 and 5).
10   Fuel injection or engine control systems malfunction (Chapters 4 and 6).

9 Engine misses at idle speed

1   Spark plugs faulty or not gapped properly (Chapter 1).
2   Faulty spark plug wires (Chapter 1).
3   Wet or damaged ignition components (Chapter 5).
4   Short circuits in ignition, coil or spark plug wires.
5   Sticking or faulty emissions systems (see Chapter 6).
6   Clogged fuel filter and/or foreign matter in fuel. Remove the fuel filter (Chapter 1) and inspect.
7   Vacuum leaks at intake manifold or hose connections. Check as described in Section 8.
8   Low or uneven cylinder compression. Check as described in Chapter 2.
9   Fuel injection or engine control systems malfunction (Chapters 4 and 6).

10 Excessively high idle speed

1   Sticking throttle linkage (Chapter 4).
2   Vacuum leaks at intake manifold or hose connections. Check as described in Section 8.
3   Fuel injection or engine control systems malfunction (Chapters 4 and 6).

11 Battery will not hold a charge

1   Alternator drivebelt defective or not adjusted properly (Chapter 1).
2   Battery cables loose or corroded (Chapter 1).
3   Alternator not charging properly (Chapter 5).
4   Loose, broken or faulty wires in the charging circuit (Chapter 5).
5   Short circuit causing a continuous drain on the battery.
6   Battery defective internally.

12 Alternator light stays on

1   Fault in alternator or charging circuit (Chapter 5).
2   Alternator drivebelt defective or not properly adjusted (Chapter 1).

13   Alternator light fails to come on when key is turned on

1   Faulty bulb (Chapter 12).
2   Defective alternator (Chapter 5).
3   Fault in the printed circuit, dash wiring or bulb holder (Chapter 12).

14   Engine misses throughout driving speed range

1   Fuel filter clogged and/or impurities in the fuel system. Check fuel filter or clean system (Chapter 4).
2   Faulty or incorrectly gapped spark plugs (Chapter 1).
3   Defective spark plug wires (Chapter 1).
4   Emissions system components faulty (Chapter 6).
5   Low or uneven cylinder compression pressures. Check as described in Chapter 2.
6   Weak or faulty ignition coil(s) (Chapter 5).
7   Weak or faulty ignition system (Chapter 5).
8   Vacuum leaks at intake manifold or vacuum hoses (see Section 8).
9   Dirty or clogged fuel injector(s) (Chapter 4).
10   Leaky EGR valve (Chapter 6).
11   Fuel injection or engine control systems malfunction (Chapters 4 and 6).

15   Hesitation or stumble during acceleration

1   Ignition system not operating properly (Chapter 5).
2   Dirty or clogged fuel injector(s) (Chapter 4).
3   Low fuel pressure. Check for proper operation of the fuel pump and for restrictions in the fuel filter and lines (Chapter 4).
4   Fuel injection or engine control systems malfunction (Chapters 4 and 6).

16   Engine stalls

1   Idle speed incorrect (Chapter 4).
2   Fuel filter clogged and/or water and impurities in the fuel system (Chapter 4).
3   Damaged or wet ignition wires.
4   Emissions system components faulty (Chapter 6).
5   Faulty or incorrectly gapped spark plugs (Chapter 1). Also check the spark plug wires (Chapter 1).
6   Vacuum leak at the intake manifold or vacuum hoses. Check as described in Section 8.
7   Fuel injection or engine control systems malfunction (Chapters 4 and 6).

17   Engine lacks power

1   Faulty or incorrectly gapped spark plugs (Chapter 1).
2   Air filter dirty (Chapter 1).
3   Faulty ignition coil(s) (Chapter 5).
4   Brakes binding (Chapters 1 and 9).
5   Automatic transmission fluid level incorrect, causing slippage (Chapter 1).
6   Fuel filter clogged and/or impurities in the fuel system (Chapter 4).
7   EGR system not functioning properly (Chapter 6).
8   Use of sub-standard fuel. Fill tank with proper octane fuel.
9   Low or uneven cylinder compression pressures. Check as described in Chapter 2.
10   Vacuum leak at intake manifold or vacuum hoses (check as described in Section 8).
11   Dirty or clogged fuel injector(s) (Chapter 4).
12   Fuel injection or engine control systems malfunction (Chapters 4 and 6).
13   Restricted exhaust system (Chapter 4).

18   Engine backfires

1   EGR system not functioning properly (Chapter 6).
2   Damaged valve springs or sticking valves (Chapter 2).
3   Vacuum leak at the intake manifold or vacuum hoses (see Section 8).

19   Engine surges while holding accelerator steady

1   Vacuum leak at the intake manifold or vacuum hoses (see Section 8).
2   Restricted air filter (Chapter 1).
3   Fuel pump or pressure regulator defective (Chapter 4).
4   Fuel injection or engine control systems malfunction (Chapters 4 and 6).

20   Pinging or knocking engine sounds when engine is under load

1   Incorrect grade of fuel. Fill tank with fuel of the proper octane rating.
2   Carbon build-up in combustion chambers. Remove cylinder head(s) and clean combustion chambers (Chapter 2).
3   Incorrect spark plugs (Chapter 1).
4   Fuel injection or engine control systems malfunction (Chapters 4 and 6).
5   Restricted exhaust system (Chapter 4).

21   Engine continues to run after being turned off

1   Idle speed too high (Chapter 4).

2   Incorrect spark plug heat range (Chapter 1).
3   Vacuum leak at the intake manifold or vacuum hoses (see Section 8).
4   Carbon build-up in combustion chambers. Remove the cylinder head(s) and clean the combustion chambers (Chapter 2).
5   Valves sticking (Chapter 2).
6   EGR system not operating properly (Chapter 6).
7   Fuel injection or engine control systems malfunction (Chapters 4 and 6).
8   Check for causes of overheating (see Section 27).

22   Low oil pressure

1   Improper grade of oil.
2   Oil pump worn or damaged (Chapter 2).
3   Engine overheating (see Section 27).
4   Clogged oil filter (Chapter 1).
5   Clogged oil strainer (Chapter 2).
6   Oil pressure gauge not working properly (Chapter 2).

23   Excessive oil consumption

1   Loose oil drain plug.
2   Loose bolts or damaged oil pan gasket (Chapter 2).
3   Loose bolts or damaged front cover gasket (Chapter 2).
4   Front or rear crankshaft oil seal leaking (Chapter 2).
5   Loose bolts or damaged valve cover gasket (Chapter 2).
6   Loose oil filter (Chapter 1).
7   Loose or damaged oil pressure switch (Chapter 2).
8   Pistons and cylinders excessively worn (Chapter 2).
9   Piston rings not installed correctly on pistons (Chapter 2).
10   Worn or damaged piston rings (Chapter 2).
11   Intake and/or exhaust valve oil seals worn or damaged (Chapter 2).
12   Worn or damaged valves/guides (Chapter 2).
13   Faulty or incorrect PCV valve allowing too much crankcase airflow.

24   Excessive fuel consumption

1   Dirty or clogged air filter element (Chapter 1).
2   Incorrect idle speed (Chapter 4).
3   Low tire pressure or incorrect tire size (Chapter 10).
4   Inspect for binding brakes (see Chapter 9).
5   Fuel leakage. Check all connections, lines and components in the fuel system (Chapter 4).
6   Dirty or clogged fuel injectors (Chapter 4).

7   Fuel injection or engine control systems malfunction (Chapters 4 and 6).
8   Thermostat stuck open or not installed.
9   Improperly operating transmission.

25 Fuel odor

1   Fuel leakage. Check all connections, lines and components in the fuel system (Chapter 4).
2   Fuel tank overfilled. Fill only to automatic shut-off.
3   Charcoal canister filter in Evaporative Emissions Control system clogged (Chapter 1).
4   Vapor leaks from Evaporative Emissions Control system lines (Chapter 6).

26 Miscellaneous engine noises

1   A strong dull noise that becomes more rapid as the engine accelerates indicates worn or damaged crankshaft bearings or an unevenly worn crankshaft. To pinpoint the trouble spot, remove the spark plug wire from one plug at a time and crank the engine over. If the noise stops, the cylinder with the removed plug wire indicates the problem area. Replace the bearing and/or service or replace the crankshaft (Chapter 2).
2   A similar (yet slightly higher pitched) noise to the crankshaft knocking described in the previous paragraph, that becomes more rapid as the engine accelerates, indicates worn or damaged connecting rod bearings (Chapter 2). The procedure for locating the problem cylinder is the same as described in Paragraph 1.
3   An overlapping metallic noise that increases in intensity as the engine speed increases, yet diminishes as the engine warms up indicates abnormal piston and cylinder wear (Chapter 2). To locate the problem cylinder, use the procedure described in Paragraph 1.
4   A rapid clicking noise that becomes faster as the engine accelerates indicates a worn piston pin or piston pin hole. This sound will happen each time the piston hits the highest and lowest points in the stroke (Chapter 2). The procedure for locating the problem piston is described in Paragraph 1.
5   A metallic clicking noise coming from the water pump indicates worn or damaged water pump bearings or pump. Replace the water pump with a new one (Chapter 2).
6   A rapid tapping sound or clicking sound that becomes faster as the engine speed increases indicates "valve tapping." This can be identified by holding one end of a section of hose to your ear and placing the other end at different spots along the valve cover. The point where the sound is loudest indicates the problem valve. If the pushrod and rocker arm components are in good shape, you likely have a collapsed valve lifter. Changing

the engine oil and adding a high viscosity oil treatment will sometimes cure a stuck lifter problem. If the problem persists, the lifters, pushrods and rocker arms must be removed for inspection (see Chapter 2).
7   A steady metallic rattling or rapping sound coming from the area of the timing chain cover indicates a worn, damaged or out-of-adjustment timing chain. Service or replace the chain and related components (Chapter 2).

Cooling system

27 Overheating

1   Insufficient coolant in system (Chapter 1).
2   Drivebelt defective or not adjusted properly (Chapter 1).
3   Radiator core blocked or radiator grille dirty and restricted (Chapter 3).
4   Thermostat faulty (Chapter 3).
5   Cooling fan not functioning properly (Chapter 3).
6   Expansion tank cap not maintaining proper pressure. Have cap pressure tested by a gas station or repair shop.
7   Defective water pump (Chapter 3).
8   Improper grade of engine oil.
9   Inaccurate temperature gauge (Chapter 12).
10   Blown cylinder head gasket (Chapter 2).

28 Overcooling

1   Thermostat faulty (Chapter 3).
2   Inaccurate temperature gauge (Chapter 12).

29 External coolant leakage

1   Deteriorated or damaged hoses. Loose clamps at hose connections (Chapter 1).
2   Water pump seals defective. If this is the case, water will drip from the weep hole in the water pump body (Chapter 3).
3   Leakage from radiator core or header tank. This will require the radiator to be professionally repaired (see Chapter 3 for removal procedures).
4   Leakage from the coolant reservoir or degas bottle.
5   Engine drain plugs or water jacket freeze plugs leaking (see Chapters 1 and 2).
6   Leak from coolant temperature switch (Chapter 3).
7   Leak from damaged gaskets or small cracks (Chapter 2).
8   Leak from oil cooler or oil cooler adapter housing (Chapter 3).

30 Internal coolant leakage

Note: *Internal coolant leaks can usually be detected by examining the oil. Check the dipstick and inside the rocker arm cover for water deposits and an oil consistency like that of a milkshake.*
1   Leaking cylinder head gasket. Have the system pressure tested or remove the cylinder head (Chapter 2) and inspect.
2   Cracked cylinder bore or cylinder head. Dismantle engine and inspect (Chapter 2).
3   Loose cylinder head bolts (tighten as described in Chapter 2).
4   Leakage from internal coolant pipe/hose (accessible only with intake manifold removed) (Chapter 2A).

31 Abnormal coolant loss

1   Overfilling system (Chapter 1).
2   Coolant boiling away due to overheating (see causes in Section 27).
3   Internal or external leakage (see Sections 29 and 30).
4   Faulty expansion tank cap. Have the cap pressure tested.
5   Cooling system being pressurized by engine compression. This could be due to a cracked head or block or leaking head gasket(s). Have the system tested for the presence of combustion gas in the coolant at a shop.

32 Poor coolant circulation

1   Inoperative water pump. A quick test is to pinch the top radiator hose closed with your hand while the engine is idling, then release it. You may be able to feel a surge of coolant if the pump is working properly (Chapter 3).
2   Restriction in cooling system. Drain, flush and refill the system (Chapter 1). If necessary, remove the radiator (Chapter 3) and have it reverse flushed or professionally cleaned.
3   Loose water pump drivebelt (Chapter 1).
4   Thermostat sticking (Chapter 3).
5   Insufficient coolant (Chapter 1).

33 Corrosion

1   Excessive impurities in the water. Soft, clean water is recommended. Distilled or rainwater is satisfactory.
2   Insufficient antifreeze solution (refer to Chapter 1 for the proper ratio of water to antifreeze).
3   Infrequent flushing and draining of system. Regular flushing of the cooling system should be carried out at the specified intervals as described in Chapter 1.

Clutch

34 Fails to release (pedal pressed to the floor - shift lever does not move freely in and out of Reverse)

1 Leak in the clutch hydraulic system. Check the master cylinder, release cylinder and lines (Chapter 8).
2 Clutch plate warped or damaged (Chapter 8).

35 Clutch slips (engine speed increases with no increase in vehicle speed)

1 Clutch plate oil soaked or lining worn. Remove clutch (Chapter 8) and inspect.
2 Clutch plate not seated. It may take 30 or 40 normal starts for a new one to seat.
3 Pressure plate worn (Chapter 8).

36 Grabbing (chattering) as clutch is engaged

1 Oil on clutch plate lining. Remove (Chapter 8) and Inspect. Correct any leakage source.
2 Worn or loose engine or transmission mounts. These units move slightly when the clutch is released. Inspect the mounts and bolts (Chapter 2).
3 Worn splines on clutch plate hub. Remove the clutch components (Chapter 8) and inspect.
4 Warped pressure plate or flywheel. Remove the clutch components and inspect.

37 Squeal or rumble with clutch fully engaged (pedal released)

Release bearing binding on transmission bearing retainer. Remove clutch components (Chapter 8) and check bearing. Remove any burrs or nicks; clean and relubricate bearing retainer before installing.

38 Squeal or rumble with clutch fully disengaged (pedal depressed)

1 Worn, defective or broken release bearing (Chapter 8).
2 Worn or broken pressure plate springs (or diaphragm fingers) (Chapter 8).

39 Clutch pedal stays on floor when disengaged

1 Linkage or release bearing binding. Inspect the linkage or remove the clutch com-

ponents as necessary.
2 Make sure proper pedal stop (bumper) is installed.

Manual transmission

Note: *All the following references are in Chapter 7A, unless noted.*

40 Noisy in Neutral with engine running

1 Input shaft bearing worn.
2 Damaged main drive gear bearing.
3 Worn countershaft bearings.
4 Worn or damaged countershaft endplay shims.

41 Noisy in all gears

1 Any of the above causes, and/or:
2 Insufficient lubricant (see the checking procedures in Chapter 1).

42 Noisy in one particular gear

1 Worn, damaged or chipped gear teeth for that particular gear.
2 Worn or damaged synchronizer for that particular gear.

43 Slips out of high gear

1 Transmission loose on clutch housing.
2 Dirt between the transmission case and engine or misalignment of the transmission.

44 Difficulty in engaging gears

1 Clutch not releasing completely (see clutch adjustment in Chapter 1).
2 Loose, damaged or out-of-adjustment shift linkage. Make a thorough inspection, replacing parts as necessary.

45 Oil leakage

1 Excessive amount of lubricant in the transmission (see Chapter 1 for correct checking procedures). Drain lubricant as required.
2 Transmission oil seal or vehicle speed sensor O-ring in need of replacement.

Automatic transmission

Note: *Due to the complexity of the automatic transmission, it's difficult for the home mechanic to properly diagnose and service this component. For problems other than the following, the vehicle should be taken to a dealer service department or a transmission shop.*

46 General shift mechanism problems

1 Common problems which may be attributed to a misadjusted shift cable are:
a) *Engine starting in gears other than Park or Neutral.*
b) *Indicator on shifter pointing to a gear other than the one actually being selected.*
c) *Vehicle moves when in Park.*
2 Refer to Chapter 7B to check the transmission range (TR) sensor adjustment.

47 Transmission will not downshift with accelerator pedal pressed to the floor

Since these transmissions are electronically controlled, your dealer or a professional shop with the proper equipment will have to diagnose the probable cause.

48 Transmission slips, shifts rough, is noisy or has no drive in forward or reverse gears

1 There are many probable causes for the above problems, but the home mechanic should be concerned with only one possibility - fluid level.
2 Before taking the vehicle to a repair shop, check the level and condition of the fluid as described in Chapter 1. Correct fluid level as necessary or change the fluid and filter if needed. If the problem persists, have a professional diagnose the problem.

49 Fluid leakage

1 Automatic transmission fluid is a deep red color. Fluid leaks should not be confused with engine oil, which can easily be blown by airflow to the transmission.
2 To pinpoint a leak, first remove all built-up dirt and grime from around the transmission. Degreasing agents and/or steam cleaning will achieve this. With the underside clean, drive the vehicle at low speeds so airflow will not blow the leak far from its source. Raise the vehicle and determine where the leek is coming from. Common areas of leakage are:
a) *Pan: Tighten the mounting bolts and/ or replace the pan gasket as necessary (see Chapter 7B).*
b) *Filler pipe: Replace the rubber seal where the pipe enters the transmission case.*
c) *Transmission oil lines: Tighten the connectors where the lines enter the transmission case and/or replace the lines.*

d) **Vent pipe:** *Transmission overfilled and/ or water in fluid (see checking procedures, Chapter 1).*

e) **Speedometer connector:** *Replace the O-ring where the speedometer sensor enters the transmission case (Chapter 7B).*

Transfer case

50 Transfer case is difficult to shift into the desired range

1 Speed may be too great to permit engagement. Stop the vehicle and shift into the desired range.
2 Shift linkage loose, bent or binding on a manual shift transfer case. Check the linkage for damage or wear and replace or lubricate as necessary (Chapter 7C).
3 Defective circuit and/or range switch on electric shift transfer case (Chapter 7C).
4 If the vehicle has been driven on a paved surface for some time, the driveline torque can make shifting difficult. Stop and shift into two-wheel drive on paved or hard surfaces.
5 Insufficient or incorrect grade of lubricant. Drain and refill the transfer case with the specified lubricant. (Chapter 1).
6 Worn or damaged internal components. Disassembly and overhaul of the transfer case may be necessary (Chapter 7C).

51 Transfer case noisy in all gears

Insufficient or incorrect grade of lubricant. Drain and refill (Chapter 1).

52 Noisy or jumps out of four-wheel drive Low range

1 Transfer case not fully engaged. Stop the vehicle, shift into Neutral and then engage 4L.
2 Shift linkage loose, worn or binding. Tighten, repair or lubricate linkage as necessary.
3 Shift fork cracked, inserts worn or fork binding on the rail. See your dealer for a new or rebuilt unit.

53 Lubricant leaks from the vent or output shaft seals

1 Transfer case is overfilled. Drain to the proper level (Chapter 1).
2 Vent is clogged or jammed closed. Clear or replace the vent.
3 Output shaft seal incorrectly installed or damaged. Replace the seal and check contact surfaces for nicks and scoring.

Driveshaft

54 Oil leak at seal end of driveshaft

Defective transmission or transfer case oil seal. See Chapter 7 for replacement procedures. While this is done, check the splined yoke for burrs or a rough condition which may be damaging the seal. Burrs can be removed with crocus cloth or a fine whetstone.

55 Knock or clunk when the transmission is under initial load (just after transmission is put into gear)

1 Loose or disconnected rear suspension components. Check all mounting bolts, nuts and bushings (see Chapter 10).
2 Loose driveshaft bolts. Inspect all bolts and nuts and tighten them to the specified torque.
3 Worn or damaged universal joint bearings. Check for wear (see Chapter 8).

56 Metallic grinding sound consistent with vehicle speed.

Pronounced wear in the universal joint bearings. Check as described in Chapter 8.

57 Vibration

Note: *Before assuming that the driveshaft is at fault, make sure the tires are perfectly balanced and perform the following test.*
1 Install a tachometer inside the vehicle to monitor engine speed as the vehicle is driven. Drive the vehicle and note the engine speed at which the vibration (roughness) is most pronounced. Now shift the transmission to a different gear and bring the engine speed to the same point.
2 If the vibration occurs at the same engine speed (rpm) regardless of which gear the transmission is in, the driveshaft is NOT at fault since the driveshaft speed varies.
3 If the vibration decreases or is eliminated when the transmission is in a different gear at the same engine speed, refer to the following probable causes.
4 Bent or dented driveshaft. Inspect and replace as necessary (see Chapter 8).
5 Undercoating or built-up dirt, etc. on the driveshaft. Clean the shaft thoroughly and recheck.
6 Worn universal joint bearings. Remove and inspect (see Chapter 8).
7 Driveshaft and/or companion flange out of balance. Check for missing weights on the shaft. Remove the driveshaft (see Chapter 8) and reinstall 180-degrees from original position, then retest. Have the driveshaft professionally balanced if the problem persists.

Axles

58 Noise

1 Road noise. No corrective procedures available.
2 Tire noise. Inspect tires and check tire pressures (Chapter 1).
3 Rear wheel bearings loose, worn or damaged (Chapter 8).

59 Vibration

See probable causes under Driveshaft. Proceed under the guidelines listed for the driveshaft. If the problem persists, check the rear wheel bearings by raising the rear of the vehicle and spinning the rear wheels by hand. Listen for evidence of rough (noisy) bearings. Remove and inspect (see Chapter 8).

60 Oil leakage

1 Pinion seal damaged (see Chapter 8).
2 Axleshaft oil seals damaged (see Chapter 8).
3 Differential inspection cover leaking. Tighten the bolts or replace the gasket as required (see Chapters 1 and 8).

Brakes

Note: *Before assuming that a brake problem exists, make sure that the tires are in good condition and inflated properly (see Chapter 1), that the front end alignment is correct and that the vehicle is not loaded with weight in an unequal manner.*

61 Vehicle pulls to one side during braking

1 Defective, damaged or oil-contaminated disc brake pads or linings on one side. Inspect as described in Chapter 9.
2 Excessive wear of brake pad or shoe material or disc/drum on one side. Inspect and correct as necessary.
3 Loose or disconnected front suspension components. Inspect and tighten all bolts to the specified torque (Chapter 10).
4 Defective brake caliper/drum brake assembly (Chapter 9).

62 Noise (high-pitched squeal with the brakes applied)

1 Disc brake pads worn out. The noise comes from the wear sensor rubbing against the disc (does not apply to all vehicles) or the actual pad backing plate itself if the material is completely worn away. Replace the pads with new ones immediately (Chapter 9). If the

pad material has worn completely away, the brake discs should be inspected for damage as described in Chapter 9.

2   Linings contaminated with dirt or grease. Replace pads or shoes.

3   Incorrect linings. Replace with correct linings.

63   Excessive brake pedal travel

1   Partial brake system failure. Inspect the entire system (Chapter 9) and correct as required.

2   Insufficient fluid in the master cylinder. Check (Chapter 1), add fluid and bleed the system if necessary (Chapter 9).

3   Rear drum brakes not adjusting properly. Make a series of starts and stops while the vehicle is in Reverse. If this does not correct the situation, remove the drums and inspect the self-adjusters (Chapter 9).

4   Problem with the anti-lock brake system (Chapter 9)

64   Brake pedal feels spongy when depressed

1   Air in the hydraulic lines. Bleed the brake system (Chapter 9).

2   Faulty flexible hoses. Inspect all system hoses and lines. Replace parts as necessary.

3   Master cylinder mounting bolts/nuts loose.

4   Master cylinder defective (Chapter 9).

5   Problem with the anti-lock brake system (Chapter 9).

65   Excessive effort required to stop vehicle

1   Power brake booster not operating properly (Chapter 9).

2   Excessively worn pads or linings. Inspect and replace if necessary (Chapter 9).

3   One or more caliper pistons or wheel cylinders seized or sticking (Chapter 9).

4   Brake pads or linings contaminated with oil or grease. Inspect and replace as required (Chapter 9).

5   New pads or shoes installed and not yet seated. It will take a while for the new material to seat against the disc/drum.

6   Problem with the anti-lock brake system (Chapter 9).

66   Pedal travels to the floor with little resistance

1   Little or no fluid in the master cylinder

reservoir caused by leaking wheel cylinder(s), leaking caliper piston(s), loose, damaged or disconnected brake lines. Inspect the entire system and correct as necessary.

2   Worn master cylinder seals (Chapter 9).

3   Problem with the anti-lock brake system (Chapter 9).

67   Brake pedal pulsates during brake application

1   Caliper improperly installed. Remove and inspect (Chapter 9).

2   Disc or drum defective. Check for excessive lateral runout and parallelism (discs) or excessive out-of-roundness (drums). Have the disc or drum resurfaced or replace it with a new one (Chapter 9).

Suspension and steering systems

68   Vehicle pulls to one side

1   Tire pressures uneven (Chapter 1).

2   Defective tire (Chapter 1).

3   Excessive wear in suspension or steering components (Chapter 10).

4   Front end in need of alignment

5   Front brakes dragging. Inspect the brakes as described in Chapters 1 and 9.

69   Shimmy, shake or vibration

1   Tire or wheel out-of-balance or out-of-round. Have professionally balanced.

2   Loose, worn or out-of-adjustment front wheel bearings (Chapter 10).

3   Shock absorbers and/or suspension components worn or damaged (Chapter 10).

70   Excessive pitching and/or rolling around corners or during braking

1   Defective shock absorbers. Replace as a set (Chapter 10).

2   Broken or weak springs and/or suspension components. Inspect as described in Chapter 10.

71   Excessively stiff steering

1   Lack of fluid in power steering fluid reservoir (Chapter 1).

2   Incorrect tire pressures (Chapter 1).

3   Lack of lubrication at steering joints (see Chapter 1).

4   Front end out of alignment.

5   Lack of power assistance (see Section 73).

72   Excessive play in steering

1   Loose front wheel bearings (Chapter 1).

2   Excessive wear in suspension or steering components (Chapter 10).

3   Steering gearbox damaged or out of adjustment (Chapter 10).

73   Lack of power assistance

1   Steering pump drivebelt faulty or not adjusted properly (Chapter 1).

2   Fluid level low (Chapter 1).

3   Hoses or lines restricted. Inspect and replace parts as necessary.

4   Air in power steering system. Bleed the system (Chapter 10).

74   Excessive tire wear (not specific to one area)

1   Incorrect tire pressures (Chapter 1).

2   Tires out-of-balance. Have professionally balanced.

3   Wheels damaged. Inspect and replace as necessary.

4   Suspension or steering components excessively worn (Chapter 10).

75   Excessive tire wear on outside edge

1   Inflation pressures incorrect (Chapter 1).

2   Excessive speed in turns.

3   Front end alignment incorrect. Have professionally aligned.

4   Suspension arm bent or twisted (Chapter 10).

76   Excessive tire wear on inside edge

1   Inflation pressures incorrect (Chapter 1).

2   Front end alignment incorrect (toe-out). Have professionally aligned.

3   Loose or damaged steering components (Chapter 10).

77   Tire tread worn in one place

1   Tires out-of-balance.

2   Damaged or buckled wheel. Inspect and replace if necessary.

3   Defective tire (Chapter 1).

Chapter 1
Tune-up and routine maintenance

Contents

Specifications

Recommended lubricants and fluids

Note: *Listed here are manufacturer recommendations at the time this manual was written. Manufacturers occasionally upgrade their fluid and lubricant specifications, so check with your local auto parts store for current recommendations.*

Engine oil
  Type .... API grade "certified for gasoline engines"
Viscosity
  2005 and earlier models
    0-degrees to 100-degrees F and higher 10W-30
    100-degrees to -20-degrees F and cooler 5W-30
  2006 and later models
    3.7L V6 engines 5W-20
    4.0L V6 engines 10W-30
Engine coolant .... 50/50 mixture of Mopar® 5 year/100,000 mile Formula antifreeze/coolant with HOAT (Hybrid Organic Additive Technology) and distilled water
Automatic transmission fluid type .... Mopar ATF +4 Type 9602 or equivalent fluid
Manual transmission lubricant type .... Mopar Manual Transmission Lubricant or equivalent
Transfer case lubricant type .... Mopar ATF+4 Type 9602 or equivalent fluid
Differential lubricant type
  Normal operation
    186FIA (model 30) front axle .... SAE 75W-140 synthetic gear lubricant
    198RBI (model 35) rear axle .... SAE 75W-140 synthetic gear lubricant
    8-1/4 rear axle .... SAE 75W-90 API GL-5 gear lubricant
    Trailer towing package rear axle .... SAE 75W-140 synthetic gear lubricant
    Limited-slip differential .... Add Mopar Friction Modifier additive
Brake fluid type .... DOT 3 brake fluid
Power steering fluid .... Mopar ATF +4 Type 9602 or equivalent fluid
Chassis grease type .... NLGI LB or GC-LB chassis grease

Capacities*

Engine oil (with filter change)
  2.4L four-cylinder .... 5 quarts (4.7L)
  3.7L V6 .... 5 quarts (4.7L)
  4.0L V6 .... 5.5 quarts (5.2L)
Cooling system
  2.4L four cylinder .... 10 quarts (9.5L)
  3.7L V6
    2004 and earlier models .... 14 quarts (13.2L)
    2005 through 2007 models .... 13 quarts (12.3L)
    2008 and later models .... 14 quarts (13.2L)
  4.0L V6 .... 14 quarts (13.2L)

Capacities* (continued)

Automatic transmission (drain and refill) .... Capacities vary - see Section 25 for procedure
Manual transmission
    NV1500 .... 2.4 pints (1.4L)
    NSG370 .... 3.17 pints (1.5L)
Transfer case
    NP 231 (Command-Trac) .... 2.95 pints (1.4L)
    NP 242 (Selec-Trac) .... 3.4 pints (1.6L)
    NV241 Gen II .... 4.2 pints (2.0L)
    MP1522 .... 3.8 pints (1.8L)
    MP3022 .... 4.0 pints (1.9L)
Differential
    Front .... 2.6 pints (1.2L)
    Rear
        198RBI (model 35) .... 3.8 pints** (1.8L)
        8-1/4 rear axle .... 4.4 pints** (2.1L)

* All capacities approximate. Add as necessary to bring to appropriate level.
**On limited-slip rear differential, include 4 oz. (188 ml) of Mopar Friction Modifier additive

Ignition system

Spark plug type and gap
    2.4L four cylinder .... Champion RE14MCC5 or equivalent @ 0.048 to 0.053 inch
    3.7L V6 engine
        2007 and earlier models .... NGK ZFR6F or equivalent @ 0.048 to 0.053 inch
        2008 and later models .... NGK ZFR6F-11G or equivalent @ 0.044 inch
    4.0L V6 .... NGK ZFR5LP-13G or equivalent @ 0.051 inch
Firing order
    2.4L four cylinder 1-3-4-2
    3.7L V6 engine 1-6-5-4-3-2
    4.0L V6 1-2-3-4-5-6

General

Disc brake pad lining thickness (minimum).. 1/8-inch
Drum brake shoe lining thickness (minimum)
    Bonded shoes.. 1/16-inch
    Riveted shoes (thickness above rivets)............................... 1/32-inch

Torque specifications

Note: *One foot-pound (ft-lb) of torque is equivalent to 12 inch-pounds (in-lbs) of torque. Torque values below approximately 15 ft-lbs are expressed in inch-pounds, because most foot-pound torque wrenches are not accurate at these smaller values.*

	Ft-lbs (unless otherwise indicated)	Nm
Automatic transmission pan bolts..	15	20
Rear differential cover bolts	30	40
Front differential cover bolts	17	23
Transfer case fill/drain plug	20	27
Spark plugs		
2.4L four-cylinder..	132 in-lbs	15
3.7L V6	20	27
4.0L V6	156 in-lbs	17.5
Engine oil drain plug		
2.4L four-cylinder..	20	27
3.7L V6	25	34
4.0L V6	20	27
Oxygen sensor		
2.4L four-cylinder and 3.7L V6.......................................	22	30
4.0L V6	30	41
Wheel lug nuts..	85 to 115	116 to 156

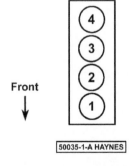

Front

50035-1-A HAYNES

Cylinder locations - four-cylinder engine

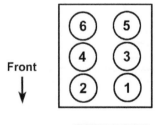

Front

50035-1-B HAYNES

Cylinder locations - V6 engine

1 Jeep Liberty & Dodge Nitro Maintenance schedule

The maintenance intervals in this manual are provided with the assumption that you, not the dealer, will be doing the work. These are the minimum maintenance intervals recommended by the factory for vehicles that are driven daily. If you wish to keep your vehicle in peak condition at all times, you may wish to perform some of these procedures even more often. Because frequent maintenance enhances the efficiency, performance and resale value of your car, we encourage you to do so. If you drive in dusty areas, tow a trailer, idle or drive at low speeds for extended periods or drive for short distances (less than four miles) in below freezing temperatures, shorter intervals are also recommended.

When your vehicle is new, it should be serviced by a factory authorized dealer service department to protect the factory warranty. In many cases, the initial maintenance check is done at no cost to the owner.

Every 250 miles or weekly, whichever comes first

Check the engine oil level (Section 4)
Check the engine coolant level (Section 4)
Check the windshield washer fluid level (Section 4)
Check the brake and clutch fluid levels (Section 4)
Check the tires and tire pressures (Section 5)
Check the automatic transmission fluid level (Section 6)
Check the power steering fluid level (Section 7)

Every 3000 miles or 3 months, whichever comes first

All items listed above, plus . . .
Change the engine oil and filter (Section 8)
Check and service the battery (Section 9)
Check the cooling system (Section 10)
Inspect and replace, if necessary, all underhood hoses (Section 11)
Inspect and replace, if necessary, the windshield wiper blades (Section 12)

Every 6000 miles or 6 months, whichever comes first

All items listed above, plus . . .
Inspect the suspension and steering components (Section 13)
Inspect the exhaust system (Section 14)
Check the manual transmission lubricant (Section 15)
Check the transfer case lubricant level (4WD models) (Section 16)
Check the differential lubricant level (Section 17)
Rotate the tires (Section 18)
Check the brakes (Section 19) *
Inspect the fuel system (Section 20)
Check the engine drivebelt (Section 21)
Check the seat belts (Section 22)

Every 15,000 or 12 months, whichever comes first

Lubricate the chassis (Section 23)
Replace the cabin air filter (Section 24a)

Every 30,000 miles or 24 months, whichever comes first

All items listed above, plus . . .
Replace the air filter (Section 24b)
Change the automatic transmission fluid and filter (Section 25)**
Change the manual transmission lubricant (Section 26)
Change the transfer case lubricant (4WD models) (Section 27)
Change the differential lubricant (Section 28)
Service the cooling system (drain, flush and refill) (Section 29)
Check the Positive Crankcase Ventilation (PCV) system (Section 30)
Check the evaporative emissions control system (Section 31)
Replace the spark plugs (Section 32)
Inspect the spark plug wires (2.4L four-cylinder model) (Section 33)

Every 60,000 miles or 48 months, whichever comes first

Replace the drivebelt (Section 21)
On 2.4L four-cylinder engines, replace the spark plug wires (Section 33)

Every 120,000 miles or 96 months, whichever comes first

On 2.4L four-cylinder and 4.0L V6 engines, replace the timing belt (Chapter 2A or Chapter 2C).

* This item is affected by "severe" operating conditions, as described below. If the vehicle is operated under severe conditions, perform all maintenance indicated with an asterisk (*) at 3000 mile/three-month intervals. Severe conditions exist if you mainly operate the vehicle . . .

   in dusty areas
   towing a trailer
   idling for extended periods and/or driving at low speeds when outside temperatures remain below freezing and most trips are less than four miles long

** If operated under one or more of the following conditions, change the automatic transmission fluid every 15,000 miles:

   in heavy city traffic where the outside temperature regularly reaches 90-degrees F or higher
   in hilly or mountainous terrain
   frequent trailer pulling

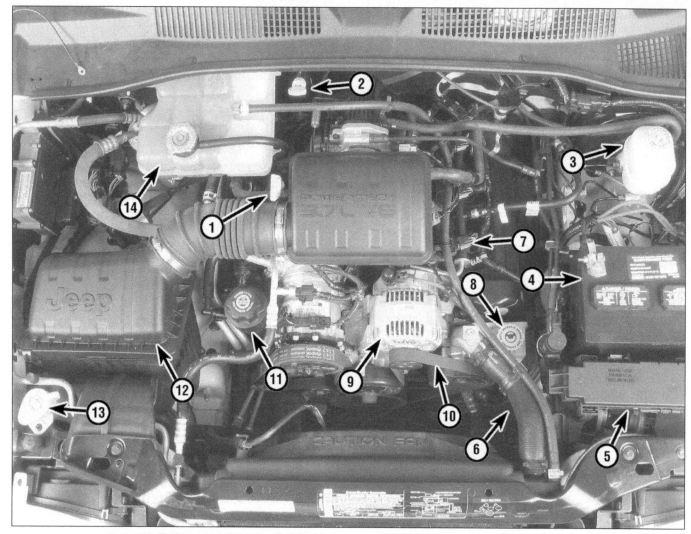

Engine compartment component locations - 2003 Jeep Liberty, 3.7L V6 engine

1   Engine oil dipstick
2   Automatic transmission fluid dipstick
3   Brake fluid reservoir
4   Battery
5   Power Distribution Center (PDC) - fuses and relays
6   Upper radiator hose
7   Ignition coils/spark plugs

8   Power steering fluid reservoir
9   Alternator
10 Drivebelt
11 Engine oil filler cap
12 Air filter housing
13 Windshield washer fluid reservoir
14 Coolant expansion tank

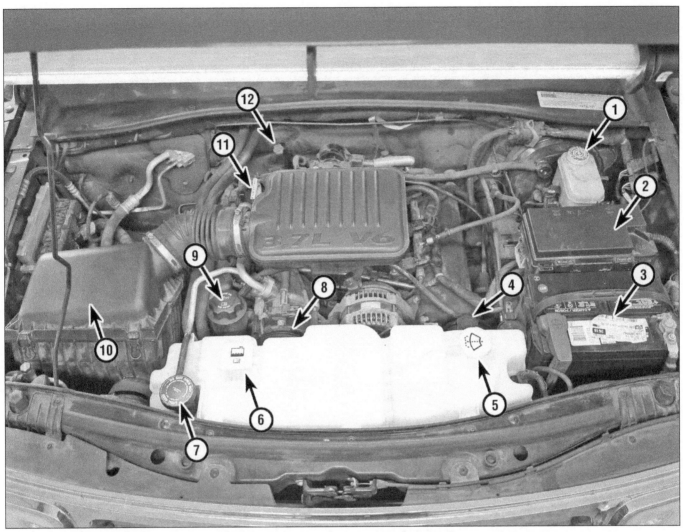

Engine compartment component locations - 2011 Dodge Nitro, 3.7L V6 engine

1   Brake fluid reservoir
2   Power Distribution Center (PDC) - fuses and relays
3   Battery
4   Power steering fluid reservoir
5   Windshield washer fluid reservoir
6   Coolant reservoir
7   Radiator cap

8   Drivebelt
9   Engine oil filler cap
10 Air filter housing
11 Engine oil dipstick
12 Automatic transmission fluid filler/dipstick tube

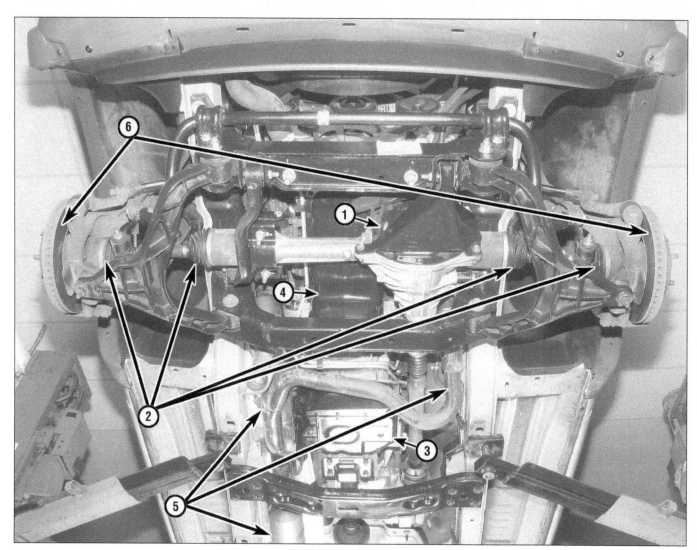

Engine compartment underside component locations

1	Front differential check/fill plug	3	Automatic transmission fluid pan	5	Exhaust system
2	Driveaxle boot (4WD models)	4	Engine oil pan	6	Front disc brake

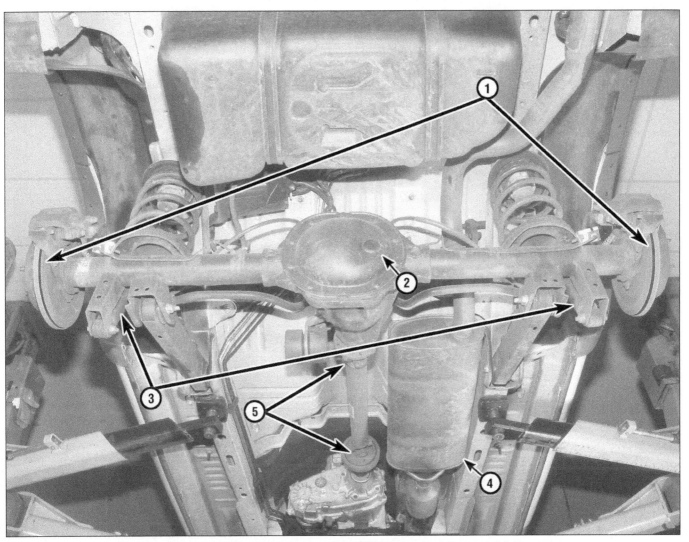

Rear underside component locations (2007 and earlier Liberty models)

1 Rear drum or disc brake	3 Rear shock absorber	5 Driveshaft universal joints
2 Differential check/fill plug	4 Exhaust system	

2   Introduction

This Chapter is designed to help the home mechanic maintain Jeep Liberty and Dodge Nitro models with the goals of maximum performance, economy, safety and reliability in mind.

Included is a master maintenance schedule, followed by procedures dealing specifically with each item on the schedule. Visual checks, adjustments, component replacement and other helpful items are included. Refer to the accompanying illustrations of the engine compartment and the underside of the vehicle for the locations of various components.

Servicing your vehicle in accordance with the mileage/time maintenance schedule and the step-by-step procedures will result in a planned maintenance program that should produce a long and reliable service life. Keep in mind that it's a comprehensive plan, so maintaining some items but not others at the specified intervals will not produce the same results.

As you service your vehicle, you will discover that many of the procedures can - and should - be grouped together because of the nature of the particular procedure you're performing or because of the close proximity of two otherwise unrelated components to one another.

For example, if the vehicle is raised for chassis lubrication, you should inspect the exhaust, suspension, steering and fuel systems while you're under the vehicle. When you're rotating the tires, it makes good sense to check the brakes since the wheels are already removed. Finally, let's suppose you have to borrow or rent a torque wrench. Even if you only need it to tighten the spark plugs, you might as well check the torque of as many critical fasteners as time allows.

The first step in this maintenance program is to prepare yourself before the actual work begins. Read through all the procedures you're planning to do, then gather up all the parts and tools needed. If it looks like you might run into problems during a particular job, seek advice from a mechanic or an experienced do-it-yourselfer.

The following maintenance intervals are based on the assumption the vehicle owner will be doing the maintenance or service work, as opposed to having a dealer service department do the work. Although the time/mileage intervals are loosely based on factory recommendations, most have been shortened to ensure, for example, that such items as lubricants and fluids are checked/changed at intervals that promote maximum engine/driveline service life. Also, subject to the preference of the individual owner interested in keeping the vehicle in peak condition at all times, and with the vehicle's ultimate resale in mind, many of the maintenance procedures may be performed more often than recommended in the following schedule. We encourage such owner initiative.

When the vehicle is new it should be serviced initially by a factory authorized dealer service department to protect the factory warranty. In many cases the initial maintenance check is done at no cost to the owner (check with your dealer service department for additional information).

3   Tune-up general information

The term tune-up is used in this manual to represent a combination of individual operations rather than one specific procedure.

If, from the time the vehicle is new, the routine maintenance schedule is followed closely and frequent checks are made of fluid levels and high wear items, as suggested throughout this manual, the engine will be kept in relatively good running condition and the need for additional work will be minimized.

More likely than not, however, there will be times when the engine is running poorly due to lack of regular maintenance. This is even more likely if a used vehicle, which has not received regular and frequent maintenance checks, is purchased. In such cases, an engine tune-up will be needed outside of the regular routine maintenance intervals.

The first step in any tune-up or diagnostic procedure to help correct a poor running engine is a cylinder compression check. A compression check (see Chapter 2C) will help determine the condition of internal engine components and should be used as a guide for tune-up and repair procedures. If, for instance, the compression check indicates serious internal engine wear, a conventional tune-up won't improve the performance of the engine and would be a waste of time and money. Because of its importance, the compression check should be done by someone with the right equipment and the knowledge to use it properly.

The following procedures are those most often needed to bring a generally poor running engine back into a proper state of tune.

Minor tune-up

Check all engine related fluids (Section 4)
Clean, inspect and test the battery
    (Section 9)
Check the cooling system (Section 10)
Check all underhood hoses (Section 11)
Check the drivebelt (Section 21)
Check the air filter (Section 24b)
Check the PCV system (Section 30)
Replace the spark plugs (Section 32)
Inspect the spark plug wires (four-cylinder
    engine) (Section 33)

Major tune-up

All items listed under Minor tune-up, plus . . .
Check the fuel system (Section 20)
Replace the air filter (Section 24b)
Replace the spark plug wires (four-
    cylinder engine) (Section 33)
Check the ignition system (Chapter 5)
Check the charging system (Chapter 5)

4   Fluid level checks (every 250 miles or weekly)

Note: *The following are fluid level checks to be done on a 250 mile or weekly basis. Additional fluid level checks can be found in specific maintenance procedures which follow. Regardless of intervals, be alert to fluid leaks under the vehicle which would indicate a fault to be corrected immediately.*

1   Fluids are an essential part of the lubrication, cooling, brake, clutch and windshield washer systems. Because the fluids gradually become depleted and/or contaminated during normal operation of the vehicle, they must be periodically replenished. See *Recommended lubricants and fluids* at the beginning of this Chapter before adding fluid to any of the following components. **Note:** *The vehicle must be on level ground when fluid levels are checked.*

Engine oil

Refer to illustrations 4.2 and 4.4

2   The engine oil level is checked with a dipstick that extends through a tube and into the oil pan at the bottom of the engine **(see illustration)**.

3   The oil level should be checked before the vehicle has been driven, or about 5 minutes after the engine has been shut off. If the oil is checked immediately after driving the vehicle, some of the oil will remain in the upper engine components, resulting in an inaccurate reading on the dipstick.

4   Pull the dipstick out of the tube and wipe all the oil from the end with a clean rag or paper towel. Insert the clean dipstick all the way back into the tube, then pull it out again. Note the oil at the end of the dipstick. Add oil as necessary to keep the level between the ADD and SAFE marks on the dipstick **(see illustration)**.

5   Do not overfill the engine by adding too much oil since this may result in oil fouled spark plugs, oil leaks or oil seal failures.

6   Oil is added to the engine after removing a cap from the valve cover on four-cylinder engines or the oil filler tube on the passenger's side of the engine on V6 engines **(see illustration 4.2)**. Always make sure the area around the opening is clean before removing the cap to prevent dirt from contaminating the engine. When adding oil into the filler tube, a funnel may help to reduce spills.

7   Checking the oil level is an important preventive maintenance step. A consistently low oil level indicates oil leakage through damaged seals, defective gaskets or past worn rings or valve guides. If the oil looks milky or has water droplets in it, the cylinder head gasket(s) may be blown or the head(s) or block may be cracked. The engine should be checked immediately. The condition of the oil should also be checked. Whenever you check the oil level, slide your thumb and index finger up the dipstick before wiping off the oil. If you see small dirt or metal particles clinging to the dipstick, the oil should be changed (see Section 8).

4.2 On 3.7L V6 engines, check or fill the engine oil, coolant and automatic transmission fluid at these locations

1 Engine oil dipstick
2 Oil filler cap
3 Coolant expansion tank
4 Automatic transmission fluid dipstick

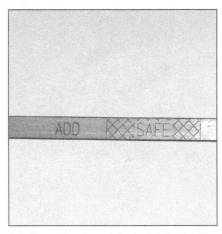

4.4 The oil level must be maintained in the SAFE range at all times - it takes one quart of oil to raise the level from the ADD to the SAFE mark

Engine coolant

Refer to illustrations 4.8 and 4.9

Warning: *Do not allow antifreeze to come in contact with your skin or painted surfaces of the vehicle. Flush contaminated areas immediately with plenty of water. Don't store new coolant or leave old coolant lying around where it's accessible to children or pets - they're attracted by its sweet smell. Ingestion of even a small amount of coolant can be fatal! Wipe up garage floor and drip pan coolant spills immediately. Keep antifreeze containers covered and repair leaks in the cooling system as soon as they are noted.*

8   On some models, a plastic coolant expansion tank located in the rear of the engine compartment is connected by a hose to the engine **(see illustration 4.2)**. As the engine heats up during operation, the coolant level in the tank rises. **Warning:** *Never remove the pressure cap on the expansion tank when the engine is warm!* On other models, a non-pressurized coolant reservoir is located at the

front of the engine compartment **(see illustration)**. On these models, coolant that escapes past the radiator cap flows into the reservoir; when the engine cools down, the coolant is drawn back into the radiator.

9   The coolant level in the tank should be checked regularly. The level in the tank/reservoir varies with the temperature of the engine. When the engine is cold, the coolant level should be at or slightly above the COLD FILL or MIN mark **(see illustrations)**. If it isn't, allow the engine to cool, then remove the cap from the tank and add a 50/50 mixture of ethylene glycol-based antifreeze and de-ionized or distilled water.

10   Drive the vehicle and recheck the coolant level. If only a small amount of coolant is required to bring the system up to the proper level, water can be used. However, repeated additions of water will dilute the antifreeze and water solution. In order to maintain the proper ratio of antifreeze and water, always top up the coolant level with the correct mixture. An

empty plastic milk jug or bleach bottle makes an excellent container for mixing coolant. Do not use rust inhibitors or additives.

11   If the coolant level drops consistently, there may be a leak in the system. Inspect the radiator, hoses, pressure cap, drain plugs and water pump (see Section 10). If no leaks are noted, have the expansion tank cap pressure tested by a service station.

12   Models with a pressurized expansion tank: If you have to remove the expansion tank cap, wait until the engine has cooled, then wrap a thick cloth around the cap and turn it to the first stop. If coolant or steam escapes, let the engine cool down longer, then remove the cap.

13   Check the condition of the coolant as well. It should be relatively clear. If it's brown or rust colored, the system should be drained, flushed and refilled. Even if the coolant appears to be normal, the corrosion inhibitors wear out, so it must be replaced at the specified intervals.

4.8 Coolant reservoir location (Nitro model shown; on some Liberty models, a non-pressurized coolant reservoir is located at the left-front corner of the engine compartment)

4.9a If the coolant level is below the COLD FILL mark, add a 50/50 mixture of ethylene glycol-based antifreeze and de-ionized or distilled water

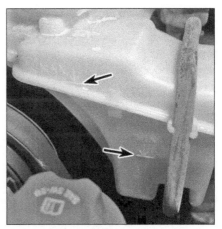

4.9b MIN and MAX marks on a non-pressurized coolant reservoir

4.14a The windshield washer fluid reservoir is located on the right front side of the engine compartment

4.14b On other models, the windshield washer fluid reservoir is located at the left front of the engine compartment

Windshield washer fluid

Refer to illustrations 4.14a and 4.14b

14   Fluid for the windshield washer system is located in a plastic reservoir in the right front corner of the engine compartment **(see illustrations)**.

15   In milder climates, plain water can be used in the reservoir, but it should be kept no more than 2/3 full to allow for expansion if the water freezes. In colder climates, use windshield washer system antifreeze, available at any auto parts store, to lower the freezing point of the fluid. Mix the antifreeze with water in accordance with the manufacturer's directions on the container. **Caution:** *Don't use cooling system antifreeze - it will damage the vehicle's paint.*

16   To help prevent icing in cold weather, warm the windshield with the defroster before using the washer.

Brake and clutch fluid

Refer to illustration 4.18

17   The brake and (if equipped) clutch mas-

ter cylinder is mounted on the upper left of the engine compartment firewall.

18   The fluid inside can be checked looking at the translucent plastic reservoir **(see illustration)**. Be sure to wipe the top of the reservoir cover with a clean rag to prevent contamination of the brake and/or clutch system before removing the cover.

19   When adding fluid, pour it carefully into the reservoir to avoid spilling it on surrounding painted surfaces. Be sure the specified fluid is used, since mixing different types of brake fluid can cause damage to the system. See *Recommended lubricants and fluids* at the front of this Chapter or your owner's manual. **Warning:** *Brake fluid can harm your eyes and damage painted surfaces, so use extreme caution when handling or pouring it. Do not use brake fluid that has been standing open or is more than one year old. Brake fluid absorbs moisture from the air. Excess moisture can cause a dangerous loss of brake performance.*

20   At this time, the fluid and master cylinder can be inspected for contamination. The system should be drained and refilled if deposits, dirt particles or water droplets are seen in the fluid.

21   After filling the reservoir to the proper level, make sure the cover or cap is on tight to prevent fluid leakage.

22   The brake fluid level in the master cylinder will drop slightly as the pads at the front wheels wear down during normal operation. If the master cylinder requires repeated additions to keep it at the proper level, it's an indication of leakage in the brake system, which should be corrected immediately. Check all brake lines and connections (see Section 19 for more information).

23   If, upon checking the master cylinder fluid level, you discover one or both reservoirs empty or nearly empty, the brake system should be bled (Chapter 9).

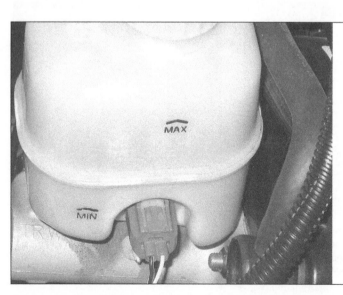

4.18 The brake fluid level is easily checked visually

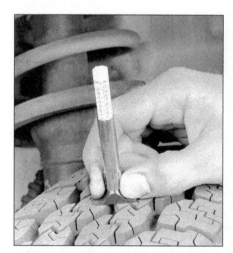

5.2 Use a tire tread depth indicator to monitor tire wear - they are available at auto parts stores and service stations and cost very little

UNDERINFLATION

OVERINFLATION

CUPPING

Cupping may be caused by:
- Underinflation and/or mechanical irregularities such as out-of-balance condition of wheel and/or tire, and bent or damaged wheel.
- Loose or worn steering tie-rod or steering idler arm.
- Loose, damaged or worn front suspension parts.

**INCORRECT TOE-IN
OR EXTREME CAMBER**

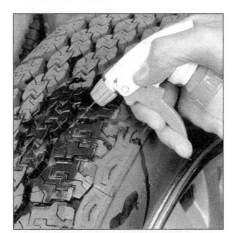

**FEATHERING DUE
TO MISALIGNMENT**

5.3 This chart will help you determine the condition of the tires, the probable cause(s) of abnormal wear and the corrective action necessary

---

5   Tire and tire pressure checks (every 250 miles or weekly)

Refer to illustrations 5.2, 5.3, 5.4a, 5.4b and 5.8

1   Periodic inspection of the tires may spare you the inconvenience of being stranded with a flat tire. It can also provide you with vital information regarding possible problems in the steering and suspension systems before major damage occurs.

2   The original tires on this vehicle are equipped with 1/2-inch wide wear bands that will appear when tread depth reaches 1/16-inch, at which point the tires can be considered worn out. Tread wear can be monitored with a simple, inexpensive device known as a tread depth indicator **(see illustration)**.

3   Note any abnormal tread wear **(see illustration)**. Tread pattern irregularities such as cupping, flat spots and more wear on one side than the other are indications of front

end alignment and/or balance problems. If any of these conditions are noted, take the vehicle to a tire shop or service station to correct the problem.

4   Look closely for cuts, punctures and embedded nails or tacks. Sometimes a tire will hold air pressure for a short time or leak down very slowly after a nail has embedded itself in the tread. If a slow leak persists, check the valve stem core to make sure it's tight **(see illustration)**. Examine the tread for an object that may have embedded itself in the tire or for a "plug" that may have begun to leak (radial tire punctures are repaired with a plug that's installed in a puncture). If a puncture is suspected, it can be easily verified by spraying a solution of soapy water onto the puncture area **(see illustration)**. The soapy solution will bubble if there's a leak. Unless the puncture is unusually large, a tire shop or service station can usually repair the tire.

5   Carefully inspect the inner sidewall of each tire for evidence of brake fluid leakage. If you see any, inspect the brakes immediately.

6   Correct air pressure adds miles to the lifespan of the tires, improves mileage and enhances overall ride quality. Tire pressure cannot be accurately estimated by looking at a tire, especially if it's a radial. A tire pressure gauge is essential. Keep an accurate gauge in the vehicle. The pressure gauges attached to the nozzles of air hoses at gas stations are often inaccurate.

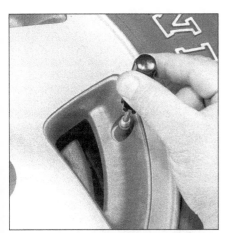

5.4a If a tire loses air on a steady basis, check the valve stem core first to make sure it's snug (special inexpensive wrenches are commonly available at auto parts stores)

5.4b If the valve stem core is tight, raise the corner of the vehicle with the low tire and spray a soapy water solution onto the tread as the tire is turned slowly - leaks will cause small bubbles to appear

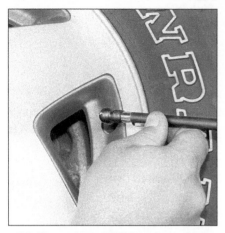

5.8 To extend the life of the tires, check the air pressure at least once a week with an accurate gauge (don't forget the spare!)

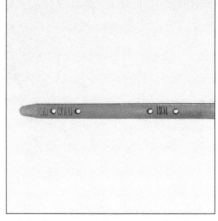

6.6a The automatic transmission fluid must be kept between the two holes in the COLD or HOT range, depending on the temperature of the fluid

6.6b Use a funnel like this one (available at auto parts stores) inserted into the transmission dipstick tube to add fluid

7   Always check tire pressure when the tires are cold. Cold, in this case, means the vehicle has not been driven over a mile in the three hours preceding a tire pressure check. A pressure rise of four to eight pounds is not uncommon once the tires are warm.

8   Unscrew the valve cap protruding from the wheel or hubcap and push the gauge firmly onto the valve stem **(see illustration)**. Note the reading on the gauge and compare the figure to the recommended tire pressure shown on the placard on the driver's side door pillar. Be sure to reinstall the valve cap to keep dirt and moisture out of the valve stem mechanism. Check all four tires and, if necessary, add enough air to bring them up to the recommended pressure.

9   Don't forget to keep the spare tire inflated to the specified pressure (refer to your owner's manual or the tire sidewall).

---

6   Automatic transmission fluid level check (every 250 miles or weekly)

Refer to illustrations 6.6a, 6.6b, 6.7a, 6.7b and 6.8

Note: *To check the transmission fluid level on models with the NAG1 transmission (4.0L V6 models), or 2007 and later Liberty/2008 and later Nitro models with a 42RLE transmission, a special dipstick (part no. 9336A) will be required. An alternative to this tool can be fabricated from a length of coathanger, welding rod or similar; it must be long enough to be inserted into the dipstick/filler tube and contact the bottom of the transmission fluid tube, while also protruding from the filler tube so it doesn't fall in. Also, the use of a scan tool to measure the temperature of the transmission fluid is recommended. If a scan tool is unavailable, an infrared heat gun pointed at the transmission fluid pan can be used to measure the approximate temperature of the fluid.*

1   The automatic transmission fluid level should be carefully maintained. Low fluid level can lead to slipping or loss of drive, while overfilling can cause foaming and loss of fluid.

2   With the parking brake set, start the engine, then move the shift lever through all the gear ranges, ending in Neutral (with the 42RLE transmissions) or PARK (with the 45RFE and NAG1 transmission). The fluid level must be checked with the vehicle level, the transmission selector in the correct position for your transmission model and the engine running at idle for at least one minute. **Note:** *Incorrect fluid level readings will result if the vehicle has just been driven at high speeds for an extended period, in hot weather in city traffic, or if it has been pulling a trailer. If any of these conditions apply, wait until the fluid has cooled (about 30 minutes).*

3   With the transmission at normal operating temperature, remove the dipstick from the filler tube. The dipstick is located at the rear of the engine compartment **(see illustration 4.2)**.

4   Determine whether the fluid is warm or hot. Wipe the fluid from the dipstick with a clean rag and push it back into the filler tube until the cap seats.

5   Pull the dipstick out again and note the fluid level.

6   On the 45RFE transmission, if the fluid is warm, the level should be between the COLD and HOT holes. If the fluid is hot, the level should be between the two HOT holes **(see illustration)**. On the 42RLE transmission, if the fluid is warm, the level should be above the MIN hole and below the circle in the cross-hatched area. If the fluid is hot, the level should be in the crosshatched area, marked OK. If additional fluid is required, add it directly into the tube using a funnel **(see illustration)**. It takes about one pint to raise the level from the bottom of the crosshatched area to respective HOT or OK zone with a hot transmission, so add the

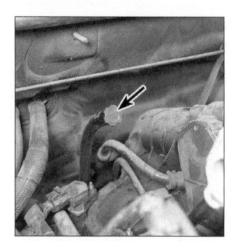

6.7a On models without a dipstick, remove this cap from the filler tube and check the fluid level with tool no. 9336A or equivalent

fluid a little at a time and keep checking the level until it's correct.

7   On models with the NAG1 transmission (4.0L V6 models) and 2007 and later Liberty/2008 and later Nitro models with a 42RLE transmission, with the transmission at normal operating temperature, remove the cap from the filler tube **(see illustration)**, insert the dipstick tool into the filler tube and pull out again, then measure the fluid level from the bottom of the tool. Determine the correct fluid level by using the temperature of the fluid and the height of the fluid on the dipstick with the accompanying chart **(see illustrations)**. If additional fluid is required, add it directly into the tube using a funnel. Add the fluid a little at a time and keep checking the level until it's correct. Once the oil level is correct, install the filler tube cap.

Note: *If using a scan tool, the transmission fluid temperature can only be read while the transmission is in Reverse or any forward gear position.*

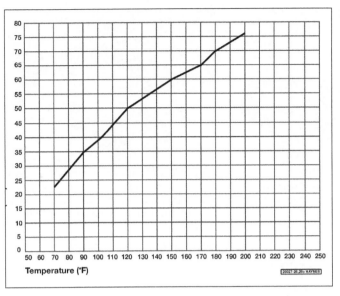

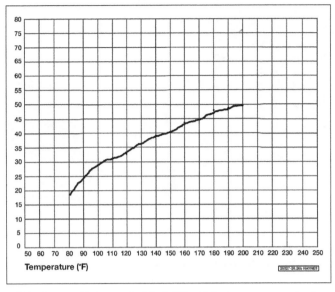

6.7b Automatic transmission fluid level-to-temperature chart -
NAG1 transmission

6.7c Automatic transmission fluid level-to-temperature chart -
2007 and later Liberty/2008 and later Nitro models
with a 42RLE transmission

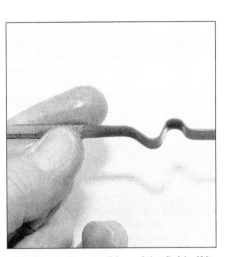

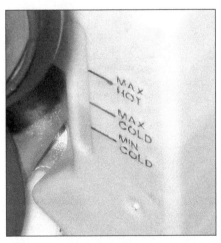

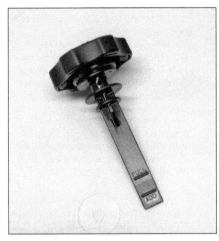

6.8 Check the condition of the fluid - if it
isn't red in color, or if it smells burnt, it
should be changed

7.6a On some models, the power steering
fluid reservoir has marks on it so the fluid
can be checked hot or cold

7.6b On other models, the power steering
fluid reservoir is not translucent and uses
a dipstick attached to the cap

8   The condition of the fluid should also be checked along with the level. If the fluid at the end of the dipstick is a dark reddish-brown color, or if it smells burnt, it should be changed **(see illustration)**. If you are in doubt about the condition of the fluid, purchase some new fluid and compare the two for color and smell.

7   Power steering fluid level check
   (every 250 miles or weekly)

Refer to illustrations 7.6a and 7.6b

1   Unlike manual steering, the power steering system relies on fluid which may, over a period of time, require replenishing.
2   The fluid reservoir for the power steering

pump is located on the pump body at the front of the engine.
3   For the check, the front wheels should be pointed straight ahead and the engine should be off.
4   Use a clean rag to wipe off the reservoir cap and the area around the cap. This will help prevent any foreign matter from entering the reservoir during the check.
5   Twist off the cap and determine the temperature of the fluid.
6   The fluid should be at the proper level, depending on whether it was checked hot or cold **(see illustrations)**. Never allow the fluid level to drop below the lower mark on the reservoir or dipstick.
7   If additional fluid is required, pour the specified type directly into the reservoir, using a funnel to prevent spills.

8   If the reservoir requires frequent fluid additions, all power steering hoses, hose connections and the power steering pump should be carefully checked for leaks.

8   Engine oil and filter change
   (every 3000 miles or 3 months)

Refer to illustrations 8.3, 8.9, 8.14 and 8.18

1   Frequent oil changes are the most important preventive maintenance procedures that can be done by the home mechanic. As engine oil ages, it becomes diluted and contaminated, which leads to premature engine wear.
2   Although some sources recommend oil filter changes every other oil change, we feel that

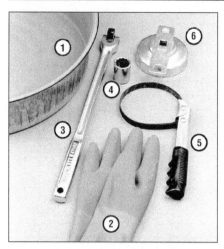

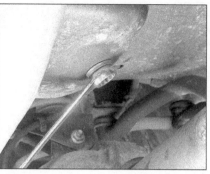

8.9 Use a proper size box-end wrench or socket to remove the oil drain plug and avoid rounding it off

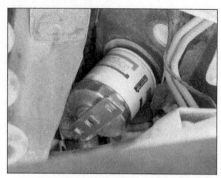

8.14 Since the oil filter is on very tight, you'll need a special wrench for removal - DO NOT use the wrench to tighten the new filter

8.3 These tools are required when changing the engine oil and filter

1   **Drain pan** - It should be fairly shallow in depth, but wide to prevent spills
2   **Rubber gloves** - When removing the drain plug and filter, you will get oil on your hands (the gloves will prevent burns)
3   **Breaker bar** - Sometimes the oil drain plug is tight, and a long breaker bar is needed to loosen it
4   **Socket** - To be used with the breaker bar or a ratchet (must be the correct size to fit the drain plug - six-point preferred)
5   **Filter wrench** - This is a metal band-type wrench, which requires clearance around the filter to be effective
6   **Filter wrench** - This type fits on the bottom of the filter and can be turned with a ratchet or breaker bar (different-size wrenches are available for different types of filters)

8.18 Lubricate the oil filter gasket with clean engine oil before installing the filter on the engine

the minimal cost of an oil filter and the relative ease with which it is installed dictate that a new filter be installed every time the oil is changed.
3   Gather together all necessary tools and materials before beginning this procedure **(see illustration)**.
4   You should have plenty of clean rags and newspapers handy to mop up any spills. Access to the underside of the vehicle may be improved if the vehicle can be lifted on a hoist, driven onto ramps or supported by jackstands. **Warning:** *Do not work under a vehicle which is supported only by a bumper, hydraulic or scissors-type jack.*
5   If this is your first oil change, get under the vehicle and familiarize yourself with the locations of the oil drain plug and the oil filter. The engine and exhaust components will be warm during the actual work, so note how they are situated to avoid touching them when working under the vehicle.
6   Warm the engine to normal operating temperature. If the new oil or any tools are needed, use this warm-up time to gather everything necessary for the job. The correct type of oil for your application can be found in *Recommended lubricants and fluids* at the beginning of this Chapter.

7   With the engine oil warm (warm engine oil will drain better and more built-up sludge will be removed with it), raise and support the vehicle. Make sure it's safely supported!
8   Move all necessary tools, rags and newspapers under the vehicle. Set the drain pan under the drain plug. Keep in mind that the oil will initially flow from the pan with some force; position the pan accordingly.
9   Being careful not to touch any of the hot exhaust components, use a wrench to remove the drain plug near the bottom of the oil pan **(see illustration)**. Depending on how hot the oil is, you may want to wear gloves while removing the plug the final few turns.
10   Allow the old oil to drain into the pan. It may be necessary to move the pan as the oil flow slows to a trickle.
11   After all the oil has drained, wipe off the drain plug with a clean rag. Small metal particles may cling to the plug and would immediately contaminate the new oil.
12   Clean the area around the drain plug opening and reinstall the plug. Tighten the plug securely with the wrench. If a torque wrench is available, use it to tighten the plug to the torque listed in this Chapter's Specifications.
13   Move the drain pan into position under the oil filter.
14   Use the filter wrench to loosen the oil filter **(see illustration)**.
15   Completely unscrew the old filter. Be careful; it's full of oil. Just as the filter is detached from the block, immediately tilt the open end up to prevent the oil inside the filter

from pouring out. Remove the filter and empty the oil inside the filter into the drain pan.
16   Compare the old filter with the new one to make sure they're the same type.
17   Use a clean rag to remove all oil, dirt and sludge from the area where the oil filter mounts to the engine. Check the old filter to make sure the rubber gasket isn't stuck to the engine. If the gasket is stuck to the engine, remove it.
18   Apply a light coat of clean oil to the rubber gasket on the new oil filter **(see illustration)**.
19   Attach the new filter to the engine, following the tightening directions printed on the filter canister or packing box. Most filter manufacturers recommend against using a filter wrench due to the possibility of over-tightening and damage to the seal.
20   Remove all tools, rags, etc. from under the vehicle, being careful not to spill the oil in the drain pan, then lower the vehicle.
21   Move to the engine compartment and locate the oil filler cap.
22   Pour the fresh oil through the filler opening. A funnel may be helpful.
23   Pour four quarts of fresh oil into the engine. Wait a few minutes to allow the oil to drain into the pan, then check the level on the oil dipstick (see Section 4 if necessary). If the oil level is above the ADD mark, start the engine, make sure the oil pressure warning light turns of after a few seconds and allow the new oil to circulate.
24   Run the engine for only about a minute and then shut it off. Immediately look under the vehicle and check for leaks at the oil pan drain plug and around the oil filter. If either is leaking, tighten with a bit more force.
25   With the new oil circulated and the filter now completely full, recheck the level on the dipstick and add more oil as necessary.
26   During the first few trips after an oil change, make it a point to check frequently for leaks and proper oil level.
27   The old oil drained from the engine cannot be reused in its present state and should be disposed of. Check with your local auto parts store, disposal facility or environmental agency to see if they will accept the oil for recycling. After the oil has cooled it can be drained into a container (capped plastic jugs, topped bottles, milk cartons, etc.) for transport to one of these disposal sites. Don't dispose of the oil by pouring it on the ground or down a drain!

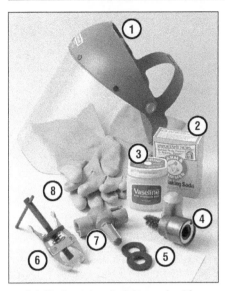

9.1 Tools and materials required for battery maintenance

1   *Face shield/safety goggles - When removing corrosion with a brush, the acidic particles can easily fly up into your eyes*
2   *Baking soda - A solution of baking soda and water can be used to neutralize corrosion*
3   *Petroleum jelly - A layer of this on the battery posts will help prevent corrosion*
4   *Battery post/cable cleaner - This wire brush cleaning tool will remove all traces of corrosion from the battery posts and cable clamps*
5   *Treated felt washers - Placing one of these on each post, directly under the cable clamps, will help prevent corrosion*
6   *Puller - Sometimes the cable clamps are very difficult to pull off the posts, even after the nut/bolt has been completely loosened. This tool pulls the clamp straight up and off the post without damage*
7   *Battery post/cable cleaner - Here is another cleaning tool that is a slightly different version of Number 4 above, but it does the same thing*
8   *Rubber gloves - Another safety item to consider when servicing the battery; remember that's acid inside the battery!*

---

9   Battery check, maintenance and charging (every 3000 miles or 3 months)

Refer to illustrations 9.1, 9.6a, 9.6b, 9.7a and 9.7b

Warning: *Certain precautions must be followed when checking and servicing the battery. Hydrogen gas, which is highly flammable, is always present in the battery cells, so keep lighted tobacco and all other open flames and sparks away from the battery. The electrolyte inside the battery is actually dilute sulfuric acid, which will cause injury if splashed on your skin or in your eyes. It will*

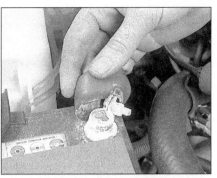

9.6a Battery terminal corrosion usually appears as light, fluffy powder

9.7a When cleaning the cable clamps, all corrosion must be removed (the inside of the clamp is tapered to match the taper on the post, so don't remove too much material)

also ruin clothes and painted surfaces. When removing the battery cables, always detach the negative cable first and hook it up last!

Maintenance

1   A routine preventive maintenance program for the battery in your vehicle is the only way to ensure quick and reliable starts. But before performing any battery maintenance, make sure that you have the proper equipment necessary to work safely around the battery **(see illustration)**.
2   There are also several precautions that should be taken whenever battery maintenance is performed. Before servicing the battery, always turn the engine and all accessories off and disconnect the cable from the negative terminal of the battery.
3   The battery produces hydrogen gas, which is both flammable and explosive. Never create a spark, smoke or light a match around the battery. Always charge the battery in a ventilated area.
4   Electrolyte contains poisonous and corrosive sulfuric acid. Do not allow it to get in your eyes, on your skin or your clothes. Never ingest it. Wear protective safety glasses when working near the battery. Keep children away from the battery.
5   Note the external condition of the battery. If the positive terminal and cable clamp on your vehicle's battery is equipped with a rubber protector, make sure that it's not torn or damaged.

9.6b Removing the cable from a battery post with a wrench - sometimes special battery pliers are required for this procedure if corrosion has caused deterioration of the nut hex (always remove the ground cable first and hook it up last!)

9.7b Regardless of the type of tool used on the battery posts, a clean, shiny surface should be the result

It should completely cover the terminal. Look for any corroded or loose connections, cracks in the case or cover or loose hold-down clamps. Also check the entire length of each cable for cracks and frayed conductors.
6   If corrosion, which looks like white, fluffy deposits **(see illustration)** is evident, particularly around the terminals, the battery should be removed for cleaning. Loosen the cable clamp bolts with a wrench, being careful to remove the ground cable first, and slide them off the terminals **(see illustration)**. Then disconnect the hold-down clamp bolt and nut, remove the clamp and lift the battery from the engine compartment.
7   Clean the cable clamps thoroughly with a battery brush or a terminal cleaner and a solution of warm water and baking soda **(see illustration)**. Wash the terminals and the top of the battery case with the same solution but make sure that the solution doesn't get into the battery. When cleaning the cables, terminals and battery top, wear safety goggles and rubber gloves to prevent any solution from coming in contact with your eyes or hands. Wear old clothes too - even diluted, sulfuric acid splashed onto clothes will burn holes in them. If the terminals have been extensively corroded, clean them up with a terminal cleaner **(see illustration)**. Thoroughly wash all cleaned areas with plain water.

8   Make sure that the battery tray is in good condition and the hold-down clamp bolts are tight. If the battery is removed from the tray, make sure no parts remain in the bottom of the tray when the battery is reinstalled. When reinstalling the hold-down clamp bolts, do not overtighten them.

9   Any metal parts of the vehicle damaged by corrosion should be covered with a zinc-based primer, then painted.

10   Information on removing and installing the battery can be found in Chapter 5. Information on jump-starting can be found at the front of this manual. For more detailed battery checking procedures, refer to the *Haynes Automotive Electrical Manual*.

Charging

Warning: *When batteries are being charged, hydrogen gas, which is very explosive and flammable, is produced. Do not smoke or allow open flames near a charging or a recently charged battery. Wear eye protection when near the battery during charging. Also, make sure the charger is unplugged before connecting or disconnecting the battery from the charger.*

Note: *The manufacturer recommends the battery be removed from the vehicle for charging because the gas that escapes during this procedure can damage the paint. Fast charging with the battery cables connected can result in damage to the electrical system.*

11   Slow-rate charging is the best way to restore a battery that's discharged to the point where it will not start the engine. It's also a good way to maintain the battery charge in a vehicle that's only driven a few miles between starts. Maintaining the battery charge is particularly important in the winter when the battery must work harder to start the engine and electrical accessories that drain the battery are in greater use.

12   It's best to use a one or two-amp battery charger (sometimes-called a "trickle" charger). They are the safest and put the least strain on the battery. They are also the least expensive. For a faster charge, you can use a higher amperage charger, but don't use one rated more than 1/10th the amp/hour rating of the battery. Rapid boost charges that claim to restore the power of the battery in one to two hours are hardest on the battery and can damage batteries not in good condition. This type of charging should only be used in emergency situations.

13   The average time necessary to charge a battery should be listed in the instructions that come with the charger. As a general rule, a trickle charger will charge a battery in 12 to 16 hours.

14   Remove all the cell caps (if equipped) and cover the holes with a clean cloth to prevent spattering electrolyte. Disconnect the cable from the negative terminal of the battery and hook the battery charger cable clamps up to the battery posts (positive to positive, negative to negative), then plug in the charger. Make sure it is set at 12-volts if it has a selector switch.

15   If you're using a charger with a rate higher than two amps, check the battery reg-ularly during charging to make sure it doesn't overheat. If you're using a trickle charger, you can safely let the battery charge overnight after you've checked it regularly for the first couple of hours.

16   If the battery has removable cell caps, measure the specific gravity with a hydrometer every hour during the last few hours of the charging cycle. Hydrometers are available inexpensively from auto parts stores - follow the instructions that come with the hydrometer. Consider the battery charged when there's no change in the specific gravity reading for two hours and the electrolyte in the cells is gassing (bubbling) freely. The specific gravity reading from each cell should be very close to the others. If not, the battery probably has a bad cell(s).

17   Some batteries with sealed tops have built-in hydrometers on the top that indicate the state of charge by the color displayed in the hydrometer window. Normally, a bright-colored hydrometer indicates a full charge and a dark hydrometer indicates the battery still needs charging.

18   If the battery has a sealed top and no built-in hydrometer, you can hook up a digital voltmeter across the battery terminals to check the charge. A fully charged battery should read 12.6 volts or higher after the surface charge has been removed.

19   Further information on the battery and jump-starting can be found in Chapter 5 and at the front of this manual.

---

10   Cooling system check (every 3000 miles or 3 months)

Refer to illustration 10.4

1   Many major engine failures can be attributed to a faulty cooling system. If the vehicle is equipped with an automatic transmission, the cooling system also cools the transmission fluid and thus plays an important role in prolonging transmission life.

2   The cooling system should be checked with the engine cold. Do this before the vehicle is driven for the day or after it has been shut off for at least three hours.

3   Remove the cap from the plastic reserve tank by turning the cap counterclockwise. If you hear a hissing sound (indicating there is still pressure in the system), wait until this stops. Thoroughly clean the cap, inside and out, with clean water. Also clean the filler neck on the tank. All traces of corrosion should be removed. The coolant inside the tank should be relatively transparent. If it is rust colored, the system should be drained and refilled (see Section 29). If the coolant level is not up to the top, add additional antifreeze/coolant mixture to the tank (see Section 4).

4   Carefully check the large upper and lower radiator hoses along with the smaller diameter heater hoses that run from the engine to the firewall. Inspect each hose along its entire length, replacing any hose that is cracked, swollen or shows signs of

Check for a chafed area that could fail prematurely.

Check for a soft area indicating the hose has deteriorated inside.

Overtightening the clamp on a hardened hose will damage the hose and cause a leak.

Check each hose for swelling and oil-soaked ends. Cracks and breaks can be located by squeezing the hose.

10.4 Hoses, like drivebelts, have a habit of failing at the worst possible time - to prevent the inconvenience of a blown radiator or heater hose, inspect them carefully as shown here

deterioration. Cracks may become more apparent if the hose is squeezed **(see illustration)**. Regardless of condition, it's a good idea to replace hoses with new ones every two years.

5   Make sure all hose connections are tight. A leak in the cooling system will usually show up as white or rust colored deposits on the areas adjoining the leak. If wire-type clamps are used at the ends of the hoses, it may be a good idea to replace them with more secure screw-type clamps.

6   Use compressed air or a soft brush to remove bugs, leaves, etc. from the front of the radiator or air conditioning condenser. Be careful not to damage the delicate cooling fins or cut yourself on them.

7   Every other inspection, or at the first indication of cooling system problems, have the cap and system pressure tested. If you don't have a pressure tester, most gas stations and repair shops will do this for a minimal charge.

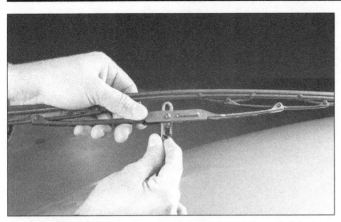

12.5a On the front wipers, press the retaining tab in and slide the wiper assembly down and out of the hook in the end of the wiper arm

12.5b On the rear wipers, lift the release lever and slide the wiper assembly down and out of the hook in the end of the wiper arm

11   Underhood hose check and replacement (every 3000 miles or 3 months)

Caution: *Replacement of air conditioning hoses must be left to a dealer service department or air conditioning shop that has the equipment to depressurize the system safely. Never remove air conditioning components or hoses until the system has been depressurized.*

General

1   High temperatures in the engine compartment can cause the deterioration of the rubber and plastic hoses used for engine, accessory and emission systems operation. Periodic inspection should be made for cracks, loose clamps, material hardening and leaks. Information specific to the cooling system hoses can be found in Section 10.
2   Some hoses are secured to the fittings with clamps. Where clamps are used, check to be sure that they haven't lost their tension, allowing the hose to leak. If clamps aren't used, make sure the hose has not expanded and/or hardened where it slips over the fitting, allowing it to leak.

Vacuum hoses

3   It's quite common for vacuum hoses, especially those in the emissions system, to be color-coded or identified by colored stripes molded into them. Various systems require hoses with different wall thickness, collapse resistance and temperature resistance. When replacing hoses, be sure the new ones are made of the same material.
4   Often the only effective way to check a hose is to remove it completely from the vehicle. If more than one hose is removed, be sure to label the hoses and fittings to ensure correct installation.
5   When checking vacuum hoses, be sure to include any plastic T-fittings in the check. Inspect the fittings for cracks and the hose where it fits over the fitting for distortion, which could cause leakage.
6   A small piece of vacuum hose (1/4-inch inside diameter) can be used as a stetho-

scope to detect vacuum leaks. Hold one end of the hose to your ear and probe around vacuum hoses and fittings, listening for the "hissing" sound characteristic of a vacuum leak. **Warning:** *When probing with the vacuum hose stethoscope, be very careful not to come into contact with moving engine components such as the drivebelt, cooling fan, etc.*

Fuel hose

Warning: *There are certain precautions that must be taken when inspecting or servicing fuel system components: Work in a well ventilated area and do not allow open flames (cigarettes, appliance pilot lights, etc.) or bare light bulbs near the work area. Mop up any spills immediately and do not store fuel soaked rags where they could ignite. On fuel-injected models, the fuel system is under high pressure, so if any fuel lines are to be disconnected, the pressure in the system must be relieved first (see Chapter 4 for more information).*
7   Check all rubber fuel lines for deterioration and chafing. Check especially for cracks in areas where the hose bends and just before fittings, such as where a hose attaches to the fuel filter.
8   High quality fuel line, specifically designed for high-pressure fuel injection applications, should be used for fuel line replacement. Never, under any circumstances, use unreinforced vacuum line, clear plastic tubing or water hose for fuel lines.
9   Spring-type clamps are commonly used on fuel lines. These clamps often lose their tension over a period of time, and can be "sprung" during removal. Replace all spring-type clamps with screw clamps whenever a hose is replaced.

Metal lines

10   Sections of metal line are often used in the fuel system. Check carefully to be sure the line has not been bent or crimped, and that cracks have not started in the line.
11   If a section of metal fuel line must be replaced, only seamless steel tubing should be used, since copper and aluminum tubing don't have the strength necessary to withstand normal engine vibration.

12   Check the metal brake lines where they enter the master cylinder and brake proportioning unit (if used) for cracks in the lines or loose fittings. Any sign of brake fluid leakage calls for an immediate thorough inspection of the brake system.

12   Wiper blade inspection and replacement (every 3000 miles or 3 months)

Refer to illustrations 12.5a and 12.5b
1   The front and rear wiper and blade assemblies should be inspected periodically for damage, loose components and cracked or worn blade elements.
2   Road film can build up on the wiper blades and affect their efficiency, so they should be washed regularly with a mild detergent solution.
3   The action of the wiping mechanism can loosen the bolts, nuts and fasteners, so they should be checked and tightened, as necessary, at the same time the wiper blades are checked.
4   If the wiper blade elements are cracked, worn or warped, they should be replaced with new wiper blade units - because the wiper blade elements are non-replaceable. Also, if the wiper blade element doesn't conform to the window curvature, the wiper blade assembly should be replaced.
5   On the front wipers, lift the wiper blade/arm assembly away from the glass for clearance, press on the release lever located under the tip of the arm, then slide the wiper blade assembly out of the hook in the end of the arm **(see illustration)**. On the rear wiper, disengage the wiper arm from the support, lift the wiper blade/arm assembly away from the glass for clearance. Pull up on the release lever located on top of the arm, then slide the wiper blade assembly out of the hook in the end of the arm **(see illustration)**.
6   Installation is the reverse of removal.
7   Wet the windshield and check for proper operation.

13.6 Push in on the CV joint boots and check for damage and leaking grease

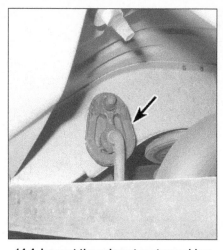

14.4 Inspect the exhaust system rubber mounts for damage

16.1 The 4WD transfer case fill plug is located on the rear side of the housing

13 Suspension, steering and driveaxle boot check (every 6000 miles or 6 months)

Refer to illustration 13.6

1   Indications of a fault in these systems are excessive play in the steering wheel before the front wheels react, excessive sway around corners, body movement over rough roads or binding at some point as the steering wheel is turned.

2   Raise the front of the vehicle periodically, support it securely on jackstands and visually check the suspension and steering components for wear.

3   Check the wheel bearings. Do this by spinning the front wheels. Listen for any abnormal noises and watch to make sure the wheel spins true (doesn't wobble). Grab the top and bottom of the tire and pull in-and-out on it. Notice any movement that would indicate a loose wheel bearing assembly. If the bearings are suspect, refer to Chapter 10 for more information.

4   From under the vehicle, check for loose bolts, broken or disconnected parts and deteriorated rubber bushings on all suspension and steering components. Look for fluid leaking from the steering gear assembly. Check the power steering hoses, belts and connections for leaks. Inspect the shock absorbers for fluid leaks, indicating the need for replacement.

5   Have an assistant turn the steering wheel from side-to-side and check the steering components for free movement, chafing and binding. If the steering doesn't react with the movement of the steering wheel, try to determine where the slack is located.

6   The front axle Constant Velocity (CV) joints on 4WD models are protected by rubber boots. Check the boots for cuts, wear and signs of leaking grease **(see illustration)**. The boots must be replaced if they are damaged, otherwise the CV joint will be contaminated and eventually fail (see Chapter 8).

14 Exhaust system check (every 6000 miles or 6 months)

Refer to illustration 14.4

1   With the engine cold (at least three hours after the vehicle has been driven), check the complete exhaust system from the manifold to the end of the tailpipe. Be careful around the catalytic converter, which may be hot even after three hours. The inspection should be done with the vehicle on a hoist to permit unrestricted access. If a hoist isn't available, raise the vehicle and support it securely on jackstands.

2   Check the exhaust pipes and connections for signs of leakage and/or corrosion indicating a potential failure. Make sure that all brackets and hangers are in good condition and tight.

3   Inspect the underside of the body for holes, corrosion, open seams, etc. which may allow exhaust gasses to enter the passenger compartment. Seal all body openings with silicone sealant or body putty.

4   Rattles and other noises can often be traced to the exhaust system, especially the hangers, mounts and heat shields **(see illustration)**. Try to move the pipes, mufflers and catalytic converter. If the components can come in contact with the body or suspension parts, secure the exhaust system with new mounts, brackets or hangers.

15 Manual transmission lubricant level check (every 6000 miles or 6 months)

1   The manual transmission has a filler plug on the passenger's side of the front transmission case. Remove the filler plug to check the lubricant level. If the vehicle is raised to gain access to the plug, be sure to

support it safely on jackstands - DO NOT crawl under a vehicle that is supported only by a jack! Be sure the vehicle is level or the check may be inaccurate.

2   Unscrew the filler plug from the transmission. If lubricant immediately starts leaking out, thread the plug back into the case - the level is correct. If lubricant doesn't leak out, completely remove the plug and reach inside the hole with your finger. The level should be even with the bottom of the fill plug hole.

3   If the level is low, add the recommended lubricant through the filler plug hole with a pump, syringe or squeeze bottle.

4   Install and tighten the plug and check for leaks after the first few miles of driving.

16 Transfer case lubricant level check (4WD models) (every 6000 miles or 6 months)

Refer to illustration 16.1

1   The transfer case lubricant level is checked by removing a filler plug from the backside of the case **(see illustration)**. If the vehicle is raised to gain access to the plug, be sure to support it safely on jackstands - DO NOT crawl under a vehicle that is supported only by a jack! Be sure the vehicle is level or the check may be inaccurate.

2   With the engine and transfer case cold, remove the plug. If lubricant immediately starts leaking out, thread the plug back into the case - the level is correct. If lubricant doesn't leak out, completely remove the plug and reach inside the hole with your finger. The level should be even with the bottom of the fill plug hole.

3   If the level is low, add the recommended lubricant through the fill plug hole with a pump, syringe or squeeze bottle.

4   Install and tighten the plug and check for leaks after the first few miles of driving.

17.1a On the rear differential, remove the differential plug with a standard screwdriver

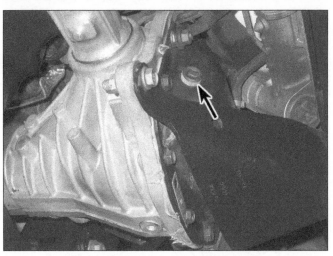

17.1b On 4WD models, remove the front differential filler plug with a 3/8-drive ratchet/extension

17   Differential lubricant level check (every 6000 miles or 6 months)

Refer to illustrations 17.1a and 17.1b
Note: *4WD vehicles have two differentials - be sure to check the lubricant level in both differentials.*

1   The differential lubricant level is check by removing a filler plug from the differential cover **(see illustrations)**. If the vehicle is raised to gain access to the plug, be sure to support it safely on jackstands - DO NOT crawl under a vehicle that is supported only by a jack! Be sure the vehicle is level or the check may be inaccurate.
2   With the differential cold, remove the fill plug. The level should be level with the bottom of the fill plug hole.
3   If the level is low, add the recommended lubricant through the filler plug hole with a pump, syringe or squeeze bottle.
4   Install the plug and check for leaks after the first few miles of driving.

18   Tire rotation (every 6000 miles or 6 months)

Refer to illustration 18.2
1   The tires should be rotated at the specified intervals and whenever uneven wear is noticed.
2   Refer to the accompanying illustration for the preferred tire rotation pattern **(see illustration)**.
3   Refer to the information in *Jacking and towing* at the front of this manual for the proper procedures to follow when raising the vehicle and changing a tire. If the brakes are to be checked, don't apply the parking brake as stated. Make sure the tires are blocked to prevent the vehicle from rolling as it's raised.
4   Preferably, the entire vehicle should be raised at the same time. This can be done on a hoist or by jacking up each corner and then lowering the vehicle onto jackstands placed under the frame rails. Always use four jackstands and make sure the vehicle is safely supported.

5   After rotation, check and adjust the tire pressures as necessary and be sure to properly tighten the lug nuts.

19   Brake check (every 6000 miles or 6 months)

Warning: *Brake system dust is hazardous to your health. Wear an approved filtering mask whenever working on the brakes. DO NOT blow it out with compressed air, inhale it or use gasoline or solvents to remove it. Use brake system cleaner only.*
Note: *For detailed information of the brake system, refer to Chapter 9.*
1   In addition to the specified intervals, the brakes should be inspected every time the wheels are removed or whenever a defect is suspected.
2   To check the brakes, raise the vehicle and place it securely on jackstands. Remove the wheels (see *Jacking and towing* at the front of the manual, if necessary).

Disc brakes

Refer to illustration 19.5
3   Disc brakes are used on the front wheels on all models as well as the rear wheels on some models. Extensive damage to the discs can occur if the pads are not replaced when needed.
4   The disc brake calipers, which contain the pads, are visible with the wheels removed. There is an outer pad and an inner pad in each caliper. All pads should be inspected.
5   Each caliper has a "window" or opening to inspect the pads. Check the thickness of the pad lining by looking into the caliper at each end and down through the inspection window at the top of the housing **(see illustration)**. If the lining material is less than the thickness listed in this Chapter's Specifications, the pads should be replaced. **Note:** *The metal backing plate is not included in this measurement.*

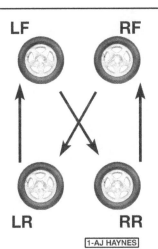

1-AJ HAYNES

18.2 The recommended tire rotation pattern for these vehicles

19.5 You will find an inspection hole like this in most calipers - placing a ruler across the hole should enable you to determine the thickness of remaining pad material

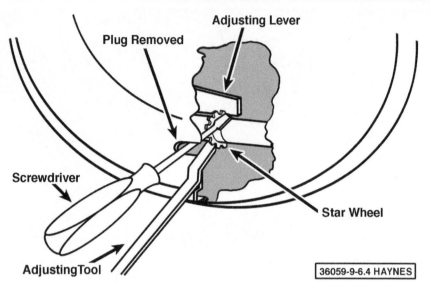

19.11 Use a thin screwdriver to push the lever away, then use an adjusting tool or another screwdriver to back off the star wheel

6   If you're unsure about the exact thickness of the remaining lining material, remove the pads for further inspection or replacement (refer to Chapter 9).

7   Before installing the wheels, check for leakage and/or damage (cracks, splitting, etc.) around the brake hose connections. Replace the hose or fittings as necessary, referring to Chapter 9.

8   Check the condition of the disc. Look for score marks, deep scratches and burned spots. If these conditions exist, the hub/disc assembly should be removed for servicing (see Chapter 9).

Rear drum brakes

Refer to illustrations 19.11, 19.13 and 19.15

9   Drum brakes are used on the rear wheels on some models. Remove the drum by pulling it off the axle and brake assembly. If this proves difficult, make sure the parking brake is released, then squirt penetrating oil around the center hub areas. Allow the oil to soak in and try to pull the drum off again.

10   If the drum still cannot be pulled off, the self-adjusting mechanism may have to be retracted slightly. This is done by first removing the small plug from the backing plate.

11   With the plug removed, insert a thin screwdriver and lift the adjusting lever off the star wheel, then use an adjusting tool or screwdriver to back off the star wheel several turns **(see illustration)**. This will move the brake shoes away from the drum. If the drum still won't pull off, tap around its inner circumference with a soft-face hammer.

12   With the drum removed, do not touch any brake dust (see the **Warning** at the beginning of this Section).

13   Note the thickness of the lining material on rear brake shoes. If the material has worn away to within 1/32-inch of the recessed rivets or 1/16-inch of the metal backing, the shoes should be replaced **(see illustration)**. The shoes should also be replaced if they're cracked, glazed (shiny surface) or contaminated with brake fluid.

14   Make sure that all the brake assembly springs are connected and in good condition.

15   Check the brake components for any signs of fluid leakage. Carefully pry back the rubber cups on the wheel cylinders located at the top of the brake shoes **(see illustration)**. Any leakage is an indication that the wheel cylinders should be overhauled immediately (see Chapter 9). Also check brake hoses and connections for signs of leakage.

16   Wipe the inside of the drum with a clean rag and brake system cleaner. Again, be careful not to breathe the dangerous dust.

17   Check the inside of the drum for cracks, score marks, deep scratches and hard spots, which will appear as small discolorations. If these imperfections cannot be removed with fine emery cloth, the drum must be taken to a machine shop equipped to resurface the drums.

18   If after the inspection process all parts are in good working condition, reinstall the brake drum. If the rear brake adjustment was changed, the rear brakes should be readjusted (see Chapter 9).

19   Install the wheels and lower the vehicle.

Parking brake

20   The parking brake is operated by a hand lever and locks the rear brake system. The easiest, and perhaps most obvious method of periodically checking the operation of the parking brake assembly is to stop the vehicle on a steep hill with the parking brake set and the transmission in Neutral. If the parking brake cannot prevent the vehicle from rolling, adjust it (see Chapter 9).

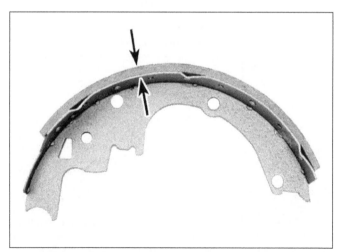

19.13 If the lining is bonded to the brake shoe, measure the lining thickness from the outer surface to the metal shoe, as shown here; if the lining is riveted to the shoe, measure from the lining outer surface to the rivet head

19.15 To check for wheel cylinder leakage, use a small screwdriver to pry the boot away from the cylinder

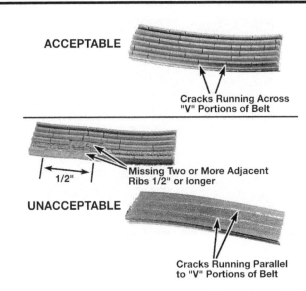

ACCEPTABLE

Cracks Running Across
"V" Portions of Belt

1/2"

Missing Two or More Adjacent
Ribs 1/2" or longer

UNACCEPTABLE

Cracks Running Parallel
to "V" Portions of Belt

21.2a Small cracks on the underside of V-ribbed belts are acceptable - lengthwise cracks, or missing pieces that cause the belt to make noise are cause for replacement

20 Fuel system check (every 6000 miles or 6 months)

Warning: *Gasoline is extremely flammable, so take extra precautions when working on any part of the fuel system. Don't smoke or allow open flames or bare light bulbs in or near the work area, and don't work in a garage where a gas-type appliance (such as a water heater or clothes dryer) is present. Since gasoline is carcinogenic, wear latex gloves when there's a possibility of being exposed to fuel, and, if you spill fuel on your skin, rinse it off immediately with soap and water. Have a Class B fire extinguisher on hand.*

1   The fuel tank is located under the rear of the vehicle.
2   The fuel system is most easily checked with the vehicle raised on a hoist so the components underneath the vehicle are readily visible and accessible.
3   If the smell of gasoline is noticed while driving or after the vehicle has been in the

sun, the system should be thoroughly inspected immediately.
4   Remove the gas tank cap and check for damage, corrosion and an unbroken sealing imprint on the gasket. Replace the cap with a new one if necessary.
5   With the vehicle raised, check the fuel tank and filler neck for punctures, cracks and other damage. The connection between the filler neck and the tank is especially critical. Sometimes a rubber filler neck will leak due to loose clamps or deteriorated rubber; problems a home mechanic can usually rectify.
6   Carefully check all rubber hoses and metal lines leading away from the fuel tank. Look for loose connections, deteriorated hoses, crimped lines and other damage. Follow the lines to the front of the vehicle, carefully inspecting them all the way. Repair or replace damaged sections as necessary.
7   If a fuel odor is still evident after the inspection, refer to Section 31.

21 Drivebelt check, adjustment and replacement (every 6000 miles or 6 months)

Refer to illustrations 21.2a, 21.2b, 21.3 and 21.5

Check

1   The drivebelt, also referred to as a serpentine belt or simply "fan" belt, is located at the front of the engine. A single serpentine drivebelt with an automatic belt tensioner drives the alternator, power steering pump and air conditioning compressor (some models). The condition and tension of the belt is critical to the operation of the engine. Because of the composition and the high stresses to which it is subjected, the drivebelt stretches and deteriorates as it gets older. It must therefore be periodically inspected. The automatic belt tensioner is non-adjustable.
2   With the engine off, open the hood and locate the drivebelt at the front of the engine. With a flashlight, check the belt for separation of the ribs from the adhesive rubber, cracking or separation of the ribs, and torn or worn ribs or cracks in the inner ridges of the ribs **(see illustration)**. Both sides of the belt should be inspected, which means you will have to twist the belt to check the underside. Use your fingers to feel the belt where you can't see it. If any of the above conditions are evident, replace the belt. On V6 engines, replace the belt if the wear indicator gap measurement exceeds 0.94 inches **(see illustration)**.

Replacement

3   On four-cylinder models, use a 1/2-inch drive ratchet to rotate the tensioner bracket counterclockwise to release the tension on the belt. On 3.7L V6 models, use a wrench to rotate the tensioner bolt clockwise to release the tension on the belt **(see illustration)**. On 4.0L V6 engines, remove the coolant recovery bottle (see Chapter 3), then insert a 3/8-inch drive ratchet into the hole on the tensioner arm to rotate the tensioner clockwise to release the tension on the belt. The

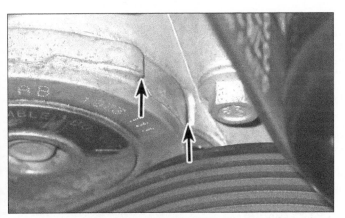

21.2b The serpentine drivebelt tensioner used on V6 models automatically keeps the proper tension on the drivebelt, but it does have limits - replace the belt if the belt wear indicator gap exceeds 0.94 inches (between arrows)

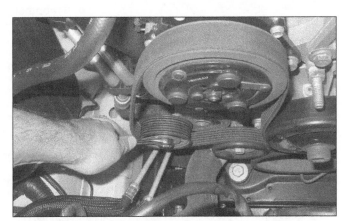

21.3 On V6 models, release the belt tension by rotating the tensioner clockwise using a wrench on the tensioner bolt

tensioner will swing down out of the way once the tension of the belt is released. Remove the belt and carefully release the tensioner.

4   Compare the old and new belts in order to make a direct comparison for length, width and design.

5   Route the new belt over the pulleys, again rotating the tensioner to allow the belt to be installed, then release the belt tensioner. When installing the new belt, make sure it is routed properly because improper routing can cause the water pump to rotate backward, causing overheating. The belt routing diagram is located on the fan shroud **(see illustration)**.

6   Before starting the engine, make sure the ribs on the belt are correctly placed in the grooves of each pulley. If not, release the tension on the belt again, and move the belt into place. It is essential that the belt be properly centered.

4.0L V6 power steering belt

Removal

Note: *The power steering belt has no tensioner.*

7   Remove the main drivebelt (see Step 3).

8   Place a plastic trim tool or equivalent between the power steering belt and the power steering pulley at the top of the pulley.

9   Using a socket and ratchet, rotate the crankshaft pulley bolt clockwise.

10   Hold the tool at a slight angle and allow the plastic tool to "walk" the belt off of the power steering pulley while slowly rotating the crankshaft pulley bolt until the belt is removed from the pulley.

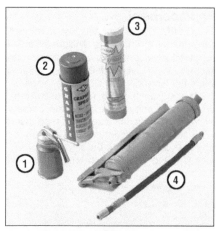

23.1 Materials required for chassis and body lubrication

*1   **Engine oil** - Light engine oil in a can like this can be used for door and hood hinges*

*2   **Graphite spray** - Used to lubricate lock cylinders*

*3   **Grease** - Grease, in a variety of types and weights, is available for use in a grease gun. Check the Specification for your requirements*

*4   **Grease gun** - A common grease gun, shown here with a detachable hose and nozzle, is needed for chassis lubrication. After use, clean it thoroughly!*

21.5 A drivebelt routing diagram is found on the radiator brace or shroud (typical)

Installation

11   Place the power steering belt around the crankshaft pulley, making sure the belt is fully seated in all the crankshaft grooves in the pulley.

12   Place the belt at the top of the power steering pulley, then position a plastic trim tool between the belt and top of the pulley. Slowly rotate the crankshaft pulley bolt while using the tool to guide the belt onto the power steering pulley.

13   Once the belt has started on the pulley, continue to "walk" the belt onto the pulley until all of the belt is seated in the grooves of the pulley.

14   If the new belt is too difficult to install using the plastic tool, insert a cable tie through the access hole on the pulley and secure the belt to the pulley. Slowly rotate the pulley until the belt is fully seated, the plastic tie may break once the belt is seated. If it tie doesn't break cut it off with a pair of diagonal cutting pliers.

15   Make sure the ribs on the belt are correctly seated in the grooves on the pulleys.

16   Install the main drivebelt (see Steps 5 and 6).

22   Seat belt check (every 6000 miles or 6 months)

Warning: *In addition to driver and passenger airbags, some later models may be equipped with pyrotechnic seat belt tensioners and/or side-impact airbags located in the seatbacks or along the inside edges of the headliner. Whenever working around these components, perform the procedure to disable the airbag system (see Chapter 12).*

1   Check the seat belts, buckles, latch plates and guide loops for any obvious damage or signs of wear.

2   Make sure the seat belt reminder light comes on when the key is turned on.

3   The seat belts are designed to lock up during a sudden stop or impact, yet allow free movement during normal driving. The retractors should hold the belt against your chest while driving and rewind the belt when the buckle is unlatched.

4   If any of the above checks reveal problems with the seat belt system, replace parts as necessary.

23   Chassis lubrication (every 15,000 miles or 12 months)

Refer to illustration 23.1

1   Refer to *Recommended lubricants and fluids* at the front of this Chapter, to obtain the necessary grease, etc. You'll also need a grease gun **(see illustration)**. Occasionally plugs will be installed rather than grease fittings. If so, grease fittings will have to be purchased and installed. Some models may have sealed front wheel bearings, tie-rod ends, balljoints and universal joints with no plugs or fittings for lubrication.

2   Look under the vehicle and locate any grease fittings.

3   For easier access under the vehicle, raise it with a jack and place jackstands under the frame. Make sure it's safely supported by the stands. If the wheels are to be removed at this interval for tire rotation or brake inspection, loosen the lug nuts slightly while the vehicle is still on the ground.

4   Before beginning, force a little grease out of the nozzle to remove any dirt from the end of the gun. Wipe the nozzle clean with a rag.

5   With the grease gun and plenty of clean rags, crawl under the vehicle and begin lubricating the components.

6   Wipe one of the grease fitting nipples clean and push the nozzle firmly over it. Pump the gun until the component is completely lubricated. On balljoints and steering connections, stop pumping when the rubber seal is firm to the touch. Do not pump too much grease into the fitting as it could rupture the seal. For all other suspension and steering components, continue pumping grease into the fitting until it oozes out of the joint between the two components. If it escapes around the grease gun nozzle, the nipple is clogged or the nozzle is not completely seated on the fitting. Resecure the gun nozzle to the fitting and try again. If necessary, replace the fitting with a new one.

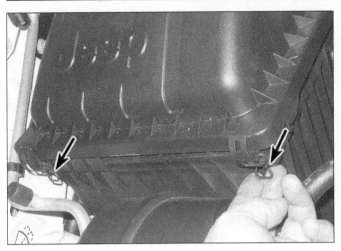

24b.2 Release the two spring clips on the air filter housing . . .

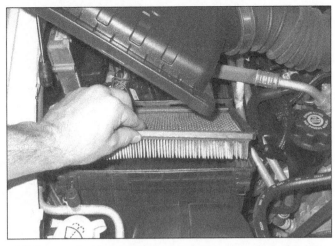

24b.3 . . . swing the housing cover slightly open, then slide it to release the locating tabs from their slots; lift the housing cover away and slide the filter element out of the housing

7   Wipe the excess grease from the components and the grease fitting. Repeat the procedure for the remaining fittings.
8   Also clean and lubricate the parking brake cable, along with the cable guides and levers. This can be done by smearing some of the chassis grease onto the cable and its related parts with your fingers.

24a Cabin air filter replacement (every 15,000 miles or 12 months)

Note: *Not all models have cabin air filters.*
1   The cabin air filter is located behind the center console and glove box.
2   Remove the glove box assembly (see Chapter 11).
3   Remove the cabin air filter cover by releasing the tabs on each end.
4   Slide the filter from the HVAC housing.
5   Installation is the reverse of removal. Make sure the arrow on the filter element faces down.

24b Air filter replacement (every 30,000 miles or 24 months)

Refer to illustrations 24b.2 and 24b.3
1   At the specified intervals, the air filter element should be replaced with a new one. The filter element sits inside of a plastic air filter housing located in the right front corner of the engine compartment. The housing cover is secured by two spring clips on the front and locating tabs on the rear.
2   Release the two spring clips on the filter housing by prying them **(see illustration)**.
3   Swing the housing cover slightly open, then slide it to release the locating tabs from their slots. Lift the housing cover away and slide the filter element out of the housing **(see illustration)**. Wipe out the inside of the housing with a clean rag.

4   Place the new filter element in the housing, making sure it seats properly.
5   Insert the housing cover locating tabs back into their slots and close the housing. Make sure the air filter housing is seated, then secure the housing with the two spring clips.

25 Automatic transmission fluid and filter change (every 30,000 miles or 24 months)

Refer to illustrations 25.4, 25.7, 25.8, 25.10 and 25.11
1   At the specified intervals, the transmission fluid should be drained and replaced. Before beginning work, purchase the specified transmission fluid (see *Recommended lubricants and fluids* at the front of this Chapter), a new filter and silicone sealant. **Note:** *On the 42RLE transmission, Mopar Lock & Seal Adhesive or equivalent, will also be needed for one specific transmission pan bolt. On the 45RFE transmission, replacement of the additional cooler return filter, requires a special filter wrench.* Since the fluid will remain hot long

after driving, perform this procedure only after the engine has cooled down completely. Other tools necessary for this job include a floor jack, jackstands to support the vehicle in a raised position, a drain pan capable of holding at least eight pints, newspapers and clean rags.
2   Raise the vehicle and support it securely on jackstands.
3   Place the drain pan underneath the transmission pan.
4   Remove the rear and side pan mounting bolts, but only loosen the front pan bolts approximately four turns. Carefully pry the transmission pan loose with a screwdriver, allowing much of the fluid to drain **(see illustration)**. Remove the remaining bolts and carefully lower the pan into the drain pan. **Note:** *On the 42RLE transmission, take notice of the pan bolt that has a sealant patch on it. Set aside this bolt for reuse and mark the bolt hole that you removed it from.*
5   Use a gasket scraper and a rag to carefully clean the mating surfaces of the transmission and pan until all traces of old sealant are removed. **Caution:** *Be very careful not gouge the delicate aluminum surface on the transmission.*

25.4 With the front bolts in place but loose, pull the rear of the pan down to drain the fluid

25.7   Remove the filter screw(s) and filter

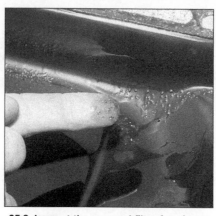

25.8   Inspect the pan and filter for pieces of metal or clutch material

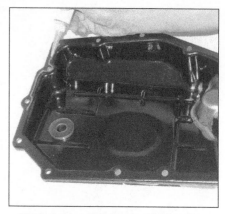

25.10   Apply a thin bead of RTV sealant on the pan

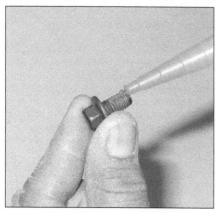

25.11   On the 42RLE transmission, re-apply sealant to the bolt located second from the front, on the left side of the pan

27.2   Remove the transfer case fill plug and drain plug from the back side of the case

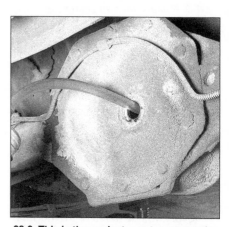

28.6   This is the easiest way to remove the lubricant: work the end of the hose to the bottom of the differential housing and draw out the old lubricant with a hand pump

6   Drain the transmission pan into the drain pan. Clean the transmission pan and pan magnet with solvent and dry it with compressed air, if available.

7   On 42RLE and 45RFE transmissions, remove the transmission filter retaining screw(s) and place the filter in the drain pan **(see illustration)**. If the filter seal or O-ring did not come off of the transmission with the filter, remove the O-ring or seal from the transmission. On NAG1 transmissions, pull the filter down and out from the electric-hydraulic control unit on the valve body, then remove the O-ring.

8   Once the pan and filter are off, you can get a good idea of the condition of your transmission by thoroughly inspecting the bottom of the pan, the filter and the fluid. **Note:** *Some models may be equipped with a magnet inside the pan. Clean the magnet and reinstall it in the cleaned fluid pan before installation.* Although normally bright red, transmission fluid may turn dark red or brown during normal use. If you find the fluid very dark colored, or if it smells burned, it usually indicates the transmission has been overheated. If you find small pieces of metal or clutch material in the pan or filter, it indicates wear or damage have

occurred to the internal parts or clutches **(see illustration)**. If you have any concerns about the condition of your transmission based on what you find in the fluid, pan and filter, it's a good idea to take your vehicle to your dealer or a transmission shop for further evaluation.

9   Install the new O-ring, filter and retaining screws, if equipped, on the transmission. **Caution:** *On the 45RFE transmission, install the seal COMPLETELY flush on the transmission oil pump. Then install the filter and retaining screw. Otherwise, transmission damage may occur.*

10   Apply a THIN bead of RTV sealant to the mating surface of the transmission drain pan **(see illustration)**.

11   Put the pan in place against the transmission and, working around the pan, tighten each bolt a little at a time to the torque listed in this Chapter's Specifications. **Note:** *On the 42RLE transmission, re-apply Mopar Lock & Seal Adhesive or equivalent to the designated pan bolt, before installing the bolt* **(see illustration)**.

12   Lower the vehicle.

13   On 42RLE transmissions, add 4 quarts (if the oil and filter were replaced) or 8.8 quarts (if the transmission was overhauled) of the specified type of automatic transmission

fluid through the filler tube. On the 45RFE transmission, add 5 quarts of the specified type of automatic transmission fluid through the filler tube (see Section 6).

14   With the transmission in Park and the parking brake set, run the engine at a fast idle, but don't race it. Move the gear selector through each range and back to Park.

15   Check the fluid level and add fluid, if necessary, until the level is within the correct range on the dipstick (see Section 6).

16   Check under the vehicle for leaks during the first few trips. Check the fluid level again when the transmission is hot (see Section 6).

26   Manual transmission lubricant change (every 30,000 miles or 24 months)

1   Raise the vehicle and support it securely on jackstands. Make sure the vehicle is level.

2   Move a drain pan, rags, newspapers and wrenches under the transmission.

3   Remove the fill plug from the right side of the front case and the drain plug from the bottom of the case.

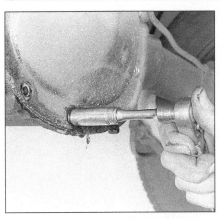

28.8a Remove the bolts from the lower edge of the cover . . .

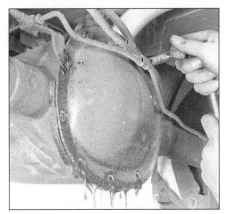

28.8b . . . then loosen the top bolts and let the lubricant drain out

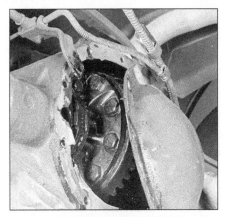

28.8c After the lubricant has drained, remove the bolts and the cover

28.10 Carefully scrape the old sealant material off to ensure a leak-free seal

4   After the lubricant has drained completely, clean the drain plug and drain hole area and reinstall the drain plug on the bottom of the case. Tighten it securely.
5   Using a hand pump, syringe, or squeeze bottle, fill the transmission with the specified lubricant until it begins to leak out through the hole.
6   Clean the fill plug and fill hole area. Install the fill plug and tighten it securely.
7   Lower the vehicle.
8   Drive the vehicle for a short distance, then check the drain and fill plugs for leakage.

27   Transfer case lubricant change (4WD models) (every 30,000 miles or 24 months)

Refer to illustration 27.2
1   Raise the vehicle and support it securely on jackstands. Make sure the vehicle is level.
2   Remove the fill plug and the drain plug from the back side of the case **(see illustration)**.
3   After the lubricant has drained completely, clean the drain plug and drain hole area and reinstall the drain plug. Tighten it securely.

4   Fill the case with the specified lubricant until it is level with the lower edge of the filler hole.
5   Clean the fill plug and fill hole area. Install the fill plug and tighten it securely.
6   Drive the vehicle for a short distance, then check the drain and fill plugs for leakage.

28   Differential lubricant change (every 30,000 miles or 24 months)

Drain

Rear differential
Refer to illustrations 28.6, 28.8a, 28.8b, 28.8c, 28.10
1   This procedure should be performed after the vehicle has been driven so the lubricant will be warm and therefore flow out of the differential more easily.
2   Raise the vehicle and support it securely on jackstands. Make sure the vehicle is level.
3   The easiest way to drain the differential is to remove the lubricant through the filler plug hole with a suction pump. If the differential cover gasket is leaking, it will be necessary to remove the cover to drain the lubricant (which will also allow you to inspect the differential).

Changing the lubricant with a suction pump
4   Remove the filler plug from the differential (see Section 17).
5   Insert the flexible hose.
6   Work the hose down to the bottom of the differential housing and pump the lubricant out **(see illustration)**.

Changing lubricant by removing the cover
7   Move a drain pan, rags, newspapers and wrenches under the vehicle.
8   Remove the bolts on the lower half of the cover. Loosen the bolts on the upper half and use them to loosely retain the cover. Allow the oil to drain into the pan, then completely remove the cover **(see illustrations)**.
9   Using a lint-free rag, clean the inside of the cover and the accessible areas of the dif-

ferential housing. As this is done, check for chipped gears and metal particles in the lubricant, indicating that the differential should be more thoroughly inspected and/or repaired.
10   Use a gasket scraper or putty knife to remove all traces of old sealant **(see illustration)**. Thoroughly clean the mating surfaces of the differential housing and the cover plate.
11   Apply a thin bead of RTV sealant to the cover flange, and install the cover.

Front differential
12   Place a drain plug under the differential. Remove the fill plug and the drain plug from the differential housing.
13   After the lubricant has drained completely, clean the drain plug and drain hole area and reinstall the drain plug. Tighten it securely.

Refill
14   Use a hand pump, syringe or squeeze bottle to fill the differential housing with the specified lubricant until it's level with the bottom of the plug hole.
15   Install the fill plug and tighten it securely.

29   Cooling system servicing (draining, flushing and refilling) (every 30,000 miles or 24 months)

Warning: *Do not allow antifreeze to come in contact with your skin or painted surfaces of the vehicle. Rinse off spills immediately with plenty of water. Antifreeze is highly toxic if ingested. Never leave antifreeze lying around in an open container or in puddles on the floor; children and pets are attracted by its sweet smell and may drink it. Check with local authorities about disposing of used antifreeze. Many communities have collection centers that will see that antifreeze is disposed of safely.*

Draining
Refer to illustration 29.3 and 29.5
1   Periodically, the cooling system should be drained, flushed and refilled to replenish the antifreeze mixture and prevent formation

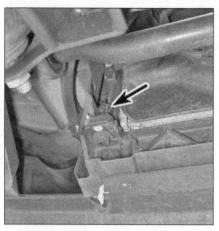

29.3 The radiator drain fitting is located on the lower left side of the radiator

29.5 V6 engine cylinder block drain location - there is one on each side of the block (right side shown); left side is behind the starter motor

of rust and corrosion, which can impair the performance of the cooling system and cause engine damage. When the cooling system is serviced, all hoses and the radiator cap should be checked and replaced if necessary.

2   Apply the parking brake and block the wheels. **Warning:** *If the vehicle has just been driven, wait several hours to allow the engine to cool down before beginning this procedure.*

3   Move a large container under the radiator drain to catch the coolant. The radiator drain fitting is located on the lower left side of the radiator **(see illustration)**. Attach a hose to the drain fitting (if possible) to direct the coolant into the container. With the expansion tank cap on, open the drain fitting (this should cause the expansion tank to drain also).

4   Allow the radiator to completely drain, then close the drain fitting.

5   On V6 models, move the container under the engine block drain plugs - there's one on each side of the block **(see illustration)**. The starter motor must be removed to access the left side drain plug (see Chapter 5). Remove the plugs and allow the coolant in the block to drain. **Note:** *Frequently, the coolant will not drain from the block after the plug is removed. This is due to a rust layer that has built up behind the plug. Insert a Phillips screwdriver into the hole to break the rust barrier.* After the engine block is completely drained, install the engine block drain plugs.

6   Check the condition of the radiator hoses, heater hoses and clamps (refer to Section 10 if necessary). Replace any damaged clamps or hoses.

Flushing

Refer to illustration 29.9

7   Once the system is completely drained, remove the thermostat from the engine (see Chapter 3). Then reinstall the thermostat housing without the thermostat. This will allow the system to be thoroughly flushed.

8   Tighten the radiator drain plug, if not already done. Turn your heating system controls to Hot, so that the heater core will be flushed at the same time as the rest of the cooling system.

9   Disconnect the upper radiator hose, then place a garden hose in the upper radiator inlet and flush the system until the water runs clear at the upper radiator hose **(see illustration)**.

10   In severe cases of contamination or clogging of the radiator, remove the radiator (see Chapter 3) and have a radiator repair facility clean and repair it if necessary.

11   Many deposits can be removed by the chemical action of a cleaner available at auto parts stores. Follow the procedure outlined in the manufacturer's instructions. **Note:** *When the coolant is regularly drained and the system refilled with the correct antifreeze/water mixture, there should be no need to use chemical cleaners or descalers.*

Refilling

12   To refill the system, install the thermostat and reconnect any radiator hoses.

13   Place the heater temperature control in the maximum heat position.

14   Make sure to use the proper coolant listed in this Chapter's Specifications. Slowly fill the radiator with the recommended mixture of antifreeze and water to the base of the filler neck. Then add coolant to the reservoir until it reaches the COLD FULL mark. Wait five minutes and recheck the coolant level in the coolant tank and radiator, adding if necessary.

15   Install the radiator and coolant tank caps.

16   Start the engine, allow it to reach normal operating temperature. Check for leaks. When the system has cooled, add coolant as necessary. The cooling system will push air out of the system, into the coolant tank - after driving the vehicle for about one hour, let the system cool and add coolant to the coolant tank as necessary.

---

30   Positive Crankcase Ventilation (PCV) valve check and replacement (every 30,000 miles or 24 months)

Refer to illustrations 30.1 and 30.2

1   The PCV valve on the 2.4L four-cylinder engine is located on the left side of the valve cover, toward the front. The PCV valve on 2006 and earlier 3.7L V6 engines is located in the neck of the oil filler tube, which is mounted to the front of the right cylinder head. On 2007 and later V6 engines, the PCV valve is located near the rear of the left cylinder head **(see illustration)**.

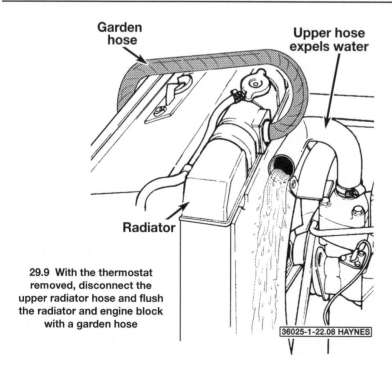

29.9 With the thermostat removed, disconnect the upper radiator hose and flush the radiator and engine block with a garden hose

Garden hose

Upper hose expels water

Radiator

36025-1-22.08 HAYNES

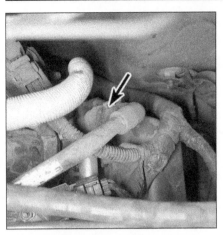

30.1 On 2007 and later V6 engines, the PCV valve is located at the rear of the left (driver's side) cylinder head

30.2 On 2006 and earlier V6 engines, rotate the PCV valve 90 degrees counterclockwise and pull straight out to remove it

31.2 The evaporative emissions control canister is located in front of the fuel tank, next to the suspension coil spring on the left rear - check the hoses and connections for damage

2   On the 2.4L four-cylinder engine, remove the hose and mounting bolt and pull the valve straight out from the valve cover. On 2006 and earlier 3.7L V6 engines, remove the hose and rotate the valve 90-degrees counterclockwise, then pull the valve straight out to remove it from the filler neck and reconnect the vacuum hose **(see illustration)**. On 2007 and later V6 engines, disconnect the hose, then unscrew the PCV valve.

3   With the engine idling, place your finger over the valve opening. If there's no vacuum at the valve, check for a plugged hose, manifold port, or the valve itself. Replace any plugged or deteriorated hoses.

4   Turn off the engine and shake the PCV valve, listening for a rattle. If the valve doesn't rattle, replace it with a new one.

5   To replace the valve, pull it from the end of the hose, noting its installed position.

6   When purchasing a replacement PCV valve, make sure it's for your particular vehicle and engine size. Compare the old valve with the new one to make sure they're the same. Make sure the new PCV valve is equipped with an O-ring. If not, remove the O-ring from the old valve and install it on the new one. Be sure the O-ring is in good condition before doing so or vacuum leaks may occur.

7   Install the PCV valve securely into position. On 2006 and earlier 3.7L V6 engines, rotate the valve 90-degrees clockwise to lock it into place on the oil filler tube. On 2007 and later V6 engines, thread the PCV valve into the right rear valve cover and tighten it securely.

8   Push the vacuum hose over the end of the valve until it's seated.

31   Evaporative emissions control system check (every 30,000 miles or 24 months)

Refer to illustration 31.2

1   The function of the evaporative emissions control system is to draw fuel vapors

from the gas tank and fuel system, temporarily store them in a charcoal canister and route them to the intake manifold during normal engine operation.

2   The most common symptom of a fault in the evaporative emissions system is a strong fuel odor in the engine compartment. If a fuel odor is detected, inspect the charcoal canister, located in front of the fuel tank, near the suspension coil spring on the left rear **(see illustration)**. Check the canister and all hoses for damage and deterioration.

3   The evaporative emissions control system is explained in more detail in Chapter 6.

32   Spark plug replacement (every 30,000 miles or 24 months)

Refer to illustrations 32.2, 32.5a, 32.5b, 32.11a, 32.11b, 32.15 and 32.20

1   Open the hood and label each spark plug wire with a number to ensure proper installation.

2   In most cases, the tools necessary for spark plug replacement include a ratchet, spark plug socket (spark plug sockets are padded inside to prevent damage to the porcelain insulators on the new plugs), various extensions and a gap gauge to check and adjust the gaps on the new plugs **(see illustration)**. For 2.4L four-cylinder models, a special plug wire removal tool is available for separating the plug wire boots from the spark plugs, but it isn't absolutely necessary. **Note:** *A torque wrench MUST be used to tighten the new plugs on the 2.4L four-cylinder model, and SHOULD be used on the 3.7L V6 model.*

3   The best approach when replacing the spark plugs is to purchase the new ones in advance, adjust them to the proper gap and replace them one at a time. When buying the new spark plugs, be sure to obtain the correct plug type for your particular engine. This information can be found on the *Emission Control*

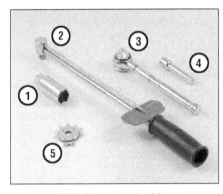

32.2 Tools required for spark plug replacement

1   **Spark plug socket** - *This will have special padding inside to protect the spark plug's porcelain insulator*

2   **Torque wrench** - *Although not mandatory, using this tool is the best way to ensure the plugs are tightened properly*

3   **Ratchet** - *Standard hand tool to fit the spark plug socket*

4   **Extension** - *Depending on model and accessories, you may need special extensions and universal joints to reach one or more of the plugs*

5   **Spark plug gap gauge** - *This gauge for checking the gap comes in a variety of styles. Make sure the gap for your engine is included*

Information label located under the hood, in your owner's manual and in the Specifications at the front of this Chapter. If differences exist between the plug specified on the emissions label and in the owner's manual, assume that the emissions label is correct.

4   Allow the engine to cool completely before attempting to remove any of the plugs. While you're waiting for the engine to cool, check the new plugs for defects and adjust the gaps.

32.5a Spark plug manufacturers recommend using a wire-type gauge when checking the gap - if the wire does not slide between the electrodes with a slight drag, adjustment is required

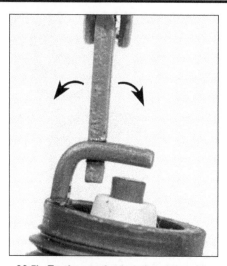

32.5b To change the gap, bend the side electrode only, as indicated by the arrows, and be very careful not to crack or chip the porcelain insulator surrounding the center electrode

32.11a Apply a thin coat of anti-seize compound to the spark plug threads

5   The gap is checked by inserting the proper-thickness gauge between the electrodes at the tip of the plug **(see illustration)**. The gap between the electrodes should be the same as the one specified on the *Emissions Control Information* label or in this Chapter's Specifications. The wire should just slide between the electrodes with a slight amount of drag. If the gap is incorrect, use the adjuster on the gauge body to bend the curved side electrode slightly until the proper gap is obtained **(see illustration)**. If the side electrode is not exactly over the center electrode, bend it with the adjuster until it is. Check for cracks in the porcelain insulator (if any are found, the plug should not be used).

6   If compressed air is available, use it to blow any dirt or foreign material away from the spark plug boots (2.4L four-cylinder models) or individual coils (3.7L V6 models).

2.4L four-cylinder models only

7   Remove the air cleaner tube, air cleaner housing and throttle body (see Chapter 4) to access the spark plugs for cylinder nos. 2 and 3.

8   By pulling only on the boot at the end of the wire (do not pull on the wire), remove the spark plug wire from one spark plug. A plug wire removal tool should be used to do this, if available.

9   If compressed air is available, use it to blow any dirt or foreign material out of the spark plug hole. The idea here is to eliminate the possibility of debris falling into the cylinder as the spark plug is removed.

10   Place the spark plug socket and extension over the plug and remove it from the engine by turning it in a counterclockwise direction. Compare the spark plug with the chart on the inside back cover of this manual to get an indication of the general running condition of the engine.

11   Apply a thin film of anti-seize compound

to the spark plug threads **(see illustration)**. Thread one of the new plugs into the hole until you can no longer turn it with your fingers, then tighten it with a torque wrench. It's a good idea to slip a short length of rubber hose over the end of the plug to use as a tool to thread it into place **(see illustration)**. The hose will grip the plug well enough to turn it, but will start to slip if the plug begins to cross-thread in the hole - this will prevent damaged threads and the accompanying repair costs.

12   Inspect the spark plug wire following the procedures outlined in Section 33.

13   Attach the plug wire, again using a twisting motion on the boot until it's seated on the spark plug.

14   Repeat the procedure for the remaining spark plugs, replacing them one at a time to prevent mixing up the spark plug wires.

V6 engines

15   On 4.0L V6 models, remove the upper intake manifold (see Chapter 2C). Disconnect the electrical connector, remove the mounting nut on the individual coil and pull the coil pack off of the spark plug **(see illustration)**.

16   If compressed air is available, use it to blow any dirt or foreign material inside the spark plug hole. The idea here is to eliminate the possibility of debris falling into the cylinder as the spark plug is removed.

17   Place the spark plug socket over the plug and remove it from the engine by turning it in a counterclockwise direction.

18   Compare the spark plug with the chart on the inside back cover of this manual to get an indication of the general running condition of the engine.

19   Apply a thin film of anti-seize compound to the spark plug threads **(see illustration 32.11a)**. Thread one of the new plugs into the hole until you can no longer turn it with your fingers, then tighten it with a torque

32.11b A length of snug-fitting rubber hose will save time and prevent damaged threads when installing the spark plugs

wrench (if available) or the ratchet. It's a good idea to slip a short length of rubber hose over the end of the plug to use as a tool to thread it into place **(see illustration 32.11b)**. The hose will grip the plug well enough to turn it, but will start to slip if the plug begins to cross-thread in the hole - this will prevent damaged threads and the accompanying repair costs.

20   Inspect the coil pack O-ring for cracks or tears. Replace the O-ring if necessary **(see illustration)**.

21   Apply a thin coat of silicone lubricant to the coil O-ring, install the coil over the new spark plug and install the mounting nut.

22   Repeat the procedure for the remaining spark plugs, replacing them one at a time to prevent mixing up the spark plug wires.

23   Remaining installation is the reverse of removal.

32.15 On 3.7L V6 models, disconnect the electrical connector, remove the mounting nut and pull the coil off each spark plug

32.20 On 3.7L V6 models, inspect the o-ring on the coil (arrow) for cracks or tears - replace if necessary

33   Spark plug wires, check and replacement (2.4L four-cylinder engine) (every 30,000 miles or 24 months)

1   The spark plug wires should be checked at the recommended intervals and whenever new spark plugs are installed in the engine.

2   The wires should be inspected one at a time to prevent mixing up the order, which is essential for proper engine operation.

3   Disconnect the plug wire from one spark plug. To do this, grab the rubber boot, twist slightly and pull the wire free. Do not pull on the wire itself, only on the rubber boot.

4   Check inside the boot for corrosion, which will look like a white crusty powder. Push the wire and boot back onto the end of the spark plug. It should be a tight fit on the plug. If it isn't, remove the wire and use a pair of pliers to carefully crimp the metal connector inside the boot until it fits securely on the end of the spark plug.

5   Using a clean rag, wipe the entire length of the wire to remove any built-up dirt and grease. Once the wire is clean, check for holes, burned areas, cracks and other damage. Don't bend the wire excessively or the conductor inside might break.

6   Disconnect the wire from the coil pack and pull the wire straight out. Pull only on

the rubber boot during removal. Check for corrosion and a tight fit in the same manner as the spark plug end. Reattach the wire to the coil pack.

7   Check the remaining spark plug wires one at a time, making sure they are securely fastened at the coil pack and the spark plug when the check is complete.

8   If new spark plug wires are required, purchase a new set for your specific engine model. Wire sets are available pre-cut, with the rubber boots already installed. Remove and replace the wires one at a time to avoid mix-ups in the firing order. The wire routing is extremely important, so be sure to note exactly how each wire is situated before removing it.

Notes

Service record

Date	Mileage/hours	Work performed

Service record

Date	Mileage/hours	Work performed

Chapter 2 Part A
2.4L four-cylinder engine

Contents

Specifications

General

Bore	3.445 inches (87.5 mm)
Stroke	3.976 inches (101 mm)
Compression ratio	9.5:1
Displacement	2.4 liters (148 cubic inches)
Firing order	1-3-4-2

Camshaft

Bearing bore diameter	1.024 to 1.025 inches (26.020 to 26.041 mm)
Bearing journal diameter	
2004 and earlier models	1.021 to 1.022 inches (25.951 to 25.970mm)
2005 models	1.022 to 1.023 inches (25.976 to 25.995 mm)
Bearing clearance	
2004 and earlier models	0.0027 to 0.0030 inch (0.0691 to 0.0710 mm)
2005 models	0.0009 to 0.0025 inch (0.025 to 0.065 mm)
Endplay	0.0019 to 0.0066 inch (0.05 to 0.17 mm)
Lobe lift	
Intake	0.324 inch (8.25 mm)
Exhaust	0.259 inch (6.60 mm)

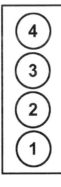

Cylinder locations
(2.4L four-cylinder)

Front

50035-1-A HAYNES

Cylinder head

Head gasket surface warpage limit.. 0.004 inch maximum (0.1 mm)
Intake and Exhaust manifold mounting surface warpage limit .... 0.006 inch per foot maximum (0.15 mm per 300 mm)

Intake and exhaust manifolds

Warpage limit .... 0.006 inch per foot maximum (0.15 mm per 300 mm)

Oil pump

Cover warpage limit .... 0.001 inch (0.025 mm)
Inner rotor thickness (minimum)
    2004 and earlier models... 0.370 inch (9.40 mm)
    2005 models .... 0.421 inch (10.699 mm)
Outer rotor thickness (minimum)... 0.370 inch (9.40 mm)
Outer rotor diameter (minimum)
    2004 and earlier models... 3.148 inches (79.95 mm)
    2005 models .... 3.383 inches (85.924 mm)
Rotor-to-pump cover clearance (maximum) .... 0.004 inch (0.10 mm)
Outer rotor-to-housing clearance (maximum)....................................... 0.015 inch (0.039 mm)
Inner rotor-to-outer rotor lobe clearance (maximum)........................... 0.008 inch (0.20 mm)
Pressure relief spring free length.. 2.39 inches (approximate)

Torque specifications*

Ft-lbs (unless otherwise indicated)   **Nm**

Note: *One foot-pound (ft-lb) of torque is equivalent to 12 inch-pounds (in-lbs) of torque. Torque values below approximately 15 foot-pounds are expressed in inch-pounds, because most foot-pound torque wrenches are not accurate at these smaller values.*

	Ft-lbs	Nm
Camshaft bearing cap bolts **(see illustration 11.5 for bolt tightening sequence)**		
M6 bolts	105 in-lbs	12
M8 bolts	21	28
Camshaft timing sprocket bolt		
2004 and earlier models..	75	101
2005 models	85	115
Crankshaft damper bolt	100	136
Cylinder head bolts **(in sequence - see illustration 15.18)**		
Step 1	25	34
Step 2	50	68
Step 3	50	68
Step 4	Tighten an additional 1/4-turn (90-degrees)	
Driveplate-to-crankshaft bolts...	70	95
Exhaust manifold-to-cylinder head bolts	16.5	22.5
Exhaust manifold-to-exhaust pipe bolts	21	28
Intake manifold bolts..	21	28
Oil pan bolts	105 in-lbs	12
Oil pump		
Attaching bolts...	21	28
Cover screws	105 in-lbs	12
Pick-up tube bolt	20	27
Relief valve cap bolt...	30	41
Timing belt		
Cover bolts...	105 in-lbs	12
Idler pulley bolt..	45	61
Tensioner assembly		
Tensioner pulley bolt	45	61
Tensioner assembly mounting bolt	45	61
Valve cover bolts		
Step 1..	40 in-lbs	4.5
Step 2..	80 in-lbs	9
Step 3..	105 in-lbs	12

* Refer to Part D for additional torque specifications

1   General information

This Part of Chapter 2 is devoted to in-vehicle engine repair procedures. Information concerning engine removal and installation can be found in Part D of this Chapter.

The following repair procedures are based on the assumption that the engine is installed in the vehicle. If the engine has been removed from the vehicle and mounted on a stand, many of the steps outlined in this Part of Chapter 2 will not apply.

The Specifications included in this Part of Chapter 2 apply only to the procedures contained in this Part.

There is one type of four-cylinder engine installed in the models covered in this book: the 2.4L dual overhead camshaft (DOHC) engine.

2   Repair operations possible with the engine in the vehicle

Many major repair operations can be accomplished without removing the engine from the vehicle.

Clean the engine compartment and the exterior of the engine with some type of degreaser before any work is done. It will make the job easier and help keep dirt out of the internal areas of the engine.

Depending on the components involved, it may be helpful to remove the hood to improve access to the engine as repairs are performed (refer to Chapter 11 if necessary). Cover the fenders to prevent damage to the paint. Special pads are available, but an old bedspread or blanket will also work.

If vacuum, exhaust, oil or coolant leaks develop, indicating a need for gasket or seal replacement, the repairs can generally be made with the engine in the vehicle. The intake and exhaust manifold gaskets, oil pan gasket, camshaft and crankshaft oil seals and cylinder head gasket are all accessible with the engine in place.

Exterior engine components, such as the intake and exhaust manifolds, the oil pan, the oil pump, the water pump, the starter motor, the alternator, the distributor and the fuel system components can be removed for repair with the engine in place.

Since the camshafts and cylinder head can be removed without pulling the engine, valve component servicing can also be accomplished with the engine in the vehicle. Replacement of the timing belt and sprockets is also possible with the engine in the vehicle.

In extreme cases caused by a lack of necessary equipment, repair or replacement of piston rings, pistons, connecting rods and rod bearings is possible with the engine in the vehicle. However, this practice is not recommended because of the cleaning and preparation work that must be done to the components involved.

3   Top Dead Center (TDC) for number one piston - locating

1   Top Dead Center (TDC) is the highest point in the cylinder that each piston reaches as it travels up-and-down when the crankshaft turns. Each piston reaches TDC on the compression stroke and again on the exhaust stroke, but TDC generally refers to piston position on the compression stroke. The cast-in timing mark arrow on the crankshaft timing belt pulley (installed on the front of the crankshaft) refers to the number one piston at TDC - when the arrow is straight up, or at "12 o'clock," and aligned with the cast-in timing mark arrow on the oil pump housing (see Section 7).

2   Positioning a specific piston at TDC is an essential part of many procedures such as camshaft(s) removal, rocker arm removal, timing belt and sprocket replacement.

3   In order to bring any piston to TDC, the crankshaft must be turned using one of the methods outlined below. When looking at the front of the engine, normal crankshaft rotation is clockwise. **Warning:** *Before beginning this procedure, be sure to set the emergency brake, place the transmission in Park or Neutral and disable the ignition system by disconnecting the primary electrical connector from the ignition coil pack.*

a) *The preferred method is to turn the crankshaft with a large socket and breaker bar attached to the crankshaft balancer hub bolt that is threaded into the front of the crankshaft.*

b) *A remote starter switch, which may save some time, can also be used. Attach the switch leads to the S (switch) and B (battery) terminals on the starter solenoid. Once the piston is close to TDC, discontinue with the remote switch and use a socket and breaker bar as described in the previous paragraph.*

c) *If an assistant is available to turn the ignition switch to the Start position in short bursts, you can get the piston close to TDC without a remote starter switch. Use a socket and breaker bar as described in Paragraph a) to complete the procedure.*

4   Remove all of the spark plugs as this will make it easier to rotate the engine by hand.

5   Insert a compression gauge (screw-in type with a hose) in the number 1 spark plug hole. Place the gauge dial where you can see it while turning the crankshaft pulley hub bolt. **Note:** *The number one cylinder is located at the front of the engine.*

6   Turn the crankshaft clockwise until you see compression building up on the gauge - you are on the compression stroke for that cylinder. If you did not see compression build up, continue with one more complete revolution to achieve TDC for the number one cylinder.

7   Remove the compression gauge. Through the number one cylinder spark plug

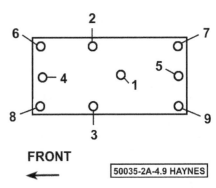

4.9 Valve cover bolt tightening sequence

hole insert a length of wooden dowel or plastic rod and slowly push it down until it reaches the top of the piston. **Caution:** *Don't insert a metal or sharp object into the spark plug hole as the piston crown may be damaged.*

8   With the dowel or rod in place on top of the piston crown, slowly rotate the crankshaft clockwise until the dowel or rod is pushed upward, stops, and then starts to move back down. At this point, rotate the crankshaft slightly counterclockwise until the dowel or rod has reached it uppermost travel. At this point the number one piston is at the TDC position.

9   Remove the upper timing belt cover (see Section 7). Then check the alignment of the camshaft timing marks **(see illustration 7.11a)**. At this point, the camshafts' timing marks should be aligned. If not repeat this procedure until alignment is correct.

10   After the number one piston has been positioned at TDC on the compression stroke, TDC for any of the remaining cylinders can be located by turning the crankshaft 180-degrees (1/2-turn) at a time and following the firing order (refer to the Specifications).

4   Valve cover - removal and installation

Removal

1   Disconnect the cable from the negative terminal of the battery (see Chapter 5).

2   Remove the upper intake manifold (see Section 5).

3   Remove the ignition coil pack from the valve cover (see Chapter 5).

4   Clearly label and detach any electrical wiring harnesses which connect to or cross over the valve cover.

5   Disconnect the PCV valve hose and breather hose from the valve cover.

6   Remove the valve cover bolts in the reverse order of the tightening sequence **(see illustration 4.9)**, then lift off the cover. If the cover sticks to the cylinder head, tap on it with a soft-face hammer or place a wood block against the cover and tap on the wood with a hammer. **Caution:** *If you have to pry*

between the valve cover and the cylinder head, be extremely careful not to gouge or nick the gasket surfaces of either part - which could cause a leak to develop after reassembly.

7   Remove the valve gaskets and spark plug well seals. Thoroughly clean the valve cover and remove all traces of old gasket material. Gasket removal solvents are available from auto parts stores and may prove helpful. After cleaning the surfaces, degrease them with a rag soaked in lacquer thinner or acetone.

Installation

Refer to illustration 4.9

8   Install a new gasket and spark plug well seals on the cover, using RTV sealant to hold them in place. Apply RTV sealant to the camshaft cap corners and the top edges of the half-round seal. Place the cover on the engine and install the cover bolts. **Note:** *Make sure the two bolts with the sealing washers are in their proper locations.*

9   Tighten the valve cover bolts in three stages in the proper sequence to the torque listed in this Chapter's Specifications **(see illustration)**.

10   The remaining steps are the reverse of removal. When finished, run the engine and check for oil leaks.

5   Intake manifold - removal and installation

1   The intake manifold is a two-piece aluminum design. **Note:** *If you are removing the intake manifold simply to access the spark plugs or valve cover, it's not necessary to remove the entire intake manifold unit - only remove the upper intake manifold. Otherwise, the intake manifold can be removed as a complete unit (upper and lower plenum sections bolted together).*

2   If removing the complete intake manifold, relieve the fuel system pressure (see Chapter 4). If only removing the upper intake manifold, this step may be skipped.

3   Disconnect the cable from the negative terminal of the battery (see Chapter 5).

4   Disconnect the electrical connector at the Intake Air Temperature (IAT) sensor and remove the air intake tube and air filter upper housing.

5   Disconnect the electrical connectors at the Idle Air Control (IAC) motor, Throttle Position Sensor (TPS) and Manifold absolute Pressure (MAP) sensor.

6   Disconnect the vacuum hoses at the intake manifold from the PCV valve, Leak Detection Pump (LDP), evaporative system purge solenoid, EGR transducer, power brake booster and, if equipped, the cruise control vacuum reservoir.

7   Disconnect the accelerator cable from its bracket and lever and, if equipped, remove the cruise control and transmission control cables.

8   Remove the EGR pipe bolts at the EGR valve and at the intake manifold, then remove the EGR pipe.

9   Remove the oil dipstick.

10   If the upper and lower intake manifolds are being removed as a unit, skip to step 11. If only the upper intake manifold is being removed, proceed as follows:

   a) Remove the bolt from the manifold support bracket at the front of the upper intake manifold.

   b) Remove the upper intake manifold attaching bolts and lift the upper manifold off.

   c) Refer to Step 17 for the inspection procedure.

   d) Use a new manifold gasket and install the upper intake manifold and tighten the upper intake manifold bolts gradually and evenly, starting with the center bolts and working outward, to the torque listed in this Chapter's Specifications.

   e) The remainder of the Installation is the reverse of removal.

11   If you are removing the complete intake manifold unit, remove the electrical connector at the Coolant Temperature Sensor (CTS) and proceed with the remaining steps.

12   Remove the fuel injector harness connectors.

13   Remove the fuel rail, fuel pressure regulator and fuel injectors as a single assembly (see Chapter 4).

14   If you're planning to *replace* or *service* the intake manifold plenum, remove the throttle body (see Chapter 4). If you're simply removing the intake manifold plenum to remove or service the cylinder head, it's not necessary to remove the throttle body from the intake manifold plenum.

15   Remove the bolts from the intake manifold support bracket.

16   Remove the intake manifold bolts and washers, then remove the intake manifold and the manifold gasket.

Inspection

17   Using a straightedge and feeler gauge, check the intake manifold mating surface for warpage.

18   Check the intake manifold surface on the cylinder head also.

19   If the warpage on either surface exceeds the limit listed in this Chapter's Specifications, the intake manifold and/or the cylinder head must be resurfaced at an automotive machine shop or, if the warpage is too excessive for resurfacing, it should be replaced.

Installation

20   Using a new manifold gasket, install the intake manifold and tighten the intake manifold-to-cylinder head bolts gradually and evenly, starting with the inner bolts and working outward, to the torque listed in this Chapter's Specifications.

21   The remainder of the installation is the reverse of removal.

6   Exhaust manifold - removal and installation

Warning: *Allow the engine to cool completely before beginning this procedure.*

Removal

1   Raise the front of the vehicle and place it securely on jackstands.

2   Disconnect the exhaust pipe from the exhaust manifold and lower the vehicle.

3   Remove the air filter housing and bracket (see Chapter 4).

4   Remove the exhaust manifold heat shield.

5   Unplug the electrical connector for the oxygen sensor, then remove the oxygen sensor from the exhaust manifold (see Chapter 6).

6   Unscrew the mounting bolts and remove the exhaust manifold and the old manifold gasket.

Inspection

7   Inspect the exhaust manifold for cracks and any other obvious damage. If the manifold is cracked or damaged in any way, replace it.

8   Using a wire brush, clean up the threads of the exhaust manifold bolts and inspect the threads for damage. Replace any bolts that have thread damage.

9   Using a scraper, remove all traces of gasket material from the mating surfaces and inspect them for wear and cracks. **Caution:** *When removing gasket material from any surface, especially aluminum, be very careful not to scratch or gouge the gasket surface. Any damage to the surface may result in a leak after reassembly. Gasket removal solvents are available from auto parts stores and may prove helpful.*

10   Using a straightedge and feeler gauge, inspect the exhaust manifold mating surface for warpage. Check the exhaust manifold surface on the cylinder head also. If the warpage on any surface exceeds the limits listed in this Chapter's Specifications, the exhaust manifold and/or cylinder head must be replaced or resurfaced at an automotive machine shop.

Installation

11   Using a new exhaust manifold gasket, install the exhaust manifold and hand-tighten the mounting bolts.

12   Raise the front of the vehicle and place it securely on jackstands.

13   Hand-tighten the exhaust pipe bolts.

14   Starting in the center and working your way outward, tighten the exhaust manifold mounting bolts to the torque listed in this Chapter's Specifications.

15   Tighten the exhaust pipe bolts to the torque listed in this Chapter's Specifications.

16   The remainder of installation is the reverse of removal.

7.7 Insert a large screwdriver or bar through the opening in the pulley and wedge it against the engine block, then loosen the bolt with a socket and breaker bar

7.8 Install a 3-jaw puller onto the damper pulley, position the center post of the puller on the crankshaft end (use the proper insert to keep from damaging the crankshaft threads), tighten the puller and remove the pulley from the crankshaft

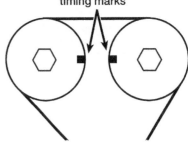
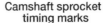

Camshaft sprocket
timing marks

50035-2A-7.11a HAYNES

7.11a Before removing the timing belt, make sure that the camshaft timing marks are aligned . . .

7 Timing belt - removal, inspection and installation

Refer to illustrations 7.7, 7.8, 7.11a and 7.11b
Caution 1: *Do not try to turn the crankshaft with a camshaft sprocket bolt and do not rotate the crankshaft counterclockwise.*
Caution 2: *Do not turn the camshafts after the timing belt has been removed. Doing so will damage the valves from contact with other valves and/or the piston(s).*

Removal

1   Position the number one piston at Top Dead Center (see Section 3).
2   Disconnect the cable from the negative terminal of the battery (see Chapter 5).
3   Remove the air filter housing and tube (see Chapter 4).
4   Remove the drivebelt (see Chapter 1).
5   Set the parking brake and block the rear wheels. Raise the front of the vehicle and support it securely on jackstands.
6   Remove the drivebelt splash shield from underneath the vehicle.
7   Loosen the large bolt in the center of the crankshaft damper pulley. It might be very tight; to break it loose, insert a large screwdriver or bar through the opening in the pulley to keep the pulley stationary and loosen the bolt with a socket and breaker bar **(see illustration)**.
8   Install a 3-jaw puller onto the damper pulley and remove the pulley from the crankshaft **(see illustration)**. Use the proper insert to keep the puller from damaging the crankshaft bolt threads. If the pulley is difficult to remove, tap the center bolt of the puller with a brass mallet to break it loose. Reinstall the bolt with a spacer so you can rotate the crankshaft later.
9   Remove the drivebelt tensioner and pulley assembly (right below the alternator).
10   Remove the lower timing belt cover bolts and remove the cover. Remove the upper timing belt cover bolts and remove the cover.

Note: *Lower the vehicle, if necessary.*
11   Before removing the timing belt, make sure that the camshaft timing marks are aligned and that the TDC mark on the crankshaft timing belt sprocket is aligned with the stationary index mark on the oil pump housing **(see illustrations)**. Note that the crankshaft sprocket TDC mark is located on the trailing edge of the sprocket tooth. If you don't align the trailing edge of the sprocket tooth with the TDC mark on the oil pump housing, the camshaft timing marks won't be aligned.
12   Insert a 6 mm Allen wrench into the belt tensioner and insert the long end of a 1/8-inch or a 3 mm Allen wrench into the small hole located on the front of the tensioner. Then, using the 6 mm Allen wrench as a lever, rotate the tensioner counterclockwise and simultaneously and lightly push in your 1/8-inch or 3 mm Allen wrench until it slides into the hole in the tensioner.
13   Remove the timing belt.

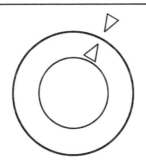

50035-2A-7.11b HAYNES

7.11b . . . and that the TDC mark on the crankshaft timing belt sprocket is aligned with the stationary index mark on the oil pump housing

Inspection

Refer to illustration 7.15
14   Rotate the tensioner pulley and idler pulley by hand and move them side-to-side to detect roughness and excess play. Replace parts as necessary (see Section 8). Visually inspect the sprockets for any signs of damage and wear.
15   Inspect the timing belt for cracks, separation, wear, missing teeth and oil contamination. Replace the belt if it's in questionable condition **(see illustration)**. If the timing belt is excessively worn or damaged on one side, it might be due to incorrect tracking (misalignment). If the belt looks like it was misaligned, be sure to replace the belt tensioner assembly.
16   Check the automatic tensioner for leaks or any obvious damage to the body.

Installation

Refer to illustration 7.18
17   Make sure that the TDC mark on the crankshaft timing belt sprocket is still aligned with the stationary index mark on the oil pump housing **(see illustrations 7.11a and 7.11b)**.

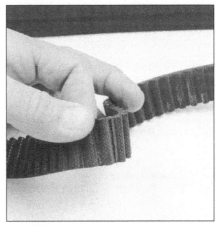

7.15 Carefully inspect the timing belt; bending it backwards will often make wear or damage more apparent

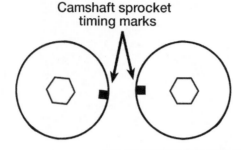

Camshaft sprocket timing marks

7.18 Before installing the timing belt, turn the exhaust camshaft sprocket clockwise so that the timing mark on the exhaust sprocket is slightly (1/2-notch) below the timing mark on the intake camshaft sprocket

50035-2A-7.18 HAYNES

18   Turn the exhaust camshaft sprocket clockwise so that the timing mark on the exhaust sprocket is slightly below the timing mark on the intake camshaft sprocket **(see illustration)**.

19   Install the timing belt as follows: Start at the crankshaft sprocket, then thread the belt onto the water pump sprocket, the idler pulley, the camshaft sprockets and then finally the tensioner **(see illustration)**. Now take up tension by moving the exhaust camshaft sprocket counterclockwise until the timing marks on the two cam sprockets are realigned.

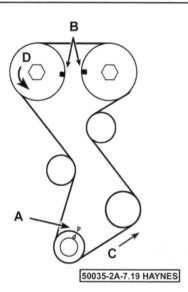

50035-2A-7.19 HAYNES

7.19 When installing the belt, start at the crankshaft sprocket and thread the belt around the water pump sprocket, the idler pulley, the cam sprockets and the tensioner pulley

A   *Crankshaft TDC marks aligned*
B   *Exhaust camshaft mark 1/2-notch below the intake camshaft mark*
C   *Install the belt in a counterclockwise direction (looking at the front of the engine)*
D   *Rotate the exhaust camshaft counterclockwise to remove the timing belt slack*

20   Remove the Allen wrench from the tensioner. If you installed a new tensioner assembly, remove the pull pin (new tensioners are locked in the "wound" position by a pull pin).

21   Rotate the crankshaft 720-degrees (two complete revolutions) and verify that the TDC marks on the crankshaft sprocket and the oil pump housing are still aligned and that the timing marks on the camshaft sprockets are still aligned. If they're not, stop right here, go back to Step 17 and install the belt again.

22   Install the upper and lower timing belt covers and tighten the fasteners to the torque listed in this Chapter's Specifications.

23   Install the drivebelt idler pulley and tighten the idler pulley bolt to the torque listed in this Chapter's Specifications.

24   Install the crankshaft vibration damper, the washer and pulley bolt. Tighten the pulley bolt to the torque listed in this Chapter's Specifications.

25   The remainder of installation is the reverse of removal.

---

8   Timing belt tensioner and pulley - removal and installation

Removal

1   Remove the timing belt (see Section 7).

8.9 When installing a camshaft sprocket, make sure the pin in the camshaft is aligned with the hole in the sprocket

2   Remove the timing belt idler pulley bolt, then remove the timing belt idler pulley.

3   Using a suitable holding tool, remove the camshaft sprocket bolts. **Caution:** *Don't allow the camshafts to turn as the bolts are loosened.* Then, using two large screwdrivers, lever the sprockets off the camshafts.

4   Remove the bolts for the rear timing belt cover and then remove the rear cover.

5   Remove the lower timing belt tensioner bolt and remove the tensioner assembly.

Installation

Refer to illustration 8.9

6   Place the timing belt tensioner in position on the front of the engine and then install the lower mounting bolt to hold it in place, but don't tighten the lower bolt yet. Install an engine bracket mounting bolt (M10) in the upper tensioner bolt hole and then screw it in five to seven turns. Tighten the lower tensioner bolt to the torque listed in this Chapter's Specifications and then remove the upper bolt.

7   Install the rear timing belt cover and tighten the cover fasteners securely.

8   Install the timing belt idler pulley and tighten the bolt to the torque listed in this Chapter's Specifications.

9   Install each camshaft sprocket, aligning the pin in the camshaft with the hole in the sprocket **(see illustration)**. Use an appropriate tool to hold the camshaft sprocket while tightening the bolt to the torque listed in this Chapter's Specifications.

10   Install the timing belt (see Section 7).

---

9   Crankshaft front oil seal - replacement

Refer to illustrations 9.2, 9.3, 9.5 and 9.6
Caution: *Do not rotate the camshafts when the timing belt is removed or damage to the engine may occur.*
Note: *The crankshaft sprocket must be installed to the same depth on the crankshaft as it was originally, or the timing belt may not track properly. Use of a factory tool (6792) is recommended, but if not available, you must accurately measure the distance between the back of the sprocket and the engine before removing the crankshaft sprocket. Use that measurement when installing the sprocket.*

1   Remove the timing belt and timing belt covers (see Sections 7 and 8).

2   Pull the crankshaft sprocket from the crankshaft with a bolt-type gear puller **(see illustration)**. Remove the Woodruff key. **Caution:** *When installing the crankshaft sprocket, it must be positioned in its original position to ensure proper timing belt alignment. Before removing the sprocket, measure the distance between the end of the crankshaft and the sprocket and record this distance. When installing the sprocket, be sure to return it to the same position on the crankshaft.*

9.2 Attach a bolt-type gear puller to the crankshaft sprocket and remove the sprocket from the crankshaft

9.3 Carefully pry the seal out of its bore

9.5 Lubricate the new seal with engine oil and drive the seal into place with a hammer and seal driver or a socket

3   Use a screwdriver or seal removal tool to pry the seal out of its bore **(see illustration)**. Take care to prevent damaging the oil pump assembly, the crankshaft and the seal bore.

4   Thoroughly clean and inspect the seal bore and sealing surface on the crankshaft. Minor imperfections can be removed with emery cloth. If there is a groove worn in the crankshaft sealing surface (from contact with the seal), installing a new seal will probably not stop the leak.

5   Lubricate the new seal with engine oil and drive the seal into place with a hammer and an appropriate size socket **(see illustration)**.

6   Use a hammer and socket to install the sprocket to the required depth. **Caution:** *Position the crankshaft sprocket with the arrow mark or "FRONT" facing out and in its original position on the crankshaft (as discussed in Step 2)* **(see illustration)**.

7   The remaining steps are the reverse of removal. Reinstall the timing belt (see Section 7).

8   Run the engine and check for oil leaks.

10   Camshaft oil seal - replacement

Refer to illustrations 10.3 and 10.5
Caution: *Do not rotate the camshafts when the timing belt is removed or damage to the engine may occur.*

1   Remove the timing belt (see Section 7)

2   Remove the camshaft sprockets and rear timing belt cover (see Section 8). Remove the exhaust camshaft target ring and sensor.

3   Note how far the seal is seated in the bore, then carefully pry it out with a small screwdriver **(see illustration)**. Don't scratch the bore or damage the camshaft in the process (if the camshaft is damaged, the new seal will end up leaking).

4   Clean the bore and coat the outer edge of the new seal with engine oil or multi-purpose grease. Also lubricate the seal lip.

5   Using a socket with an outside diameter slightly smaller than the outside diameter of the seal **(see illustration)**, carefully drive the new seal into place with a hammer. Make sure it's installed squarely and driven in to the same depth as the original. If a socket isn't available,

a short section of pipe will also work.

6   Install the rear timing belt cover part and camshaft sprockets (see Section 8).

7   Install the timing belt (see Section 7).

8   Run the engine and check for oil leaks at the camshaft seal.

11   Camshafts - removal, inspection and installation

Removal

Refer to illustration 11.5

1   Remove the valve cover (see Section 4).

2   Remove the Camshaft Position Sensor (CMP) (see Chapter 6) and the camshaft target magnet.

3   Remove the timing belt (see Section 7).

4   Remove the camshaft sprockets and the rear timing belt cover (see Section 8).

5   The camshaft bearing caps are identified with their numbered location in the cylinder head. They must be returned to their original locations. Remove the front and rear bearing caps at each end of the camshafts.

9.6 Position the crankshaft sprocket with the mark facing out, and install it onto the crankshaft in its original position

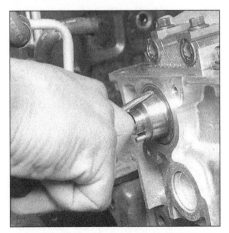

10.3 Carefully pry the camshaft seal out of the bore - DO NOT nick or scratch the camshaft or seal bore

10.5 Gently tap the new seal into place with the spring side toward the engine

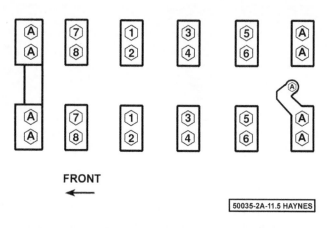

FRONT

50035-2A-11.5 HAYNES

11.5 First, remove the outside bearing caps (A), then loosen the remaining camshaft bearing cap bolts a little at a time in the sequence shown

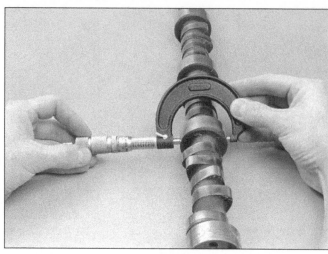

11.9 Measure the camshaft bearing journal diameters with a micrometer

Remove the remaining camshaft bearing caps. Loosen the bolts a little at a time to prevent distorting the camshafts, in the sequence shown **(see illustration)**. When the bearing caps have all been loosened enough for removal, they may still be difficult to remove. Use the bearing cap bolts for leverage and move the cap back and forth to loosen the cap from the cylinder head. If they are still difficult to remove you can tap them gently with a soft face mallet until they can be lifted off. **Caution:** *Store them in order so they can be returned to their original locations, with the same side facing forward.*

6　　Carefully lift the camshafts out of the cylinder head. Mark the camshafts INTAKE and EXHAUST. They cannot be interchanged.

7　　Remove the front seal from each camshaft. **Note:** *It would be a good idea to inspect the rocker arms and lash adjusters at this time* (see Section 12).

Inspection

Refer to illustration 11.9

8　　Clean the camshaft(s) and the gasket surface. Inspect the camshaft for wear and/or damage to the lobe surfaces, bearing journals, and seal contact surfaces. Inspect the camshaft bearing surfaces in the cylinder head and bearing caps for scoring and other damage.

9　　Measure the camshaft bearing journal diameters **(see illustration)**. Measure the inside diameter of the camshaft bearing surfaces in the cylinder head, using a telescoping gauge (temporarily install the bearing caps). Subtract the journal measurement from the bearing measurement to obtain the camshaft bearing oil clearance. Compare this clearance with the value listed this Chapter's Specifications. Replace worn components as required.

10　　Replace the camshaft if it fails any of the above inspections. **Note:** *If the lobes are worn, replace the rocker arms and lash adjusters along with the camshaft.* The cylinder head may need to be replaced, if the camshaft bearing surfaces in the head are damaged or excessively worn.

11　　Clean and inspect the cylinder head (see Chapter 2C).

Camshaft endplay measurement

12　　Lubricate the camshaft(s) and cylinder head bearing journals with clean engine oil.

13　　Place the camshaft in its original location in the cylinder head. **Note:** *Do not install the rocker arms for this check.* Install the rear bearing cap and tighten the bolts to the torque listed in this Chapter's Specifications.

14　　Install a dial indicator on the cylinder head and place the indicator tip on the camshaft at the sprocket end.

15　　Use a screwdriver to carefully pry the camshaft fully to the rear until it stops. Zero the dial indicator and pry the camshaft fully to the front. The amount of indicator travel is the camshaft endplay. Compare the endplay with the tolerance given in this Chapter's Specifications. If the endplay is excessive, check the camshaft and cylinder head bearing journals for wear. Replace as necessary.

Installation

Refer to illustrations 11.17 and 11.20

16　　Install the valve lash adjusters and rocker arms (see Section 11).

17　　Clean the camshaft and bearing journals and caps. Liberally coat the journals, lobes, and thrust portions of the camshaft with assembly lube or engine oil **(see illustration)**.

18　　Carefully install the camshafts in the cylinder head in their original location. Temporarily install the camshaft sprockets and rotate the camshafts so that their timing marks align **(see illustration 7.11a)**. Make sure the crankshaft is positioned with the crankshaft sprocket timing mark at three teeth BTDC. **Caution:** *If the pistons are at TDC when tightening the camshaft bearing caps, damage to the engine may occur.*

19　　Install the bearing caps, except for the No. 1 and No. 6 (left side) end caps. Tighten the bolts in several steps, in the proper sequence **(see illustration 11.5)** to the torque listed in this Chapter's Specifications.

20　　Apply a small bead of anaerobic sealant

11.17 Prior to installing each camshaft, lubricate the bearing journals, thrust surfaces and lobes with assembly lube or clean engine oil

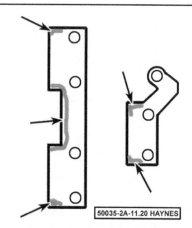

50035-2A-11.20 HAYNES

11.20 Apply a bead of anaerobic sealant to the indicated areas on the No. 1 and No. 6 bearing caps, then install these bearing caps

(approximately 1/8-inch) to the No. 1 and No. 6 (left side) bearing caps (see illustration). Install the bearing caps and tighten the bolts to the torque listed in this Chapter's Specifications.

21 Install new camshaft oil seals (see Section 10).

22 Install the timing belt, covers, and related components (see Sections 7 and 8).

23 Install the valve cover (see Section 4).

12 Rocker arms - removal, inspection and installation

Removal

1 Remove the valve cover (see Section 4).

2 Remove both camshafts (see Section 11).

3 Once the camshafts have been removed, the rocker arms can be lifted off. **Caution:** *Each rocker arm must be placed back in the same location it was removed from, so mark each rocker arm or place them in a container (such as an egg carton) so they won't get mixed up.* The lash adjusters can remain in the head at this time, unless they are being replaced (see Section 13).

Inspection

4 Inspect the rocker arm tip, roller and lash adjuster pocket for wear. Replace them if evidence of wear or damage is found.

5 Carefully inspect each lash adjuster for signs of wear and damage, particularly on the ball tip that contacts the rocker arm. The lash adjusters become clogged as they age, so it's a good idea to replace them if you're concerned about their condition or if the engine is making valve "tapping" noises.

Installation

6 Installation is the reverse of removal. When reinstalling the rocker arms, make sure that you install them at the same locations from which they were removed.

13 Valve lash adjusters - removal, inspection and installation

1 Remove the rocker arms (see Section 12).

2 If the lash adjusters aren't already removed from the head, lift them out now. **Caution:** *Be sure to keep the adjusters in order so they can be placed back in the same location in the cylinder head it was removed from.*

3 Inspect each adjuster carefully for signs of wear and damage, particularly on the ball tip that contacts the rocker arm. Since the lash adjusters can become clogged, we recommend replacing them if you're concerned about their condition or if the engine is exhibiting valve "tapping" noises.

4 The lash adjusters must be partially full of engine oil - indicated by little or no plunger action when the adjuster is depressed. If there's excessive plunger travel, place the lash adjuster into clean engine oil and pump the plunger until the plunger travel is elimi-

14.4 This is what the air hose adapter that threads into the spark plug hole looks like - they're generally available from auto parts stores

nated. **Note:** *If the plunger still travels within the lash adjuster when full of oil it's defective and the lash adjuster must be replaced.*

5 When re-starting the engine after replacing the adjusters, the adjusters will normally make "tapping" noises. After warm-up, raise the speed of the engine from idle to 3,000 rpm for one minute. If the adjuster(s) do not become silent, replace the defective ones.

14 Valve springs, retainers and seals - replacement

Refer to illustrations 14.4, 14.7, 14.8, 14.13 and 14.15

Note: *Broken valve springs and defective valve stem seals can be replaced without removing the cylinder heads. Two special tools and a compressed air source are normally required to perform this operation, so read through this Section carefully and rent or buy the tools before beginning the job.*

1 Remove the spark plug from the cylinder that has the defective component. If all of the valve stem seals are being replaced, all of the spark plugs should be removed.

2 Turn the crankshaft until the piston in the affected cylinder is at top dead center on the compression stroke (refer to Section 3). If you're replacing all of the valve stem seals, begin with cylinder number one and work on the valves for one cylinder at a time. Move from cylinder-to-cylinder following the firing order sequence (see this Chapter's Specifications).

3 Remove the camshaft(s) and the rocker arm(s) (see Sections 11 and 12).

4 Thread an adapter into the spark plug hole **(see illustration)** and connect an air hose from a compressed air source to it. Most auto parts stores can supply the air hose adapter. **Note:** *Many cylinder compression gauges utilize a screw-in fitting that may work with your air hose quick-disconnect fitting.*

5 Apply compressed air to the cylinder. **Warning:** *The piston may be forced down by compressed air, causing the crankshaft to*

14.7 Use needle-nose pliers (shown) or a small magnet to remove the valve spring keepers - be careful not to drop them down into the engine!

turn suddenly. If the wrench used when positioning the number one piston at TDC is still attached to the bolt in the crankshaft nose, remove it as it could cause damage or injury when the crankshaft moves.

6 Stuff clean shop rags into the various oil-drain holes in the cylinder head, to prevent parts and tools from falling into the engine.

7 Use a valve spring compressor to compress the spring. Remove the keepers with small needle-nose pliers or a magnet **(see illustration)**.

8 Remove the spring retainer and valve spring, then remove the valve guide seal/spring seat assembly **(see illustration)**. **Caution:** *If air pressure fails to hold the valve in the closed position during this operation, the valve face and/or seat is probably damaged. If so, the cylinder head will have to be removed for additional repair operations.*

9 Wrap a rubber band or tape around the top of the valve stem so the valve won't fall into the combustion chamber, then release the air pressure.

10 Inspect the valve stem for damage.

14.8 Remove the valve guide seal with a pair of pliers

Rotate the valve in the guide and check the end for eccentric movement, which would indicate that the valve is bent.

11   Move the valve up-and-down in the guide and make sure it doesn't bind. If the valve stem binds, either the valve is bent or the guide is damaged. In either case, the head will have to be removed for repair.

12   Pull up on the valve stem to close the valve, reapply air pressure to the cylinder to retain the valve in the closed position, then remove the tape or rubber band from the valve stem.

13   Lubricate the valve stem with engine oil and install a new valve guide seal/spring seat assembly. Tap into place with a deep socket **(see illustration)**.

14   Install the spring in position over the valve.

15   Install the valve spring retainer. Compress the valve spring and carefully position the keepers in the groove. Apply a small dab of grease to the inside of each keeper to hold it in place if necessary **(see illustration)**.

16   Remove the pressure from the spring tool and make sure the keepers are seated.

17   Disconnect the air hose and remove the adapter from the spark plug hole.

18   The remainder of the installation is the reverse of removal.

19   Start and run the engine, then check for oil leaks and unusual sounds coming from the valve cover area.

---

15   Cylinder head - removal and installation

Warning: *Allow the engine to cool completely before beginning this procedure.*

Removal

Refer to illustrations 15.4 and 15.12

1   Position the number one piston at Top

14.13 Gently tap the new seal into place with a hammer and a deep socket

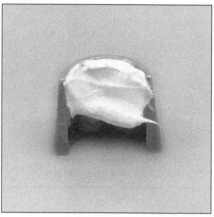

14.15 Apply a small dab of grease to each keeper before installation to hold it in place on the valve stem until the spring is released

Dead Center (see Section 3).

2   Disconnect the cable from the negative terminal of the battery (see Chapter 5).

3   Drain the cooling system (see Chapter 1).

4   Remove the intake manifold (see Section 5). Cover the intake ports with duct tape to keep out debris **(see illustration)**.

5   Remove the spark plug wires, spark plugs and ignition coil (see Chapters 1 and 5).

6   Remove the upper radiator and heater hoses (see Chapter 1). Remove the heater tube support bracket.

7   Detach the power steering pump bolts from the engine (leaving the hoses attached) and set the pump aside (see Chapter 10). Remove the accessory bracket(s).

8   Raise the front of the vehicle and place it securely on jackstands. Disconnect the exhaust pipe from the exhaust manifold (see Chapter 4) and lower the vehicle.

9   Remove the timing belt (see Section 7).

10   Remove the valve cover, camshafts and

rocker arms (see Sections 4, 11 and 12).

11   Loosen the cylinder head bolts 1/4-turn at a time, in the reverse of the tightening sequence **(see illustration 15.18)** until they can be removed by hand. **Note:** *Write down the location of the different length bolts so they will be reinstalled in the correct location.*

12   Carefully lift the cylinder head straight up and place the head on wood blocks to prevent damage to the sealing surfaces. If the head sticks to the engine block, dislodge it by placing a wood block against the head casting and tapping the wood with a hammer or by prying the head with a prybar placed carefully on a casting protrusion **(see illustration)**. **Caution:** *The cylinder head is aluminum, so you must be very careful not to gouge the sealing surfaces.*

Inspection

13   Special gasket removal solvents that soften gaskets and make removal much eas-

15.4 Cover the intake ports with duct tape to keep out debris before removing the cylinder head (typical)

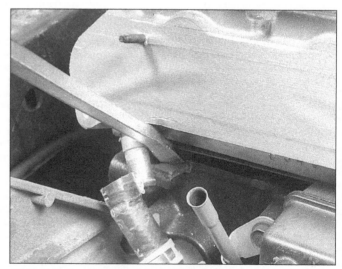

15.12 If the head sticks to the engine block, dislodge it by placing a wood block against the head casting and tapping the wood with a hammer or by prying the head with a prybar placed carefully on a casting protrusion (typical)

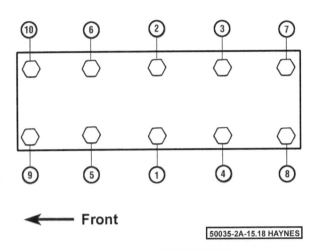

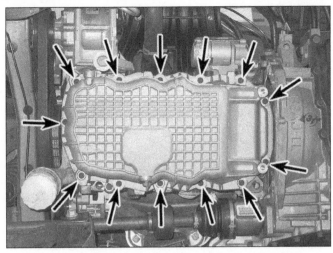

15.18 Cylinder head bolt TIGHTENING sequence

16.11a Using a criss-cross pattern, loosen and remove the oil pan bolts (typical) . . .

ier are available at auto parts stores. Remove all traces of old gasket material from the block and head. Do not allow anything to fall into the engine.

14 Using a straightedge and feeler gauge, check the gasket mating surfaces for warpage. If the warpage on either surface exceeds the limit listed in this Chapter's Specifications, the cylinder head must be resurfaced at an automotive machine shop. The machine shop can also clean and inspect the cylinder head components.

15 Clean and inspect all threaded bolts and be sure the threaded holes in the block are clean and dry. Inspect the threads on the cylinder head bolts with a straight edge. If any of the threads don't touch the straight edge, the bolts are stretched, and should be replaced.

Installation

Refer to illustrations 15.18

16 Place a new gasket and the cylinder head in position on the engine block. The part number on the gasket should be alongside the number 1 cylinder bore and must face UP.

17 Apply clean engine oil to the cylinder head bolt threads prior to installation. The four short bolts (4.33 inch long) are to be installed in each corner of the cylinder head.

18 Tighten the cylinder head bolts in several stages in the recommended sequence **(see illustration)** to the torque listed in this Chapter's Specifications.

19 Reinstall the timing belt (see Section 7).

20 Reinstall the remaining parts in the reverse order of removal.

21 Refill the cooling system and change the engine oil and filter (see Chapter 1). Rotate the crankshaft clockwise slowly by hand through six complete revolutions. Recheck the camshaft timing marks (see Section 7).

22 Start the engine and run it until normal operating temperature is reached. Check for leaks and proper operation.

16 Oil pan - removal and installation

Removal

Refer to illustrations 16.11a, 16.11b, 16.11c, 16.12a, 16.12b and 16.13

1 Disconnect the cable from the negative terminal of the battery (see Chapter 5).

2 Remove the air filter housing (see Chapter 4).

3 Raise the vehicle and support it securely on jackstands.

4 Drain the engine oil (see Chapter 1).

5 Disconnect the exhaust pipe at the exhaust manifold (see Chapter 4).

6 Remove the structural collar that connects the rear of the engine oil pan to the front of the transmission bellhousing.

7 On 4WD models, support the front axle assembly with a floor jack, remove the front axle assembly mounting bolts (see Chapter 8), then lower the axle housing as much

as possible.

8 Loosen the engine mount through-bolts/nuts.

9 Support the engine from above with an engine hoist (but not on the intake manifold), take the weight off the engine mounts with the hoist and remove the engine mount through-bolts (see Section 19).

10 Raise the engine sufficiently to allow access for oil pan removal.

11 Remove the mounting bolts and lower the oil pan from the vehicle **(see illustrations)**. If the pan is stuck, tap it with a soft-face hammer **(see illustration)** or place a wood block against the pan and tap the wood block with a hammer. **Caution:** *If you're wedging something between the oil pan and the engine block to separate the two, be extremely careful not to gouge or nick the gasket surface of either part; an oil leak could result.*

12 Remove the oil pump pickup tube and

16.11b . . . then lower the pan carefully (there might still be some residual oil in the pan) (typical)

16.11c If the pan is stuck, tap it with a soft-face hammer or place a wood block against the pan and tap the wood block with a hammer (typical)

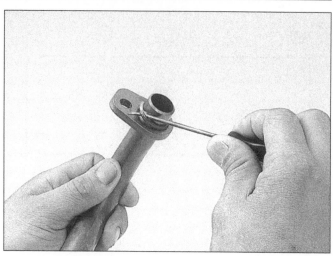

16.12a Remove the bolt and remove the oil pump pick-up tube and screen assembly - clean both the tube and screen thoroughly before reassembly

16.12b Replace the seal at the oil pump pick-up tube mounting flange with a new one

screen assembly **(see illustration)** and clean both the tube and screen thoroughly. Install the pick-up tube and screen with a new seal **(see illustration)**.

13   Thoroughly clean the oil pan and sealing surfaces on the block and pan **(see illustration)**. Use a scraper to remove all traces of old gasket material. Gasket removal solvents are available at auto parts stores and may prove helpful. Check the oil pan sealing surface for distortion. Straighten or replace as necessary. After cleaning and straightening (if necessary), wipe the gasket surfaces of the pan and block clean with a rag soaked in lacquer thinner or acetone.

Installation

14   Apply a 1/8-inch bead of RTV sealant at the cylinder block-to-oil pump assembly joint at the oil pan flange. Install a new oil pan gasket.

15   Place the oil pan into position and install

the bolts finger-tight. Working side-to-side from the center out, tighten the bolts to the torque listed in this Chapter's Specifications.

16   Lower the engine and install the through-bolts in the engine mounts (see Section 19). Tighten them securely.

17   Install the structural collar as follows:

   a) Place the collar in position between the transmission and the oil pan and then hand-start the collar-to-transmission bolts.

   b) Install and then hand-tighten the collar-to-oil pan bolts.

   c) Tighten the collar-to-transmission bolts.

   d) Tighten the collar-to-oil pan bolts.

18   The remainder of installation is the reverse of removal.

19   Refill the crankcase with the correct quantity and grade of oil, run the engine and check for leaks.

20   Road test the vehicle and check for leaks again.

17   Oil pump - removal, inspection and installation

Removal

Refer to illustrations 17.3a, 17.3b, 17.5a, 17.5b and 17.5c

1   Remove the timing belt, timing belt covers and the crankshaft front seal (see Sections 7, 8 and 9).

2   Remove the oil pan and pick-up tube/screen assembly (see Section 16).

3   Remove the bolts and detach the oil pump assembly from the engine **(see illustration). Caution:** *If the pump doesn't come off by hand, tap it gently with a soft-faced hammer or pry on a casting boss* **(see illustrations)**.

4   Remove the crankshaft oil seal from the oil pump, by prying it out with a screwdriver wrapped in a rag. **Caution:** *To prevent an oil leak after the new seal is installed, be very*

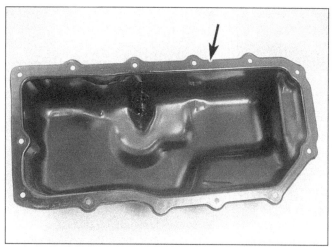

16.13 Thoroughly clean the oil pan and sealing surfaces on the engine block and oil pan (arrow) with a scraper to remove all traces of old gasket material (typical)

17.3a Remove the oil pump assembly mounting bolts and remove the assembly (typical)

17.3b If the pump doesn't come off by hand, tap it gently with a soft-faced hammer or pry gently on a casting protrusion (typical)

17.5a Remove the cover mounting screws (typical) . . .

17.5b . . . and the cover (typical)

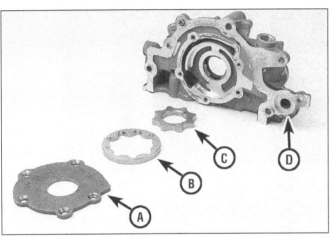

17.5c Arrangement of oil pump components (typical)

A	Cover	C	Inner rotor
B	Outer rotor	D	Oil pump body

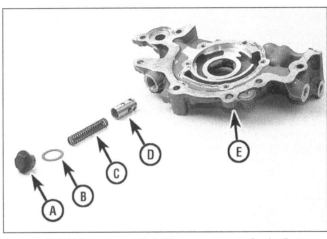

17.7 Oil pressure relief valve components (typical)

A	Cap	C	Spring	E	Oil pump
B	Gasket	D	Relief valve		body

careful not to scratch or otherwise damage the seal-bore in the pump body.

5   Remove the mounting screws and remove the cover **(see illustrations)**. Remove the inner and outer rotor from the body **(see illustration)**. **Caution:** *Be very careful with these parts. Close tolerances are critical in creating the correct oil pressure. Any nicks or other damage will require replacement of the complete pump assembly.*

Inspection
Refer to illustrations 17.7, 17.8a, 17.8b, 17.8c, 17.8d and 17.8e

6   Clean all components including the block surfaces with solvent, then inspect all surfaces for excessive wear and/or damage.

7   To disassemble the relief valve, remove the cap bolt, washer, spring and relief valve **(see illustration)**. Check the oil pressure relief valve piston sliding surface and valve spring. If either the spring or the valve is damaged, they must be replaced as a set. If no damage is found reassemble the relief valve

parts. **Caution:** *Make sure to install the relief valve into the pump body with the grooved end going in first - otherwise engine damage may occur. Coat the parts with oil and reinstall them in the oil pump body.* Tighten the

cap bolt to this Chapter's Specifications.

8   Check the clearance of the oil pump components with a micrometer and a feeler gauge **(see illustrations)** and compare the results to this Chapter's Specifications.

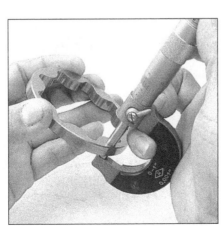

17.8a Measure the outer rotor thickness

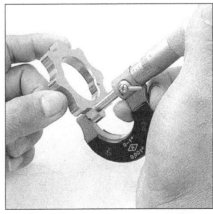

17.8b Measure the inner rotor thickness

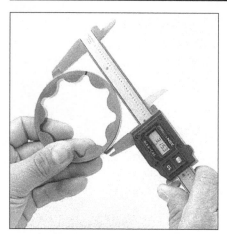

17.8c Use a caliper and measure the outer diameter of the outer rotor

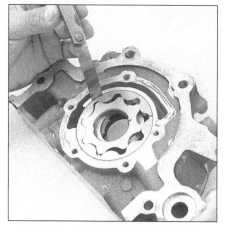

17.8d Use a feeler gauge and measure the outer rotor-to-case clearance

17.8e Place a precision straightedge over the rotors and measure the clearance between the rotors and the cover

Installation

Refer to illustrations 17.10a and 17.10b

9   Lubricate the housing and the inner and outer rotors with clean engine oil and install both rotors in the body. **Caution:** *When installing the inner rotor, the chamfer faces the pump cover.* Prime the oil pump by filling the rotor cavities with clean engine oil. Install the cover and tighten the cover screws to the torque listed in this Chapter's Specifications.

10   Install a new O-ring in the oil discharge passage **(see illustration)**. Apply sealant to the oil pump body **(see illustration)**, and raise the pump assembly to the block. Align the flat parts of the rotor with the flat parts of the crankshaft. Tighten the bolts to the torque listed in this Chapter's Specifications.

11   Install the crankshaft seal (see Section 9).

12   Install the pick-up tube/strainer assembly and oil pan (see Section 16).

13   Install the crankshaft sprocket (see Section 9)

14   Install the rear timing belt cover (see Section 8).

15   Install the timing belt (see Section 7).

16   Install a new oil filter and engine oil (see

Chapter 1).

17   Start the engine and check for oil pressure and leaks.

18   Recheck the engine oil level.

---

18   Flywheel/driveplate - removal and installation

Removal

Refer to illustration 18.5

1   Raise the vehicle and support it securely on jackstands.

2   Remove the transmission assembly (see Chapter 7).

3   If the vehicle has a manual transmission, remove the pressure plate and clutch disc (see Chapter 8). Now is a good time to check/replace the clutch components.

4   To ensure correct alignment during reinstallation, mark the position of the flywheel/driveplate to the crankshaft before removal.

5   Remove the bolts that secure the flywheel/driveplate to the crankshaft **(see illustration)**. A tool is available at most auto parts stores to hold the flywheel/driveplate while

17.10a Replace the O-ring seal in the oil pump body - apply clean engine oil to the seal

loosening the bolts. If the tool is not available, wedge a screwdriver in the ring gear teeth to jam the flywheel.

6   Remove the flywheel/driveplate from the crankshaft. Since the flywheel is fairly heavy, be sure to support it while removing the last bolt.

17.10b Apply a bead of anaerobic sealant to the housing sealing surface as shown

18.5 Mark the relative position of the flywheel or driveplate to the crankshaft and, using an appropriate tool to hold the flywheel, remove the bolts (typical)

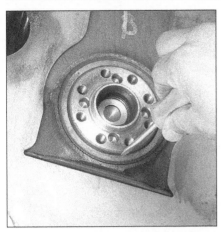

19.5 Carefully pry the crankshaft seal out of the bore - DO NOT nick or scratch the crankshaft or seal bore

7   Clean the flywheel to remove grease and oil. Inspect the flywheel (see Chapter 8).
8   Clean and inspect the mating surfaces of the flywheel/driveplate and the crankshaft. If the crankshaft rear main seal is leaking, replace it before reinstalling the flywheel/driveplate (see Section 19).

Installation

9   Position the flywheel/driveplate against the crankshaft. Align the previously applied match marks. Before installing the bolts, apply thread locking compound to the threads.
10   Hold the flywheel/driveplate with the holding tool, or wedge a screwdriver in the ring gear teeth to keep the flywheel/driveplate from turning as you tighten the bolts to the torque listed in this Chapter's Specifications.
11   The remainder of installation is the reverse of the removal procedure.

19   Rear main oil seal - replacement

Refer to illustrations 19.5 and 19.7

1   The one-piece rear main oil seal is pressed into a bore machined into the rear main bearing cap and engine block.
2   Remove the transmission (see Chapter 7)
3   If equipped, remove the clutch components (see Chapter 8)
4   Remove the flywheel or driveplate (see Section 18).
5   Observe that the oil seal is installed flush with the outer surface of the block. Using the chamfered edge on the crankshaft, pry out the old seal with a 3/16-inch flat blade

screwdriver **(see illustration)**. **Caution:** *To prevent an oil leak after the new seal is installed, be very careful not to scratch or otherwise damage the crankshaft sealing surface or the bore in the engine block.*
6   Clean the crankshaft and seal bore in the block thoroughly and de-grease these areas by wiping them with a rag soaked in lacquer thinner or acetone. If there is a burr or scratch on the crankshaft chamfer edge, sand it with 400-grit sandpaper. Do not lubricate the lip or outer diameter of the new seal - it must be installed as it comes from the manufacturer.
7   Position the new seal onto the crankshaft. **Note:** *When installing the new seal, the words THIS SIDE OUT on the seal must face out, toward the rear of the vehicle. If the seal is not marked, the seal lip (that seals against the crankshaft) must face in.* Using an appropriate size driver and pilot tool, gently and gradually drive the seal into the cylinder block until it is flush with the outer surface of the block **(see illustration)**. **Caution:** *If the seal is driven-in past the flush-position, there will be an oil leak.*
8   The remainder of installation is the reverse of removal.

20   Engine mounts - check and replacement

1   Engine mounts seldom require attention, but broken or deteriorated mounts should be replaced immediately or the added strain placed on the driveline components may cause damage or wear.

Check

2   During the check, the engine must be raised slightly to remove the weight from the mounts.
3   Raise the vehicle and support it securely on jackstands, then position a jack under the engine oil pan. Place a large wood block between the jack head and the oil pan to prevent oil pan damage, then carefully raise the engine just enough to take the weight off the mounts. **Warning:** *DO NOT place any part of your body under the engine when only a jack supports it.*
4   Check the mounts to see if the rubber is cracked, hardened or separated from the metal backing. Sometimes the rubber will split right down the center.
5   Check for relative movement between the mount plates and the engine or frame (use a large screwdriver or pry bar to attempt to move the mounts). If movement is noted, lower the engine and tighten the mount fasteners.

19.7 Position the new seal with the words THIS SIDE OUT facing out, toward the rear of the engine. If the seal is not marked, the seal lip (that seals against the crankshaft) must face in. Gently and gradually drive the seal into the cylinder block until it is flush with the outer surface of the block. Do not drive it past flush or there will be an oil leak - the seal must be flush

6   Rubber preservative may be applied to the mounts to slow deterioration.

Replacement

7   Disconnect the cable from the negative terminal of the battery, then raise the vehicle and support it securely on jackstands (if not already done).
8   Place a floor jack under the engine (with a wood block between the jack head and oil pan) and raise the engine slightly to relieve the weight from the mounts.
9   Remove the through-bolt from the engine mount.
10   Remove the engine mount from the side of the engine block. **Note:** *Do not disconnect more than one mount at a time, except during engine removal.*
11   Install the new engine mount assembly to the engine block and tighten the bolts securely.
12   Slowly lower the engine, while watching for proper mount-to-frame alignment, and then install the through-bolt. Hand-tighten the nut on the through-bolt.
13   Completely lower the engine and tighten the through-bolt securely.
14   If necessary, repeat the process for the motor mount on the other side of the engine.
15   The remainder of installation is the reverse of removal.

Notes

Chapter 2 Part B
3.7L V6 engine

Contents

Specifications

General

Displacement	3.7 liters (226 cubic inches)
Bore and stroke	3.66 x 3.40 inches (93.0 x 90.8 mm)
Cylinder numbers (front to rear)	
Left bank	1-3-5
Right bank	2-4-6
Firing order	1-6-5-4-3-2

Camshaft

Endplay	0.003 to 0.0079 inch (0.075 to 0.200 mm)

Valve lift
2004 and earlier models

Intake	0.472 inch (12.0 mm)
Exhaust	0.4292 (10.90 mm)

2005 and later models

Intake and exhaust	0.472 inch (12.0 mm)
Camshaft bearing oil clearance	
Standard	0.001 to 0.0026 inch (0.025 to 0.065 mm)
Service limit	0.0026 inch (0.065 mm)
Camshaft journal diameter	1.0227 to 1.0235 inch (25.975 to 25.995 mm)
Camshaft bore diameter	1.0245 to 1.0252 inch (26.02 to 26.04 mm)

Front

Cylinder identification diagram

50035-1-B HAYNES

Timing chain

Idler gear endplay.. 0.004 to 0.010 inch (0.101 to 0.254 mm)

Oil pump

Cover warpage limit (maximum).. 0.001 inch (0.025 mm)
Inner and outer rotor thickness.. 0.4731 inch (12.02 mm)
Outer rotor diameter (minimum) .... 3.382 inches (85.925 mm)
Outer rotor-to-housing clearance (maximum)............................ 0.0093 inch (0.235 mm)
Inner rotor-to-outer rotor lobe clearance (maximum)................... 0.006 inch (0.150 mm)
Oil pump housing-to-rotor side clearance (maximum)............... 0.0038 inch (0.095 mm)

Torque specifications

Note: *One foot-pound (ft-lb) of torque is equivalent to 12 inch-pounds (in-lbs) of torque. Torque values below approximately 15 foot-pounds are expressed in inch-pounds, because most foot-pound torque wrenches are not accurate at these smaller values.*

	Ft-lbs (unless otherwise indicated)	**Nm**
Camshaft sprocket bolts (non-oiled)...	90	122
Camshaft bearing cap bolts..	100 in-lbs	11
Crankshaft pulley/vibration damper bolt	130	175
Cylinder head bolts (in sequence - **see illustration 11.19a**)		
Step one **(do this step twice)**		
M10 bolts (1 through 8)..	20	27
Step two		
M8 bolts (9 through 12)..	120 in-lbs	13.5
Step three **(do this step twice)**		
M10 bolts (1 through 8)..	Turn an additional 90-degrees	
Step four		
M8 bolts (9 through 12)..	19	26
Driveplate bolts..	70	95
Engine mount bracket-to-block bolts	45	61
Engine mount through bolt/nut	45	61
Exhaust manifold bolts	18	25
Exhaust manifold heat shield nuts		
Step 1...	72 in-lbs	8
Step 2...	Loosen 45-degrees	
Intake manifold bolts..	105 in-lbs	12
Oil pan bolts...	130 in-lbs	14.5
Oil pan drain plug	25	34
Oil pick-up tube mounting bolt/nut..	21	28
Oil pump mounting bolts...	21	28
Oil pump cover screws..	105 in-lbs	12
Timing chain cover bolts...	43	58
Timing chain guide bolts...	21	28
Timing chain guide access plugs..	15	20
Timing chain idler sprocket bolt	25	34
Timing chain tensioner arm pivot bolt.......................................	21	28
Timing chain tensioner (secondary)	21	28
Timing chain tensioner (primary)	21	28
Transmission-to-oil pan support brace	40	54
Valve cover bolts	105 in-lbs	12
Water outlet housing...	105 in-lbs	12

3.5 A compression gauge can be used in the number one plug hole to assist in finding TDC

3.8 Align the groove in the damper with the TDC mark on the timing chain cover

1 General information

This Part of Chapter 2 is devoted to in-vehicle repair procedures for the 3.7L V6 single overhead camshaft (SOHC) engine. These engines utilize a cast iron engine block with six cylinders arranged in a "V" shape at a 90-degree angle between the two banks. The overhead camshaft aluminum cylinder heads are equipped with replaceable valve guides and seats. Stamped steel rocker arms with an integral roller bearing actuate the valves.

All information concerning engine removal and installation and engine overhaul can be found in Part D of this Chapter.

The following repair procedures are based on the assumption that the engine is installed in the vehicle. If the engine has been removed from the vehicle and mounted on a stand, many of the steps outlined in this Part of Chapter 2 will not apply.

The Specifications included in this Part of Chapter 2 apply only to the procedures contained in this Part. Information concerning engine removal and overhaul or replacement can be found in Chapter 2, Part D.

2 Repair operations possible with the engine in the vehicle

Many major repair operations can be accomplished without removing the engine from the vehicle.

Clean the engine compartment and the exterior of the engine with some type of degreaser before any work is done. It will make the job easier and help keep dirt out of the internal areas of the engine.

Depending on the components involved, it may be helpful to remove the hood to improve access to the engine as repairs are performed (refer to Chapter 11, if necessary). Cover the fenders to prevent damage to the paint. Special pads are available, but an old bedspread or blanket will also work.

If vacuum, exhaust, oil or coolant leaks develop, indicating a need for gasket or seal replacement, the repairs can generally be made with the engine in the vehicle. The intake and exhaust manifold gaskets, oil pan gasket, crankshaft oil seals and cylinder head gaskets are all accessible with the engine in place.

Exterior engine components, such as the intake and exhaust manifolds, the oil pan, the oil pump, the water pump (see Chapter 3), the starter motor, the alternator and the fuel system components (see Chapter 4) can be removed for repair with the engine in place.

Since the cylinder heads can be removed without pulling the engine, valve component servicing can also be accomplished with the engine in the vehicle. Replacement of the camshafts, timing chains and sprockets are also possible with the engine in the vehicle.

In extreme cases caused by a lack of necessary equipment, repair or replacement of piston rings, pistons, connecting rods and rod bearings is possible with the engine in the vehicle. However, this practice is not recommended because of the cleaning and preparation work that must be done to the components involved.

3 Top Dead Center (TDC) for number one piston - locating

Refer to illustrations 3.5 and 3.8

1 Top Dead Center (TDC) is the highest point in the cylinder that each piston reaches as it travels up the cylinder bore. Each piston reaches TDC on the compression stroke and again on the exhaust stroke, but TDC generally refers to piston position on the compression stroke.

2 Positioning the piston(s) at TDC is an essential part of many procedures such as valve timing, camshaft and timing chain/sprocket removal.

3 Before beginning this procedure, be sure to place the transmission in Neutral and apply the parking brake or block the rear wheels. Also, disable the ignition system by discon-necting the primary electrical connectors at the ignition coil packs and remove the spark plugs (see Chapter 1).

4 In order to bring any piston to TDC, the crankshaft must be turned using one of the methods outlined below. When looking at the front of the engine, normal crankshaft rotation is clockwise.

a) *The preferred method is to turn the crankshaft with a socket and ratchet attached to the bolt threaded into the front of the crankshaft. Turn the bolt in a clockwise direction only. Never turn the bolt counterclockwise.*

b) *A remote starter switch, which may save some time, can also be used. Follow the instructions included with the switch. Once the piston is close to TDC, use a socket and ratchet as described in the previous paragraph.*

c) *If an assistant is available to turn the ignition switch to the Start position in short bursts, you can get the piston close to TDC without a remote starter switch. Make sure your assistant is out of the vehicle, away from the ignition switch, then use a socket and ratchet as described in Paragraph (a) to complete the procedure.*

5 Install a compression pressure gauge in the number one spark plug hole (refer to Chapter 2D). It should be a gauge with a screw-in fitting and a hose at least six inches long **(see illustration)**.

6 Rotate the crankshaft using one of the methods described above while observing for pressure on the compression gauge. The moment the gauge shows pressure indicates that the number one cylinder has begun the compression stroke.

7 Once the compression stroke has begun, TDC for the compression stroke is reached by bringing the piston to the top of the cylinder.

8 Continue turning the crankshaft until the notch in the crankshaft damper is aligned with the TDC mark on the timing chain cover **(see illustration)**. At this point, the number one cylinder is at TDC on the compression

stroke. If the marks are aligned but there was no compression, the piston was on the exhaust stroke. Continue rotating the crankshaft 360-degrees (1-turn). **Note:** *If a compression gauge is not available, you can simply place a blunt object over the spark plug hole and listen for compression as the engine is rotated. Once compression at the No.1 spark plug hole is noted, the remainder of the Step is the same.*

9   After the number one piston has been positioned at TDC on the compression stroke, TDC for any of the remaining cylinders can be located by turning the crankshaft 90 degrees and following the firing order (refer to the Specifications). Rotating the engine 90 degrees past TDC #1 will put the engine at TDC compression for cylinder #6.

---

4   Valve cover - removal and installation

Refer to illustrations 4.3, 4.5 and 4.6

Removal

1   Disconnect the cable from the negative terminal of the battery.
2   Remove the air intake duct and the throttle body resonator (see Chapter 4).
3   Unclip the fuel injector wiring harness from the studs on the valve cover **(see illustration)**.
4   Disconnect the electrical connectors from the fuel injectors on the side from which the valve cover is to be removed (see Chapter 4). If both valve covers are to be removed, disconnect all of the connectors from the fuel injectors **(see illustration 4.3)**.
5   Remove the left side breather tube **(see illustration)**.
6   Remove the valve cover studs and bolts **(see illustration)**. Make a note of the stud locations before removal to ensure correct positioning during installation.
7   Detach the valve cover. **Note:** *If the cover sticks to the cylinder head, use a block*

of wood and a hammer to dislodge it. *If the cover still won't come loose, pry on it carefully, but don't distort the sealing flange.*

Installation

8   The mating surfaces of each cylinder head and valve cover must be perfectly clean when the covers are installed. The valve cover gasket can be reused if it isn't hardened, cracked or otherwise damaged. **Caution:** *Do not use harsh cleaners when cleaning the valve covers or damage to the covers may occur.*
9   Clean the mounting bolt or stud threads with a wire brush if necessary to remove any corrosion and restore damaged threads. Use a tap to clean the threaded holes in the heads.
10   Place the valve cover and gasket in position, then install the stud/nuts and bolts in the correct location from which they where removed. Tighten the bolts in several steps to the torque listed in this Chapter's Specifications.
11   Complete the installation by reversing the removal procedure. Start the engine and check carefully for oil leaks.

---

5   Rocker arms and hydraulic lash adjusters - removal, inspection and installation

Refer to illustrations 5.5, 5.7 and 5.8
Note 1: *A special valve spring compressor available from most aftermarket specialty tool manufacturers will be required for this procedure. The only other alternative to accomplishing this task without the use of this special tool is to remove the timing chains and the camshafts, which requires major disassembly of the engine and surrounding components.*
Note 2: *This engine is a non-freewheeling engine and the pistons must be down in the cylinder bore before the valve and spring assembly can be compressed to allow rocker arm removal.*

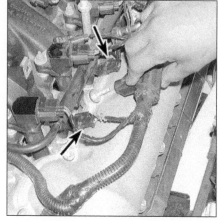

4.3 Detach the wiring harness from the valve cover studs, then disconnect the fuel injector electrical connectors and position the harness aside

1   Before beginning this procedure, be sure to place the transmission in Park and apply the parking brake or block the rear wheels. Also, disable the ignition system by disconnecting the primary electrical connectors at the ignition coils and remove the spark plugs (see Chapter 1).
2   Remove the valve cover(s) (see Section 4).
3   Before the rocker arms and lash adjusters are removed, arrange to label and store them, so they can be kept separate and reinstalled on the same valve they were removed from.
4   Rotate the engine with a socket and ratchet in a clockwise direction only by the crankshaft pulley/vibration damper bolt until the piston(s) are positioned correctly to remove the rocker arms from the corresponding cylinders as follows:

 a) Remove the rocker arms from cylinders **No. 2 and 6** with the No.1 piston at TDC on the compression stroke (see Section 3).

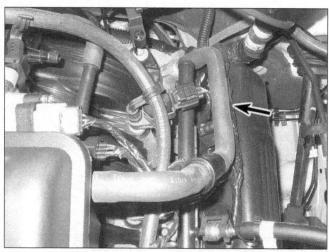

4.5 Remove the left side breather tube

4.6 Remove the valve cover studs and bolts (left side shown)

5.5 Using a special type valve spring compressor, depress the valve spring just enough to remove the rocker arm

5.7 Pull the lash adjuster up and out of its bore to remove it from the cylinder head

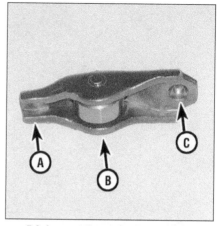

5.8 Inspect the rocker arms at the following locations

A Valve stem seat
B Roller
C Lash adjuster pocket

b) *Rotate the crankshaft another 180-degrees (1/2-turn) from TDC on the compression stroke to bring the No. 1 piston to BDC on the combustion stroke. Remove the rocker arms from cylinder* **No. 1** *with the No. 1 piston at Bottom Dead Center (BDC) on the combustion stroke.*

c) *Rotate the crankshaft another 180-degrees (1/2-turn) from BDC on the combustion stroke to bring the No.1 piston to TDC on the exhaust stroke. Remove the rocker arms from cylinders* **No. 3 and 5** *with the No. 1 piston at TDC on the exhaust stroke.*

d) *Rotate the crankshaft another 180-degrees (1/2-turn) from TDC on the exhaust stroke to bring the No.1 piston to BDC on the intake stroke. Remove the rocker arms from cylinder* **No. 4** *with the No. 1 piston at BDC on the intake stroke.*

5   Hook the valve spring compressor around the base of the camshaft. Depress the valve spring just enough to release tension on the rocker arm to be removed. Once tension on the rocker arm is relieved, the rocker arm can be removed by simply pulling it out **(see illustration)**.

6   If you're removing or replacing only a few of the rockers arms or lash adjusters, locate the cylinder number of the rocker arm or lash adjuster you wish to remove in Step 4, then rotate the crankshaft to the corresponding position. Remember to keep the rocker arm and lash adjuster for each valve together so they can be reinstalled in the same locations. Refer to Section 3 as necessary to help position the designated cylinder at TDC.

7   Once the rocker arms are removed, the lash adjusters can be pulled out of the cylinder head and stored with the corresponding rocker arm **(see illustration)**.

8   Inspect each rocker arm for wear, cracks and other damage. Make sure the rollers turn freely and show no signs of wear, also check the pivot area for wear, cracks and galling **(see illustration)**.

9   Inspect the lash adjuster contact surfaces for wear or damage. Make sure the lash adjusters move up and down freely in their bores on the cylinder head without excessive side to side play.

10   Installation is the reverse of removal with the following exceptions: Always install the lash adjuster first and make sure they're at least partially full of oil before installation. This is indicated by little or no lash adjuster plunger travel.

6   Valve springs, retainers and seals - replacement

Refer to illustrations 6.5, 6.7a, 6.7b, 6.8, 6.13 and 6.15

Note: *Broken valve springs and defective valve stem seals can be replaced without removing the cylinder heads. Two special tools and a compressed air source are normally required to perform this operation, so read through this Section carefully. The special type of valve spring compressor required for rocker arm and valve spring removal may not be available at all tool rental yards, so check on the availability before beginning the job.*

1   Remove the valve cover(s) (see Section 4).

2   Refer to Section 5 and remove the rocker arms from the affected cylinder head.

3   Remove the spark plug (see Chapter 1) from the cylinder that has the defective component. If all of the valve stem seals are being replaced, all of the spark plugs and rockers arms should be removed.

4   Turn the crankshaft until the piston in the affected cylinder is at Top Dead Center on the compression stroke (refer to Section 3). If you're replacing all of the valve stem seals, begin with cylinder number one and work on the valves for one cylinder at a time. Move from cylinder to cylinder following the firing order sequence (see this Chapter's Specifications).

5   Thread a long adapter into the spark plug hole and connect an air hose from a compressed air source to it **(see illustration)**. Most auto parts stores can supply the air hose adapter. **Note:** *Because of the length of the spark plug wells, it will be necessary to use a long spark plug adapter with a length of hose attached (as used on many cylinder compression gauges) utilizing a quick-disconnect fitting to hook to your air source.*

6   Apply 90 to 100 psi of compressed air to the cylinder. **Warning:** *The piston may be forced down by the compressed air, causing the crankshaft to turn suddenly. If the wrench*

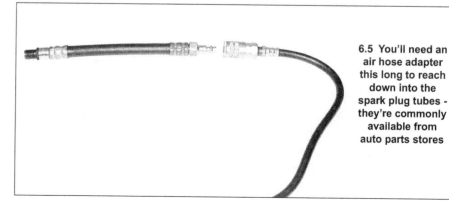

6.5 You'll need an air hose adapter this long to reach down into the spark plug tubes - they're commonly available from auto parts stores

6.7a Compress the valve spring enough to release the valve stem locks . . .

6.7b . . . and lift them out with a magnet or needle-nose pliers

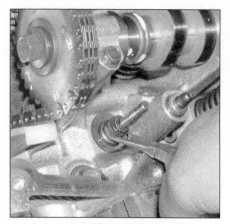

6.8 A pair of pliers will be required to remove the valve stem seal from the valve guide

used when positioning the number one piston at TDC is still attached to the bolt in the crankshaft nose, it could cause damage or injury when the crankshaft moves.

7   Stuff shop rags into the cylinder head holes around the valves to prevent parts and tools from falling into the engine, then use a valve spring compressor to compress the spring. Remove the valve stem locks with small needle-nose pliers or a magnet **(see illustrations)**. **Note:** *The valves should be held in place by the air pressure. If the valve faces or seats are in poor condition, leaks may prevent air pressure from retaining the valves. If the valves cannot hold air, the cylinder head should be removed and taken to a machine shop for a valve job.*

8   Remove the spring retainer and valve spring, then remove the valve stem seal **(see illustration)**. **Note:** *Label and store the valve springs so that they can be reinstalled on the same valve that they were removed from.*

9   Wrap a rubber band or tape around the top of the valve stem so the valve won't fall into the combustion chamber, then release the air pressure.

10   Inspect the valve stem for damage. Rotate the valve in the guide and check the

end for eccentric movement, which would indicate that the valve is bent.

11   Move the valve up-and-down in the guide and make sure it doesn't bind. If the valve stem binds, either the valve is bent or the guide is damaged. In either case, the head will have to be removed for repair.

12   Reapply air pressure to the cylinder to retain the valve in the closed position, then remove the tape or rubber band from the valve stem.

13   Lubricate the valve stems with engine oil and install the valve spring seat/valve seal assembly over the top of the valve stems. Using the stem of the valves as a guide, slide the seals down to the top of each valve guide. Using a hammer and a deep socket or seal installation tool, gently tap each seal into place until it's completely seated on the guide **(see illustration)**. Don't twist or cock the seals during installation or they won't seal properly on the valve stems. Make sure the garter spring (if equipped) is still in place around the top of the seal.

14   Install the spring and retainer in position over the valve with the colored mark (if equipped) on the spring facing UP. Compress the valve spring assembly only enough to

install the keepers in the valve stem.

15   Position the keepers in the valve stem groove. Apply a small dab of grease to the inside of each keeper to hold it in place if necessary **(see illustration)**. Remove the pressure from the spring tool and make sure the keepers are seated.

16   Disconnect the air hose and remove the adapter from the spark plug hole.

17   Repeat the above procedure on the remaining cylinders, following the firing order sequence (see the Specifications). Bring each piston to top dead center on the compression stroke before applying air pressure.

18   Refer to Section 5 and install the rocker arms.

19   Refer to Section 4 and install the valve covers.

20   Refer to Section 1 and install the spark plug(s) and the ignition coils.

21   Start and run the engine, then check for oil leaks and unusual sounds coming from the valve cover area.

7   Timing chain and sprockets - removal, inspection and installation

Warning: *Wait until the engine is completely cool before beginning this procedure.*

Note 1: *Special tools are necessary to complete this procedure. Read through the entire procedure and obtain the special tools before beginning work.*

Note 2: *The 3.7L V6 engine utilizes three timing chains to produce proper valve timing. The primary timing chain runs around the crankshaft sprocket and the idler gear sprocket. This chain synchronizes the crankshaft and pistons with the idler gear, while two secondary timing chains run around the rear of the idler gear and up to the camshaft sprockets to synchronize the valve timing with the crankshaft. Refer to Step 24 for the in-vehicle timing chain inspection procedure prior to timing chain removal.*

6.13 Using a deep socket and hammer, gently tap the new seals onto the valve guide only until seated

6.15 Apply a small dab of grease to each valve stem lock as shown here before installation - it will hold them in place on the valve stem as the spring is released

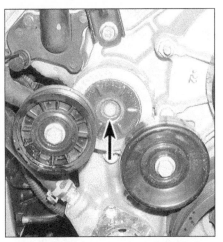

7.7 Drivebelt tensioner retaining bolt

7.11 When the engine is at TDC on the exhaust stroke, the crankshaft sprocket is aligned with the mark on the timing chain cover and the "V6" marks on the camshaft sprockets should be pointing to the 12 o'clock position

Removal

Refer to illustrations 7.7, 7.11, 7.13, 7.14, 7.15, 7.16a, 7.16b, 7.17 and 7.18

1   Disconnect the cable from the negative terminal of the battery.

2   Drain the cooling system (see Chapter 1).

3   Refer to Chapter 3 and remove the cooling fan, the accessory drivebelt and the fan shroud from the engine compartment.

4   Detach the heater hoses and the lower radiator hose from the timing chain cover and position them aside.

5   Unbolt the power steering pump and set it aside without disconnecting the fluid lines (see Chapter 10).

6   Remove the alternator (see Chapter 5). Also unbolt the air conditioning compressor (if equipped) and position it aside without disconnecting the refrigerant lines.

7   Remove the accessory drivebelt tensioner from the timing chain cover **(see illustration)**.

8   Remove the valve covers (see Section 4) and the spark plugs (see Chapter 1).

9   Remove all of the rocker arms following the procedure outlined in Section 5. **Note:** *This step is not absolutely necessary, but it will help make alignment of the camshaft sprockets easier upon installation and also eliminate any possibility of the pistons contacting the valves during this procedure, since the 3.7L V6 is an interference engine.*

10   Remove the camshaft position sensor (CPS) from the right cylinder head (see Chapter 6).

11   Position the number one piston at TDC on the exhaust stroke (see Section 3). Visually confirm the engine is at TDC on the exhaust stroke, by verifying that the timing mark on the timing chain cover and the "V6" marks on the camshaft sprockets are pointing straight up in the 12 o'clock position **(see illustration 3.8 and the accompanying illustration)**.

12   Remove the crankshaft damper/pulley (see Section 12).

13   Remove the timing chain cover and the

7.13 Timing chain cover retaining bolts

7.14 Locking the primary timing chain tensioner in the retracted position

7.15 Secondary timing chain tensioner mounting bolts

water pump as an assembly **(see illustration)**. Note that various types and sizes of bolts are used. They must be reinstalled in their original locations. Mark each bolt or make a sketch to help remember where they go.

14   Cover the oil pan opening with shop rags to prevent any components from falling

into the engine. Collapse the primary timing chain tensioner with a pair of locking pliers and install a locking pin into the holes in the tensioner body to keep it in the retracted position **(see illustration)**.

15   Remove the secondary timing chain tensioners **(see illustration)**.

7.16a Removing the left camshaft sprocket bolt while holding the sprocket with a pin spanner wrench - note the chain guide access plug

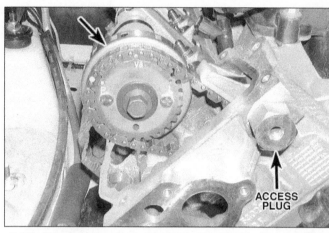

7.16b The right camshaft sprocket is identified by the camshaft position sensor ring which is fastened to the rear of the sprocket - be extremely careful not to damage or place a magnetic object of any kind near the camshaft position sensor ring or a "no start" condition may occur after installation

16   Remove the camshaft sprocket retaining bolts **(see illustrations)**. Pull the camshaft sprockets off the camshaft hubs one at a time. Lower the sprocket(s) into the cylinder head opening until the chain can be displaced from around the sprocket, then remove the camshaft sprockets from the engine and let the secondary chains fall down between the timing chain guides. **Warning 1:** *If the rocker arms* **were not** *removed as suggested in Step 9, it will be necessary to hold the camshafts from rotating with a set of locking pliers while the sprocket is being removed. Work on one camshaft and sprocket at a time starting with the left sprocket and proceeding to the right sprocket. After the sprocket is removed from the camshaft(s), let the camshaft slowly rotate to its neutral position. This is typically 15-degrees clockwise on the left camshaft sprocket and 45-degrees counterclockwise on the right camshaft. Pressure from the valve springs will make the cam-*

shafts rotate as the sprockets are removed. Sudden movement of the camshafts may allow the valves to strike the pistons. **Warning 2:** *Never install the locking pliers on a camshaft lobe as damage to the camshaft will occur. When using locking pliers, always rotate the camshaft with locking pliers by the base shaft.* **Warning 3:** *Do not rotate the crankshaft or camshafts separately after the secondary timing chains are loosened or removed with the rocker arms installed in the engine as piston or valve damage may occur. The only exception to this rule is when the camshafts must be rotated slightly, to realign the camshaft sprockets with the camshafts during installation.* **Caution:** *The right camshaft sprocket is identified by the camshaft position sensor ring which is fastened to the rear of the sprocket. Be extremely careful not to damage or place a magnetic object of any kind near the camshaft position sensor ring or a no start condition may occur after installation.*

17   Remove the idler sprocket bolt, then detach the idler sprocket, the crankshaft sprocket, the primary timing chain and the secondary timing chains as an assembly **(see illustration)**.
18   Remove the cylinder head access plugs **(see illustrations 7.16a and 7.16b)**. Also remove the oil fill tube (if not already removed) from the front of the right cylinder head. Detach the timing chain guides and tensioner arms **(see illustration)**.

Inspection

19   Inspect the camshaft and crankshaft sprockets for wear on the teeth and keyways.
20   Inspect the chains for cracks or excessive wear of the rollers.
21   Inspect the facings of the primary chain tensioner, secondary chain guides and tensioner arms for excessive wear. If any of the components show signs of excessive wear or the chain guides are grooved in excess

7.17 Remove the primary timing chain and the secondary chains as an assembly from the engine

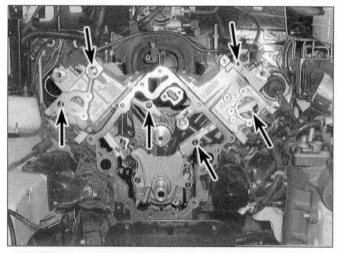

7.18 Timing chain guide and tensioner arm pivot bolts (arrows)

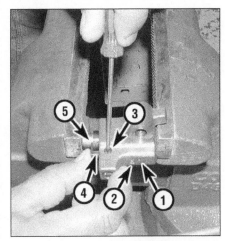

7.27 Locking the secondary tensioner(s) in the retracted position - note the "identification" mark on the side of the tensioner, as they are not interchangeable

1   Insert locking pin
2   Identification mark
3   Ratchet pawl
4   Ratchet
5   Tensioner piston

of 0.039 inch deep, they must be replaced.

22   If any of the timing chain guides are excessively grooved or melted, the tensioner lube jet may be clogged. Be sure to remove the jet and clean it with a small metal pick, and then blow compressed air through it to remove any debris or foreign material.

23   Inspect the idler sprocket bushing, shaft and spline joint for wear (see Chapter 2D).

24   Inspect the tensioner piston and ratchet assembly on both of the secondary chain tensioners. If it appears there has been heavy contact between the piston, and ratchet assembly, replace the tensioner arm and secondary timing chain. **Note:** *Secondary timing chain stretch can be checked by rotating the engine clockwise until the pistons in the secondary tensioners reach their maximum travel or extension. Using a machinist's ruler or a dial caliper, measure the piston protrusion or extension from the stepped ledge on the piston to the tensioner housing on each tensioner. If the maximum extension of either tensioner piston exceeds 0.590 inch, the secondary timing chains are worn beyond their limits and should be replaced.*

Installation

Refer to illustrations 7.27, 7.31, 7.33, 7.39 and 7.48

25   If removed, install the timing chain guides and tensioner pivot arms back onto the engine and tighten the bolts to the torque listed in this Chapter's Specifications. Apply several drops of medium strength thread-locking compound to the tensioner pivot arm bolts before installing them. Note that the silver bolts retain the guides to the cylinder head and the black colored bolts retain the guides to the engine block.

26   If the primary timing chain tensioner was removed or replaced, install it back onto

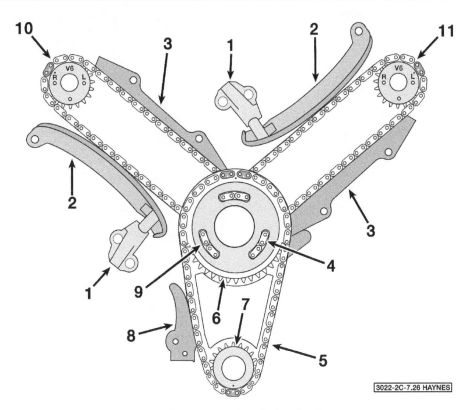

7.31 Timing chain installation details

1	Secondary timing chain tensioner	7	Crankshaft sprocket
2	Secondary tensioner arm	8	Primary chain tensioner
3	Chain guide	9	Two plated links on left camshaft chain
4	Two plated links on right camshaft chain	10	Right camshaft sprocket and secondary chain
5	Primary chain	11	Left camshaft sprocket and secondary chain
6	Idler sprocket		

the engine in the locked position and tighten the lower two bolts to the torque listed in this Chapter's Specifications **(see illustration 15.3)**.

27   Working on one secondary timing chain tensioner at a time, compress the tensioner piston in a vise until the stepped edge is flush with the tensioner body **(see illustration)**. Insert a small scribe or other suitable tool into the side of the tensioner body and push the spring loaded ratchet pawl away from the ratchet mechanism, then push the ratchet down into the tensioner body until it's approximately 0.080 inch away from the tensioner body. Insert the end of a paper clip into the hole on the front of the tensioner to lock the tensioner in place.

28   After the two secondary tensioners have been compressed and locked into place, install them on the engine and tighten the bolts to the torque listed in this Chapter's Specifications. Make sure the tensioner with the "R" mark is installed on the right secondary chain and the tensioner with "L" mark is installed on the left secondary chain. The secondary chain tensioners **cannot** be switched with one another. Also make sure the plate

behind the left secondary chain tensioner is installed correctly.

29   If you purchased a new timing chain verify that you have the correct timing chain for your vehicle by counting the number of links the chain has and comparing the new chain with the old chain. Also compare the position of the colored links in the new chain with the position of the colored links in the old chain.

30   Note that the idler sprocket has three sprockets and a gear incorporated into it. The front or forward facing sprocket (the largest of the three) is for the primary timing chain, the second or middle sprocket is for the left secondary chain, the third or rear sprocket is for the right secondary chain and the gear is for the counterbalance shaft. The next 5 Steps will involve assembling the timing chains onto the idler sprocket on a workbench.

31   Place the idler sprocket on a workbench with the mark on the front in the 12 o'clock position. Loop the right camshaft chain over the rear gear (farthest away from the primary chain gear) on the idler sprocket and position it so that the two plated links on the chain are visible through the lower (4 o'clock) window in the idler sprocket **(see illustration)**.

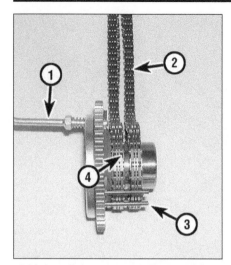

7.33 Install the secondary chain holding tool onto the idler sprocket with the plated links on the right camshaft chain in the 4 o'clock position and the left camshaft chain in the 8 o'clock position (typical)

1   *Secondary chain holding tool*
2   *Right camshaft timing chain*
3   *Secondary chain holding tool retaining pins*
4   *Left camshaft timing chain*

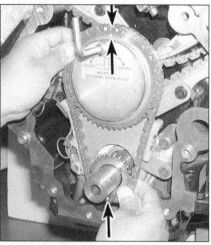

7.39 Align the timing mark on the idler sprocket gear with the timing mark on the counterbalance shaft, then install the idler sprocket, the crankshaft sprocket, the timing chains and the chain holding tool as an assembly onto the engine with the marks aligned as shown - feed the secondary chains up through the timing chain guides and loop them over the camshaft hubs, then push the primary chain assembly the rest of the way on the engine until it's seated against the block

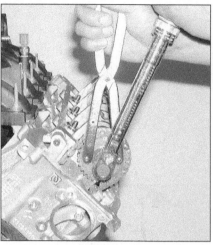

7.48 Tightening the left camshaft sprocket bolt

32   Loop the left camshaft chain over the front of the idler sprocket and position it over the middle gear so that the two plated links on the chain are visible through the lower (8 o'clock) window in the idler sprocket. **Note:** *After the left chain is in position, the two plated links on the right chain will no longer be visible through the 4 o'clock window in the idler sprocket, so be sure that the right camshaft chain is installed correctly before installing the left camshaft chain.*

33   After the secondary (camshaft) chains have been installed properly on the idler sprocket, install the special secondary chain holding tool onto the idler sprocket **(see illustration)**. This tool serves as a third hand, to secure the camshaft chains to the idler sprocket during the installation of the idler sprocket onto the engine.

34   Install the primary timing chain onto the primary chain gear of the idler sprocket and align the double plated links with the mark on the front of the sprocket. The mark on the idler sprocket should still be in the 12 o'clock position.

35   Insert the teeth of the crankshaft sprocket into the primary timing chain with the mark on the crankshaft sprocket pointing down in the 6 o'clock position and aligned with the single plated link on the chain.

36   Lubricate the idler shaft and bushing with clean engine oil.

37   Install the idler sprocket, the crankshaft sprocket and the timing chains with the chain holding tool as an assembly onto the engine. Slide the crankshaft sprocket over the keyway on the crankshaft and position the idler

sprocket partially over the idler shaft (just enough to hold the primary chain in place). Then feed the secondary chains up through the chain guides and the cylinder head. **Note:** *It may be easier to bend a hook in the end of a coat hanger to help pull the secondary chains up through the timing chain guides and the cylinder head opening.*

38   Loop the secondary chains over the camshaft hubs and secure them with rubber bands to remove the slack from the chains.

39   Align the timing mark on the idler sprocket gear with the timing mark on the counterbalance shaft gear, then push the idler sprocket, primary timing chain and the crankshaft sprocket assembly back on to the engine until they're fully seated against the block **(see illustration)**.

40   Thoroughly clean the idler sprocket bolt. Make sure all oil is removed from the bolt threads before installation, as over-tightening of bolts may occur if oil is not removed, then lubricate the idler sprocket washer with small amount of clean engine oil making sure not to get oil on the bolt threads. Remove the secondary timing chain holding tool from the idler sprocket, then install the idler sprocket retaining bolt and tighten it to the torque listed in this Chapter's Specifications.

41   Align the "L" mark on left camshaft sprocket with the plated link on the left camshaft chain and position the camshaft sprocket over the camshaft hub **(see illustration 7.31)**. The camshaft may have to be rotated slightly to align the dowel pin on the camshaft with the slot on the sprocket.

42   Align the "R" mark on right camshaft

sprocket with the plated link on the right camshaft chain and position the camshaft sprocket over the camshaft hub. The camshaft may have to be rotated slightly to align the dowel pin on the camshaft with the slot on the sprocket. **Note:** *If the rocker arms **were not** removed as suggested in Step 9, it will be necessary to rotate and hold the camshafts with a set of locking pliers while the sprocket is being installed. This is typically 15-degrees counterclockwise on the left camshaft sprocket and 45-degrees clockwise on the right camshaft (the exact opposite of removal). Work on one camshaft and sprocket at a time starting with the left sprocket and proceeding to the right sprocket and never rotate the camshaft by a camshaft lobe or damage to the camshaft will occur.*

43   Thoroughly clean the camshaft sprocket bolts. Make sure all oil is removed from the bolt threads before installation, as over-tightening of bolts may occur if oil is not removed, then lubricate the bolt washers with small amount of clean engine oil making sure not to get oil on the threads.

44   Install the camshaft sprocket bolts finger tight.

45   Verify that all the plated timing chain links are aligned with their corresponding marks.

46   Remove the locking pins from the primary timing chain tensioner and the secondary timing chain tensioners. **Caution:** *Do not manually extend the tensioners; doing so will only over-extend the tensioners and lead to premature timing chain wear.*

47   Rotate the engine two complete revolutions and re-verify the position of the timing marks again. The idler sprocket mark should be located in the 12 o'clock position and the crankshaft sprocket mark should be located in the 6 o'clock position with the "V6" marks on the camshaft sprockets located in the 12 o'clock position.

48   Using a spanner wrench to hold the sprockets from turning, tighten the camshaft sprocket bolts to the torque listed in this Chapter's Specifications **(see illustration)**.

8.6 The timing chain tensioner wedge is pushed down between the chain strands to secure the secondary chain and the tensioner in place while the camshaft is removed - this wedge is fabricated from a block of wood and a piece of wire

8.9 Verify that the camshaft bearing caps are marked to ensure correct reinstallation - do not mix-up the caps from the left cylinder head with the caps from the right cylinder head

49 Remove all traces of old sealant or gasket material from the timing chain cover and the engine block.

50 Place the timing cover and gasket in position on the engine and install the bolts in their original locations and tighten the bolts to the torque listed in this Chapter's Specifications.

51 The remainder of installation is the reverse of removal. Be sure to use pipe sealant on the cylinder head plugs to prevent oil leaks.

8 Camshafts - removal, inspection and installation

Note 1: *Special tools are necessary to complete this procedure. Read through the entire procedure and obtain the special tools before beginning work.*

Note 2: *The camshafts should always be thoroughly inspected before installation and camshaft endplay should always be checked prior to camshaft removal.*

Removal

Refer to illustrations 8.6 and 8.9

1 Disconnect the cable from the negative terminal of the battery.

2 Remove the valve covers (see Section 4).

3 Rotate the engine with a socket and ratchet (in a clockwise direction only) by the crankshaft pulley/vibration damper bolt until the "V6" marks on the camshaft sprockets are located in the 12 o'clock position (see illustration 7.11).

4 Using a permanent marker, apply alignment marks to the secondary timing chain links on either side of the "V6" marks on both camshaft sprockets to help aid the installation process (4 marks total).

5 Using a spanner wrench to hold the camshaft sprockets from turning, loosen the camshaft sprocket bolts several turns, then retighten the bolts by hand until they're

snug up against the sprocket. If the camshaft sprockets have rotated during the bolt-loosening process, rotate the engine clockwise until the "V6" marks on the cam sprockets are realigned in the 12 o'clock position.

6 Install a timing chain tensioner wedge through the opening in the top of the cylinder head and force the wedge down between the narrowest section of the secondary chain (see illustration). If both camshafts are to be removed, two timing chain wedges will be necessary (one for the left camshaft chain and one for the right camshaft chain). The wedge is used to secure the chain and the secondary tensioner in place while the camshaft is removed. Caution 1: *Failure to use a timing chain wedge will allow the secondary tensioner to over-extend and require removal of the timing chain cover to reset the tensioners.* Caution 2: *Never force the wedge past the narrowest section of the secondary timing chain as damage to the tensioner will occur.* If a timing chain wedge is not available, they may be fabricated using a block of wood that is 3/8 to 1/2-inch thick and a piece of wire to pull the wedge out of the cylinder head after installation.

7 Remove the Camshaft Position Sensor (CPS) (see Chapter 6).

8 Remove the camshaft sprocket retaining bolt(s) and detach the camshaft sprocket(s) from the camshaft hub(s). Disengage the camshaft chain(s) from the sprocket(s) and remove the camshaft sprocket(s) from the engine.

9 Make note of the markings on the camshaft bearing caps. The caps are marked from 1 to 4 with arrow marks on the caps indicating the front of the engine (see illustration). If both camshafts are being removed, use a permanent marker to mark each bearing cap on the right cylinder head with an "R" and each bearing cap on the left cylinder head with an "L" to indicate from which cylinder head they came. Loosen the camshaft bearing caps bolts 1/2-turn at a time beginning with the bearing caps on the

ends, then working inward. Caution: *Keep the caps in order. They must go back in the same location they were removed from.*

10 Detach the bearing caps. Note: *The rocker arms may slide out of position. Mark the rocker arms so that they will be installed in the same location.*

11 Remove the camshaft(s) from the cylinder head. Mark the camshaft(s) "Left" or "Right" to indicate which cylinder head it came from.

Inspection

Refer to illustrations 8.14, 8.15 and 8.18

12 Inspect the camshaft sprockets for wear on the teeth.

13 Inspect the chains for cracks or excessive wear of the rollers. If any of the components show signs of excessive wear they must be replaced.

14 Visually check the camshaft bearing surfaces on the cylinder head(s) for pitting, score marks, galling and abnormal wear. If the bearing surfaces are damaged, the cylinder head may have to be replaced (see illustration).

8.14 Inspect the cam bearing surfaces in each cylinder head for pits, score marks and abnormal wear - if wear or damage is noted, the cylinder head must be replaced

15   Measure the outside diameter of each camshaft bearing journal and record your measurements **(see illustration)**. Compare them to the journal outside diameter specified in this Chapter, then measure the inside diameter of each corresponding camshaft bearing and record the measurements. Subtract each cam journal outside diameter from its respective cam bearing bore inside diameter to determine the oil clearance for each bearing. Compare the results to the specified journal-to-bearing clearance. If any of the measurements fall outside the standard specified wear limits in this Chapter, either the camshaft or the cylinder head, or both, must be replaced.

16   Check camshaft runout by placing the camshaft back into the cylinder head and set up a dial indicator on the center journal. Zero the dial indicator. Turn the camshaft slowly and note the dial indicator readings. Runout should not exceed 0.0010 inch. If the measured runout exceeds the specified runout, replace the camshaft.

17   Check the camshaft endplay by placing a dial indicator with the stem in line with the camshaft and touching the snout. Push the camshaft all the way to the rear and zero the dial indicator. Next, pry the camshaft to the front as far as possible and check the reading on the dial indicator. The distance it moves is the endplay. If it's greater than the Specifications listed in this Chapter, check the bearing caps for wear. If the bearing caps are worn, the cylinder head must be replaced.

18   Compare the camshaft lobe height by measuring each lobe with a micrometer **(see illustration)**. Measure of each of the intake lobes and write the measurements and relative positions down on a piece of paper. Then measure of each of the exhaust lobes and record the measurements and relative positions also. This will let you compare all of the intake lobes to one another and all of the exhaust lobes to one another. If the difference between the lobes exceeds 0.005 inch the camshaft should be replaced. Do not compare intake lobe heights to exhaust lobe heights, as lobe lift may be different. Only compare intake lobes to intake lobes and exhaust lobes to exhaust lobes for this comparison.

Installation

19   Apply moly-based engine assembly lubricant to the camshaft lobes and journals and install the camshaft(s) into the cylinder head with the dowel pins in the 10 o'clock position. If the old camshafts are being used, make sure they're installed in the exact location from which they came.

20   Install the bearing caps and bolts and tighten them hand tight.

21   Tighten the bearing cap bolts 1/2-turn at a time, to the torque listed in this Chapter's Specifications, starting with the middle bolts and working outward.

22   Engage the camshaft sprocket teeth with the camshaft drive chain links so that the "V6" mark on the sprocket(s) is between the two marks made in Step 5 during removal, then position the sprocket over the dowel on the

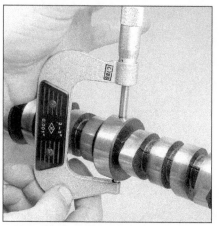

8.15 Measure the outside diameter of each camshaft journal and the inside diameter of each bearing to determine the oil clearance measurement

camshaft hub. At this point the chain marks and the "V6" marks on the sprocket should be pointing up in the 12 o'clock position.

23   Thoroughly clean the camshaft sprocket bolts. Make sure all oil is removed from the bolt threads before installation, as over-tightening of bolts may occur if oil is not removed, then lubricate the bolt washers with small amount of clean engine oil making sure not to get oil on the threads.

24   Install the camshaft sprocket bolts and tighten them to the torque listed in this Chapter's Specifications **(see illustration 7.48)**.

25   Remove the timing chain wedge(s).

26   Install the camshaft position sensor (CPS) (see Chapter 6).

27   Install the valve covers (see Section 4).

28   Connect the cable to the negative terminal of the battery.

---

9   Intake manifold - removal and installation

Warning: *The engine must be completely cool before beginning this procedure.*

Removal

1   Relieve the fuel pressure (see Chapter 4).

2   Disconnect the cable from the negative terminal of the battery.

3   Refer to Chapter 4 and remove the air intake duct from the throttle body.

4   Label and disconnect the vacuum hoses leading to the intake manifold for the PCV valve, power brake booster, cruise control servo and evaporative emission control system.

5   Disconnect all of the electrical connectors leading to the intake manifold for the various sensors, the ignition coils and the Idle Air Control motor (IAC).

6   Disconnect the electrical connectors for the alternator and the air conditioning compressor.

7   Unbolt the ground straps attached to the intake manifold and the throttle body.

8.18 Measuring cam lobe height with a micrometer, make sure you move the micrometer to get the highest reading (top of cam lobe)

8   Remove the engine oil dipstick nut from the stud on the intake manifold, then follow the dipstick tube to the rear of the engine block and remove the bolt securing the tube at the back of the block. Pull the dipstick tube up and out of the engine block to remove it.

9   Detach the throttle cable and the cruise control cable from the throttle body and the throttle cable bracket (see Chapter 4).

10   Remove the ignition coils (see Chapter 5).

11   Remove the fuel rails, then remove the throttle body and mounting bracket (see Chapter 4).

12   Refer to Chapter 1 and drain the cooling system, then remove the heater hoses from the front cover and heater core tubes.

13   Refer to Chapter 6 and remove the coolant temperature sensor. This step is necessary to allow clearance for the intake manifold as it is removed. On 2007 and later models, disconnect and remove the EGR tube connecting the intake manifold to the EGR valve at the rear of the left cylinder head.

14   Remove the intake manifold mounting nuts and bolts in the reverse order of the tightening sequence **(see illustration 9.20)**.

15   The manifold will probably be stuck to the cylinder heads and force may be required to break the gasket seal. **Caution:** *Don't pry between the manifold and the heads or damage to the gasket sealing surfaces may occur, leading to vacuum leaks.*

16   Label and detach any remaining hoses which would interfere with the removal of the intake manifold.

17   Lift the manifold up level with the vehicle and remove it from the engine.

Installation

Refer to illustration 9.20

18   Clean and inspect the intake manifold-to-cylinder head sealing surfaces. Inspect the gaskets on the manifold for tears or cracks replacing them if necessary. The gaskets can be reused if not damaged.

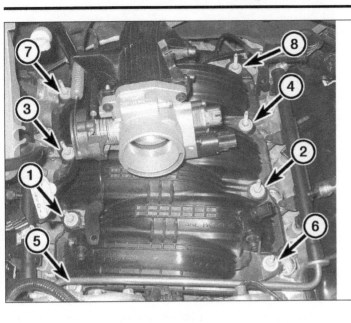

9.20 Intake manifold bolt TIGHTENING sequence

19   Position the manifold on the engine making sure the gaskets and manifold are aligned correctly over the cylinder heads, then install the intake manifold nuts and bolts hand tight.
20   Following the recommended tightening sequence, tighten the bolts to the torque listed in this Chapter's Specifications **(see illustration)**.
21   The remainder of the installation is the reverse of the removal procedure.
22   Fill the cooling system, run the engine and check for fuel, vacuum and coolant leaks.

10   Exhaust manifold - removal and installation

Warning: *The engine must be completely cool before beginning this procedure.*

Removal

1   Disconnect the cable from the negative terminal of the battery.
2   Block the rear wheels, set the parking brake, raise the front of the vehicle and support it securely on jackstands.
3   Unbolt the exhaust pipe from the exhaust manifold.
4   Remove the exhaust manifold heat shields.
5   Remove the exhaust manifold mounting bolts and remove the manifold from the vehicle.

Installation

6   Clean the mating surfaces to remove all traces of old gasket material, then inspect the manifold for distortion and cracks. Warpage can be checked with a precision straightedge held against the mating flange. If a feeler gauge thicker than 0.030-inch can be inserted between the straightedge and flange surface, take the manifold to an automotive machine shop for resurfacing.

7   Place the exhaust manifold in position with a new gasket and install the mounting bolts finger tight.
8   Starting in the middle and working out toward the ends, tighten the mounting bolts in several increments, to the torque listed in this Chapter's Specifications. Tighten the heat shield bolts to the torque listed in this Chapter's Specifications.
9   Install the remaining components in the reverse order of removal.
10   Start the engine and check for exhaust leaks between the manifold and cylinder head and between the manifold and exhaust pipe.

11   Cylinder head - removal and installation

Caution: *The engine must be completely cool before beginning this procedure.*
Note: *The following procedure describes how to remove the cylinder heads with the camshaft(s) and the exhaust manifold(s) still attached to the cylinder head.*

Removal

1   Remove the intake manifold (see Section 9).
2   Refer to Section 7 and remove the timing chains, sprockets and the timing chain guides.
3   Raise the front of the vehicle and support it securely on jackstands.
4   Unbolt the exhaust pipes from the exhaust manifolds.
5   Label and remove any remaining items attached to the cylinder head, such as coolant fittings, ground straps, cables, hoses, wires or brackets.
6   Using a breaker bar and the appropriate sized socket, loosen the twelve cylinder head bolts in 1/4-turn increments until they can be removed by hand. Loosen the bolts in

the reverse order of the tightening sequence **(see illustration 11.19a)** to avoid warping or cracking the head.
7   Lift the cylinder head off the engine block with the camshaft in place and the exhaust manifold attached. If it's stuck, very carefully pry up at the front end of the cylinder head, beyond the gasket surface, at a casting protrusion.
8   Remove all external components from the head to allow for thorough cleaning and inspection. **Note:** *See Chapter 2, Part C, for cylinder head inspection and servicing procedures.*

Installation

Refer to illustrations 11.19a and 11.19b

9   The mating surfaces of the cylinder head and block must be perfectly clean when the head is installed.
10   Use a gasket scraper to remove all traces of carbon and old gasket material from the cylinder head and engine block being careful not to gouge the aluminum, then clean the mating surfaces with lacquer thinner or acetone. If there's oil on the mating surfaces when the head is installed, the gasket may not seal correctly and leaks could develop. When working on the block, stuff the cylinders with clean shop rags to keep out debris. Use a vacuum cleaner to remove material that falls into the cylinders.
11   Check the block and head mating surfaces for nicks, deep scratches and other damage. If damage is slight, it can be removed with a file; if it's excessive, machining may be the only alternative.
12   Use a tap of the correct size to chase the threads in the head bolt holes, then clean the holes with compressed air - make sure that nothing remains in the holes. **Warning:** *Wear eye protection when using compressed air!*
13   With a straight-edge, check each cylinder head bolt for stretching. If all threads do not contact the straight-edge, replace the bolt.
14   Check the cylinder head for warpage (see Chapter 2D). Check the head gasket, intake and exhaust manifold surfaces.
15   Install any components that were removed from the head such as the lash adjusters, the exhaust manifold and the camshaft back onto the cylinder head.
16   Position the new cylinder head gasket over the dowel pins on the block noting which direction on the gasket faces up.
17   Carefully set the head over the dowels on the block without disturbing the gasket.
18   Before installing the M10 head bolts (bolts 1 through 8), apply a small amount of clean engine oil to the threads and hardened washers (if equipped). The chamfered side of the washers must face the bolt heads. Before installing the M8 head bolts (bolts 9 through 12) apply a small amount of thread sealant to the bolt threads.

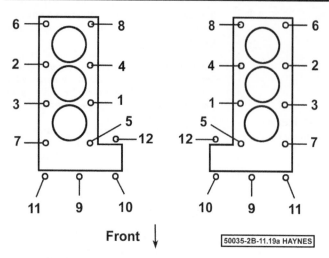

50035-2B-11.19a HAYNES

11.19a Cylinder head bolt TIGHTENING sequence

11.19b Using an angle measurement gauge during the final stages of tightening

19   Install the bolts in their original locations and tighten them finger tight. Then tighten them, following the proper sequence, to the torque and angle of rotation listed in this Chapter's Specifications **(see illustrations)**.
20   Install the timing chain guides, the timing chains and the timing chain sprockets as described in Section 7. The remaining installation steps are the reverse of removal.
21   Refill the cooling system and check the engine oil, adding if necessary (see Chapter 1).
22   Start the engine and check for oil and coolant leaks.

12   Crankshaft pulley/vibration damper - removal and installation

Refer to illustrations 12.4, 12.5 and 12.6

1   Disconnect the cable from the negative terminal of the battery.
2   Refer to Chapter 3 and remove the cooling fans and shroud assembly.
3   Remove the drivebelts (see Chapter 1)

and position the belt tensioner away from the crankshaft pulley.
4   Use a strap wrench around the crankshaft pulley to hold it while using a breaker bar and socket to remove the crankshaft pulley center bolt **(see illustration)**.
5   Pull the damper off the crankshaft with a puller **(see illustration)**. **Caution:** *The jaws of the puller must only contact the hub of the pulley - not the outer ring.* **Note:** *A long Allen-head bolt should be screwed into the crankshaft nose for the puller's tapered tip to push against to prevent damage to the crankshaft threads.*
6   Check the surface on the pulley hub that the oil seal rides on. If the surface has been grooved from long-time contact with the seal, a press-on sleeve may be available to renew the sealing surface **(see illustration)**. This sleeve is pressed into place with a hammer and a block of wood and is commonly available at auto parts stores for various applications.
7   Lubricate the pulley hub with clean engine oil. Align the slot in the pulley with the key on the crankshaft and push the crankshaft pulley on the crankshaft as far as it will go. Use

a vibration damper installation tool to press the pulley the rest of the way onto the crankshaft.
8   Install the crankshaft pulley retaining bolt and tighten it to the torque listed in this Chapter's Specifications.
9   The remainder of installation is the reverse of the removal.

13   Crankshaft front oilseal - replacement

Refer to illustrations 13.2 and 13.4

1   Remove the crankshaft pulley from the engine (see Section 12).
2   Carefully pry the seal out of the cover with a seal removal tool or a large screwdriver **(see illustration)**. **Caution:** *Be careful not to scratch, gouge or distort the area that the seal fits into or an oil leak will develop.*
3   Clean the bore to remove any old seal material and corrosion. Position the new seal in the bore with the seal lip (usually the side with the spring) facing IN (toward the engine).

12.4 Use a strap wrench to hold the crankshaft pulley while removing the center bolt (a chain-type wrench may be used if you wrap a section of old drivebelt or a rag around the crankshaft pulley first)

12.5 The use of a three jaw puller will be necessary to remove the crankshaft pulley - always place the puller jaws around the pulley hub, not the outer ring

12.6 If the sealing surface of the pulley hub has a wear groove from contact with the seal, repair sleeves are available at most auto parts stores

13.2 Pry the seal out very carefully with a seal removal tool or screwdriver, being careful not to nick or gouge the seal bore or the crankshaft

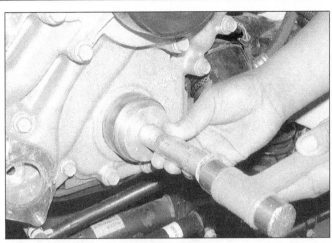

13.4 Use a seal driver or large-diameter socket to drive the new seal into the cover

A small amount of oil applied to the outer edge of the new seal will make installation easier - but don't overdo it!

4   Drive the seal into the bore with a seal driver or a large socket and hammer until it's completely seated **(see illustration)**. Select a socket that's the same outside diameter as the seal and make sure the new seal is pressed into place until it bottoms against the cover flange.

5   Lubricate the seal lips with engine oil and reinstall the crankshaft pulley.

6   The remainder of installation is the reverse of the removal. Run the engine and check for oil leaks.

14 Oil pan - removal and installation

Refer to illustrations 14.4 and 14.9

Removal

1   Disconnect the cable from the negative terminal of the battery.

2   Apply the parking brake and block the rear wheels. Raise the front of the vehicle and place it securely on jackstands.

3   Drain the engine oil (see Chapter 1).

4   If equipped, remove the skidplate, then remove the transmission-to-oil pan support brace at the rear of the pan **(see illustration)**.

5   Remove the transmission oil cooler line bracket.

6   On 4WD models, remove the front axle assembly (see Chapter 8)

7   Remove the engine mount through-bolts (see Section 18).

8   Using an engine hoist, raise the engine until the engine fan is almost in contact with the fan shroud.

9   Remove the bolts and nuts, noting the stud locations, then carefully separate the oil pan from the block. Don't pry between the block and the pan or damage to the sealing surfaces and gasket could occur and oil leaks may develop. Instead, tap on the side of the oil pan with a rubber mallet if necessary to break the gasket seal **(see illustration)**.

10   Remove the two nuts and one bolt that secures the oil pump pick-up tube and windage tray. Drop the pick-up tube into the oil pan, then remove the pick-up tube, windage tray and oil pan as a unit.

Installation

11   Clean the oil pan with solvent and remove any gasket material from the block and the pan mating surfaces. Clean the mating surfaces with lacquer thinner or acetone and make sure the bolt holes in the block are clear. Check the oil pan flange for distortion, particularly around the bolt holes.

12   Inspect the oil pan gasket for cuts and tears, replacing it if necessary. If the gasket is in good condition, it can be reused. **Note:** *The oil pan gasket and the windage tray are a one piece design, therefore must be replaced together.*

13   Place the pick-up tube and oil pan gasket/windage tray in the oil pan and position the oil pan toward the engine. **Note:** *Always use a new O-ring on the pick-up tube and tighten the pick-up tube-to-oil pump bolt first.*

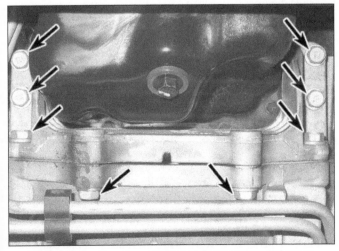

14.4 Remove the oil pan-to-transmission structural cover

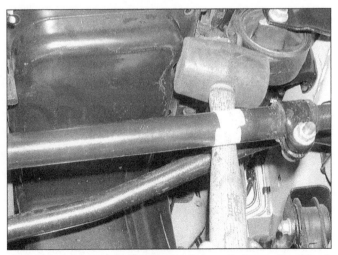

14.9 If the oil pan is stuck to the gasket, gently tap on the side of the oil pan to break the gasket seal

15.3 Oil pump housing/primary timing chain tensioner
retaining bolts

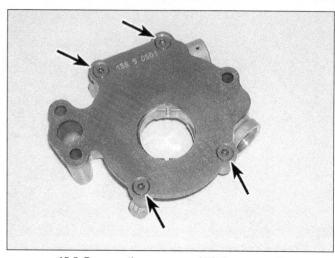

15.5 Remove the screws and lift the cover off

14   After the fasteners are installed, tighten all the bolts in several steps, in a criss-cross pattern starting from the center and working out to the ends, to the torque listed in this Chapter's Specifications.

15   Place the transmission-to-oil pan structural cover in position and install the vertically mounted bolts. Torque the vertical mounted bolts to 10 in-lbs, then install the horizontally mounted bolts. Torque the horizontal mounted bolts to 40 ft-lbs, then re-torque the vertical mounted bolts to 40 ft-lbs (see illustration 14.4).

16   The remaining steps are the reverse of the removal procedure.

17   Refill the engine with oil (see Chapter 1), replace the filter, run it until normal operating temperature is reached and check for leaks.

15   Oil pump - removal, inspection and installation

Removal

Refer to illustrations 15.3

1   Refer to Section 7 and remove the timing chains and sprockets.

2   Remove the oil pan, windage tray and pick-up tube (see Section 14).

3   Remove the primary timing chain tensioner/oil pump bolts (see illustration), then remove the tensioner.

4   Remove the remaining oil pump bolts, then gently pry the oil pump housing outward enough to clear the flats on the crankshaft and remove it from the engine.

Inspection

Refer to illustrations 15.5, 15.7a, 15.7b, 15.7c, 15.7d, 15.7e and 15.7f

5   Remove the screws holding the front cover on the oil pump housing (see illustration).

6   Clean all components with solvent, then inspect them for wear and damage. **Caution:** *The oil pressure relief valve and spring are an integral part of the oil pump housing. Removal of the relief valve and spring from the oil pump housing will damage the oil pump and require replacement of the entire oil pump assembly.*

7   Check the clearance of the following oil pump components with a feeler gauge and a micrometer or dial caliper (see illustrations) and compare the measurement to the clear-

ance specifications listed in this Chapter's Specifications.

a) Cover flatness
b) Outer rotor diameter and thickness
c) Inner rotor thickness
d) Outer rotor-to-body clearance
e) Inner rotor-to-outer rotor tip clearance
f) Cover-to-inner rotor side clearance
g) Cover-to-outer rotor side clearance

If any clearance is excessive, replace the entire oil pump assembly.

8   Fill the oil pump rotor cavities with clean engine oil to prime it. Assemble the oil pump and tighten all fasteners to the torque listed in this Chapter's Specifications.

Installation

9   To install the pump, turn the flats in the rotor so they align with the flats on the crankshaft and push the oil pump back into position against the block.

10   Position the primary timing chain tensioner over the oil pump and install the pump-to-block bolts. Tighten the oil pump/primary timing chain tensioner bolts to the torque listed in this Chapter's Specifications.

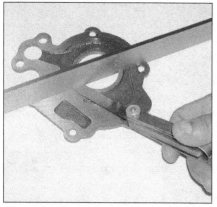

15.7a Place a straightedge across the oil pump cover and check it for warpage with a feeler gauge

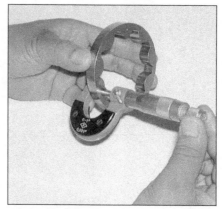

15.7b Use a micrometer or dial caliper to check the thickness and the diameter of the outer rotor

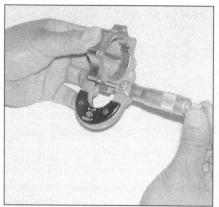

15.7c Use a micrometer or dial caliper to check the thickness of the inner rotor

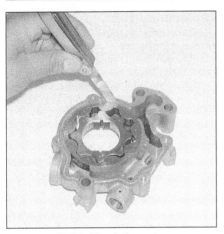

15.7d Check the outer rotor-to-housing clearance

15.7e Check the clearance between the tips of the inner and outer rotors

15.7f Using a straightedge and feeler gauge, check the side clearance between the surface of the oil pump and the inner and outer rotors

11   The remainder of installation is the reverse of removal.

16 Driveplate - removal and installation

1   Raise the vehicle and support it securely on jackstands, then refer to Chapter 7B and remove the transmission. **Warning:** *The engine must be supported from above with an engine hoist or three-bar support fixture before working underneath the vehicle with the transmission removed.*
2   Now would be a good time to check and replace the transmission front pump seal.
3   Use paint or a center-punch to make alignment marks on the driveplate and crankshaft to ensure correct alignment during reinstallation.
4   Remove the bolts that secure the driveplate to the crankshaft. If the crankshaft turns, jam a large screwdriver or prybar through the

driveplate to keep the crankshaft from turning, then remove the mounting bolts.
5   Pull straight back on the driveplate to detach it from the crankshaft.
6   Installation is the reverse of removal. Be sure to align the matching paint marks. Use thread locking compound on the bolt threads and tighten them in several steps, in a criss-cross pattern, to the torque listed in this Chapter's Specifications.

17 Rear main oil seal - replacement

Refer to illustrations 17.2 and 17.5

1   These models use a one-piece rear main seal that is sandwiched between the engine block and the lower main bearing cap assembly, or "bed plate" as it's often referred to. Replacing this seal requires removal of the transmission (see Chapter 7) and driveplate (see Section 16).
2   The seal can be removed by prying it out

of the engine block with a screwdriver, being careful not to nick the crankshaft surface **(see illustration)**. Wrap the screwdriver tip with tape to avoid damage. Be sure to note how far it's recessed into the housing bore before removal so the new seal can be installed to the same depth.
3   Thoroughly clean the seal bore in the block with a shop towel. Remove all traces of oil and dirt.
4   Lubricate the seal lip with clean engine oil and install the seal over the end of the crankshaft. Make sure the lip of the seal points toward the engine.
5   Preferably, a seal installation tool (available at most auto parts stores) should be used to press the new seal back into place. If the proper seal installation tool is unavailable, use a large socket, section of pipe or a blunt tool and carefully drive the new seal squarely into the seal bore and flush with the rear of the engine block **(see illustration)**.
6   The remainder of installation is the reverse of the removal procedure.

17.2 Pry the seal out very carefully with a seal removal tool or screwdriver - if the crankshaft is damaged, the new seal will leak!

17.5 The rear oil seal can be pressed into place with a seal installation tool, a section of pipe or a blunt object shown here - in any case be sure the seal is installed squarely into the seal bore and flush with the rear of the engine block

18   Engine mounts - check and replacement

Refer to illustration 18.12

1   There are three powertrain mounts on the vehicle covered by this manual; left and right engine mounts attached to the engine block and to the frame and a rear mount attached to the transmission and the frame. The rear transmission mount is covered in Chapter 7B. Engine mounts seldom require attention, but broken or deteriorated mounts should be replaced immediately or the added strain placed on the driveline components may cause damage or wear.

Check

2   During the check, the engine must be raised slightly to remove the weight from the mounts.

3   Raise the vehicle and support it securely on jackstands, then position a jack under the engine oil pan. Place a large wood block between the jack head and the oil pan, then carefully raise the engine just enough to take the weight off the mounts. **Warning:** *DO NOT place any part of your body under the engine when it's supported only by a jack!*

4   Check for relative movement between the inner and outer portions of the mount (use a large screwdriver or prybar to attempt to move the mounts). If movement is noted, lower the engine and tighten the mount fasteners.

5   Check the mounts to see if the rubber is cracked, hardened or separated from the metal casing which would indicate a need for replacement.

6   Rubber preservative should be applied to the mounts to slow deterioration.

Replacement

7   Disconnect the cable from the negative terminal of the battery.

8   Remove the engine cooling fan and shroud (see Chapter 3).

9   Raise the front of the vehicle and support it securely on jackstands. **Caution:** *Refer to Chapter 3 and remove the cooling fan and fan clutch before raising the engine, or the fan assembly could be damaged.*

10   Support the engine with a lifting device from above. **Caution:** *Do not connect the lifting device to the intake manifold.* Raise the engine just enough to take the weight off the engine mounts. If you're removing the driver's side engine mount, removal of the oil filter will be necessary.

11   Working below the vehicle, remove the engine mount-to-frame support bracket through-bolt.

12   Remove the mount-to-engine block bolts, then raise the engine until the mounts can be maneuvered past the frame bracket and removed from the vehicle **(see illustration)**.

13   To remove the engine mount support brackets, simply remove the bolts from the frame and remove the brackets from the engine compartment.

14   After the engine mounts have been installed onto the engine, lower the engine while guiding the engine mount and through-bolt into the frame support bracket. Install the through-bolt and tighten it to the torque listed in this Chapter's Specifications.

15   The remainder of installation is the reverse of removal. Remove the engine hoist and the jackstands and lower the vehicle.

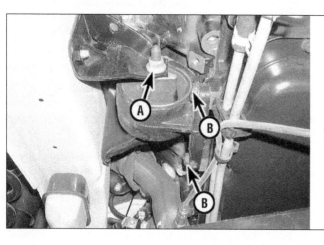

18.12 The engine mounts are secured by the through-bolt (A) and the mount-to-engine block bolts (B) - the two upper mount to block bolts are not visible in this photo

Notes

Notes

Chapter 2 Part C
4.0L V6 engine

Contents

Specifications

General

Displacement	244 cubic inches
Bore	3.780 inches
Stroke	3.583 inches
Compression ratio	10.2:1
Cylinder numbers (drivebelt end-to-transmission end)	
Right bank	1-3-5
Left bank	2-4-6
Firing order	1-2-3-4-5-6
Oil pressure	
At idle speed	5 psi (minimum)
At 3,000 rpm	45 to 105 psi

Camshaft

Bore diameter	1.6944 to 1.6953 inches
Bearing journal diameter	1.6905 to 1.6913 inches
Bearing clearance	
Standard	0.003 to 0.0047 inch
Maximum	0.0059 inch
Endplay	0.002 to 0.02 inch

Crankshaft main bearing

Main journal diameter	2.716 to 2.717 inches
Clearance	0.0013 to 0.0024 inch
Bearing clearance wear limit	0.0027 inch
Endplay	
Standard	0.002 to 0.010 inch
Maximum	0.013 inch

Cylinder head

Gasket thickness (compressed) .... 0.059 inch
Valve seat width
    Intake.. 0.031 to 0.067 inch
    Exhaust .... 0.05 to 0.067 inch
Valve seat runout (maximum)... 0.002 inch

Valves

Stem-to-guide clearance
    Intake
        Standard .... 0.0009 to 0.0026 inch
        Maximum .... 0.011 inch
    Exhaust
        Standard .... 0.002 to 0.0037 inch
        Maximum .... 0.0146 inch

Oil pump

Cover warpage limit.. 0.001 inch
Inner and outer rotor thickness
    Minimum.. 0.563 inch
    Maximum... 0.564 inch
Rotor-to-pump cover clearance .... 0.003 inch
Outer rotor-to-housing clearance... 0.015 inch
Inner rotor-to-outer rotor lobe clearance... 0.008 inch

Torque specifications   Ft-lbs (unless otherwise indicated)

Note: *One foot-pound (ft-lb) of torque is equivalent to 12 inch-pounds (in-lbs) of torque. Torque values below approximately 15 ft-lbs are expressed in inch-pounds, since most foot-pound torque wrenches are not accurate at these smaller values.*

Camshaft sprocket bolt
    Step 1 75
    Step 2 .... Tighten an additional 90-degrees
Crankshaft balancer bolt.. 70
Cylinder head bolts* (in sequence - see illustration 13.27)
    Step 1 45
    Step 2 65
    Step 3 65
    Step 4 .... Tighten an additional 90-degrees
Drivebelt idler sprocket bolt 21
Driveplate-to-crankshaft bolts... 70
Exhaust manifold-to-cylinder head bolts... 21
Exhaust manifold heat shield nut... 105 in-lbs
Exhaust crossover bolts 21
Intake manifold (upper) retaining bolts** .... 105 in-lbs
Intake manifold (lower)-to-cylinder head bolts.................................... 21
Oil cooler connector bolt.. 40
Oil pan drain plug 20
Oil pan bolts... 21
Oil pump pick-up tube mounting bolts 21
Oil pump cover (plate) screws... 105 in-lbs
Oil pump-to-engine block fasteners... 21
Rear main oil seal retainer bolts .... 105 in-lbs
Rocker arm shaft bolts (in sequence - see illustration 11.15)................. 23
Timing belt cover bolts
    M6 bolts.. 105 in-lbs
    M8 bolts 21
    M10 bolts 40
Valve cover-to-cylinder head bolts .... 90 in-lbs
Water pump bolts .... See Chapter 3

*Use new bolts
**Apply a non-hardening thread-locking compound to the bolt threads before installation.

1 General information

Chapter 2C is devoted to in-vehicle repair procedures for the 4.0L V6 engine.

The 4.0 liter engine also uses a "V"-shape design with 60-degrees between the two banks, a timing belt, an aluminum cylinder block with "cast-in-place" iron cylinder sleeves or liners, single overhead camshafts with hydraulic lash adjusters and four valves per cylinder.
Caution: *This engine is of an interference design and severe engine damage will occur if the timing belt breaks.*

The cylinders are numbered from front to rear on both engines. The right bank is numbered 1, 3, 5 and the left bank is numbered 2, 4, 6. The firing order is 1–2–3–4–5–6.

Information concerning engine removal and installation can be found in Chapter 2D. The following repair procedures are based on the assumption that the engine is installed in the vehicle. If the engine has been removed from the vehicle and mounted on a stand, many of the steps outlined in Chapter 2C do not apply.

2 Repair operations possible with the engine in the vehicle

Many major repair operations can be done without removing the engine from the vehicle.

Clean the engine compartment and the exterior of the engine with degreaser before any work is done. It'll make the job easier and help keep dirt out of internal parts of the engine.

It may be helpful to remove the hood to improve engine access when repairs are performed (see Chapter 11). Cover the fenders to prevent damage to the paint. Special pads are available, but an old bedspread or blanket will also work.

If vacuum, exhaust, oil, or coolant leaks develop, indicating a need for gasket or seal replacement, the repairs can generally be done with the engine in the vehicle. The intake and exhaust manifold gaskets, oil pan gasket, crankshaft oil seals, and cylinder head gaskets are all accessible with the engine in the vehicle.

Exterior engine components, such as the intake and exhaust manifolds, the oil pan, the oil pump, the water pump, the starter motor, the alternator, and fuel system components can be removed for repair with the engine in the vehicle.

Cylinder heads can be removed without pulling the engine. Valve component servicing can also be done with the engine in the vehicle. Replacement of the timing belt and sprockets is also possible with the engine in the vehicle, as is camshaft and valve train removal and installation.

Repair or replacement of piston rings, pistons, connecting rods, and rod bearings

is possible with the engine in the vehicle, however, this practice is not recommended because of the cleaning and preparation work that must be done to the components.

3 Top Dead Center (TDC) for number one piston - locating

1 Top Dead Center (TDC) is the highest point in the cylinder that each piston reaches as it travels up the cylinder bore. Each piston reaches TDC on the compression stroke and again on the exhaust stroke, but TDC generally refers to piston position on the compression stroke.
2 Positioning the piston(s) at TDC is an essential part of certain procedures such as camshaft and timing belt/sprocket removal.
3 Before beginning this procedure, be sure to place the transaxle in Neutral and apply the parking brake or block the rear wheels. Disconnect the cable from the negative terminal of the battery (see Chapter 5). Remove the ignition coils (see Chapter 5) and the spark plugs (see Chapter 1).
4 Install a compression pressure gauge in the number one spark plug hole (refer to Chapter 2D). It should be a gauge with a screw-in fitting and a hose at least six inches long.
5 Rotate the crankshaft using a socket and breaker bar on the crankshaft pulley bolt while observing for pressure on the compression gauge. The moment the gauge shows pressure, indicates that the number one cylinder has begun the compression stroke.
6 Once the compression stroke has begun, TDC for the compression stroke is reached by bringing the piston to the top of the cylinder.
7 This engine is not equipped with external components (crankshaft pulley, flywheel, timing hole, etc.) that are marked to identify the position of number 1 TDC. Therefore, the only method to double-check the location of TDC number 1 is to remove the valve cover (see Section 4) and note the rocker arm position, or use a degree wheel and a positive stop timing device threaded into the spark plug hole for cylinder number 1.
8 After the number one piston has been positioned at TDC on the compression stroke, TDC for any of the remaining cylinders can be located by turning the crankshaft 120-degrees and following the firing order (refer to the Specifications). For example, rotating the engine 120-degrees past TDC number 1 will put the engine at TDC compression for cylinder number 2.

4 Valve covers - removal and installation

Removal

1 Disconnect the cable from the negative terminal of the battery (see Chapter 5).
2 Remove the engine cover.

3 Remove the upper intake manifold (see Section 5).
Note: *Cover open ports on the intake to prevent debris from entering the engine.*
4 Remove the ignition coils from the valve cover side you are working on (see Chapter 5).
5 Remove the upper intake manifold brackets from both sides of the engine, and move the engine wiring harness to the side.
6 Loosen the valve cover bolts and remove both valve covers.

Installation

7 The mating surfaces of each cylinder head and valve cover must be perfectly clean when the covers are installed. Use a gasket scraper to remove all traces of sealant and old gasket material, then clean the mating surfaces with brake system cleaner. If there's sealant or oil on the mating surfaces when the cover is installed, oil leaks may develop.
8 Inspect spark plug tube seals; if damaged, carefully remove the seals using an appropriate pry tool. Position the new seal with the part number facing the valve cover, then use a socket that contacts the outer edge to drive the seal in place.
9 Install the valve cover and bolts, then tighten the bolts to the torque listed in this Chapter's specifications.
10 The remainder of installation is the reverse of removal.

5 Intake manifolds - removal and installation

Warning: *Wait until the engine is completely cool before beginning this procedure.*

Removal

1 If you will be removing the lower intake manifold, relieve the fuel system pressure (see Chapter 4).
2 Disconnect the cable from the negative terminal of the battery (see Chapter 5).
3 If equipped, remove the engine cover.
4 If you will be removing the lower intake manifold, drain the cooling system (see Chapter 1).

Upper intake manifold

5 Disconnect the electrical connectors to the electronic throttle body (ETC), IAT sensor and MAP sensor.
6 Loosen the clamp and disconnect the air hose from the throttle body, then remove the air filter housing (see Chapter 4).
7 Disconnect the PCV valve hose, EVAP purge solenoid and the power brake booster hoses.
8 Disconnect and remove the EGR tube (see Chapter 6).
9 Disconnect the wiring harness retainers from the intake support brackets and move the harness out of the way.
10 Remove the fasteners for the intake manifold support brackets at the side of the intake manifold.

11 Remove the upper intake manifold bolts and lift the upper intake manifold off of the lower manifold.

12 Cover the exposed intake ports to prevent any debris from entering the engine.

Lower intake manifold

13 Disconnect the upper radiator hose from the thermostat housing.

14 Remove the upper intake manifold (see Steps 5 through 12).

15 Disconnect the fuel injectors electrical connectors (see Chapter 4).

16 Disconnect the coolant temperature sensor and knock sensor electrical connectors (see Chapter 3 and Chapter 6).

17 Disconnect the heater hose from the intake manifold and the coolant overflow hose at the thermostat housing.

18 Disconnect the fuel line, then remove the fuel rail and injectors as an assembly (see Chapter 4).

19 Remove the lower intake manifold bolts, and remove the lower intake manifold.

Installation

Lower intake manifold

Refer to illustration 5.24

Note: *The mating surfaces of the cylinder heads, cylinder block, and the intake manifold must be perfectly clean when the lower intake manifold is installed. Gasket removal solvents are available at most auto parts stores and may be helpful when removing old gasket material that's stuck to the cylinder heads, cylinder block and lower intake manifold (the lower intake manifold is made of aluminum - aggressive scraping can cause damage). Be sure to follow the instructions printed on the solvent container.*

20 Use a gasket scraper to remove all traces of sealant and old gasket material, then clean the mating surfaces with lacquer thinner or acetone. If there's old sealant or oil on the mating surfaces when the lower intake manifold is installed, oil or vacuum leaks may develop. Use a vacuum cleaner to remove gasket material that falls into the intake ports or the valley between the cylinder banks.

21 Apply a 1/4-inch bead of RTV sealant (or equivalent) to the cylinder head-to-engine block junctions.

22 Position the lower intake gaskets onto the cylinder head surface.

23 Carefully lower the lower intake manifold into place and install the mounting bolts finger-tight.

24 Tighten the mounting bolts in steps, following the tightening sequence **(see illustration)**, to the torque listed in this Chapter's Specifications.

25 Reconnect the hoses and harnesses in the reverse of removal.

26 Install the upper intake manifold; and the remaining steps are the reverse of the removal procedure.

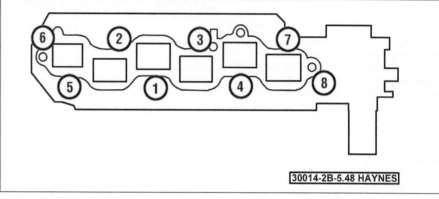

5.24 Lower intake manifold tightening sequence

Upper intake manifold

27 Install a new upper manifold gasket to the intake manifold.

28 Install the upper intake manifold onto the lower intake manifold and hand tighten the bolts.

29 Tighten the mounting bolts in several steps, starting in the center and working in a circular pattern to the torque listed in this Chapter's Specifications.

30 Installation of the remaining components is the reverse of removal.

31 Refill the cooling system, if drained (see Chapter 1), start the engine and check for leaks and proper operation.

6   Exhaust manifolds - removal and installation

Removal

1 Disconnect the cable from the negative terminal of the battery (see Chapter 5).

2 Raise the vehicle, support it securely on jackstands, and remove the engine splash shield from the passenger's side.

3 Disconnect the upstream and downstream oxygen sensor connectors.

Note: *If the manifold is being replaced, remove both oxygen sensors (see Chapter 6).*

4 Unbolt the exhaust pipe from the exhaust manifold.

5 Unhook the exhaust system hangers and lower the pipe assembly.

6 Remove the bolts and remove the upper heat shield.

7 Remove the bolts attaching the exhaust manifold to the cylinder head, and remove the exhaust manifold.

Installation

8 Clean the mating surfaces to remove all traces of old gasket material, then inspect the exhaust manifolds for distortion and cracks. Check for warpage with a precision straightedge held against the mating surface. If a feeler gauge thicker than 0.030-inch can be inserted between the straightedge and the mating surface, take the exhaust manifold(s) to an automotive machine shop for resurfacing.

9 Place the exhaust manifold in position with a new gasket and install the mounting bolts finger tight.

Note: *Be sure to identify the exhaust manifold gasket by the correct cylinder designation and the position of the exhaust ports on the gasket.*

10 Starting in the middle and working out toward the ends, tighten the bolts to the torque listed in this Chapter's Specifications.

11 Installation of the remaining components is the reverse of removal.

12 Start the engine and check for exhaust leaks between the exhaust manifolds and the cylinder heads and between the exhaust manifolds, crossover pipe and catalytic converter.

7   Crankshaft pulley - removal and installation

Removal

1 Disconnect the cable from the negative terminal of the battery (see Chapter 5).

2 Remove the air filter housing (see Chapter 4).

3 Remove the coolant recovery bottle/washer fluid reservoir (see Chapter 3).

4 Remove the cooling fan shroud fasteners and disconnect the electrical connector to the electric fan then remove the shroud (see Chapter 3).

5 Remove the fan/viscous fan drive assembly (see Chapter 3).

6 Remove the drivebelt and power steering drivebelt (see Chapter 1).

7 Remove the crankshaft pulley-to-crankshaft bolt.

8 Insert adapter #9020 or equivalent to the end of crankshaft. Install a three-jaw puller to the inner hub, pressing against the adapter on the end of the crankshaft to pull the crankshaft balancer off the crankshaft.

Caution: *Do not attach the puller to the outer edge of the pulley or damage to the pulley may result.*

Caution: *Because the pulley is recessed, adapter #9020 may be needed between the puller bolt and the crankshaft (to prevent damage to the bore and threads in the end of the crankshaft).*

Installation

9 Install the crankshaft pulley with a special installation tool that threads to the crankshaft in place of the crankshaft pulley bolt (available at most automotive parts stores). Be sure to apply clean engine oil or multi-purpose grease to the seal contact surface of the damper hub (if it isn't lubricated, the seal lip could be damaged and oil leakage would result). If the tool isn't available, the crankshaft pulley bolt and several washers used as spacers may be used, as long as the crankshaft pulley bolt torque is not exceeded.

10 Remove the tool, install the crankshaft pulley bolt and tighten it to the torque listed in this Chapter's Specifications.

11 Installation of the remaining components is the reverse of removal.

8 Timing belt - removal, installation and adjustment

Warning: *Wait until the engine is completely cool before beginning this procedure.*
Caution: *This is **not** a freewheeling engine; if the belt is broken or improperly installed, the engine will be severely damaged.*
Caution: *Do not rotate the crankshaft or camshafts separately during this procedure (with the timing belt removed), as damage to the valves may occur.*
Note: *Several special tools are required to complete these procedures, so read through the entire Section and obtain the special tools before beginning work.*

Removal

Refer to illustrations 8.14 and 8.18

1 Disconnect the cable from the negative terminal of the battery (see Chapter 5).

2 Remove the air filter housing (see Chapter 4).

3 Remove the coolant recovery bottle/washer fluid reservoir (see Chapter 3).

4 Remove the cooling fan shroud fasteners and disconnect the electrical connector to the electric fan then remove the shroud (see Chapter 3).

5 Remove the fan/viscous van drive assembly (see Chapter 3).

6 Position the number one piston at TDC on the compression stroke (see Section 3).

7 Remove the drivebelts and drivebelt tensioner (see Chapter 1).

8 Remove the air conditioning compressor mounting nuts/bolts and secure the compressor out of the way using wire.

9 Remove the accessory drive bracket fasteners and remove the bracket with idler pulleys attached, from the front of the engine.

10 Remove the bolts from the power steering pump and set the pump and bracket aside (see Chapter 10).

11 Remove the crankshaft pulley (see Section 7).

12 Remove the front timing belt cover fasteners and remove the cover. Note the various type and sizes of bolts by making a dia-

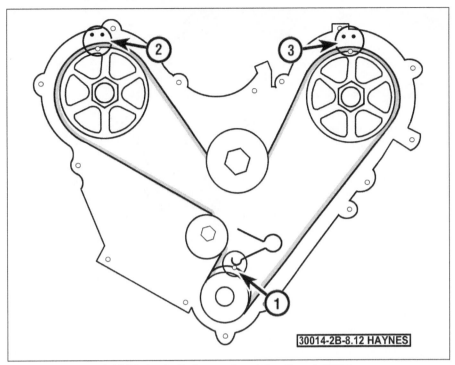

8.14 Timing belt alignment mark locations (at TDC):

1	Crankshaft pointer aligns with the TDC mark on the oil pump	2	Right crankshaft timing mark aligns between the marks on the rear cover	3	Left camshaft timing mark aligns between the marks on the rear cover

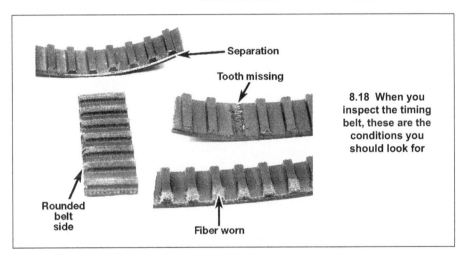

8.18 When you inspect the timing belt, these are the conditions you should look for

gram or taking careful notes while the timing belt cover is being removed. The bolts must be reinstalled in their original locations.

13 Raise the vehicle and support it securely on jackstands then remove the oil cooler hose mounting bolt and secure the hose out of the way.

14 Make sure that the number one piston is still at TDC on the compression stroke by verifying that the timing marks on all three timing belt sprockets are aligned with their respective alignment marks **(see illustration)**.

15 Remove the hydraulic tensioner mounting bolts and tensioner from the side of the rear cover.

16 Loosen the bolt on the timing belt tensioner pulley and push the pulley away from the belt, then retighten the bolt.

17 Check to see that the timing belt is marked with an arrow to show which side faces out. If there isn't a mark, paint one on (only if the same belt will be reinstalled). Slide the timing belt off the sprockets and check the condition of the tensioner.

18 Inspect the timing belt **(see illustration)**. Look at the backside (the side without the teeth): If it's cracked or peeling, or it's hard, glossy and inflexible, and leaves no indentation when pressed with your fingernail, replace the belt. Look at the drive side: If teeth are missing, cracked or excessively worn, replace the belt.

9.3 Use a hook tool and pry the seal from the oil pump

9.4 Another way of removing an old oil seal is to screw a self-tapping screw partially into the seal, then use pliers as a lever to pull it from the engine

9.5 Drive the seal squarely into the cover using a socket and hammer

Installation

Caution: *Before starting the engine, carefully rotate the crankshaft by hand through at least two full revolutions (use a socket and breaker bar on the crankshaft pulley center bolt). If you feel any resistance, STOP! There is something wrong - most likely, valves are contacting the pistons. You must find the problem before proceeding. Check your work and see if any updated repair information is available.*

19   Place the hydraulic timing belt tensioner in a bench vise and slowly compress the tensioner until a 3 mm Allen wrench can be inserted through the housing to prevent the plunger from expanding.

Note: *This should take about 5 minutes to bleed down.*

20   Install the belt on the crankshaft sprocket first, and keep the belt tight on the tension side.

21   Install the belt on the left camshaft sprocket, the water pump pulley, and the right camshaft sprocket and timing belt tensioner. Be careful not to nudge the camshaft sprocket(s) or crankshaft gear off the timing marks. Install the timing belt with the arrow pointing away from the engine.

22   Align the factory-made white lines on the timing belt with the punch mark on each of the camshaft sprockets and the crankshaft sprocket. Make sure all three sets of timing marks are properly aligned **(see illustration 8.13)**.

Adjustment

23   Insert the tensioner into the side of the rear cover and tighten the bolts to the torque listed in this Chapter's Specifications.

24   Once the belt is in place, press the tensioner pulley toward the belt, remove the retaining pin from the tensioner and allow the plunger to extend against the pulley bracket - the tensioner will automatically apply the proper amount of tension to the belt.

25   Slowly turn the crankshaft clockwise two full revolutions, returning the number one piston to TDC on the compression stroke.

Caution: *If excessive resistance is felt while turning the crankshaft, it's an indication that the pistons are coming into contact with the valves. Go back over the procedure to correct the situation before proceeding.*

26   Make sure all the timing marks are still aligned properly **(see illustration 8.13)**. Tighten the tensioner bolt to the torque listed in this Chapter's Specifications while keeping the tensioner steady with your hand.

27   Check the deflection of the timing belt by observing the force the tensioner pulley applies to the timing belt. If the belt seems loose, replace the tensioner spring.

28   Installation of the remaining components is the reverse of removal. Refer to the appropriate Sections.

9   Crankshaft front oil seal - replacement

Refer to illustrations 9.3, 9.4 and 9.5

1   Remove the crankshaft balancer (see Section 7).

2   Remove the timing belt and crankshaft sprocket (see Section 8).

3   Use a screwdriver or hook tool to carefully pry out the seal **(see illustration)**.

Note: *Be careful not to damage the oil pump cover bore where the seal is seated or the nose and sealing surface of the crankshaft.*

4   Another procedure for removing the seal is to drill a small hole on each side of the seal and place a self-tapping screw in each hole **(see illustration)**. Use these screws as a means of pulling the seal out without having to pry on it.

5   Apply clean engine oil or multi-purpose grease to the outer edge of the new seal, then install it in the cover with the lip (spring side) facing IN. Drive the seal into place with a large socket and a hammer **(see illustration)**. Make sure the seal enters the bore squarely and stop when the front face is at the proper depth.

Note: *If a large socket isn't available, a piece of pipe will also work.*

6   Check the surface on the balancer hub that the oil seal rides on. If the surface has been grooved from long-time contact with the seal, the balancer will need to be replaced.

7   Lubricate the balancer hub with clean engine oil and install the crankshaft balancer (see Section 7).

8   The remainder of installation is the reverse of the removal.

10   Camshaft oil seal - replacement

Refer to illustrations 10.15, 10.19a, 10.19b and 10.19c

Warning: *Wait until the engine is completely cool before beginning this procedure.*

1   Disconnect the cable from the negative terminal of the battery (see Chapter 5).

2   Remove the air filter housing (see Chapter 4).

3   Remove the coolant recovery bottle/washer fluid reservoir (see Chapter 3).

4   Remove the cooling fan shroud fasteners and disconnect the electrical connector to the electric fan, then remove the shroud (see Chapter 3).

5   Remove the fan/viscous fan drive assembly (see Chapter 3).

6   Remove the upper intake manifold (see Section 5).

7   Remove the drivebelts and drivebelt tensioner (see Chapter 1).

8   Remove the alternator (see Chapter 5).

9   Remove the air conditioning compressor mounting nuts/bolts and secure the compressor out of the way using wire. **Warning:** *Do not disconnect the refrigerant lines.*

10   Rotate the engine to TDC (see Section 3) and remove the timing belt (see Section 8).

11   Remove the valve covers (see Section 4).

12   Remove the rocker arm assembly (see Section 12).

13   Place a wrench onto the flats on the timing belt sprocket, then use a ratchet and

socket to remove the camshaft sprocket bolt and sprocket.

Note: *Don't mix up the camshaft sprockets. They must be installed on the same camshaft they were removed from.*

14 If you're working on the right cylinder head, remove the EGR valve (see Chapter 6).

15 Remove the rear timing cover fasteners and timing cover **(see illustration)**. Discard the O-ring seals. Note the various type and sizes of bolts by making a diagram or taking careful notes while the timing belt cover is being removed. The bolts must be reinstalled in their original locations and the O-ring seals replaced with new ones.

16 Remove the camshaft thrust plate fasteners and remove the thrust plate(s) from the rear of the cylinder head.

17 Carefully slide the camshaft out of the cylinder (towards the rear) about 3-1/2 inches.

18 Using a small screwdriver or punch, drive the seal out of the cylinder head towards the rear.

19 There are several ways to install the new seal. Fabricate a seal installation tool **(see illustrations)** or use a very large socket with an inside diameter large enough to clear the nose of the camshaft and carefully drive the seal into place **(see illustration)**. Remove the sprocket positioning pin from the nose of the cam, if necessary, to prevent damaging the pin.

20 Install the pulley onto the camshaft and hold the pulley with a large wrench, tightening the bolt to the torque in this Chapter's Specifications.

21 Installation of the remaining components is the reverse of removal.

---

11 Rocker arms and hydraulic valve lash adjusters - removal, inspection and installation

1 A noisy lash adjuster can be isolated when the engine is idling. Hold a mechanic's

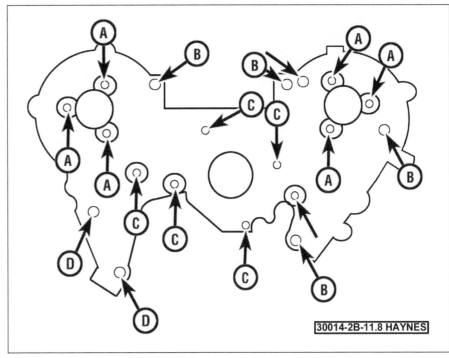

10.15 Rear timing belt cover fastener locations; the bolts letter coded for are type and size

A	M8 bolts (apply thread sealant)	B	M10 bolts	D	M10 stud/nut
		C	M6 bolts		

stethoscope or a length of hose near each valve while listening at the other end.

2 The most likely causes of noisy valve lifters are dirt trapped inside the lifter and lack of oil flow, viscosity, or pressure. Before condemning the lifters, check the oil for fuel contamination, correct level, cleanliness, and correct viscosity.

Note: *To prevent possible damage and keep air from entering the lash adjuster, keep the rocker arm assembly upright, keep the parts in order and do not allow the assembly to rest on the lash adjusters.*

Removal

3 Remove the valve covers (see Section 4) and mark the rocker arms with paint or a marker to aid installation.

4 Remove the rocker arm bolts in the reverse of the tightening sequence **(see illustration 11.15)**.

Disassembly

Refer to illustration 11.6

5 Identify the intake and exhaust rocker arms, as they are different.

10.19a Fabricate a seal installation tool from a piece of pipe and a large washer . . .

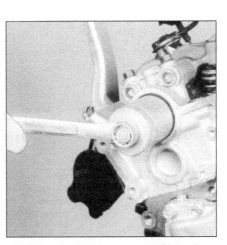

10.19b . . . and press a new seal into place with a section of pipe and a bolt of the proper size and thread pitch (don't let the camshaft turn as the bolt is tightened)

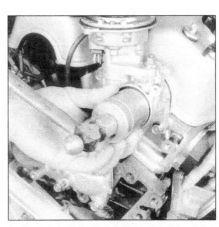

10.19c As a last resort, you can also drive a cam seal into place with a hammer and a large socket, but make sure you don't damage the sprocket positioning pin on the end of the camshaft

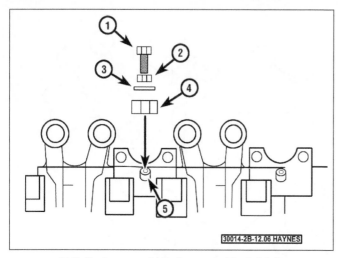

11.6 Rocker arm shaft pin removal tool details

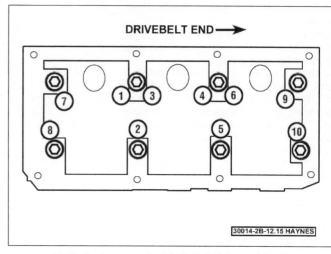

11.15 Rocker arm shaft bolt tightening sequence

1	4 mm bolt	3	Washer	5	Rocker shaft
2	Nut	4	10 mm nut		pin

6 To remove the rocker arm shafts, the dowel pins must be pulled from the shaft supports. Use a 4 mm screw with a nut threaded to the top of the screw, a washer followed by a spacer, and thread this assembly into the pin, then loosen the nut on the screw to pull the pin out **(see illustration)**. Once the pins are removed, they must be replaced.

7 Remove the rocker shaft.

Note: *The identification notches face towards the front of the engine on the front head, and toward the back of the engine on the rear head.*

Inspection

8 Check the rocker arms for scuffing or wear on the rollers and shaft.

9 Check that the swivel pad on the lash adjuster is in place and is not broken.

10 Replace the assembly if any rocker arm shows signs of wear.

Assembly

11 Position the shafts with the notches facing up before installing the rocker arms and pedestals onto the shafts.

12 Press the new dowel pins through the pedestals until they bottom out against the rocker shaft.

Installation

Refer to illustration 11.15

13 Rotate camshaft gears clockwise until the number one cylinder intake valves start to open.

14 Install the rocker arm shaft assembly with the mark facing towards the front of the engine for the front head, and towards the back of the engine for the rear head.

15 Tighten the assembly bolts in sequence **(see illustration)** to the torque listed in this Chapter's Specifications.

16 Install the valve covers (see Section 4).

Bleeding

Refer to illustration 11.20

Note: *Use this procedure to manually bleed air from a lash adjuster.*

17 Start and run the engine until it reaches normal operating temperature.

18 Remove the valve cover (see Section 4).

19 Rotate the engine until the rocker arm in question is on the low side of the camshaft lobe.

20 Insert a small wire through the air bleed hole in the rocker arm while lightly pushing the check ball down **(see illustration)**; there will be one for the intake rockers, and two for exhaust rockers.

Caution: *If the tip of the wire breaks off inside the lash adjuster, the rocker arm will need to be replaced.*

21 While holding the check ball down inside the adjuster, press the rocker arm towards the valve and hold it down for a few seconds.

22 Slowly release the rocker arm first, then remove the wire.

23 Repeat Steps 19 through 22 until all air or play is eliminated. If no difference is made, the rocker arm(s) must be replaced.

24 Installation is the reverse of removal.

12 Camshaft – removal, inspection and installation

Warning: *Wait until the engine is completely cool before beginning this procedure.*

Caution: *The timing system is complex, and severe engine damage will occur if you make any mistakes. Do not attempt this procedure unless you are highly experienced with this type of repair. If you are at all unsure of your abilities, be sure to consult an expert. Double-check all your work and be sure everything is correct before you attempt to start the engine.*

Caution: *Once the valve covers are removed, the magnetic timing wheels are exposed. The magnetic timing wheels on the camshafts must not come in contact with any type of magnet or magnetic field. If contact is made, the timing wheels will need to be replaced.*

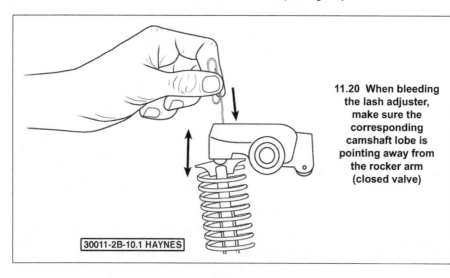

11.20 When bleeding the lash adjuster, make sure the corresponding camshaft lobe is pointing away from the rocker arm (closed valve)

12.15 Use a micrometer to measure cam lobe height

Removal

1 Disconnect the cable from the negative terminal of the battery (see Chapter 5).
2 Remove the drivebelt (see Chapter 1).
3 Drain the engine oil and coolant (see Chapter 1).
4 Remove the air filter housing and resonator (see Chapter 4).
5 Remove the intake and exhaust manifolds (see Sections 5 and 6).
6 Disconnect all wires and vacuum hoses from the cylinder heads. Label them to simplify reinstallation.
7 Disconnect the ignition coils and remove the spark plugs (see Chapter 1). Label the ignition coils to simplify reinstallation.
8 Remove the valve covers (see Section 4).
9 Remove the cylinder heads (see Section 14).
10 Remove the rocker arm shaft assembly (see Section 12).
11 Place a wrench onto the flats on the timing belt sprocket, then use a ratchet and socket to remove the camshaft sprocket bolt and slide the sprocket from the end of the camshaft.
Note: *Don't mix up the camshaft sprockets. They must be installed on the same camshaft from which they were removed.*
12 Remove the camshaft thrust plate fasteners and remove the thrust plate from the rear of the cylinder head.
13 Carefully slide the camshaft out of the cylinder head.

Inspection

Refer to illustration 12.15

14 Check the camshaft bearing surfaces for pitting, score marks, galling, and abnormal wear. If the bearing surfaces are damaged, the cylinder head will have to be replaced.
15 Compare the camshaft lobe height by measuring each lobe with a micrometer **(see illustration)**. Measure each of the intake

lobes and record the measurements and relative positions. Then measure each of the exhaust lobes and record the measurements and relative positions also. This will let you compare all of the intake lobes to one another and all of the exhaust lobes to one another. If the difference between the lobes exceeds 0.005 inch, the camshaft should be replaced. Do not compare intake lobe heights to exhaust lobe heights as lobe lift may be different. Only compare intake lobes to intake lobes and exhaust lobes to exhaust lobes for this comparison.
16 Check the rocker arms and shafts for abnormal wear, pits, galling, score marks, and rough spots. Don't attempt to restore rocker arms by grinding the pad surfaces. Replace defective parts.

Installation

Caution: *Before starting the engine, carefully rotate the crankshaft by hand through at least two full revolutions (use a socket and breaker bar on the crankshaft pulley center bolt). If you feel any resistance, STOP! There is something wrong - most likely, valves are contacting the pistons. You must find the problem before proceeding.*
17 Lubricate the camshaft bearing journals and lobes with moly-base grease or engine assembly lube, then install them carefully in the cylinder head. Don't scratch the bearing surfaces with the cam lobes!
18 Install a new thrust plate gasket and the thrust plate. Install the bolts then tighten them to the torque listed in this Chapter's Specifications.
19 Make sure the mark on the crankshaft sprocket is still aligned with its mark on the oil pump. Slide the camshaft sprockets onto the camshafts and align the marks on the sprockets with their corresponding marks on the cylinder heads. Install the bolts and tighten them to the torque listed in this Chapter's Specifications.
20 Install the timing belt (see Section 8).
21 Installation of the remaining components is the reverse of removal.

---

13 Cylinder heads - removal and installation

Warning: *Wait until the engine is completely cool before beginning this procedure.*

Removal

1 Relieve the fuel system pressure (see Chapter 4), then disconnect the cable from the negative terminal of the battery (see Chapter 5).
2 Remove the drivebelt (see Chapter 1).
3 Drain the engine oil and coolant (see Chapter 1).
4 Remove the air filter housing and resonator (see Chapter 4).
5 Remove the intake manifolds (see Section 5).
6 Disconnect all wires and vacuum hoses

from the cylinder heads. Label them to simplify reinstallation.
7 Disconnect the ignition coils and remove the spark plugs (see Chapter 1).
8 Remove the exhaust manifold(s) (see Section 6).
9 Remove the valve cover(s) (see Section 4).
10 Remove the drivebelts, drivebelt tensioner and idler pulley (see Chapter 1).
11 Remove the power steering pump (see Chapter 10).
12 Remove the timing belt covers, rotate the engine to TDC and verify the timing marks are aligned **(see illustration 8.14),** then remove the timing belt (see Section 8).
13 Remove the EGR valve (see Chapter 12).
14 Place a wrench onto the flats on the timing belt sprocket, then use a ratchet and socket to remove the camshaft sprocket bolt and sprocket.
Note: *Don't mix up the camshaft sprockets. They must be installed in their original locations.*
15 Remove the camshaft thrust plate fasteners and remove the thrust plate from the rear of the cylinder head.
16 Carefully slide the camshaft out of the cylinder (towards the rear) about 3-1/2 inches and remove the rear timing cover-to-cylinder head bolts.
17 Loosen the cylinder head bolts in the reverse order of the tightening sequence **(see illustration 13.27)**.
Note: *When removing the rear cylinder head, the front four cylinder head bolts can't be completely removed until after the cylinder head is removed. Loosen the bolts and raise them up, then keep them in that position using rubber bands or mechanic's wire.*
18 Lift the cylinder head off the block.
19 If resistance is felt, dislodge the cylinder head by striking it with a wood block and hammer. If prying is required, pry only on a casting protrusion - be very careful not to damage the cylinder head or block!
20 Have the cylinder head inspected and serviced by a qualified automotive machine shop.

Installation

Refer to illustration 13.27

21 The mating surfaces of each cylinder head and the engine block must be perfectly clean when the cylinder head is installed.
22 Carefully use a gasket scraper to remove all traces of carbon and old gasket material, then clean the mating surfaces with brake system cleaner. If there's oil on the mating surfaces when the cylinder head is installed, the gasket may not seal correctly and leaks may develop.
23 When working on the engine block, it's a good idea to cover the lifter valley with shop rags to keep debris out of the engine. Use a shop rag or vacuum cleaner to remove any debris that falls into the cylinders.
24 Check the engine block and cylinder head mating surfaces for nicks, deep scratches, and other damage. If damage is slight, it can be removed with a file; if it's excessive, machining may be the only alternative.

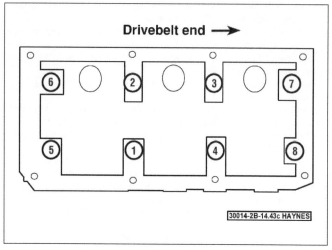

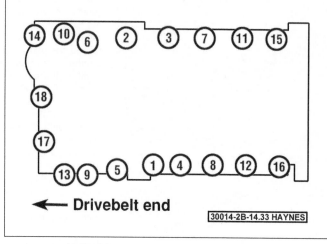

13.27 Cylinder head bolt TIGHTENING sequence

14.15 Oil pan bolt TIGHTENING sequence

25   Position the new gasket over the dowel pins in the engine block. Some gaskets are marked TOP or FRONT to ensure correct installation.
26   Carefully position the cylinder head on the engine block without disturbing the gasket.
27   Install NEW cylinder head bolts and tighten them in the recommended sequence **(see illustration)** to the torque steps listed in this Chapter's Specifications.
Caution: *Do not use a torque wrench for steps requiring additional rotation or turns; apply a paint mark to the bolt head or use a torque-angle gauge (available at most automotive parts stores) and a socket and breaker bar.*
28   Installation of the remaining components is the reverse of removal.
29   Change the engine oil and filter (see Chapter 1).
30   Refill the cooling system (see Chapter 1). Start the engine and check for leaks and proper operation.

14   Oil pan - removal and installation

Removal

1   Disconnect the cable from the negative terminal of the battery (see Chapter 5).
2   Raise the front of the vehicle and support it securely on jackstands. Apply the parking brake and block the rear wheels to keep it from rolling off the stands.
3   Drain the engine oil (see Chapter 1).
4   Remove the lower splash shield fasteners and remove the splash shield.
5   Remove the dipstick tube bracket mounting bolt then using a twisting motion, pull the dipstick tube out of the upper oil pan.
6   Disconnect the exhaust pipe-to-exhaust manifold bolts and lower the exhaust (see Section 6).
7   Remove the stabilizer bar (see Chapter 10).
8   Remove the engine mount through-bolts (see Section 19).

9   Using an engine hoist, raise the engine until the engine fan is almost in contact with the fan shroud.
10   Remove the front axle assembly (see Chapter 8).
11   Remove the oil pan bolts around the perimeter of the pan, carefully separate the oil pan from the engine block and remove the gasket.

Installation

Refer to illustration 14.15
12   Clean the pan(s) with solvent and remove all old sealant and gasket material from the engine block and pan mating surfaces. Clean the mating surfaces with lacquer thinner or acetone and make sure the bolt holes in the engine block are clear. Check the oil pan flange(s) for distortion, particularly around the bolt holes. If necessary, place the pan(s) on a wood block and use a hammer to flatten and restore the gasket surface.
13   Apply a bead of RTV sealant to the line between the engine block, oil pump housing and the rear main oil seal retainer. Install a new gasket on the oil pan flange.
14   Place the oil pan in position on the engine block and install the bolts.
15   Tighten the bolts in sequence **(see illustration)** to the torque listed in this Chapter's Specifications.
16   Installation of the remaining components is the reverse of removal.
17   Refill the engine with oil (see Chapter 1). Start and run the engine until normal operating temperature is reached, then check for leaks.

15   Oil pump - removal, inspection and installation

Removal

1   Disconnect the cable from the negative terminal of the battery (see Chapter 5).

2   Raise the front of the vehicle and support it securely on jackstands. Apply the parking brake and block the rear wheels to keep it from rolling off the stands.
3   Drain the engine oil (see Chapter 1).
4   Remove the under-vehicle splash shield.
5   Drain the cooling system (see Chapter 1).
6   Remove the oil pan (see Section 14).
7   Remove the timing belt and crankshaft sprocket (see Section 8).
8   Remove the pick-up tube, then remove the oil pump mounting bolts and the oil pump. Discard the pick-up tube O-ring.

Inspection

Refer to illustrations 15.10, 15.11, 15.13, 15.14 and 15.15
9   Clean all parts thoroughly in solvent and carefully inspect the rotors, pump cover, and housing for nicks, scratches, or burrs. Replace the assembly if it is damaged.
10   Use a straightedge and a feeler gauge to measure the oil pump cover for warpage **(see illustration)**. If it's warped more than the limit listed in this Chapter's Specifications, the pump should be replaced.
11   Measure the thickness of the outer rotor **(see illustration)**. If the thickness is less than the value listed in this Chapter's Specifications, the pump should be replaced.
12   Measure the thickness of the inner rotor. If the thickness is less than the value listed in this Chapter's Specifications, the pump should be replaced.
13   Insert the outer rotor into the oil pump housing and measure the clearance between the rotor and housing **(see illustration)**. If the measurement is more than the maximum allowable clearance listed in this Chapter's Specifications, the pump should be replaced.
14   Install the inner rotor in the oil pump assembly and measure the clearance between the lobes on the inner and outer rotors **(see illustration)**. If the clearance is more than the value listed in this Chap-

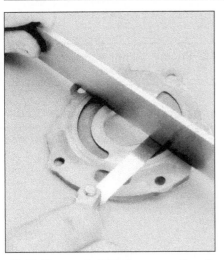

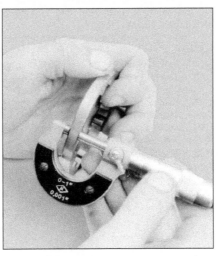

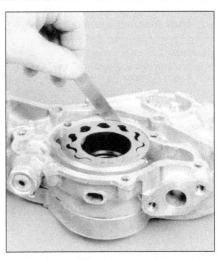

15.10 Place a straightedge across the oil pump cover and check it for warpage with a feeler gauge

15.11 Use a micrometer to measure the thickness of the outer rotor

15.13 Check the outer rotor-to-housing clearance with a feeler gauge

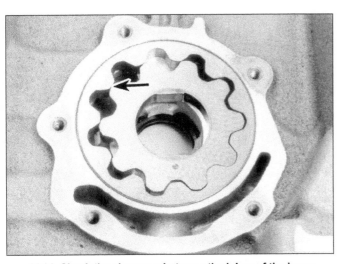

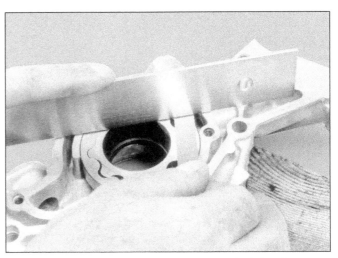

15.14 Check the clearance between the lobes of the inner and outer rotors

15.15 Using a straightedge and feeler gauge, check the clearance between the surface of the oil pump cover and the rotors

ter's Specifications, the pump should be replaced.

15   Place a straightedge across the face of the oil pump assembly **(see illustration)**. If the clearance between the pump surface and the rotors is greater than the limit listed in this Chapter's Specifications, the pump should be replaced.

Installation

16   Install the pump cover then coat the cover bolts with thread sealer and tighten the bolts to the torque listed in this Chapter's Specifications.

Note: *Only apply thread sealer to the first 1/4-inch of cover bolt threads.*

17   Install the pump and tighten the bolts to the torque listed in this Chapter's Specifications.

18   Place a new O-ring seal onto the pick-up tube and install the tube.

19   Install the timing belt and crankshaft sprocket (see Section 8).

20   Installation of the remaining components is the reverse of removal.

21   Refill the engine with oil and change the oil filter (see Chapter 1).

16 Oil cooler - removal and installation

Warning: *Wait until the engine is completely cool before beginning this procedure.*

Removal

1   Disconnect the cable from the negative terminal of the battery (see Chapter 5).

2   Drain the coolant (see Chapter 1).

3   Raise the vehicle and support it securely on jackstands.

4   Drain the engine oil and remove the oil filter (see Chapter 1).

5   Disconnect the coolant hoses from the inlet and outlet ports.

6   Remove the oil cooler connector bolt from the center of the cooler and remove the oil cooler. Discard the oil cooler O-ring.

Installation

7   Lubricate the oil cooler connector bolt and O-ring with clean engine oil.

8   Position the flat side of the oil cooler parallel to the adapter and install the oil cooler onto the adapter. Install the connector bolt and tighten it to the torque listed in this Chapter's Specifications.

9   Install new oil filter and refill the engine with oil (see Chapter 1).

10   Refill the cooling system (see Chapter 1). Run the engine until normal operating temperature is reached, and check for leaks.

17 Driveplate - removal and installation

This procedure is essentially the same as for the 3.7L engines. Refer to Chapter 2B, Section 16 and follow the procedure outlined there, but use the torque listed in this Chapter's Specifications.

18   Rear main oil seal - replacement

Caution: *The rear crankshaft oil seal and retainer are an assembly. When install a new seal assembly, do not separate the seal protector from the crankshaft oil seal before installation it onto engine. Damage to the seal lip will occur if the seal protector is removed.*

1   Remove the engine oil pan (see Section 15).

2   Lower the weight of the engine back onto the engine mounts.

3   Remove the transmission (see Chapter 7B).

4   Remove the rear seal retaining bolts then remove the seal retainer.

5   Clean the crankshaft oil seal and retainer surfaces.

6   Mount special tool #8225 to the oil pan surface using the oil pan mounting bolts.

7   Install the oil seal retainer over the end of the crankshaft and onto the engine block firmly press the retainer into place making sure it's also seated against the special tool.

8   Install the mounting bolt and tighten the bolts to the torque value listed in this Chapter's Specifications.

9   Remaining installation is the reverse of removal.

10   Refill the engine with oil (see Chapter 1).

19   Engine mounts - check and replacement

This procedure is essentially the same as for the 3.7L engines. Refer to Chapter 2B, Section 18 and follow the procedure outlined there, but use the torque listed in this Chapter's Specifications.

Chapter 2 Part D
General engine overhaul procedures

Contents

Specifications

2.4L four cylinder engine

General

Displacement	2.4 liters (148 cubic inches)
Bore and Stroke	3.445 x 3.976 inches (87.5 x 101.0 mm)
Compression ratio	9.5:1
Cylinder compression pressure	
Minimum	100 psi (690 kPa)
Maximum variation between cylinders	25 percent
Oil pressure (engine warmed-up)	
At idle	4 psi* (25 kPa)
At 3000 rpm	25 to 80 psi (170 to 550 kPa)

* **Note:** *If the oil pressure at idle is less than the specified minimum, don't run the engine at 3000 rpm.*

Torque specifications

Note: *One foot-pound (ft-lb) of torque is equivalent to 12 inch-pounds (in-lbs) of torque. Torque values below approximately 15 foot-pounds are expressed in inch-pounds, because most foot-pound torque wrenches are not accurate at these smaller values.*

	Ft-lbs (unless otherwise indicated)	Nm
Balance shaft assembly		
Chain cover	105 in-lbs	12
Carrier-to-block bolts	40	54
Gear cover bolt/stud	105 in-lbs	12
Sprocket bolt	21	28
Tensioner bolts	105 in-lbs	12
Connecting rod cap bolts (new bolts)		
Step 1	20	27
Step 2	Tighten an additional 1/4-turn	
Main bearing/bedplate bolts (bolts lightly lubed with engine oil) **(see illustration 11.19a)**		
Step one		
Tighten bolts 11, 17 and 20	Hand-tighten until bedplate contacts block mating surface	
Step two		
Tighten bolts 1 through 10	30	41
Step three		
Tighten bolts 1 through 10	Tighten an additional 1/4-turn	
Step four		
Tighten bolts 11 through 20	20	27

3.7L V6 engine

General

Displacement..	3.7 liters (226 cubic inches)
Bore and Stroke	3.66 x 3.4 inches (93.0 x 86.4 mm)
Compression ratio	9.6:1
Cylinder compression pressure...	170 to 225 psi (1,172 to 1,551 kPa)
Maximum variation between cylinders	25 percent
Oil pressure (engine warmed-up)	
At curb idle..	4 psi minimum* (25 kPa)
At 3000 rpm	25 to 110 psi (170 to 758 kPa)

* **Note:** *If oil pressure at idle is less than the specified minimum, do NOT run engine at 3000 rpm.*

Torque specifications

	Ft-lbs (unless otherwise indicated)	**Nm**

Note: *One foot-pound (ft-lb) of torque is equivalent to 12 inch-pounds (in-lbs) of torque. Torque values below approximately 15 foot-pounds are expressed in inch-pounds, because most foot-pound torque wrenches are not accurate at these smaller values.*

Bedplate fasteners **(see illustration 11.19b)**		
2002 models		
Step 1		
Tighten bolts 4, 7 and 6	Hand-tighten until bedplate contacts block mating surface	
Step 2		
Tighten bolts 11 through 18..	20	27
Step 3		
Tighten bolts 1 through 10..	15	20
Step 4		
Tighten bolts 19 through 23..	72 in-lbs	8
Step 5		
Tighten bolts 11 through 18..	Tighten an additional 66 degrees	
Step 6		
Tighten bolts 4, 6 and 7	Tighten an additional 42 degrees	
Step 7		
Tighten bolts 1, 2, 3, 5, 8, 9 and 10	Tighten an additional 36 degrees	
Step 8		
Tighten bolts 19 through 23..	Tighten an additional 32 degrees	
2003 and later models		
Step 1		
Tighten bolts 4,7 and 6	Hand-tighten until bedplate contacts block mating surface	
Step 2		
Tighten bolts 1 through 10..	40	54
Step 3		
Tighten bolts 11 through 18..	60 in-lbs	7
Step 4		
Tighten bolts 11 through 18..	Tighten an additional 1/4-turn	
Step 5		
Tighten bolts 19 through 23..	20	27
Connecting rod cap bolts (new bolts)		
Step one	20	27
Step two	Tighten an additional 90-degrees	

4.0L V6 engine

General

Displacement..	244 cubic inches
Bore	3.780 inches
Stroke	3.583 inches
Compression ratio	10.2:1
Oil pressure	
At idle speed	5 psi (minimum)
At 3,000 rpm	45 to 105 psi

Torque specifications

	Ft-lbs (unless otherwise indicated)

Note: *One foot-pound (ft-lb) of torque is equivalent to 12 inch-pounds (in-lbs) of torque. Torque values below approximately 15 foot-pounds are expressed in inch-pounds, because most foot-pound torque wrenches are not accurate at these smaller values.*

Connecting rod bearing bolts**	
Step 1	20
Step 2	Tighten an additional 90-degrees
Main bearing cap bolts (in sequence, see illustration 11.30e, 11.30f and 11.30g)	
Step 1 (inner bolts)..	15
Step 2 (inner bolts)..	Tighten an additional 90-degrees
Step 3 (outer bolts)..	20
Step 4 (outer bolts)..	Tighten an additional 90-degrees
Step 5 (outer bolts)..	21

** *Use new bolts.*

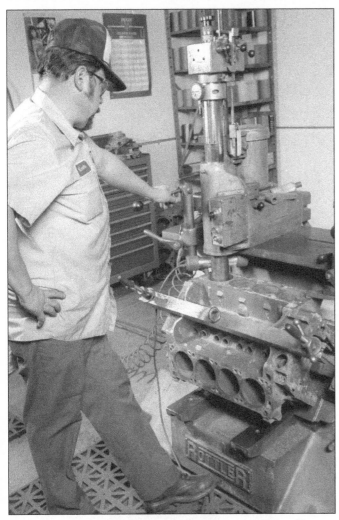

1.1 An engine block being bored. An engine rebuilder will use special machinery to recondition the cylinder bores

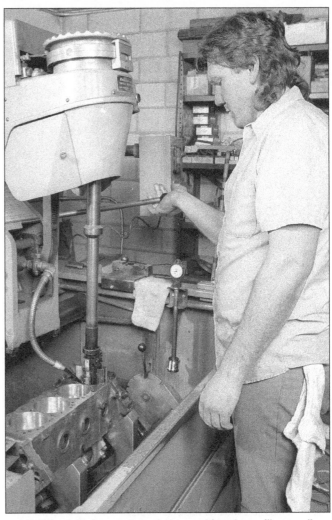

1.2 If the cylinders are bored, the machine shop will normally hone the engine on a machine like this

1   General information - engine overhaul

Refer to illustrations 1.1, 1.2, 1.3, 1.4, 1.5 and 1.6

Included in this portion of Chapter 2 is the general information and diagnostic testing procedures for determining the overall mechanical condition of you engine.

The information ranges from advice concerning preparation for an overhaul and the purchase of replacement parts to detailed, step-by-step procedures covering removal and installation.

The following Sections have been written to help you determine whether your engine needs to be overhauled and how to remove and install it once you've determined it needs to be rebuilt. For information concerning in-vehicle engine repair, see Chapters 2A, 2B or 2C.

The Specifications included in this Part are general in nature and include only those necessary for testing the oil pressure and checking the engine compression. Refer to Chapters 2A, 2B or 2C for additional engine Specifications.

It's not always easy to determine when, or if, an engine should be completely overhauled, because a number of factors must be considered.

High mileage is not necessarily an indication that an overhaul is needed, while low mileage doesn't preclude the need for an overhaul. Frequency of servicing is probably the most important consideration. An engine that's had regular and frequent oil and filter changes, as well as other required maintenance, will most likely give many thousands of miles of reliable service. Conversely, a neglected engine may require an overhaul very early in its service life.

Excessive oil consumption is an indication that piston rings, valve seals and/or valve guides are in need of attention. Make sure that oil leaks aren't responsible before deciding that the rings and/or guides are bad. Perform a cylinder compression check to determine the extent of the work required (see Section 3). Also check the vacuum readings under various conditions (see Section 4).

Check the oil pressure with a gauge installed in place of the oil pressure sending unit and compare it to this Chapter's Specifications (see Section 2). If it's extremely low, the bearings and/or oil pump are probably worn out.

Loss of power, rough running, knocking or metallic engine noises, excessive valve train noise and high fuel consumption rates may also point to the need for an overhaul, especially if they're all present at the same time. If a complete tune-up doesn't remedy the situation, major mechanical work is the only solution.

An engine overhaul involves restoring the internal parts to the specifications of a new engine. During an overhaul, the piston rings are replaced and the cylinder walls are reconditioned (rebored and/or honed) **(see illustrations 1.1 and 1.2)**. If a rebore is done by an automotive machine shop, new oversize pistons will also be installed. The main bearings, connecting rod bearings and camshaft bearings are generally replaced with new ones and, if necessary, the crankshaft may be reground to restore the journals **(see**

1.3 A crankshaft having a main bearing journal ground

1.4 A machinist checks for a bent connecting rod, using specialized equipment

illustration 1.3). Generally, the valves are serviced as well, since they're usually in less-than-perfect condition at this point. While the engine is being overhauled, other components, such as the starter and alternator, can be rebuilt as well. The end result should be a like new engine that will give many trouble free miles. **Note:** *Critical cooling system components such as the hoses, drivebelts, thermostat and water pump should be replaced with new parts when an engine is overhauled. The radiator should be checked carefully to ensure that it isn't clogged or leaking* (see Chapter 3). *If you purchase a rebuilt engine or short block, some rebuilders will not warranty their engines unless the radiator has been professionally flushed. Also, we don't recommend overhauling the oil pump - always install a new one when an engine is rebuilt.*

Overhauling the internal components on today's engines is a difficult and time-consuming task which requires a significant amount of specialty tools and is best left to a professional engine rebuilder **(see illustrations 1.4, 1.5 and 1.6)**. A competent engine rebuilder will handle the inspection of your old parts and offer advice concerning the reconditioning or replacement of the original engine, never purchase parts or have machine work done on other components until the block has been thoroughly inspected by a professional machine shop. As a general rule, time is the primary cost of an overhaul, especially since the vehicle may be tied up for a minimum of two weeks or more. Be aware that some engine builders only have the capability to rebuild the engine you bring them while other rebuilders have a large inventory of rebuilt exchange engines in stock. Also be aware that many machine shops could take as much as two weeks time to completely rebuild your engine depending on shop workload. Sometimes it makes more sense to simply exchange your engine for another engine that's already rebuilt to save time.

2   Oil pressure check

Refer to illustration 2.2

1   Low engine oil pressure can be a sign of an engine in need of rebuilding. A "low oil pressure" indicator (often called an "idiot light") is not a test of the oiling system. Such indicators only come on when the oil pressure is dangerously low. Even a factory oil pressure gauge in the instrument panel is only a relative indication, although much better for driver information than a warning light. A better test is with a mechanical (not electrical) oil pressure gauge.

2   Locate the oil pressure indicator sending unit on the engine block:

a)   On 2.4L four-cylinder engines, the oil pressure sending unit is located at the left (driver's side) end of the rear side of the block.

b)   On V6 engines, the oil pressure sending unit is located just above the oil filter, at the left front of the block) **(see illustration)**.

1.5 A bore gauge being used to check the main bearing bore

3   Unscrew the oil pressure sending unit and screw in the hose for your oil pressure gauge. If necessary, install an adapter fitting. Use Teflon tape or thread sealant on the

1.6 Uneven piston wear like this indicates a bent connecting rod

2.2 On V6 engines, the oil pressure sending unit is located at the front of the block, just above the oil filter

3.6 Use a compression gauge with a threaded fitting for the spark plug hole, not the type that requires hand pressure to maintain the seal (3.7L V6 engine shown)

4.4 A simple vacuum gauge can tell you a lot about an engine's condition (3.7L V6 engine shown)

threads of the adapter and/or the fitting on the end of your gauge's hose.

4   Connect an accurate tachometer to the engine, according to the tachometer manufacturer's instructions.

5   Check the oil pressure with the engine running (normal operating temperature) at the specified engine speed, and compare it to this Chapter's Specifications. If it's extremely low, the bearings and/or oil pump are probably worn out.

3   Cylinder compression check

Refer to illustration 3.6

1   A compression check will tell you what mechanical condition the upper end (pistons, rings, valves, head gaskets) of your engine is in. Specifically, it can tell you if the compression is down due to leakage caused by worn piston rings, defective valves and seats or a blown head gasket. **Note:** *The engine must be at normal operating temperature and the battery must be fully charged for this check.*

2   Disable the ignition system by unplugging the primary (low voltage) electrical connector(s) from the ignition coil(s). The fuel pump circuit should also be disabled (see Chapter 4, Section 2).

3   Clean the area around the spark plugs before you remove them (compressed air should be used, if available). The idea is to prevent dirt from getting into the cylinders as the compression check is being done.

4   Remove all of the spark plugs from the engine (see Chapter 1).

5   Block the throttle wide open.

6   Install the compression gauge in the number one spark plug hole **(see illustration)**.

7   Crank the engine over at least seven compression strokes and watch the gauge. The compression should build up quickly in a healthy engine. Low compression on the first

stroke, followed by gradually increasing pressure on successive strokes, indicates worn piston rings. A low compression reading on the first stroke, which doesn't build up during successive strokes, indicates leaking valves or a blown head gasket (a cracked head could also be the cause). Deposits on the undersides of the valve heads can also cause low compression. Record the highest gauge reading obtained.

8   Repeat the procedure for the remaining cylinders and compare the results to this Chapter's Specifications.

9   Add some engine oil (about three squirts from a plunger-type oil can) to each cylinder, through the spark plug hole, and repeat the test.

10   If the compression increases after the oil is added, the piston rings are definitely worn. If the compression doesn't increase significantly, the leakage is occurring at the valves or head gasket. Leakage past the valves may be caused by burned valve seats and/or faces or warped, cracked or bent valves.

11   If two adjacent cylinders have equally low compression, there's a strong possibility that the head gasket between them is blown. The appearance of coolant in the combustion chambers or the crankcase would verify this condition.

12   If one cylinder is about 20-percent lower than the others, and the engine has a slightly rough idle, a worn exhaust lobe on the camshaft could be the cause.

13   If the compression is unusually high, the combustion chambers are probably coated with carbon deposits. If that's the case, the cylinder head(s) should be removed and decarbonized.

14   If compression is way down or varies greatly between cylinders, it would be a good idea to have a leak-down test performed by an automotive repair shop. This test will pinpoint exactly where the leakage is occurring and how severe it is.

4   Vacuum gauge diagnostic checks

Refer to illustrations 4.4 and 4.6

A vacuum gauge provides valuable information about the condition of internal engine components. You can check for worn rings or cylinder walls, leaking head or intake manifold gaskets, restricted exhaust, stuck or burned valves, weak valve springs, improper ignition or valve timing and ignition problems.

Unfortunately, vacuum gauge readings are easy to misinterpret, so they should be used in conjunction with other tests to confirm the diagnosis.

Both the absolute readings and the rate of needle movement are important for accurate interpretation. Most gauges measure vacuum in inches of mercury (in-Hg). The following references to vacuum assume the diagnosis is being performed at sea level. As elevation increases (or atmospheric pressure decreases), the reading will decrease. For every 1,000 foot increase in elevation above approximately 2000 feet, the gauge readings will decrease about one inch of mercury.

Connect the vacuum gauge directly to intake manifold vacuum, not to ported (throttle body) vacuum **(see illustration 4.4)**. Be sure no hoses are left disconnected during the test or false readings will result.

Before you begin the test, allow the engine to warm up completely. Block the wheels and set the parking brake. With the transmission in Park, start the engine and allow it to run at normal idle speed. **Warning:** *Carefully inspect the fan blades for cracks or damage before starting the engine. Keep your hands and the vacuum gauge clear of the fan and do not stand in front of the vehicle or in line with the fan when the engine is running.*

Read the vacuum gauge; an average, healthy engine should normally produce about 17 to 22 inches (57.4 to 74.3 kPa) of vacuum with a fairly steady needle **(see illustration 4.6)**. Refer to the following vacuum gauge readings and what they indicate about the engine's condition:

a) *A low steady reading usually indicates a leaking gasket between the intake manifold and throttle body, a leaky vacuum hose, late ignition timing or incorrect camshaft timing. Check ignition timing with a timing light and eliminate all other possible causes, utilizing the tests provided in this Chapter before you remove the timing chain cover to check the timing marks.*

b) *If the reading is three to eight inches below normal and it fluctuates at that low reading, suspect an intake manifold gasket leak at an intake port or a faulty fuel injector.*

c) *If the needle has regular drops of about two-to-four inches at a steady rate, the valves are probably leaking. Perform a compression check or leak-down test to confirm this.*

d) *An irregular drop or down-flick of the needle can be caused by a sticking valve or an ignition misfire. Perform a compression check or leak-down test and read the spark plugs.*

e) *A rapid vibration of about four inches Hg vibration at idle combined with exhaust smoke indicates worn valve guides. Perform a leak-down test to confirm this. If the rapid vibration occurs with an increase in engine speed, check for a leaking intake manifold gasket or head gasket, weak valve springs, burned valves or ignition misfire.*

f) *A slight fluctuation, say one inch up and down, may mean ignition problems. Check all the usual tune-up items and, if necessary, run the engine on an ignition analyzer.*

g) *If there is a large fluctuation, perform a compression or leak-down test to look for a weak or dead cylinder or a blown head gasket.*

h) *If the needle moves slowly through a wide range, check for a clogged PCV system, incorrect idle fuel mixture, carburetor/throttle body or intake manifold gasket leaks.*

i) *Check for a slow return after revving the engine by quickly snapping the throttle open until the engine reaches about 2,500 rpm and let it shut. Normally the reading should drop to near zero, rise above normal idle reading (about 5 in.-Hg over) and then return to the previous idle reading. If the vacuum returns slowly and doesn't peak when the throttle is snapped shut, the rings may be worn. If there is a long delay, look for a restricted exhaust system (often the muffler or catalytic converter). An easy way to check this is to temporarily disconnect the exhaust ahead of the suspected part and re-test.*

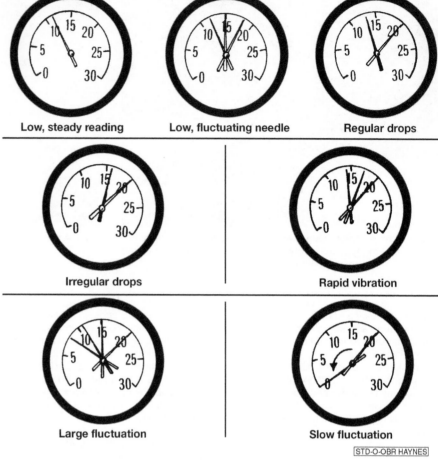

Low, steady reading Low, fluctuating needle Regular drops

Irregular drops   Rapid vibration

Large fluctuation   Slow fluctuation

STD-O-OBR HAYNES

4.6 Typical vacuum gauge readings

5 Engine rebuilding alternatives

The do-it-yourselfer is faced with a number of options when purchasing a rebuilt engine. The major considerations are cost, warranty, parts availability and the time required for the rebuilder to complete the project. The decision to replace the engine block, piston/connecting rod assemblies and crankshaft depends on the final inspection results of your engine. Only then can you make a cost effective decision whether to have your engine overhauled or simply purchase an exchange engine for your vehicle.

Some of the rebuilding alternatives include:

Individual parts - If the Inspection procedures reveal that the engine block and most engine components are in reusable condition, purchasing individual parts may be the most economical alternative. The block, crankshaft and piston/connecting rod assemblies should all be inspected carefully. Even if the block shows little wear, the cylinder bores should be surface honed.

Short block - A short block consists of an engine block with a crankshaft, camshaft (cam-in-block engines) and piston/connecting rod assemblies already installed. All new

bearings are incorporated and all clearances will be correct. The existing valve train components, cylinder head(s) and external parts can be bolted to the short block with little or no machine-shop work necessary.

Long block - A long block consists of a short block plus an oil pump, oil pan, cylinder head(s), valve cover(s) and valve train components, timing sprockets and chain or gears and timing cover. All components are installed with new bearings, seals and gaskets incorporated throughout. The installation of manifolds and external parts is all that's necessary.

Low mileage used engines - Some companies now offer low mileage used engines which is a very cost effective way to get your vehicle up and running again. These engines often come from vehicles that have been totaled in accidents or come from other countries that have a higher vehicle turn over rate. A low mileage used engine also usually has a similar warranty like the newly remanufactured engines.

Give careful thought to which alternative is best for you and discuss the situation with local automotive machine shops, auto parts dealers and experienced rebuilders before ordering or purchasing replacement parts.

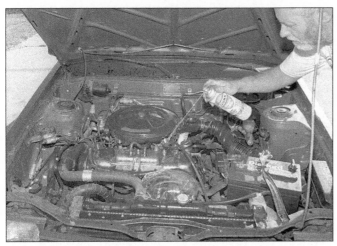

6.1 After tightly wrapping water-vulnerable components, use a spray cleaner on everything, with particular concentration on the greasiest areas, usually around the valve cover and lower edges of the block. If one section dries out, apply more cleaner (typical)

6.2 Depending on how dirty the engine is, let the cleaner soak in according to the directions and then hose off the grime and cleaner. Get the rinse water down into every area you can get at; then dry important components with a hair dryer or paper towels (typical)

6 Engine removal - methods and precautions

Refer to illustrations 6.1, 6.2, 6.3, 6.4 and 6.5

If you've decided that an engine must be removed for overhaul or major repair work, several preliminary steps should be taken.

Locating a suitable place to work is extremely important. Adequate work space, along with storage space for the vehicle, will

6.3 Get an engine hoist, and get a load-leveler device (like the one shown here) that can change the angle of the engine for easier removal and installation from/to the transmission. An engine sling can be used if a load-leveler device is not available

be needed. If a shop or garage isn't available, at the very least a flat, level, clean work surface made of concrete or asphalt is required. Cleaning the engine compartment and engine before beginning the removal procedure will help keep your tools clean and organized **(see illustrations 6.1 and 6.2)**.

An engine hoist or A-frame will also be necessary. Make sure the equipment is rated in excess of the combined weight of the engine and accessories. Safety is of primary importance, considering the potential hazards involved in lifting the engine out of the vehicle.

If the engine is being removed by a novice, a helper should be available. Advice and aid from someone more experienced would also be helpful. There are many instances when one person cannot simultaneously perform all of the operations required when lifting the engine out of the vehicle.

Plan the operation ahead of time. Arrange for or obtain all of the tools and equipment you'll need prior to beginning the job **(see illustrations 6.3, 6.4 and 6.5)**. Some of the equipment necessary to perform engine removal and installation safely and with relative ease are (in addition to an engine

6.4 Get an engine stand sturdy enough to firmly support the engine while you're working on it. Stay away from three-wheeled models; they have a tendency to tip over more easily - four-wheeled models are recommended

hoist) a heavy duty floor jack, complete sets of wrenches and sockets as described in the front of this manual, wooden blocks and plenty of rags and cleaning solvent for mop-

6.5 On models equipped with a manual transmission, a clutch alignment tool is necessary

ping up spilled oil, coolant and gasoline. If the hoist must be rented, make sure that you arrange for it in advance and perform all of the operations possible without it beforehand. This will save you money and time.

Plan for the vehicle to be out of use for quite a while. A machine shop will be required to perform some of the work which the do-it-yourselfer can't accomplish without special equipment. These shops often have a busy schedule, so it would be a good idea to consult them before removing the engine in order to accurately estimate the amount of time required to rebuild or repair components that may need work.

7   Engine - removal and installation

Refer to illustrations 7.5 and 7.27

Warning 1: *The air conditioning system is under high pressure! Have a dealer service department or other qualified facility discharge the system before disconnecting any system hoses or fittings.*

Warning 2: *Wait until the engine is completely cool before beginning this procedure.*

Warning 3: *Gasoline is flammable, so take extra precautions when you work on any part of the fuel system. Don't smoke or allow open flames or bare light bulbs near the work area, and don't work in a garage where a gas-type appliance (such as a water heater or a clothes dryer) is present. Since fuel is carcinogenic, wear latex gloves when there's a possibility of being exposed to fuel, and, if you spill any fuel on your skin, rinse it off immediately with soap and water. Mop up any spills immediately and do not store fuel-soaked rags where they could ignite. The fuel system is under constant pressure, so, if any fuel lines are to be disconnected, the fuel pressure in the system must be relieved first (see Chapter 4 for more information). When you perform any kind of work on the fuel system, wear safety glasses and have a Class B type fire extinguisher on hand.*

Removal

1   Refer to Chapter 4 and relieve the fuel system pressure, then disconnect the cable from the negative terminal of the battery.

Warning: *These models are equipped with airbags. The airbag is armed and can deploy (inflate) anytime the battery is connected. To prevent accidental deployment (and possible injury), disconnect the cable from the negative terminal of the battery whenever working near airbag components, such as the crash sensors at the front of the vehicle. After the battery is disconnected, wait at least 2 minutes before beginning work (the system has a back-up capacitor that must fully discharge). For more information see Chapter 12.*

2   Cover the fenders and cowl and remove the hood (see Chapter 11). Special pads are available to protect the fenders, but an old bedspread or blanket will also work.

3   Remove the air cleaner assembly (see Chapter 4).

4   Drain the cooling system (see Chapter 1).

5   Label the vacuum lines, emissions system hoses, wiring connectors, ground strap (at the cowl) and fuel lines, to ensure correct reinstallation **(see illustration)**, then detach them. If there's any possibility of confusion, make a sketch of the engine compartment and clearly label the lines, hoses and wires.

6   Label and detach all coolant hoses from the engine.

7   Remove the drivebelt(s) (see Chapter 1).

8   Remove the cooling fan(s), shroud, coolant expansion tank and radiator (see Chapter 3).

9   Remove the fuel rail(s) (see Chapter 4).

10   Disconnect the throttle cable (and TV cable/cruise control cable, if equipped) from the engine (see Chapter 4).

11   On 3.7L V6 models, remove the intake manifold (see Chapter 2B).

12   Unbolt the power steering pump (see Chapter 10). Leave the lines/hoses attached and make sure the pump is kept in an upright position in the engine compartment (use wire or rope to restrain it out of the way).

13   On air-conditioned vehicles, unbolt the compressor (see Chapter 3) and set it aside. Do not disconnect the hoses.

14   Drain the engine oil (see Chapter 1) and remove the oil filter. Remove the oil dipstick tube fastener, then remove the tube from the oil pan.

15   Remove the starter motor (see Chapter 5).

16   Remove the alternator (see Chapter 5).

17   Unbolt the exhaust system from the engine (see Chapter 4).

18   Remove the oil pan-to-transmission structural cover.

19   If you're working on a vehicle with an automatic transmission, refer to Chapter 7B and remove the torque converter-to-driveplate fasteners. Also unclip the automatic transmission fluid cooler lines from any studs on the engine oil pan, on models so equipped.

20   If you're working on a 3.7L V6 model with a manual transmission, refer to Chapter 2B and remove the engine mounts. Another option would be to remove the transmission to allow the engine to clear the vehicle frame during the engine removal and installation processes (see Chapter 7A).

21   Support the transmission with a jack. Position a block of wood between the jack and transmission to prevent damage to the transmission. Special transmission jacks with safety chains are available - use one if possible.

22   Install lifting brackets to the engine attachment points for equal weight distribution.

23   Roll the hoist into position and connect the load-leveler or sling device to it. Take up the slack in the sling or chain, but don't lift the engine. **Warning:** *DO NOT place any part of your body under the engine when it's supported only by a hoist or other lifting device.*

24   Remove the transmission-to-engine bolts.

25   If the engine motor mounts are still in

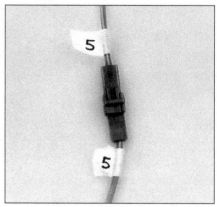

7.5 Label both ends of each wire before unplugging the connector

place, remove the engine mount through-bolts.

26   Recheck to be sure that nothing is still connecting the engine to the transmission or vehicle. Disconnect anything still remaining.

27   Raise the engine slightly. Carefully work it forward to separate it from the transmission. If you're working on a vehicle with an automatic transmission, be sure the torque converter stays in the transmission (clamp a pair of vise-grips to the housing to keep the converter from sliding out). If you're working on a vehicle with a manual transmission, the input shaft must be completely disengaged from the clutch. Slowly raise the engine out of the engine compartment **(see illustration)**. Check carefully to make sure that nothing is hanging up.

28   Remove the flywheel/driveplate and mount the engine on an engine stand.

Installation

29   Check the engine and transmission mounts. If they're worn or damaged, replace them. **Note:** *On 3.7L V6 models with a manual transmission, if the transmission was not removed, leave the engine mounts off of the engine; they must be installed after the engine is in place.*

7.27 Pull the engine forward as far as possible to clear the transmission, then lift the engine high enough to clear the body (typical)

30   If you're working on a vehicle with a manual transmission, install the clutch and pressure plate (Chapter 7A). Now is a good time to install a new clutch.

31   Carefully lower the engine into the engine compartment - make sure the engine mounts line up. If you're working on a vehicle with an automatic transmission, guide the torque converter into the crankshaft following the procedure outlined in Chapter 7B. If you're working on a vehicle with a manual transmission, apply a dab of high-temperature grease to the input shaft and guide it into the crankshaft pilot bearing until the bellhousing is flush with the engine block.

32   Install the transmission-to-engine bolts and tighten them securely. **Caution:** *DO NOT use the bolts to force the transmission and engine together.*

33   On 3.7L V6 models with a manual transmission, install the motor mounts (see Chapter 2B).

34   Reinstall the remaining components in the reverse order of removal.

35   Add coolant, oil, power steering and transmission fluid as needed (see Chapter 1).

36   Run the engine and check for leaks and proper operation of all accessories, then install the hood and test drive the vehicle.

37   If any air conditioning system hoses or fittings were disconnected, have the system recharged and leak tested.

8   Engine overhaul - disassembly sequence

1   It's much easier to disassemble and work on the engine, if it's mounted on a portable engine stand. A stand can often be rented quite cheaply from an equipment rental yard. Before the engine is mounted on a stand, the flywheel/driveplate should be removed from the engine.

2   If a stand isn't available, it's possible to disassemble the engine with it blocked up on the floor. Be extra careful not to tip or drop the engine when working without a stand.

3   If you're going to obtain a rebuilt engine, all external components must come off first to be transferred to the replacement engine, just as they will if you're doing a complete engine overhaul yourself. These include:

   Accessory brackets
   Clutch and flywheel/driveplate
   Emissions control components
   Engine mounts
   Fuel Injection components
   Ignition coil(s)
   Intake/exhaust manifolds
   Oil filter
   Spark plug wires and spark plugs
   Thermostat and housing assembly
   Water pump

Note: *When removing the external components from the engine, pay close attention to details that may be helpful or important during installation. Note the installed position of gaskets, seals, spacers, pins, brackets, washers, bolts and other small items.*

4   If you're going to obtain a short block (assembled engine block, crankshaft, pistons and connecting rods), then remove the timing belt/chains, cylinder head(s), oil pan, oil pump pick-up tube, oil pump and water pump from your engine so that you can turn in your old short block to the rebuilder as a core. See *Engine rebuilding alternatives* for additional information regarding the different possibilities to be considered.

5   On 2.4L four-cylinder engines, a vibration-reducing balance shaft assembly is mounted to the bottom of the block. The balance shaft assembly must be removed to allow disassembly of the pistons/connecting rods and/or crankshaft (see Section 9).

9   Balance shaft assembly - removal and installation (2.4L four-cylinder engine)

Removal

1   Remove the balance shaft assembly chain cover.

2   Remove the balance shaft sprocket bolt.

3   Loosen the balance shaft chain tensioner bolts.

4   Push the balance shaft inward, and pull or pry the balance shaft sprocket off of the shaft, so that the sprocket hangs freely. Remove the chain and sprocket.

5   Remove the balance shaft carrier-to-block bolts and remove the carrier assembly.

Installation

Refer to illustrations 9.7, 9.8 and 9.11

6   Install the carrier assembly and tighten the bolts to the torque listed in this Chapter's Specifications.

7   Remove the bolt from the other balance shaft and remove the gear cover. Rotate the crankshaft to the No. 1 top-dead-center (TDC) position, so that the No. 1 TDC timing mark on the crankshaft sprocket is aligned with the parting line of the main bearing cap **(see illustration)**.

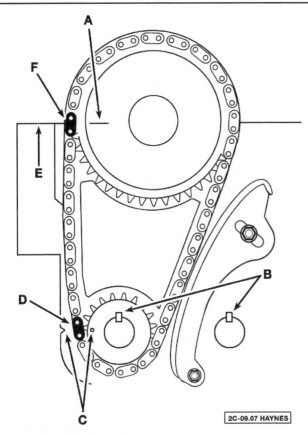

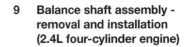

2C-09.07 HAYNES

9.7 Rotate the crankshaft so that the No. 1 TDC mark on the crankshaft sprocket is aligned with the parting line of the main bearing cap/bedplate. When the balance shaft chain and balance shaft driven-sprocket are installed, they must also be aligned as shown

A	*Mark on crankshaft sprocket*
B	*Balance shaft keyways pointing up*
C	*Alignment marks on counter balance sprocket and counter balance carrier case*

D	*Plated link on balance shaft chain*
E	*Parting line of main bearing cap/ bedplate*
F	*Plated link on balance shaft chain*

ENGINE BEARING ANALYSIS

Debris

Babbitt bearing embedded with debris from machinings

Microscopic detail of debris

Microscopic detail of gouges

Overplated copper alloy bearing gouged by cast iron debris

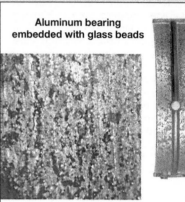

Aluminum bearing embedded with glass beads

Microscopic detail of glass beads

Damaged lining caused by dirt left on the bearing back

Misassembly

Result of a lower half assembled as an upper - blocking the oil flow

Excessive oil clearance is indicated by a short contact arc

Polished and oil-stained backs are a result of a poor fit in the housing bore

Result of a wrong, reversed, or shifted cap

Overloading

Damage from excessive idling which resulted in an oil film unable to support the load imposed

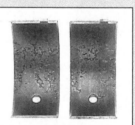

Damaged upper connecting rod bearings caused by engine lugging; the lower main bearings (not shown) were similarly affected

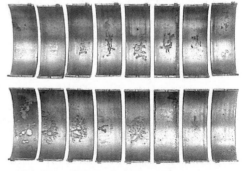

The damage shown in these upper and lower connecting rod bearings was caused by engine operation at a higher-than-rated speed under load

Misalignment

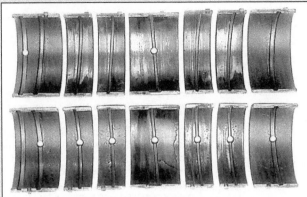

A warped crankshaft caused this pattern of severe wear in the center, diminishing toward the ends

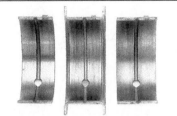

A poorly finished crankshaft caused the equally spaced scoring shown

A bent connecting rod led to the damage in the "V" pattern

A tapered housing bore caused the damage along one edge of this pair

Lubrication

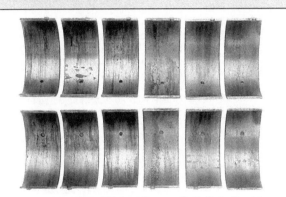

Result of dry start: The bearings on the left, farthest from the oil pump, show more damage

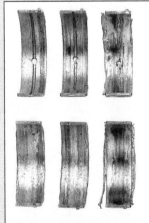

Result of a low oil supply or oil starvation

Severe wear as a result of inadequate oil clearance

Corrosion

Microscopic detail of corrosion

Corrosion is an acid attack on the bearing lining generally caused by inadequate maintenance, extremely hot or cold operation, or inferior oils or fuels

Microscopic detail of cavitation

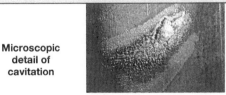

Example of cavitation - a surface erosion caused by pressure changes in the oil film

Damage from excessive thrust or insufficient axial clearance

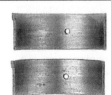

Bearing affected by oil dilution caused by excessive blow-by or a rich mixture

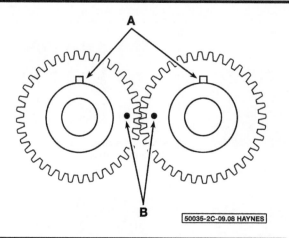

9.8 Align the balance shaft keyways and balance shaft gear marks as shown

A *Balance shaft keyways pointing up*
B *Alignment marks on balance shaft gears*

8   Align both balance shaft keyways at the 12 o'clock position (pointing toward the top of the engine) so that both balance shaft gear timing dots are pointed to each other **(see illustration)**. Install the gear cover and tighten the bolt/stud to the torque listed in this Chapter's Specifications.
9   Install the chain so that the each of the plated links are aligned with each of the timing marks on the crankshaft and balance shaft sprockets, then slide the sprocket onto the balance shaft **(see illustration 9.7)**.
10   Check the timing marks, plated links and keyways for correct alignment, then install both of the balance bolts and tighten to the torque listed in this Chapter's Specifications.

Note: *To prevent the crankshaft from rotating, wedge a wooden block between a crankshaft counterweight and the block.*
11   Position the chain tensioner and loosely install the tensioner bolts. Install the chain guide, making sure the tab is in the slot. Position a 0.039-inch thick x 2.75-inch long spacer-shim between the tensioner and chain **(see illustration)**. Push firmly on the tensioner, then tighten the top tensioner arm bolt, then tighten the lower tensioner pivot bolt - each to the torque listed in this Chapter's Specifications. Remove the spacer-shim.
12   Install the chain cover and tighten the bolts to the torque listed in this Chapter's Specifications

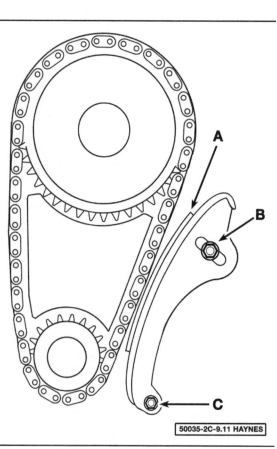

9.11 To adjust the balance shaft chain tensioner, position a 0.039-inch (1 mm) thick by 2.75-inch (70 mm) long shim spacer between the chain and tensioner, push on the tensioner firmly, then tighten the tensioner arm bolt and tensioner pivot bolt

A   *0.039 (1 mm) thick shim*
B   *Tensioner adjuster bolt*
C   *Tensioner pivot bolt*

10   Pistons/connecting rods - removal and installation

Refer to illustrations 10.1, 10.3, 10.4a, 10.4b and 10.6
Note: *Prior to removing the piston/connecting rod assemblies, remove the cylinder head(s), the oil pan and the oil pump by referring to the appropriate Sections in Chapter 2A, 2B or 2C. On 2.4L four-cylinder engines, the balance shaft carrier assembly must be removed (see Section 9).*
1   Use your fingernail to feel if a ridge has formed at the upper limit of ring travel (about 1/4-inch down from the top of each cylinder). If carbon deposits or cylinder wear have produced ridges, they must be completely removed with a special tool **(see illustration)**. Follow the manufacturer's instructions provided with the tool. Failure to remove the ridges before attempting to remove the piston/connecting rod assemblies may result in piston breakage.
2   After the cylinder ridges have been removed, turn the engine upside-down so the crankshaft is facing up.
3   Before the main bearing cap assembly and connecting rods are removed, check the connecting rod endplay with feeler gauges. Slide them between the first connecting rod and the crankshaft throw until the play is removed **(see illustration)**. Repeat this procedure for each connecting rod. The endplay is equal to the thickness of the feeler gauge(s). Check with an automotive machine shop for the endplay service limit. If the play exceeds the service limit, new connecting rods will be required. If new rods (or a new crankshaft) are installed, the endplay may fall under the minimum listed in this Chapter's Specifications. If it does, the rods will have to be machined to restore it. If necessary, consult an automotive machine shop for advice.
4   Check the connecting rods and caps for identification marks. If they aren't plainly marked, use a permanent marking pen or scratch awl to make the appropriate number

10.1 A ridge reamer is required to remove the ridge from the top of each cylinder - do this before removing the pistons!

10.3 Check the connecting rod side clearance (endplay) with a feeler gauge as shown here

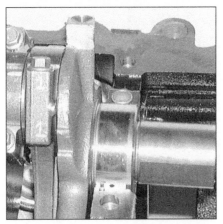

10.4a Use a permanent marking pen or scratch awl to identify each connecting rod and each cap - using any type of punch can damage the connecting rod cap

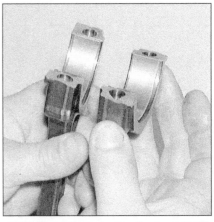

10.4b The rod cap parting line on 3.7L V6 engines is fractured at the factory (not machined) - therefore it is imperative that the rod caps be placed in the exact same position from which they were removed

on each rod and each cap (1, 2, 3, etc., depending on the cylinder they're associated with) **(see illustration)**. The marks will enable you to keep each rod cap in the same position on its corresponding connecting rod during the installation steps. **Caution:** *Using a center punch to mark the connecting rods, can damage the connecting rod cap. On the 3.7L V6 engines, it is ABSOLUTELY NECESSARY that each rod cap be installed in the same position on the same connecting rod from which it was removed, because the connecting rod cap parting line is a fractured-fit instead of a machined-fit* **(see illustration)**.

5   Loosen each of the connecting rod cap fasteners 1/2-turn at a time until they can be removed by hand. Remove the number one connecting rod cap and bearing insert. Don't drop the bearing insert out of the cap.

6   Remove the bearing insert and push the connecting rod/piston assembly out through the top of the engine. Use a wooden or plastic hammer handle to push on the upper bearing surface in the connecting rod. If resistance is felt, double-check to make sure that the entire ridge was removed from the cylinder. **Note:** *On 3.7L V6 engines, the rod bolts*

are removed with the caps. So, it might be helpful to make a set of connecting rod guides. Find two studs that fit the rods at your local hardware store. Or, use two of the old rod bolts (since they must be replaced anyway), and cut the heads of the bolts off with a hacksaw. Slip a short length of hose over the end of each stud/bolt to make a pair of connecting rod guides. Screw the guides into the connecting rods during removal and installation of the pistons/connecting rods from the engine block **(see illustration)**.

7   Repeat the procedure for the remaining cylinders. **Note:** *If the connecting rod caps are secured by bolts (instead of nuts), discard the old rod cap bolts. Use new bolts when reassembling the engine.*

8   After removal, reassemble the connecting rod caps and bearing inserts in their respective connecting rods and install the cap bolts finger tight. Leaving the old bearing inserts in place until reassembly will help prevent the connecting rod bearing surfaces from being accidentally nicked or gouged.

9   The pistons and connecting rods are

now ready for inspection and overhaul at an automotive machine shop.

Piston ring installation

Refer to illustrations 10.12, 10.13, 10.14, 10.18a, 10.18b, 10.20 and 10.21

10   Before installing the new piston rings, the ring end gaps must be checked. It's assumed that the piston ring side clearance has been checked and verified correct.

11   Lay out the piston/connecting rod assemblies and the new ring sets so the ring sets will be matched with the same piston and cylinder during the end gap measurement and engine assembly.

12   Insert the top (number one) ring into the first cylinder and square it up with the cylinder walls by pushing it in with the top of the piston **(see illustration)**. The ring should be near the bottom of the cylinder, at the lower limit of ring travel.

13   To measure the end gap, slip feeler gauges between the ends of the ring until a gauge equal to the gap width is found **(see illustration)**. The feeler gauge should slide

10.6 On 3.7L V6 engine, it will be necessary to fabricate a connecting rod guide tool before removing and installing the piston and connecting rod assemblies

10.12 When checking piston ring end gap, the ring must be square in the cylinder bore - use a piston to push the ring down near the bottom of the ring travel

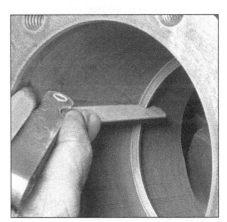

10.13 With the ring square in the cylinder, measure the end gap with feeler gauges

10.14 If the end gap is too small, clamp a fine file in a vise and file ring ends a little at a time and re-measure in the bore

10.18a Installing the spacer/expander in the oil control ring groove

10.18b Slip the oil ring side rails in by hand - do not use a ring expander

between the ring ends with a slight amount of drag. Check with an automotive machine shop for the correct end gap for your engine. If the gap is larger or smaller than specified, double-check to make sure you have the correct rings before proceeding.

14 If the gap is too small, it must be enlarged or the ring ends may come in contact with each other during engine operation, which can cause serious damage to the engine. The end gap can be increased by filing the ring ends very carefully with a fine file. Mount the file in a vise equipped with soft jaws, slip the ring over the file with the ends contacting the file face and slowly move the ring to remove material from the ends. When performing this operation, file only by pushing the ring from the outside end of the file towards the vise **(see illustration)**.

15 Repeat the procedure for each ring that will be installed in the first cylinder and for each ring in the remaining cylinders.

16 Remember to keep rings, pistons and cylinders matched up.

17 Once the ring end gaps have been checked/corrected, the rings can be installed on the pistons.

18 The oil control ring (lowest one on the piston) is usually installed first. It's composed of three separate components. Slip the spacer/expander into the groove **(see illustration)**. If an anti-rotation tang is used on the spacer/expander, make sure it's inserted into the drilled hole in the piston ring groove. Next, install the upper side rail in the same manner **(see illustration)**. Don't use a piston ring installation tool on the oil ring side rails, as they may be damaged. Instead, place one end of the side rail into the groove between the spacer/expander and the ring land, hold it firmly in place and slide a finger around the piston while pushing the rail into the groove. Finally, install the lower side rail.

19 After the three oil ring components have been installed, check to make sure that both the upper and lower side rails can be rotated smoothly inside the ring grooves.

20 The number two (middle) ring is installed next. It's usually stamped with a mark that

must face up, toward the top of the piston. Do not mix up the top and middle rings, as they have different cross-sections **(see illustration)**. **Note:** *Always follow the instructions printed on the ring package or box - different manufacturers may require different approaches.*

21 Use a piston ring installation tool and make sure the identification mark is facing the top of the piston, then slip the ring into the middle groove on the piston **(see illustration)**. Don't expand the ring any more than necessary to slide it over the piston.

22 Install the number one (top) ring in the same manner. Make sure the mark is facing up. Be careful not to confuse the number one and number two rings **(see illustrations 10.20 and 10.21)**.

23 Repeat the procedure for the remaining pistons and rings.

Installation

24 Before installing the piston/connecting rod assemblies, the cylinder walls must be perfectly clean, the top edge of each cylinder bore must be chamfered, and the crankshaft must be in place.

25 Remove the cap from the end of the number one connecting rod (refer to the marks made during removal). Remove the original bearing inserts and wipe the bearing surfaces of the connecting rod and cap with

a clean, lint-free cloth. They must be kept spotlessly clean. Connecting rods should be fitted with new cap bolts before installation.

Connecting rod bearing oil clearance check

Refer to illustrations 10.28, 10.33, 10.35 and 10.39

26 Clean the back side of the new upper bearing insert, then lay it in place in the connecting rod. Make sure the tab on the bearing fits into the recess in the rod. Don't hammer the bearing insert into place and be very careful not to nick or gouge the bearing face. Don't lubricate the bearing at this time.

27 Clean the back side of the other bearing insert and install it in the rod cap. Again, make sure the tab on the bearing fits into the recess in the cap, and don't apply any lubricant. It's critically important that the mating surfaces of the bearing and connecting rod are perfectly clean and oil free when they're assembled.

28 Position the piston ring gaps at 90-degree intervals around the piston **(see illustration)**.

29 Lubricate the piston and rings with clean engine oil and attach a piston ring compressor to the piston. Leave the piston skirt pro-

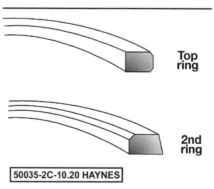

10.20 Typical example of top and middle piston ring profiles

Top ring

2nd ring

50035-2C-10.20 HAYNES

10.21 Use a piston ring installation tool to install the second and top rings - be sure the mark on the piston ring(s) is facing toward the top of the piston (the appearance of the marks may vary)

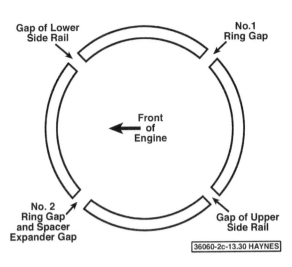

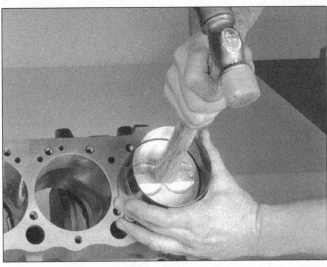

10.28 Position the piston ring end gaps as shown

10.33 Use a plastic or wooden hammer handle to push the piston into the cylinder

truding about 1/4-inch to guide the piston into the cylinder. The rings must be compressed until they're flush with the piston.

30 Rotate the crankshaft until the number one connecting rod journal is at BDC (bottom dead center) and apply a liberal coat of engine oil to the cylinder walls.

31 Gently insert the piston/connecting rod assembly into the number one cylinder bore and rest the bottom edge of the ring compressor on the engine block. **Caution:** *On 3.7L V6 engines, reinstall the connecting rod guides that were used when removing the connecting rods (see illustration 10.6). The pistons will have a weight designation mark or arrow on top of the piston, or a raised "F" near the piston pin bore. The appropriate mark must face the front of the engine. The connecting rods also have a mark or oil slinger slot, which must face the front of the engine. If the mark on the piston, or the mark on the connecting rod does not face the front of the engine, the piston and connecting rod have been assembled incorrectly.*

32 Tap the top edge of the ring compressor to make sure it's contacting the block around its entire circumference.

33 Gently tap on the top of the piston with the end of a wooden or plastic hammer handle **(see illustration)**, while guiding the end of the connecting rod into place on the crankshaft journal. The piston rings may try to pop out of the ring compressor just before entering the cylinder bore, so keep some downward pressure on the ring compressor. Work slowly, and if any resistance is felt as the piston enters the cylinder, stop immediately. Find out what's hanging up and fix it before proceeding. Do not, for any reason, force the piston into the cylinder - you might break a ring and/or the piston.

34 Once the piston/connecting rod assembly is installed, the connecting rod bearing oil clearance must be checked before the rod cap is permanently installed.

35 Cut a piece of the appropriate size Plastigage slightly shorter than the width of the connecting rod bearing and lay it in place on the

number one connecting rod journal, parallel with the journal axis **(see illustration)**.

36 Clean the connecting rod cap bearing face and install the rod cap. Make sure the mating mark on the cap is on the same side as the mark on the connecting rod **(see illustration 10.4a)**.

37 Install the old rod bolts or nuts, at this time, and tighten them to the torque listed in this Chapter's Specifications, working up to it in three steps. **Note:** *Use a thin-wall socket to avoid erroneous torque readings that can result if the socket is wedged between the rod cap and the bolt or nut. If the socket tends to wedge itself between the fastener and the cap, lift up on it slightly until it no longer contacts the cap. DO NOT rotate the crankshaft at any time during this operation.*

38 Remove the fasteners and detach the rod cap, being very careful not to disturb the Plastigage. Discard the cap bolts at this time as they cannot be reused. **Note:** *You MUST use new connecting rod bolts.*

39 Compare the width of the crushed Plastigage to the scale printed on the Plastigage envelope to obtain the oil clearance **(see illustration)**. The connecting rod oil clearance is usually about 0.001 to 0.002 inch. Consult an automotive machine shop for the clearance specified for the rod bearings on your engine.

40 If the clearance is not as specified, the bearing inserts may be the wrong size (which means different ones will be required). Before deciding that different inserts are needed, make sure that no dirt or oil was between the bearing inserts and the connecting rod or cap when the clearance was measured. Also, recheck the journal diameter. If the Plastigage was wider at one end than the other, the journal may be tapered. If the clearance still exceeds the limit specified, the bearing will have to be replaced with an undersize bearing. **Caution:** *When installing a new crankshaft, always use a standard size bearing.*

10.35 Place Plastigage on each connecting rod bearing journal parallel to the crankshaft centerline

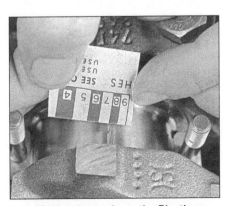

10.39 Use the scale on the Plastigage package to determine the bearing oil clearance - be sure to measure the widest part of the Plastigage and use the correct scale; it comes with both standard and metric scales

B

Backlash - The amount of play between two parts. Usually refers to how much one gear can be moved back and forth without moving gear with which it's meshed.

Bearing Caps - The caps held in place by nuts or bolts which, in turn, hold the bearing surface. This space is for lubricating oil to enter.

Bearing clearance - The amount of space left between shaft and bearing surface. This space is for lubricating oil to enter.

Bearing crush - The additional height which is purposely manufactured into each bearing half to ensure complete contact of the bearing back with the housing bore when the engine is assembled.

Bearing knock - The noise created by movement of a part in a loose or worn bearing.

Blueprinting - Dismantling an engine and reassembling it to EXACT specifications.

Bore - An engine cylinder, or any cylindrical hole; also used to describe the process of enlarging or accurately refinishing a hole with a cutting tool, as to bore an engine cylinder. The bore size is the diameter of the hole.

Boring - Renewing the cylinders by cutting them out to a specified size. A boring bar is used to make the cut.

Bottom end - A term which refers collectively to the engine block, crankshaft, main bearings and the big ends of the connecting rods.

Break-in - The period of operation between installation of new or rebuilt parts and time in which parts are worn to the correct fit. Driving at reduced and varying speed for a specified mileage to permit parts to wear to the correct fit.

Bushing - A one-piece sleeve placed in a bore to serve as a bearing surface for shaft, piston pin, etc. Usually replaceable.

C

Camshaft - The shaft in the engine, on which a series of lobes are located for operating the valve mechanisms. The camshaft is driven by gears or sprockets and a timing chain. Usually referred to simply as the cam.

Carbon - Hard, or soft, black deposits found in combustion chamber, on plugs, under rings, on and under valve heads.

Cast iron - An alloy of iron and more than two percent carbon, used for engine blocks and heads because it's relatively inexpensive and easy to mold into complex shapes.

Chamfer - To bevel across (or a bevel on) the sharp edge of an object.

Chase - To repair damaged threads with a tap or die.

Combustion chamber - The space between the piston and the cylinder head, with the piston at top dead center, in which air-fuel mixture is burned.

Compression ratio - The relationship between cylinder volume (clearance volume) when the piston is at top dead center and cylinder volume when the piston is at bottom dead center.

Connecting rod - The rod that connects the crank on the crankshaft with the piston. Sometimes called a con rod.

Connecting rod cap - The part of the connecting rod assembly that attaches the rod to the crankpin.

Core plug - Soft metal plug used to plug the casting holes for the coolant passages in the block.

Crankcase - The lower part of the engine in which the crankshaft rotates; includes the lower section of the cylinder block and the oil pan.

Crank kit - A reground or reconditioned crankshaft and new main and connecting rod bearings.

Crankpin - The part of a crankshaft to which a connecting rod is attached.

Crankshaft - The main rotating member, or shaft, running the length of the crankcase, with offset throws to which the connecting rods are attached; changes the reciprocating motion of the pistons into rotating motion.

Cylinder sleeve - A replaceable sleeve, or liner, pressed into the cylinder block to form the cylinder bore.

D

Deburring - Removing the burrs (rough edges or areas) from a bearing.

Deglazer - A tool, rotated by an electric motor, used to remove glaze from cylinder walls so a new set of rings will seat.

E

Endplay - The amount of lengthwise movement between two parts. As applied to a crankshaft, the distance that the crankshaft can move forward and back in the cylinder block.

F

Face - A machinist's term that refers to removing metal from the end of a shaft or the face of a larger part, such as a flywheel.

Fatigue - A breakdown of material through a large number of loading and unloading cycles. The first signs are cracks followed shortly by breaks.

Feeler gauge - A thin strip of hardened steel, ground to an exact thickness, used to check clearances between parts.

Free height - The unloaded length or height of a spring.

Freeplay - The looseness in a linkage, or an assembly of parts, between the initial application of force and actual movement. Usually perceived as slop or slight delay.

Freeze plug - See Core plug.

G

Gallery - A large passage in the block that forms a reservoir for engine oil pressure.

Glaze - The very smooth, glassy finish that develops on cylinder walls while an engine is in service.

H

Heli-Coil - A rethreading device used when threads are worn or damaged. The device is installed in a retapped hole to reduce the thread size to the original size.

I

Installed height - The spring's measured length or height, as installed on the cylinder head. Installed height is measured from the spring seat to the underside of the spring retainer.

J

Journal - The surface of a rotating shaft which turns in a bearing.

K

Keeper - The split lock that holds the valve spring retainer in position on the valve stem.

Key - A small piece of metal inserted into matching grooves machined into two parts fitted together - such as a gear pressed onto a shaft - which prevents slippage between the two parts.

Knock - The heavy metallic engine sound, produced in the combustion chamber as a result of abnormal combustion - usually detonation. Knock is usually caused by a loose or worn bearing. Also referred to as detonation, pinging and spark knock. Connecting rod or main bearing knocks are created by too much oil clearance or insufficient lubrication.

L

Lands - The portions of metal between the piston ring grooves.

Lapping the valves - Grinding a valve face and its seat together with lapping compound.

Lash - The amount of free motion in a gear train, between gears, or in a mechanical assembly, that occurs before movement can begin. Usually refers to the lash in a valve train.

Lifter - The part that rides against the cam to transfer motion to the rest of the valve train.

M

Machining - The process of using a machine to remove metal from a metal part.

Main bearings - The plain, or babbit, bearings that support the crankshaft.

Main bearing caps - The cast iron caps, bolted to the bottom of the block, that support the main bearings.

O

O.D. - Outside diameter.

Oil gallery - A pipe or drilled passageway in the engine used to carry engine oil from one area to another.

Oil ring - The lower ring, or rings, of a piston; designed to prevent excessive amounts of oil from working up the cylinder walls and into the combustion chamber. Also called an oil-control ring.

Oil seal - A seal which keeps oil from leaking out of a compartment. Usually refers to a dynamic seal around a rotating shaft or other moving part.

O-ring - A type of sealing ring made of a special rubberlike material; in use, the O-ring is compressed into a groove to provide the sealing action.

Overhaul - To completely disassemble a unit, clean and inspect all parts, reassemble it with the original or new parts and make all adjustments necessary for proper operation.

P

Pilot bearing - A small bearing installed in the center of the flywheel (or the rear end of the crankshaft) to support the front end of the input shaft of the transmission.

Pip mark - A little dot or indentation which indicates the top side of a compression ring.

Piston - The cylindrical part, attached to the connecting rod, that moves up and down in the cylinder as the crankshaft rotates. When the fuel charge is fired, the piston transfers the force of the explosion to the connecting rod, then to the crankshaft.

Piston pin (or wrist pin) - The cylindrical and usually hollow steel pin that passes through the piston. The piston pin fastens the piston to the upper end of the connecting rod.

Piston ring - The split ring fitted to the groove in a piston. The ring contacts the sides of the ring groove and also rubs against the cylinder wall, thus sealing space between piston and wall. There are two types of rings: Compression rings seal the compression pressure in the combustion chamber; oil rings scrape excessive oil off the cylinder wall.

Piston ring groove - The slots or grooves cut in piston heads to hold piston rings in position.

Piston skirt - The portion of the piston below the rings and the piston pin hole.

Plastigage - A thin strip of plastic thread, available in different sizes, used for measuring clearances. For example, a strip of plastigage is laid across a bearing journal and mashed as parts are assembled. Then parts are disassembled and the width of the strip is measured to determine clearance between journal and bearing. Commonly used to measure crankshaft main-bearing and connecting rod bearing clearances.

Press-fit - A tight fit between two parts that requires pressure to force the parts together. Also referred to as drive, or force, fit.

Prussian blue - A blue pigment; in solution, useful in determining the area of contact between two surfaces. Prussian blue is commonly used to determine the width and location of the contact area between the valve face and the valve seat.

R

Race (bearing) - The inner or outer ring that provides a contact surface for balls or rollers in bearing.

Ream - To size, enlarge or smooth a hole by using a round cutting tool with fluted edges.

Ring job - The process of reconditioning the cylinders and installing new rings.

Runout - Wobble. The amount a shaft rotates out-of-true.

S

Saddle - The upper main bearing seat.

Scored - Scratched or grooved, as a cylinder wall may be scored by abrasive particles moved up and down by the piston rings.

Scuffing - A type of wear in which there's a transfer of material between parts moving against each other; shows up as pits or grooves in the mating surfaces.

Seat - The surface upon which another part rests or seats. For example, the valve seat is the matched surface upon which the valve face rests. Also used to refer to wearing into a good fit; for example, piston rings seat after a few miles of driving.

Short block - An engine block complete with crankshaft and piston and, usually, camshaft assemblies.

Static balance - The balance of an object while it's stationary.

Step - The wear on the lower portion of a ring land caused by excessive side and back-clearance. The height of the step indicates the ring's extra side clearance and the length of the step projecting from the back wall of the groove represents the ring's back clearance.

Stroke - The distance the piston moves when traveling from top dead center to bottom dead center, or from bottom dead center to top dead center.

Stud - A metal rod with threads on both ends.

T

Tang - A lip on the end of a plain bearing used to align the bearing during assembly.

Tap - To cut threads in a hole. Also refers to the fluted tool used to cut threads.

Taper - A gradual reduction in the width of a shaft or hole; in an engine cylinder, taper usually takes the form of uneven wear, more pronounced at the top than at the bottom.

Throws - The offset portions of the crankshaft to which the connecting rods are affixed.

Thrust bearing - The main bearing that has thrust faces to prevent excessive endplay, or forward and backward movement of the crankshaft.

Thrust washer - A bronze or hardened steel washer placed between two moving parts. The washer prevents longitudinal movement and provides a bearing surface for thrust surfaces of parts.

Tolerance - The amount of variation permitted from an exact size of measurement. Actual amount from smallest acceptable dimension to largest acceptable dimension.

U

Umbrella - An oil deflector placed near the valve tip to throw oil from the valve stem area.

Undercut - A machined groove below the normal surface.

Undersize bearings - Smaller diameter bearings used with re-ground crankshaft journals.

V

Valve grinding - Refacing a valve in a valve-refacing machine.

Valve train - The valve-operating mechanism of an engine; includes all components from the camshaft to the valve.

Vibration damper - A cylindrical weight attached to the front of the crankshaft to minimize torsional vibration (the twist-untwist actions of the crankshaft caused by the cylinder firing impulses). Also called a harmonic balancer.

W

Water jacket - The spaces around the cylinders, between the inner and outer shells of the cylinder block or head, through which coolant circulates.

Web - A supporting structure across a cavity.

Woodruff key - A key with a radiused backside (viewed from the side).

11.1 Checking the crankshaft endplay with a dial indicator

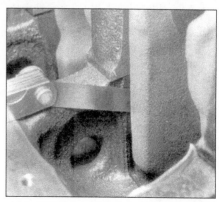

11.3 Checking crankshaft endplay with a feeler gauge

Final installation

41   Carefully scrape all traces of the Plasti-gage material off the rod journal and/or bearing face. Be very careful not to scratch the bearing - use your fingernail or the edge of a plastic card.

42   Make sure the bearing faces are perfectly clean, then apply a uniform layer of clean moly-base grease or engine assembly lube to both of them. You'll have to push the piston into the cylinder to expose the face of the bearing insert in the connecting rod.

43   Slide the connecting rod back into place on the journal, install the rod cap, install the nuts or new bolts and tighten them to the torque listed in this Chapter's Specifications. Again, work up to the torque in three steps. **Caution:** *If the connecting rod caps are secured to the rods with bolts (instead of nuts), install new connecting rod cap bolts. Do NOT reuse old bolts - they have stretched and cannot be reused.*

44   Repeat the entire procedure for the remaining pistons/connecting rods.

45   The important points to remember are:

a) *Keep the back sides of the bearing inserts and the insides of the connecting rods and caps perfectly clean when assembling them.*

b) *Make sure you have the correct piston/rod assembly for each cylinder.*

c) *The mark on the piston must face the front of the engine.*

d) *Lubricate the cylinder walls liberally with clean oil.*

e) *Lubricate the bearing faces when installing the rod caps after the oil clearance has been checked.*

46   After all the piston/connecting rod assemblies have been correctly installed, rotate the crankshaft a number of times by hand to check for any obvious binding.

47   As a final step, check the connecting rod endplay again (see Step 3). If it was correct before disassembly and the original crankshaft and rods were reinstalled, it should still be correct. If new rods or a new crankshaft were installed, the endplay may be inadequate. If so, the rods will have to be

removed and taken to an automotive machine shop for resizing.

11   Crankshaft - removal and installation

Removal

Refer to illustrations 11.1, 11.3 and 11.5
Note: *The crankshaft can be removed only after the engine has been removed from the vehicle. It's assumed that the flywheel or driveplate, crankshaft pulley, timing chain(s), oil pan, oil pump body, oil filter and piston/connecting rod assemblies have already been removed. On 2.4L four-cylinder engines, the balance shaft carrier assembly must be removed (see Section 9). The rear main oil seal retainer must be unbolted and separated from the block before proceeding with crankshaft removal.*

1   Before the crankshaft is removed, measure the endplay. Mount a dial indicator with the indicator in line with the crankshaft and touching the end of the crankshaft as shown **(see illustration)**.

2   Pry the crankshaft all the way to the rear and zero the dial indicator. Next, pry the crankshaft to the front as far as possible and check the reading on the dial indicator. The distance traveled is the endplay. A typical crankshaft endplay will fall between 0.003 and 0.010-inch (0.076 to 0.254 mm). If it's greater than that, check the crankshaft thrust surfaces for wear after its removed. If no wear is evident, new main bearings should correct the endplay.

3   If a dial indicator isn't available, feeler gauges can be used. Gently pry the crankshaft all the way to the front of the engine. Slip feeler gauges between the crankshaft and the front face of the thrust bearing or washer to determine the clearance **(see illustration)**.

4   Loosen the main bearing cap/bedplate fasteners 1/4-turn at a time each, until they can be removed by hand.

5   Carefully and evenly pry the main bearing caps/bedplate off of the engine **(see**

illustration). **Caution:** *Use the cast-in pry points on the bedplate. Do not pry on the block or bedplate mating surfaces, because damage to those surfaces can occur. Try not to drop the bearing inserts if they come out with the caps.*

6   Carefully lift the crankshaft out of the engine. It may be a good idea to have an assistant available, since the crankshaft is quite heavy and awkward to handle. With the bearing inserts in place inside the engine block and main bearing caps/bedplate, reinstall the main bearing caps/bedplate onto the engine block and tighten the fasteners finger tight.

Installation

7   Crankshaft installation is the first step in engine reassembly. It's assumed at this point that the engine block and crankshaft have been cleaned, inspected and repaired or reconditioned.

8   Position the engine block with the bottom facing up.

9   Remove the mounting fasteners and lift off the main bearing caps/bedplate.

10   If they're still in place, remove the original bearing inserts from the block and from

11.5 On 3.7L V6 engines, pry the bedplate off the engine block at the casting protrusion only, then remove the bedplate to access the crankshaft - be careful not to drop the lower bearing inserts when removing the bedplate

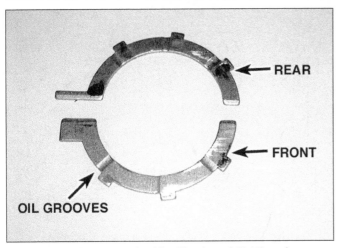

11.14a Thrust bearing identification - 3.7L V6 engine

11.14b Insert the thrust washer into the machined surface between the crankshaft and the upper bearing saddle, then rotate it down into the block until its flush with the parting line of the main bearing saddle - make sure the oil grooves on the thrust bearings face the crankshaft

the main bearing caps/bedplate. Wipe the bearing surfaces of the block and main bearing caps/bedplate with a clean, lint-free cloth. They must be kept spotlessly clean. This is critical for determining the correct bearing oil clearance.

Main bearing oil clearance check

Refer to illustrations 11.14a, 11.14b, 11.17, 11.18a, 11.18b, 11.19a, 11.19b, 11.9c, 11.9d, 11.9e and 11.21

11 Without mixing them up, clean the back sides of the new upper main bearing inserts (with grooves and oil holes) and lay one in each main bearing saddle in the block. Each upper bearing has an oil groove and oil hole in it. **Caution:** *The oil holes in the block must line up with the oil holes in the upper bearing inserts.* On 2.4L four-cylinder engines, install the grooved thrust bearing into the block saddle that has the

machined surface. **Note:** *On V6 engines, the two piece thrust washers will be installed later - after the crankshaft is laid in position in the block.* Clean the back sides of the lower main bearing inserts (without grooves) and lay them in the main bearing cap/bedplate saddles. Make sure the tab on the bearing insert fits into the recess in the block or main bearing caps/bedplate. **Caution:** *Do not hammer the bearing insert into place and don't nick or gouge the bearing faces. DO NOT apply any lubrication at this time.*

12 Clean the faces of the bearing inserts in the block and the crankshaft main bearing journals with a clean, lint-free cloth.

13 Check or clean the oil holes in the crankshaft, as any dirt here can go only one way - straight through the new bearings.

14 Once you're certain the crankshaft is clean, carefully lay it in position in the cylinder block. On V6 engines, install the two-piece

thrust washers in the upper bearing saddles that have the machined surfaces. The grooves in the thrust washers must face the crankshaft (so that the plain sides will face against the main bearing saddle) **(see illustrations)**.

15 Before the crankshaft can be permanently installed, the main bearing oil clearance must be checked.

16 Cut several strips of the appropriate size of Plastigage (they must be slightly shorter than the width of the main bearing journal).

17 Place one piece on each crankshaft main bearing journal, parallel with the journal axis, as shown **(see illustration)**.

18 Clean the faces of the bearing inserts in the main bearing caps/bedplate. Hold the bearing inserts in place and install the main bearing cap/bedplate assembly onto the crankshaft and cylinder block **(see illustrations)**. DO NOT disturb the Plastigage.

11.17 Place the Plastigage onto the crankshaft main bearing journals as shown - parallel to the crankshaft centerline

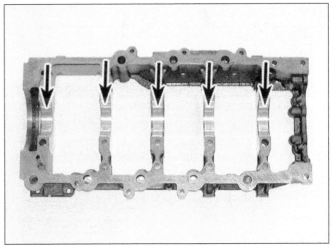

11.18a On 3.7L V6 engines, the bearings are laid into the corresponding saddles in the bedplate, then the bedplate is flipped 180 degrees . . .

11.18b . . . and set over the crankshaft onto the dowels on the engine block

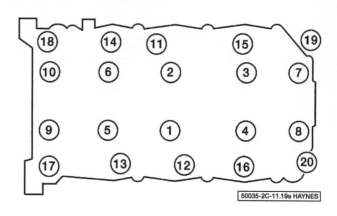

11.19a 2.4L four cylinder main bearing cap/bedplate tightening sequence

19   Apply clean engine oil to all bolt threads prior to installation, then install all bolts finger-tight. Tighten the main bearing cap/bedplate fasteners in the sequence shown **(see illustrations)** progressing in several steps, to the torque listed in this Chapter's Specifications. On 4.0L engines, only install and tighten the inner bolts. It is not necessary to install the main bearing cap outer bolts or side bolts at this time. DO NOT rotate the crankshaft at any time during this operation.

20   Remove the bolts in the reverse order of the tightening sequence and carefully lift the main bearing caps/bedplate straight up and off the block. Do not disturb the Plastigage or rotate the crankshaft.

21   Compare the width of the crushed Plastigage on each journal to the scale printed on the Plastigage envelope to determine the main bearing oil clearance **(see illustration)**, A typical main bearing oil clearance should fall between 0.0015 to 0.0023-inch (0.038 to 0.058 mm). Check with an automotive machine shop for the oil clearance for your engine.

22   If the clearance is not acceptable, the bearing inserts may be the wrong size (which means different ones will be required). Before deciding if different inserts are needed, make sure that no dirt or oil was between the bearing inserts and the main bearing caps/bedplate or block when the clearances were measured. If the Plastigage was wider at one end than the other, the crankshaft journal may be tapered. If the clearance still exceeds the limit specified, the bearing insert(s) will have to be replaced with an undersize bearing insert(s). **Caution:** *When installing a new crankshaft, always install a standard bearing insert set.*

23   Carefully scrape all traces of the Plastigage material off the main bearing journals and/or the bearing insert faces. Be sure to remove all residue from the oil holes. Use your fingernail or the edge of a plastic card -don't nick or scratch the bearing faces.

Final installation

Refer to illustrations 11.28a and 11.28b

24   Carefully lift the crankshaft out of the cylinder block.

25   Clean the bearing insert faces in the cylinder block, then apply a thin, uniform layer of moly-base grease or engine assembly lube to each of the bearing surfaces. Be sure to coat the thrust faces of the thrust bearing/washers as well. *See Steps 11 and 14 for thrust washer/bearing locations.*

26   Make sure the crankshaft journals are clean, then lay the crankshaft back in place in the cylinder block.

27   Clean the bearing insert faces and then apply the same lubricant to them.

28   On 2.4L four-cylinder engines, apply Mopar Bedplate Sealant or equivalent to the bedplate sealing area on the block **(see illustration)**. On 3.7L V6 engines, apply Mopar Engine RTV Sealant or equivalent to the bedplate sealing area on the block **(see illustration)**. Hold the bearing inserts in place and install the main bearing caps/bedplate on the crankshaft and cylinder block. Tap the main bearing caps/bedplate into place with a brass or a soft-face hammer.

29   Using NEW main bearing cap bolts, apply clean engine oil to the bolt threads,

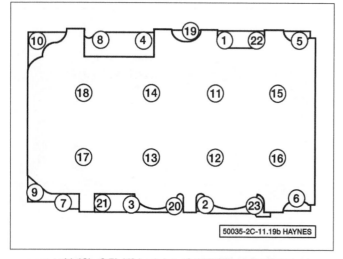

11.19b 3.7L V6 bedplate tightening sequence

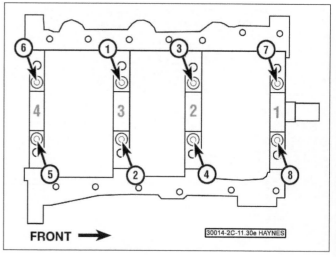

11.19c Main bearing cap inner bolt tightening sequence - 4.0L engines

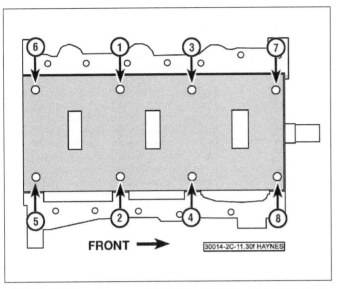

11.19d Main bearing cap outer bolt and windage tray tightening sequence - 4.0L engines

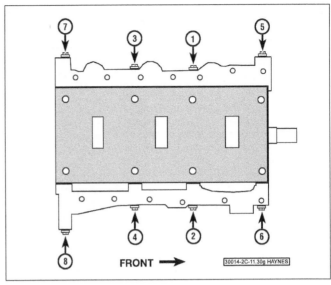

11.19e Main bearing cap side bolt tightening sequence - 4.0L engines

wipe off any excess oil and then install the fasteners finger-tight.

30 Push the crankshaft forward using a wooden wedge between the block and the crankshaft counterweight to seat the thrust bearing. Do not use the wedge against the bedplate. Once the crankshaft is pushed fully forward to seat the thrust bearing, leave the screwdriver in position so that pressure stays on the crankshaft until after all main bearing cap/bedplate fasteners have been tightened.

31 Tighten the main bearing cap/bedplate fasteners in the indicated sequence **(see illustration 11.19a, 11.9b or 11.9c)**, to the torque listed in *Step 1* of this Chapter's Specifications. Remove the screwdriver or prybar.

32 Tighten the main bearing cap/bedplate fasteners in the indicated sequence **(see illustration 11.19a, 11.9b, 11.9c, 11.9d or 11.19e)** and to the torque and angle

listed in the remaining steps of this Chapter's Specifications.

33 Recheck crankshaft endplay with a feeler gauge or a dial indicator. The endplay should be correct if the crankshaft thrust faces aren't worn or damaged and if new bearings have been installed.

34 Rotate the crankshaft a number of times by hand to check for any obvious binding.

35 Install the new rear main oil seal (see Chapter 2A, 2B or 2C).

36 On 2.4L four-cylinder engines, reinstall the balance shaft carrier assembly (see Section 9).

12 Engine overhaul - reassembly sequence

1 Before beginning engine reassembly, make sure you have all the necessary new parts, gaskets and seals as well as the

11.21 Compare the width of the crushed Plastigage to the scale on the envelope, taking the measurement at the widest point of the Plastigage

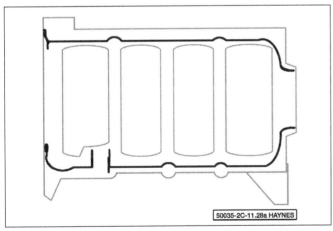

11.28a On 2.4L four-cylinder engines, apply Mopar Bedplate sealant or equivalent to the block-to-bedplate mating surface as shown

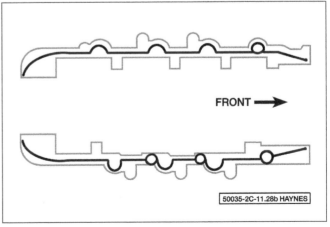

11.28b On 3.7L V6 engines, apply Mopar RTV sealant or equivalent to the block-to-bedplate mating surface as shown

following items on hand:

Common hand tools
A 1/2-inch drive torque wrench
New engine oil
Gasket sealant
Thread locking compound

2   If you obtained a short block it will be necessary to install the cylinder head, the oil pump and pick-up tube, the oil pan, the water pump, the timing belt and timing cover, and the valve cover (see Chapter 2A, 2B or 2C). In order to save time and avoid problems, the external components must be installed in the following general order:

Thermostat and housing cover
Water pump
Intake and exhaust manifolds
Carburetor or fuel injection components
Emission control components
Spark plug wires and spark plugs
Ignition coils
Oil filter
Engine mounts and mount brackets

Clutch and flywheel (manual transmission)
Driveplate (automatic transmission)

13   Initial start-up and break-in after overhaul

Warning: *Have a fire extinguisher handy when starting the engine for the first time.*
1   Once the engine has been installed in the vehicle, double-check the engine oil and coolant levels.
2   With the spark plugs out of the engine and the ignition system and fuel pump disabled, crank the engine until oil pressure registers on the gauge or the light goes out.
3   Install the spark plugs, hook up the plug wires and restore the ignition system and fuel pump functions.
4   Start the engine. It may take a few moments for the fuel system to build up pressure, but the engine should start without a great deal of effort.
5   After the engine starts, it should be

allowed to warm up to normal operating temperature. While the engine is warming up, make a thorough check for fuel, oil and coolant leaks.
6   Shut the engine off and recheck the engine oil and coolant levels.
7   Drive the vehicle to an area with minimum traffic, accelerate from 30 to 50 mph, then allow the vehicle to slow to 30 mph with the throttle closed. Repeat the procedure 10 or 12 times. This will load the piston rings and cause them to seat properly against the cylinder walls. Check again for oil and coolant leaks.
8   Drive the vehicle gently for the first 500 miles (no sustained high speeds) and keep a constant check on the oil level. It is not unusual for an engine to use oil during the break-in period.
9   At approximately 500 to 600 miles, change the oil and filter.
10   For the next few hundred miles, drive the vehicle normally. Do not pamper it or abuse it.
11   After 2000 miles, change the oil and filter again and consider the engine broken in.

Notes

Notes

Chapter 3
Cooling, heating and air conditioning systems

Contents

Specifications

General

Coolant capacity	See Chapter 1
Drivebelt tension	See Chapter 1
Radiator cap pressure rating	
2007 and earlier models	18 to 21 psi (124 to 145kPa)
2008 and later models	10 to 18 psi (69 to 124kPa)
Thermostat opening temperature	
2007 and earlier models	195-degrees F (90-degrees C)
2008 and later models	176-degrees F (80-degrees C)

Torque specifications

Note: *One foot-pound (ft-lb) of torque is equivalent to 12 inch-pounds (in-lbs) of torque. Torque values below approximately 15 foot-pounds are expressed in inch-pounds, because most foot-pound torque wrenches are not accurate at these smaller values.*

	Ft-lbs (unless otherwise indicated)	Nm
Viscous fan drive, nut-to-water pump		
3.7L V6 engine	70	95
4.0L V6 engine	37	50
Fan blade-to-viscous drive clutch bolts		
2004 and earlier models	20	28
2005 through 2010 models	17	23
Electric fan-to-shroud mounting bolts		
All except 4.0L engine	80 in-lbs	9
4.0L V6 engine	50 in-lbs	6
Thermostat housing bolts		
2.4L four-cylinder	20	28
3.7L V6	105 in-lbs	12
4.0L V6 engine	25	35
Water pump bolts		
2.4L four-cylinder	105 in-lbs	12
3.7L V6	40	54
4.0L V6 engine	105 in-lbs	12

1   General information

Engine cooling system

All vehicles covered by this manual employ a pressurized engine cooling system with thermostatically controlled coolant circulation. An impeller-type water pump mounted on the front of the block pumps coolant through the engine. The coolant flows around each cylinder and toward the rear of the engine. Cast-in coolant passages direct coolant around the intake and exhaust ports, near the spark plug areas and in close proximity to the exhaust valve guides.

A wax-pellet type thermostat is located in a housing near the front of the engine. During warm up, the closed thermostat prevents coolant from circulating through the radiator. As the engine nears normal operating temperature, the thermostat opens and allows hot coolant to travel through the radiator, where it's cooled before returning to the engine.

The cooling system is sealed by a pressure-type cap on the expansion tank or radiator, which raises the boiling point of the coolant. If the system pressure exceeds the cap pressure-relief value, the excess pressure in the system forces the spring-loaded valve inside the cap off its seat. This allows either excess pressure or coolant to escape through an overflow tube into a separate coolant recovery chamber, which is part of the expansion tank (models with the pressure cap on the expansion tank), or into a coolant reservoir (models with the pressure cap on the radiator).

Heating system

The heating system consists of a blower fan and heater core located in the heater/air conditioning unit under the dash, the hoses connecting the heater core to the engine cooling system and the heater/air conditioning control head on the instrument panel. Hot engine coolant is circulated through the heater core. When the heater mode is activated, a flap door opens to expose the heater box to the passenger compartment. A fan switch on the control head activates the blower motor, which forces air through the core, heating the air.

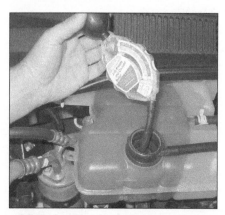

2.5 An inexpensive hydrometer can be used to test the condition of your coolant

Air conditioning system

The air conditioning system consists of a condenser mounted in front of the radiator, an evaporator mounted adjacent to the heater core under the dash, a compressor mounted on the engine, an accumulator mounted near the firewall and the plumbing that connects all of these components.

A blower fan forces the warmer air of the passenger compartment through the evaporator core (sort of a radiator-in-reverse), transferring the heat from the air to the refrigerant. The liquid refrigerant boils off into low-pressure vapor, taking the heat with it when it leaves the evaporator.

2   Antifreeze/coolant - general information

Refer to illustration 2.5
Warning 1: *Use ethylene-glycol type antifreeze only. Do not use propylene-glycol type antifreeze solutions.*
Warning 2: *Do not allow antifreeze to come in contact with your skin or painted surfaces of the vehicle. Rinse off spills immediately with plenty of water. Antifreeze is highly toxic if ingested. Never leave antifreeze lying around in an open container or in puddles on the floor; children and pets are attracted by its sweet smell and may drink it. Check with local authorities about disposing of used antifreeze. Many communities have collection centers which will see that antifreeze is disposed of safely. Never dump used anti-freeze on the ground or pour it into drains.*
Note: *Non-toxic antifreeze is now manufactured and available at local auto parts stores, but even these types should be disposed of properly.*

1   Keep antifreeze containers covered and repair leaks in your cooling system as soon as they are noticed.
2   The cooling system should be filled with a water/ethylene-glycol based antifreeze solution, which will prevent freezing down to at least -20°F, or lower if local climate requires it. It also provides protection against corrosion and increases the coolant boiling point.
3   The cooling system should be drained, flushed and refilled at the specified intervals (see Chapter 1). Old or contaminated antifreeze solutions are likely to cause damage and encourage the formation of rust and scale in the system. Use distilled water with the antifreeze.
4   Before adding antifreeze, check all hose connections, because antifreeze tends to leak through very minute openings. Engines don't normally consume coolant, so if the level goes down, find the cause and correct it.
5   The exact mixture of antifreeze-to-water which you should use depends on the relative weather conditions. The mixture should contain at least 50-percent antifreeze, but should never contain more than 70-percent antifreeze. Consult the mixture ratio chart on the antifreeze container before adding coolant. Hydrometers are available at most auto

parts stores to test the coolant **(see illustration)**. Use antifreeze which meets the vehicle manufacturer's specifications.

3   Thermostat - check and replacement

Warning: *Do not remove the expansion tank cap, drain the coolant or replace the thermostat until the engine has cooled completely.*
Note: *The following check applies to 2.4L four-cylinder engines only. On 3.7L V6 engines, the thermostat is located at the lower radiator hose and the thermostat also allows a 50-percent flow of coolant when the engine is cool. Therefore, the check portion of this Section does not apply to the 3.7L V6 engine.*

Check (2.4L four-cylinder engine only)

1   Before assuming the thermostat is to blame for a cooling system problem, check the coolant level, drivebelt tension (Chapter 1) and temperature gauge (or light) operation.
2   If the engine seems to be taking a long time to warm up (based on heater output or temperature gauge operation), the thermostat is probably stuck open. Replace the thermostat with a new one.
3   If the engine runs hot, use your hand to check the temperature of the upper radiator hose. If the hose isn't hot, but the engine is, the thermostat is probably stuck closed, preventing the coolant inside the engine from escaping to the radiator. Replace the thermostat. **Caution:** *Don't drive the vehicle without a thermostat. The computer may stay in open-loop, potentially causing oil-sludge, excessive exhaust emissions/fuel consumption.*
4   If the upper radiator hose is hot, it means that the coolant is flowing and the thermostat is open. Consult the *Troubleshooting* section at the front of this manual for cooling system diagnosis.

Replacement (2.4L and 3.7L models)

Refer to illustrations 3.8, 3.14a and 3.14b
5   Disconnect the cable from the negative terminal of the battery.
6   Drain the cooling system (see Chapter 1). If the coolant is relatively new or in good condition (see Section 2), save it and reuse it.
7   On 2.4L four-cylinder engines, follow the upper radiator hose to the engine to locate the thermostat housing. On 3.7L V6 engines, follow the lower radiator hose to the engine to locate the thermostat housing.
8   Squeeze the tabs on the hose clamp to loosen it from the hose(s), then reposition the clamp several inches back up the hose. Detach the hose(s) from the thermostat housing **(see illustration)**. **Note:** *Special hose clamps pliers are available at most auto parts stores. If the hose is stuck, grasp it near the end with a pair of adjustable pliers and twist it to break the seal, then pull it off. If the hose is old or deteriorated, cut it off and install a new one.*

3.8 Thermostat housing location on 3.7L V6 engines - view here is from below at the front of the engine

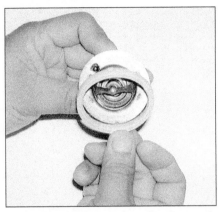

3.14a On 3.7L V6 engines, align the notch on the thermostat (arrow) with the rubber tab on the inner O-ring groove . . .

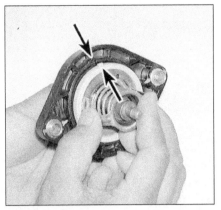

3.14b . . . then align the tab (lower arrow) on the outer diameter of the O-ring with the notch in the thermostat housing (upper arrow) and insert the thermostat into the thermostat housing

9   If the outer surface of the thermostat housing that mates with the hose is deteriorated (corroded, pitted, etc.) it may be damaged further by hose removal. If it is, the thermostat housing will have to be replaced.

10   Remove the thermostat housing from the engine. If the housing is stuck, tap it with a soft-face hammer to jar it loose. Be prepared for some coolant to spill as the gasket seal is broken.

11   Note how the thermostat is installed (which end is facing up, or out, and the position of the air bleed "jiggle valve") and remove it from the housing.

12   On 2.4L four-cylinder engines, stuff a rag into the engine opening, then remove all traces of old gasket material and sealant from the housing and cover with a gasket scraper. Remove the rag from the opening and clean the gasket mating surfaces with chemical gasket remover or lacquer thinner or acetone. On 3.7L V6 engines, simply remove the rubber O-ring from around the thermostat.

13   On 2.4L four-cylinder engines, install the new thermostat in the machined groove on the engine. Make sure the correct end faces out - the spring end is normally directed into the engine. Apply a thin, uniform layer of RTV sealant to both sides of the new gasket and position it over the thermostat on the engine.

14   On 3.7L V6 engines, install a new O-ring around the thermostat. Make sure to align the rubber tab on the inside of the O-ring groove with the notch on the thermostat **(see illustration)**. Then align the rubber tab on the outside of the O-ring with the notch on the thermostat housing and insert the thermostat and O-ring into the thermostat housing **(see illustration)**.

15   Install the thermostat housing and bolts onto the engine. Tighten the bolts to the torque listed in this Chapter's Specifications.

16   Reattach the hose(s) to the fitting(s) and tighten the hose clamp(s) securely.

17   Refill the cooling system (see Chapter 1).

18   Start the engine and allow it to reach normal operating temperature, then check for leaks and proper thermostat operation.

4.0L engines

19   Disconnect the cable from the negative battery terminal (see Chapter 5).

20   Drain the cooling system (see Chapter 1). If the coolant is relatively new or in good condition, save it and reuse it. Read the Warning in Section 2.

21   Remove the air filter housing (see Chapter 4).

22   Remove the upper intake manifold (see Chapter 2C).

23   Remove the radiator tube mounting nuts, then loosen the radiator hose clamp at the thermostat and disconnect the hose from the thermostat housing.

24   Remove the thermostat housing bolts and housing from the lower intake manifold, then remove the thermostat from the manifold noting the location of the jiggle valve on the thermostat.

25   Remove all traces of the old gasket from the mating surfaces and clean them thoroughly.

26   Install the new thermostat, with the jiggle pin in the 12 o'clock position, and the spring end directed into the engine.

27   Install a new gasket or O-ring, making sure that it is oriented in the same way as the original.

Note: *It is standard practice to use a thin layer of RTV sealant when installing flat replacement gaskets. However, if the gasket is designed with a raised crushable sealing surface (not flat), or if it is an O-ring, no RTV sealant is necessary.*

28   Install the thermostat housing cover, tightening the fasteners to the torque listed in this Chapter's Specifications.

29   Reattach the hose and tighten the hose clamp securely, if removed. Install all components that were previously removed.

30   Remaining installation is the reverse of removal.

31   Reconnect the battery (see Chapter 5).

32   Refill the cooling system (see Chapter 1).

33   Start the engine and allow it to reach normal operating temperature, then check for leaks and proper thermostat operation (as described in Section 2).

4   Engine cooling fan(s) - check and component replacement

Warning: *To avoid possible injury or damage, DO NOT operate the engine with a damaged fan. Do not attempt to repair fan blades - replace a damaged fan with a new one.*

Note: *Most of the models covered by this manual are equipped with a single electric cooling fan only. However some 3.7L V6 engines are equipped with a heavy-duty cooling package which consists of an additional mechanical clutch-type cooling fan and a larger-capacity radiator.*

Check

Electric cooling fan (all models)

1   On vehicles equipped with air conditioning, with the air conditioning ON, the electric fan motor will start when the engine coolant temperature exceeds 203° F (95°C) or the battery temperature sensor exceeds 100° F (38° C). With the air conditioning OFF, or on vehicles NOT equipped with air conditioning, the electric fan motor will start when the engine coolant temperature exceeds 176° F (80° C) or the battery temperature sensor exceeds 16° F (-9° C).

2   If the electric cooling fan does not appear to operate properly, check the fuses in the interior compartment fuse block and in the power distribution center in the engine compartment. If any of the fuses are blown, see Step 4.

3   If no fuses are blown, disconnect the electrical connector from the cooling fan motor. Connect a fused jumper wire between one of the two terminals on the cooling fan connector (the fan motor side of the connector) and the positive battery terminal. Connect another jumper wire between the other terminal and ground. When the connections are made, the cooling fan should operate. Polarity may be critical, so, if the cooling fan doesn't operate, try swapping the jumper wires and test the operation again. If the cool-

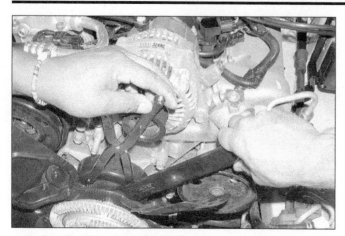

4.10 On 3.7L V6 engines with the heavy duty cooling system, use a special pin spanner wrench to hold the water pump pulley while the mechanical fan drive nut is loosened

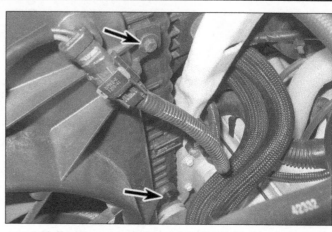

4.12 Fan shroud mounting nuts/bolts (right side shown)

ing fan does not operate, it is faulty and the fan/shroud assembly should be replaced.

4   Because the electric cooling fan on the these models is directly controlled by the Powertrain Control Module (PCM), cooling fan relay and/or air conditioning system, all other tests on this system must be performed by a dealer service department or other qualified repair facility with a specialized scan tool.

Mechanical cooling fan (3.7L and 4.0L V6 engines with heavy duty cooling package)

5   Disconnect the cable from the negative terminal of the battery.

6   Rock the fan back and forth by hand to check for excessive bearing play and visually inspect the clutch assembly for substantial fluid leakage. If problems are noted, replace the clutch assembly.

7   With the engine cold (and not running), turn the fan blades by hand. The fan should turn freely but never more than five turns.

8   With the engine completely warmed up, turn off the ignition switch and disconnect the negative battery cable from the battery. Turn the fan by hand. More drag should be evident. If the fan turns easily or it rotates the same number of turns it did when the clutch was cold, replace the fan clutch.

Replacement

Refer to illustrations 4.10, 4.12, 4.13 and 4.15

9   Disconnect the cable from the negative terminal of the battery.

10   On V6 engines with the heavy duty cooling package, a special spanner wrench obtainable at most auto parts stores may be required to hold the water pump pulley while a large wrench is used to loosen the fan drive nut on the mechanical cooling fan **(see illustration)**. Sometimes it is possible to hold the water pump pulley by applying considerable hand pressure to the serpentine belt while the large nut is turned counter-clockwise (right-hand threads), but it may require the tool if the fan drive nut is excessively tight. Carefully lower the fan and clutch assembly into the fan shroud. Be very careful not to damage the radiator fins while doing so.

11   Disconnect the electrical connector from the electric fan motor.

12   Remove the fan shroud mounting bolts **(see illustration)**.

13   Lift the shroud up until it clears the slots in the bottom bracket of the radiator, then lift the fan/clutch and shroud or the electric cooling fan and shroud assembly out of the engine compartment together **(see illustration)**. **Note 1:** *If difficulty is encountered during the removal of the shroud, it will be necessary to*

drain the cooling system and remove the upper radiator hose to help facilitate removal of the shroud and fan. **Note 2:** *The electric cooling fan and the shroud are assembled as a unit and if defective must be replaced as a unit.*

14   Carefully inspect the fan blades for damage and defects. Replace it if necessary.

15   If applicable, the mechanical fan may be unbolted from the fan clutch, if necessary **(see illustration)**. If the fan clutch is stored, keep it in a vertical position (as if still mounted on the engine) to prevent the silicone fluid from leaking out.

16   Installation is the reverse of removal. Be sure to tighten the fan and clutch mounting nuts evenly and securely.

---

5   Coolant expansion tank/coolant reservoir - removal and installation

Warning: *Wait until the engine is completely cool before beginning this procedure.*

1   Drain the cooling system (see Chapter 1).

2   Disconnect the hoses from the expansion tank.

Four cylinder models

Refer to illustration 5.3

3   Remove the expansion tank mounting

4.13 Lift the fan/shroud assembly out, being careful not to damage the fins on the radiator

4.15 If applicable and if necessary, remove the four bolts retaining the mechanical fan to the fan clutch

5.3 The coolant expansion tank is retained to the firewall with two mounting nuts

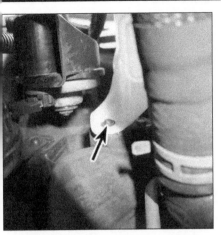

5.8a Coolant reservoir/washer fluid reservoir left lower mounting screw

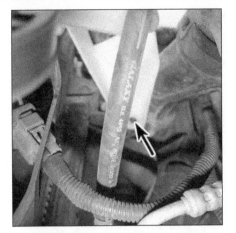

5.8b Coolant reservoir/washer fluid reservoir right lower mounting screw

5.8c Coolant reservoir/washer fluid reservoir right upper mounting screws

nuts from the firewall **(see illustration)**. Remove the tank.

4   Installation is the reverse of removal.

5   Refill the cooling system (see Chapter 1).

V6 models

Non-pressurized tank (coolant reservoir)

Refer to illustrations 5.8a, 5.8b & 5.8c

Note: *The coolant reservoir and windshield washer reservoir are combined into one assembly.*

6   Disconnect the cable from the negative terminal of the battery.

7   Disconnect the washer hoses and washer pump electrical connector.

8   Remove the fasteners along the upper and lower edges of the assembly **(see ilus-trations)**.

9   Remove the coolant reservoir/windshield washer reservoir assembly from the the radi-ator support.

Pressurized tank (expansion tank)

10   Remove the expansion tank-to-radiator bracket nuts.

11   Remove the expansion tank from the radiator bracket.

All systems

12   Installation is the reverse of removal.

13   Refill the cooling system (see Chapter 1).

---

6   Radiator - removal and installation

Warning: *Wait until the engine is completely cool before beginning this procedure.*

Removal

Refer to illustrations 6.7, 6.10, 6.11a and 6.11b

1   Disconnect the cable from the negative terminal of the battery.

2   Disconnect the electrical connector from the electric fan motor.

3   Drain the cooling system (see Chapter 1). If the coolant is relatively new or in good condition, save it and reuse it.

4   On V6 models, remove the coolant res-ervoir/windshield washer reservoir assembly (see Section 5).

5   If the vehicle is not equipped with a

mechanical cooling fan, the electric cooling fan/shroud assembly can be removed with the radiator as a unit. If the vehicle is equipped with a mechanical cooling fan, remove the mechanical fan (see Section 4).

6   Loosen the hose clamps, then detach the radiator hoses from the fittings on the radiator. If they're stuck, grasp each hose near the end with a pair of adjustable pliers and twist it to break the seal, then pull it off - be careful not to distort the radiator fittings. If the hoses are old or deteriorated, cut them off and install new ones.

7   Remove the radiator support **(see illus-tration)**. On later models, you will have to remove the grille (see Chapter 11).

8   If the vehicle is equipped with an auxiliary automatic transmission cooler, remove the cooler mounting bolts and set the cooler aside.

9   If the vehicle is equipped with an internal cooler in the radiator, detach the cooler lines from the rear of the radiator. Use a drip pan to catch spilled fluid. Plug the lines and fittings.

10   If equipped, remove the power steering cooler mounting bolts and set the cooler aside **(see illustration)**.

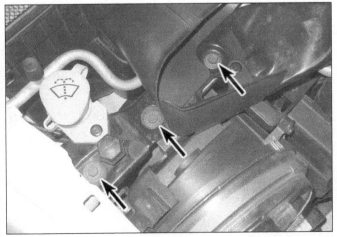

6.7 Remove the radiator support mounting bolts (right side shown)

6.10 If equipped, remove the power steering cooler mounting bolts, then set the cooler aside

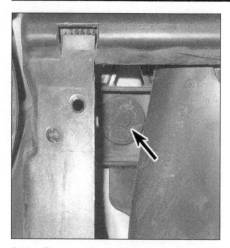

6.11a Remove the radiator mounting bolts (right side shown) (view from front of vehicle) . . .

6.11b . . . and carefully lift out the radiator

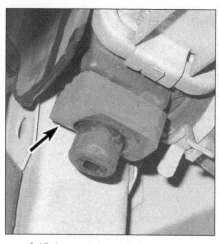

6.15 Inspect the radiator mounts for damage

11   Remove the radiator mounting bolts and carefully lift out the radiator **(see illustrations)**. Don't spill coolant on the vehicle or scratch the paint.
12   Whenever the radiator is removed from the vehicle, make note of the location of all rubber seals and their location. They should be put back exactly as originally installed to ensure proper cooling.

Installation

Refer to illustration 6.15

13   With the radiator removed, it can be inspected for leaks and damage. If it needs repair, have a radiator shop or dealer service department perform the work as special techniques are required.
14   Bugs and dirt can be removed from the front of the radiator with a garden hose, followed by compressed air and a soft brush. Don't bend the cooling fins as this is done. When blowing out the core, direct the hose or air line only from the engine side out.
15   Inspect the radiator mounts for deterioration and make sure there's nothing in them when the radiator is installed **(see illustration)**.
16   Installation is the reverse of the removal procedure.

17   After installation, fill the cooling system with the proper mixture of antifreeze and water (see Chapter 1).
18   Start the engine and check for leaks. Allow the engine to reach normal operating temperature, indicated by the upper radiator hose becoming hot. Recheck the coolant level and add more if required.
19   If you're working on an automatic transmission-equipped vehicle, check and add fluid as needed.

---

7   Water pump - check

Refer to illustrations 7.3 and 7.4

1   A failure in the water pump can cause serious engine damage due to overheating.
2   There are several ways to check the operation of the water pump while it's installed on the engine. If the pump is defective, it should be replaced with a new or rebuilt unit.
3   Water pumps are equipped with weep or vent holes. If a failure occurs in the pump seal, coolant will leak from the hole. In most cases you'll need a flashlight to find the hole on the water pump from underneath to check for leaks **(see illustration)**.

4   If the water pump shaft bearings fail there may be a howling sound at the front of the engine while it's running. Shaft wear can be felt if the water pump pulley is rocked up and down **(see illustration)**. Don't mistake drivebelt slippage, which causes a squealing sound, for water pump bearing failure.
5   It is possible for a water pump to be bad, even if it doesn't howl or leak water. Sometimes the fins on the back of the impeller can corrode away until the pump is no longer effective. The only way to check for this is to remove the pump for examination.

---

8   Water pump - removal and installation

Refer to illustration 8.12
Warning: *Wait until the engine is completely cool before beginning this procedure.*

Removal

Four-cylinder and 3.7L V6 engines

1   Disconnect the cable from the negative terminal of the battery. Drain the cooling system (see Chapter 1). If the coolant is relatively new or in good condition, save it and reuse it.
2   On V6 models, remove the coolant reservoir/windshield washer reservoir assembly (see Section 5).
3   Remove the cooling fan(s) and shroud (see Section 4).
4   Remove the drivebelts (see Chapter 1) and the water pump pulley.
5   On 2.4L four-cylinder engines, remove the timing belt, timing belt idler pulley, camshaft sprockets and rear timing belt cover (see Chapter 2A).
6   On 3.7L V6 engines, detach the lower radiator hose from the pump. If it is stuck, grasp it near the end with a pair of adjustable pliers and twist it to break the seal, then pull off. If the hose is deteriorated, cut it off and install a new one.

7.3 The water pump weep hole will drip coolant when the seal on the pump shaft fails (typical)

7.4 Grasp the water pump flange and try to rock the shaft back and forth to check for play (typical)

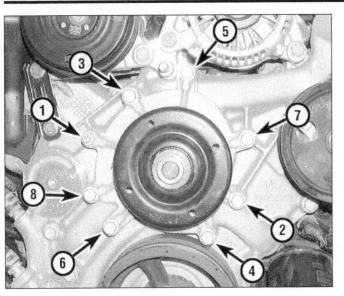

8.12 Water pump mounting bolts on a 3.7L V6 engine - during installation, tighten the bolts in the sequence shown

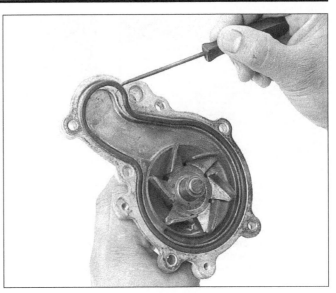

8.17 Be sure to install a new O-ring in the water pump groove (four-cylinder engine)

4.0L engines

7   Remove the air filter housing (see Chapter 4).

8   Remove the right engine mount through-bolt (see Chapter 2C).

9   Using an engine hoist, raise the engine slightly.

10   With the engine raised high enough to access the compressor mounting bolts, secure the engine in this position using a jack, then remove the air conditioning compressor mounting bolts. Warning: Do not disconnect the refrigerant lines.

11   Remove the accessory drivebelt bracket and timing belt (see Chapter 2C).

All engines

12   Remove the bolts and detach the water pump from the engine. Note the locations of the various lengths and different types of bolts as they're removed to ensure correct installation **(see illustration)**.

Installation

Refer to illustration 8.17

13   Clean the bolt threads and the threaded holes in the engine to remove corrosion and sealant.

14   Compare the new pump to the old one to make sure they're identical. If the old pump is being reused, check the impeller blades on the backside for corrosion. If any fins are missing or badly corroded, replace the pump with a new one.

15   Remove all traces of old gasket material from the engine with a gasket scraper.

16   Clean the engine and new water pump mating surfaces with brake system cleaner.

Four-cylinder and 3.7L V6 engines

17   On 2.4L four-cylinder engines, apply a thin coat of RTV sealant to the new water pump O-ring and position it in the groove on the back of the water pump **(see illustration)**. **Caution:** *Make sure the O-ring is correctly seated in the water pump groove to avoid a coolant leak.*

18   On 3.7L V6 engines, apply a thin layer of RTV sealant to the gasket mating surface of the new pump, and position the gasket on the pump. Apply a thin layer of RTV to the engine-side of the gasket and slip a couple of bolts through the pump mounting holes to hold the gasket in place.

19   Carefully attach the pump and O-ring/gasket to the engine and thread the bolts into the holes finger tight. If installing a new pump on a 3.7L V6 engine, spin the pump by hand to ensure that there is no interference between the impeller and the timing cover.

20   Install the remaining bolts and tighten them in a criss-cross pattern to the torque listed in this Chapter's Specifications in 1/4-turn increments. Turn the water pump by hand to make sure it rotates freely.

4.0L engines

21   Install a new O-ring seal in the groove that lines the water pump body, then apply a thin film of RTV sealant to hold the seal in place during installation. **Caution:** *Make sure that the O-ring seal is correctly seated in the water pump groove to avoid a coolant leak.*

22   Carefully mate the pump to the engine.

23   Install the water pump mounting bolts and tighten them to the torque listed in this Chapter's Specifications. **Caution:** *Don't overtighten the mounting bolts; doing so will damage the pump.*

24   Install the timing belt and accessory drivebelt bracket (see Chapter 2C).

All engines

25   Reinstall all parts removed for access to the pump. **Caution:** *Make sure the serpentine*

drivebelt is installed as originally routed (see Chapter 1) or overheating could result.

26   Refill the cooling system and check the drivebelt tension (see Chapter 1). Run the engine and check for leaks.

9   Heater and air conditioning blower motor - circuit check

Warning: *These models are equipped with airbags. The airbag is armed and can deploy (inflate) anytime the battery is connected. To prevent accidental deployment (and possible injury), disconnect the negative battery cable whenever working near airbag components. After the battery is disconnected, wait at least 2 minutes before beginning work (the system has a back-up capacitor that must fully discharge). For more information see Chapter 12.*

Note: *The circuit checking procedure outlined below applies to manually operated heating and air conditioning systems only. Automatic heating and air conditioning systems cannot be tested using conventional equipment. Due to the use of an integrated electronic control module that can only be tested with specialized equipment it will be necessary to take these vehicles to dealer service department or other qualified repair shop.*

Circuit check

Refer to illustrations 9.4, 9.6, 9.7a and 9.7b

1   Check the fuse and all connections in the circuit for looseness and corrosion. Make sure the battery is fully charged.

2   With the transmission in Park, the parking brake securely applied, turn the ignition switch to the On position (engine not running).

3   The blower motor is located under the

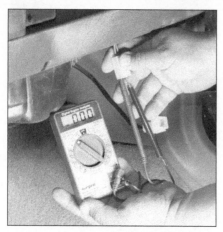

9.4 Disconnect the electrical connector to the blower motor and attach a voltmeter to the harness side of the connector

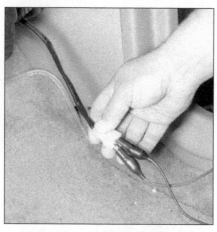

9.6 Supply a ground connection and fused battery voltage directly to the blower side of the connector to see if the blower operates

9.7a Remove the two screws and remove the blower motor resistor from the case

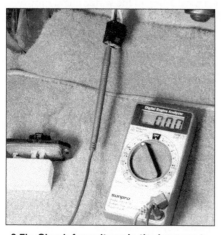

9.7b Check for voltage in the harness to the resistor, at each blower switch speed

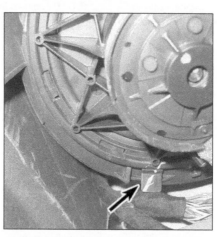

10.3 The blower motor is mounted under the instrument panel, near the glove box - disconnect the electrical connector, then release the blower motor locking tab

10.4 If the motor is being replaced, detach the fan clip, remove the fan and install it on the shaft of the new motor

glove compartment area of the dash, near the firewall. Bend back the clip on one of the blower motor mounting screws and release the wiring harness, pulling it down for easier disconnection of the connector.

4   Connect a voltmeter to the blower motor connector **(see illustration)**.

5   Move the blower switch through each of its positions and note the voltage readings. Changes in voltage indicate that the motor speeds will also vary as the switch is moved to the different positions. Slower speeds will deliver less voltage to the blower, and the HIGH position will bypass the resistor to supply a full 12 volts.

6   If there is voltage, but the blower motor does not operate, connect a jumper wire between the motor ground terminal and a good chassis ground. Connect a fused jumper wire between the battery positive terminal and the positive terminal on the motor side of the connector **(see illustration)**. If the motor now works, remove the jumper wire. If the motor stops working when the ground

wire is removed, check for bad ground and re-test. If the motor still doesn't work, the blower motor is probably faulty.

7   If there's no voltage at the blower motor, remove the resistor block connector and check it for voltage **(see illustrations)**. If there's voltage at any of the connector terminals, check the resistor block for continuity and the wiring between the resistor block and the motor for an open or short.

8   If there's no voltage to the resistor, remove the heater control panel (see Section 11) and, with the ignition ON, check for voltage at the connector for the blower motor switch.

9   If there's no voltage at the blower switch, check the wiring between the fuse panel and the switch for an open or a short.

10   If there's voltage, connect one end of a jumper wire to the terminal of the switch connector with voltage. Connect the other end of the jumper to each of the terminals that supply voltage to the resistor block. If the motor now operates normally, replace the switch.

10 Blower motor - removal and installation

Refer to illustrations 10.3 and 10.4
Warning: *These models are equipped with airbags. The airbag is armed and can deploy (inflate) anytime the battery is connected. To prevent accidental deployment (and possible injury), disconnect the negative battery cable whenever working near airbag components. After the battery is disconnected, wait at least 2 minutes before beginning work (the system has a back-up capacitor that must fully discharge). For more information see Chapter 12.*

1   Disconnect the cable from the negative terminal of the battery.

2   Disconnect the blower motor connector (see Section 9).

3   Release locking tab, then rotate and remove the motor from the housing **(see illustration)**.

4   Detach the fan retainer clip from the motor shaft **(see illustration)**.

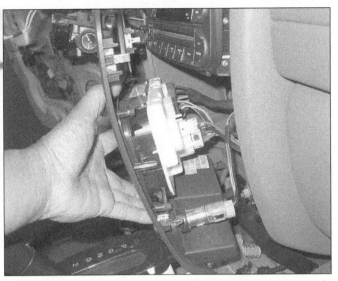

11.3a Pull the air conditioning and heater control panel out and disconnect the vacuum and electrical connectors

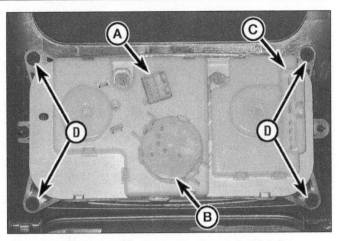

11.3b Air conditioning and heater control details

A   Temperature control/rear defogger connection
B   Vacuum connector
C   Blower speed switch connection
D   Mounting screws

11.10 Use a hand-held vacuum pump to test the one-way check valve (arrow) in the vacuum line to the control panel

A   Vacuum line to the check valve - attach hand-held vacuum pump
B   Check valve

5   Slip the fan off the old motor shaft, mount it on the new motor and install the retainer clip.
6   Place the blower motor in position, and rotate it, making sure all of the locking tabs are in their slots.
7   Reconnect the electrical connector and battery cable. Test the motor operation and make sure the air seal around the motor is not leaking.

11 Air conditioning and heater control assembly - removal and installation

Warning: *These models are equipped with airbags. The airbag is armed and can deploy (inflate) anytime the battery is connected. To prevent accidental deployment (and possible injury), disconnect the negative battery cable whenever working near airbag components. After the battery is disconnected, wait at least 2 minutes before beginning work (the system has*

a back-up capacitor that must fully discharge). For more information, see Chapter 12.
1   Disconnect the negative cable from the battery.

2007 and earlier models
Refer to illustrations 11.3a and 11.3b
2   Refer to Chapter 11 and remove the center trim bezel.
3   Pull the air conditioning and heater control panel forward and disconnect the electrical connectors and vacuum plug **(see illustrations)**.
4   Remove the four mounting screws and detach the air conditioning and heater control panel.
5   To install the control, reverse the removal procedure.

2008 and later models
Note: *The A/C heater control assembly is mounted directly to the center trim bezel.*
6   Remove the center trim bezel from the instrument panel and disconnect the electrical connectors to the panel (see Chapter 11).

7   Remove the A/C heater control assembly mounting screws and separate the control assembly from the trim bezel.
8   Installation is the reverse of removal. **Note:** *When installing a new control assembly or when reinstalling the original control assembly, the A/C heater control assembly will automatically perform the Actuator Calibration function when the ignition key is first turned on. If control assembly is from another vehicle, the Actuator Calibration function will not be automatically performed. The Actuator Calibration must be done following the diagnostic guidelines using a scan tool and software.*

Control panel vacuum tests (2007 and earlier models)
Refer to illustration 11.10
9   Vacuum is supplied to the climate control system by a line attached to the intake manifold. A one-way check valve located behind the glove box, is used in this line. With the engine OFF, remove the check valve (note its direction).
10   Apply 8 inches of vacuum with a hand-held pump on the side of the check valve that was attached to the engine vacuum line **(see illustration)**. Vacuum should flow through.
11   With vacuum applied to the control panel side of the valve, vacuum should hold. If not, replace the check valve.
12   Reconnect the check valve in the vacuum line in the proper direction and plug the hose back onto the intake manifold.
13   With the control panel pulled out and the vacuum harness disconnected **(see illustration 11.3b)** apply 8 inches of vacuum in turn to each of the ports on the connector.
14   If any of the circuits don't hold vacuum, follow the lines to their destination to isolate the source of the leak and repair it. **Note:** *The vacuum lines are all colored differently, according to their destination, making it easier to trace vacuum lines during troubleshooting.*

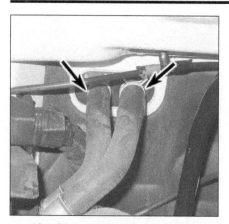

**12.4 Disconnect the heater hoses
at the firewall**

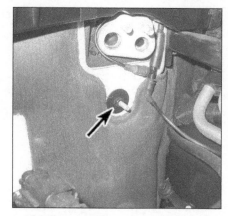

**12.5 Remove this nut securing the air
conditioning/heater unit to the firewall**

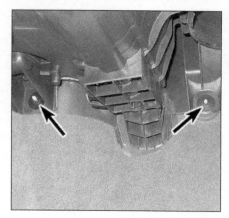

**12.8 Remove the air conditioning/heater
unit mounting nuts from inside the vehicle**

12   Heater core - removal and installation

Warning 1: *These models are equipped with airbags. The airbag is armed and can deploy (inflate) anytime the battery is connected. To prevent accidental deployment (and possible injury), disconnect the negative battery cable whenever working near airbag components. After the battery is disconnected, wait at least 2 minutes before beginning work (the system has a back-up capacitor that must fully discharge). For more information see Chapter 12.*
Warning 2: *The air conditioning system is under high pressure. Do not loosen any hose fittings or remove any components until after the system has been discharged and recovered. Air conditioning refrigerant must be properly discharged into an EPA-approved recovery/recycling unit at a dealer service department or an automotive air conditioning repair facility. Always wear eye protection when disconnecting air conditioning system fittings.*
Warning 3: *Wait until the engine is completely cool before beginning this procedure.*

Removal

Refer to illustrations 12.4, 12.5, 12.8 and 12.10
Note: *Heater core removal is a difficult task for the home mechanic. It can be done with slow, careful attention-to-detail, but many fasteners and wiring connectors are difficult to get at behind the instrument panel. The entire instrument panel must be removed to allow the heater/air conditioning unit to be removed from the car.*

1   On models with air conditioning, have the system discharged and recovered (see **Warning** above).
2   Disconnect the cable from the negative terminal of the battery.
3   Drain the cooling system (see Chapter 1). Remove the expansion tank (see Section 5).
4   Disconnect the heater hoses at the heater core at the right side of the engine compartment at the firewall **(see illustration)** and tape or plug all openings.
5   On models with air conditioning, detach the accumulator and mounting bracket and disconnect both air conditioning lines from the evaporator (see Section 14). Also remove the nut securing the air conditioning/heater unit to the firewall **(see illustration)**.

6   Remove the instrument panel (see Chapter 11). Remove the accumulator at the firewall to access the two nuts on the engine side of the firewall securing the HVAC housing. On later models, disconnect the straps securing the instrument panel to the top of the HVAC housing.
7   From inside the vehicle, disconnect the electrical connector from the blower motor resistor (see Section 10). Disconnect the electrical connector and vacuum hoses from the heater/air conditioning unit.
8   Remove the air conditioning/heater unit mounting nuts located in the passenger compartment floor area **(see illustration)**.
9   Remove the heater/air conditioning unit from the interior of the vehicle.
10   Remove the two heater core retaining screws, if equipped, and release the four retaining tabs to lift the heater core out from the heater/air conditioning unit **(see illustration)**.

Installation

11   Position the heater core tabs in the housing, and install the screws.
12   When installing the heater core, cement the seal in place to keep it from moving when the blower assembly is installed.
13   Reinstall the remaining components in the reverse order of removal. **Note:** *When installing the blower/evaporator housing, avoid pinching any wiring between the housing fresh air inlet and the instrument panel on the right side of the housing.*
14   Refill the cooling system (see Chapter 1).
15   On models with air conditioning, have the system evacuated, recharged and leak-tested.
16   Start the engine and check for proper operation.

13   Air conditioning and heating system - check and maintenance

Air conditioning system

Refer to illustrations 13.1 and 13.5
Warning: *The air conditioning system is under high pressure. Do not loosen any hose fittings or remove any components until after the sys-*

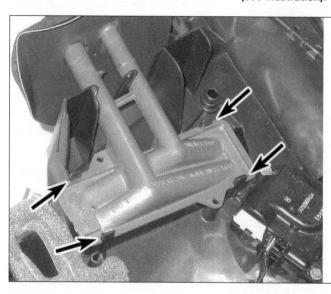

12.10 Remove the heater core retaining screws and release the tabs, then lift the heater core out from the heater/air conditioning unit

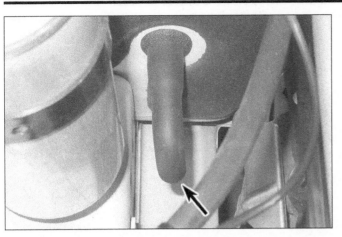

13.1 Check the evaporator housing drain tube for blockage - the view here is from the engine compartment (passenger's side), looking down at the firewall (next to the accumulator)

13.5 The air conditioning relay is located in the power distribution center which is in the left front corner of the engine compartment

tem has been discharged and recovered. Air conditioning refrigerant should be properly discharged into an EPA-approved recovery/recycling unit at a dealer service department or an automotive air conditioning repair facility. Always wear eye protection when disconnecting air conditioning system fittings.

Caution 1: *All models covered by this manual use environmentally friendly R-134a. This refrigerant (and its appropriate refrigerant oils) are not compatible with R-12 refrigerant system components and must never be mixed or the components will be damaged.*

Caution 2: *When replacing entire components, additional refrigerant oil should be added equal to the amount that is removed with the component being replaced. Be sure to read the can before adding any oil to the system, to make sure it is compatible with the R-134a system.*

1   The following maintenance checks should be performed on a regular basis to ensure that the air conditioning continues to operate at peak efficiency.

a) *Inspect the condition of the compressor drivebelt. If it is worn or deteriorated, replace it (see Chapter 1).*

b) *Check the drivebelt tension and, if necessary, adjust it (see Chapter 1).*

c) *Inspect the system hoses. Look for cracks, bubbles, hardening and deterioration. Inspect the hoses and all fittings for oil bubbles or seepage. If there is any evidence of wear, damage or leakage, replace the hose(s).*

d) *Inspect the condenser fins for leaves, bugs and any other foreign material that may have embedded itself in the fins. Use a "fin comb" or compressed air to remove debris from the condenser.*

e) *Make sure the system has the correct refrigerant charge.*

f) *If you hear water sloshing around in the dash area or have water dripping on the carpet, check the evaporator housing drain tube* **(see illustration)** *and insert a piece of wire into the opening to check for blockage.*

2   It's a good idea to operate the system for about 10 minutes at least once a month, particularly during the winter. Long-term non-use can cause hardening, and subsequent failure, of the seals.

3   Leaks in the air conditioning system are best spotted when the system is brought up to temperature and pressure, by running the engine with the air conditioning ON for five minutes. Shut the engine off and inspect the air conditioning hoses and connections. Traces of oil usually indicate refrigerant leaks.

4   Because of the complexity of the air conditioning system and the special equipment necessary to service it, in-depth troubleshooting and repairs are not included in this manual. However, simple checks and component replacement procedures are provided in this Chapter.

5   If the air conditioning system doesn't operate at all, check the fuse panel and the air conditioning relay, located in the relay box in the engine compartment **(see illustration)**.

6   The most common cause of poor cooling is simply a low system refrigerant charge. If a noticeable drop in cool air output occurs, the following quick check will help you determine if the refrigerant level is low. For more complete information on the air conditioning system, refer to the *Haynes Automotive Heating and Air Conditioning Manual*.

Checking the refrigerant charge

7   Warm the engine up to normal operating temperature.

8   Place the air conditioning temperature selector at the coldest setting and put the blower at the highest setting. Open the doors (to make sure the air conditioning system doesn't cycle off as soon as it cools the passenger compartment).

9   With the compressor engaged, the clutch will make an audible click and the center of the clutch will rotate. Feel the evaporator inlet pipe between the fixed orifice tube and the evaporator with one hand while placing your other hand on the metal portion of the hose between the evaporator and the condenser.

10   The pipe leading from the fixed orifice to the evaporator should be much colder than on the condenser side of the fixed orifice. If it isn't, the system charge is probably low. The earliest warning that a system is low on refrigerant is the air temperature coming out of the ducts inside the vehicle. If the air isn't as cold as it used to be, the system probably needs a charge. Further inspection or testing of the system requires special tools and techniques and is beyond the scope of this manual.

11   If the inlet pipe has frost accumulation or feels cooler than the accumulator surface, the refrigerant charge is low.

Adding refrigerant

Refer to illustrations 13.12, 13.15a, 13.15b and 13.18

12   Buy an automotive charging kit at an auto parts store. A charging kit includes a can of refrigerant, a tap valve and a short section of hose that can be attached between the tap valve and the system low side service valve **(see illustration). Caution:** *Although the system will hold more than one can of refrigerant, don't add more than one can (you could overfill the system).*

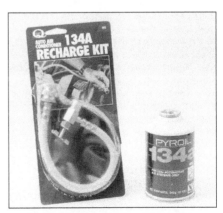

13.12 A basic charging kit for 134a systems is available at most auto parts stores - it must say 134a (not R-12) and so should the can of refrigerant

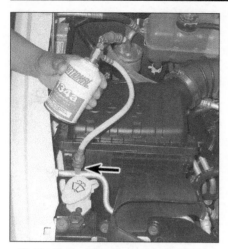

13.15a Add refrigerant to the system at the low-pressure port (early model shown)

13.15b Low-side charging port - later models

13.18 If you have an accurate thermometer, you can place it in the center air conditioning duct inside the vehicle

13   Hook up the charging kit by following the manufacturer's instructions. **Warning:** *DO NOT hook the charging kit hose to the system high side!* The fittings on the charging kit are designed to fit **only** on the low side of the system.

14   Back off the valve handle on the charging kit and screw the kit onto the refrigerant can, making sure first that the O-ring or rubber seal inside the threaded portion of the kit is in place. **Warning:** *Wear protective eyewear when dealing with pressurized refrigerant cans.*

15   Remove the dust cap from the low-side charging port (by the windshield washer tank) and attach the quick-connect fitting on the kit hose **(see illustrations)**.

16   Warm up the engine and turn on the air conditioner. Keep the charging kit hose away from the fan and other moving parts. The charging process requires the compressor to be running. If the clutch cycles off, remove the cycling switch connector from the accumulator **(see illustration 14.3)** and attach a jumper wire to the terminals on the harness side. This will keep the compressor engaged.

17   Turn the valve handle on the kit until the stem pierces the can, then back the handle out to release the refrigerant. You should be able to hear the rush of gas. Add refrigerant to the low side of the system until both the accumulator surface and the evaporator inlet pipe feel about the same temperature . Allow stabilization time between each addition.

18   If you have an accurate thermometer, you can place it in the center air conditioning duct inside the vehicle **(see illustration)** and keep track of the outlet air temperature. A charged system that is working properly should cool to approximately 40-degrees F. If the ambient (outside) air temperature is very high, say 110-degrees F, or if the relative humidity is high, the duct air temperature may be as high as 60- to 70-degrees F, but generally the air conditioning is 30 to 50-degrees F cooler than the ambient air.

19   When the can is empty, turn the valve handle to the closed position and release the

connection from the low-side port. Replace the dust cap.

20   Remove the charging kit from the can and store the kit for future use with the piercing valve in the UP position to prevent inadvertently piercing the can on the next use.

Eliminating air conditioning odors

Refer to Illustration 13.24

21   Unpleasant odors that often develop in air conditioning systems are caused by the growth of a fungus, usually on the surface of the evaporator core. The warm, humid environment there is a perfect breeding ground for mildew to develop.

22   The evaporator core on most vehicles Is difficult to access, and factory dealerships have a lengthy, expensive process for eliminating the fungus by opening up the evaporator case and using a powerful disinfectant and rinse on the core until the fungus is gone. You can service your own system at home, but it takes something much stronger than basic household germ-killers or deodorizers.

23   Aerosol disinfectants for automotive air conditioning systems are available in most auto parts stores, but remember when shopping for them that the most effective treatments are also the most expensive. The basic procedure for using these sprays is to start by running the system in the Recirculating mode for ten minutes with the blower on its highest speed. Use the highest heat mode to dry out the system - keep the compressor from engaging by disconnecting the wiring connector at the compressor (see Section 15).

24   Make sure that the disinfectant can comes with a long spray hose. Work the nozzle through the insulating foam surrounding the evaporator outlet pipe so that it protrudes inside the evaporator housing **(see illustration)** and then spray according to the manufacturer's recommendations. Try to cover the whole surface of the evaporator core, by aiming the spray up, down and sideways. Follow the manufacturer's recommendations for the length of spray and waiting time between applications.

25   Once the evaporator has been cleaned, the best way to prevent the mildew from coming back again is to make sure your evaporator housing drain tube is clear **(see illustration 13.1)**.

Heating system

26   If the carpet under the heater core is damp, or if antifreeze vapor or steam is coming through the vents, the heater core is leaking. Remove it (see Section 12) and install a new unit (most radiator shops will not repair a leaking heater core).

27   If the air coming out of the heater vents isn't hot, the problem could stem from any of the following causes:

a)   *The thermostat is stuck open, preventing the engine coolant from warming up enough to carry heat to the heater core. Replace the thermostat (see Section 3).*

b)   *There is a blockage in the system, preventing the flow of coolant through the heater core. Feel both heater hoses at the firewall. They should be hot. If one of them is cold, there is an obstruction in one of the hoses or in the heater core. Detach the hoses and back flush the heater core with a water hose. If the heater core is clear but circulation is still restricted, remove the two hoses and flush them out with a water hose.*

c)   *If flushing fails to remove the blockage from the heater core, the core must be replaced (see Section 12).*

---

14   Air conditioning system accumulator - removal and installation

---

Warning: *The air conditioning system is under high pressure. Do not loosen any hose fittings or remove any components until after the system has been discharged and recovered. Air*

13.24 To disinfect the evaporator housing, insert the nozzle of the disinfectant can through the insulating foam that surrounds the outlet tube of the evaporator at the engine side of the firewall

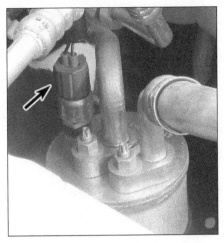

14.3 Disconnect the clutch-cycling pressure switch from the accumulator

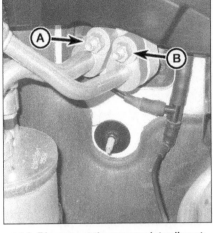

14.6 Disconnect the accumulator line at the firewall, where it goes to the evaporator

A   Accumulator line
B   Liquid line

conditioning refrigerant should be properly discharged into an EPA-approved recovery/recycling unit at a dealer service department or an automotive air conditioning repair facility. Always wear eye protection when disconnecting air conditioning system fittings.

1   The accumulator holds liquid refrigerant and moisture and is located near the engine firewall. Have the air conditioning system discharged and recovered (see **Warning** above).
2   Disconnect the cable from the negative terminal of the battery.

2007 and earlier models

Refer to illustration 14.3 and 14.6

3   Disconnect the electrical connector from the pressure switch near the top of the accumulator **(see illustration)**, which is located on the right side of the engine compartment, next to the fenderwell.
4   Disconnect the refrigerant suction line from the compressor by unbolting it **(see illustration 15.3)**. **Note:** *The refrigerant lines do not detach from the accumulator - when the accumulator is replaced, the lines must be replaced along with it.*
5   Plug the open fittings to prevent entry of dirt and moisture.
6   Disconnect the accumulator line where it goes through the firewall to the evaporator **(see illustration)**.
7   Remove the accumulator and bracket from the firewall. If a new accumulator is being installed, remove the Schrader valve (under the pressure switch) from the old accumulator. Add 3 oz (90 ml) of new refrigerant oil (a type designated as compatible with refrigerant R-134a) to the new accumulator.

2008 and later models

8   Remove the grille (see Chapter 11).
9   Remove the air seal from the left side of the radiator assembly.
10   Working from the bottom of the accumu-

lator, remove the accumulator-to-condenser mounting bolt.
11   Disconnect the line from the A/C condenser to the accumulator and cover the line.
12   Remove the accumulator from the vehicle.

All models

13   Installation is the reverse of removal. **Note:** *New R-134a compatible O-rings should be used in each fitting during reassembly.*
14   Take the vehicle back to the shop that discharged it. Have the air conditioning system evacuated, charged and leak tested.

---

15   Air conditioning system compressor - removal and installation

---

Refer to illustrations 15.3, 15.6a and 15.6b
Warning: *The air conditioning system is under high pressure. Do not loosen any hose fittings*

or remove any components until after the system has been discharged and recovered. Air conditioning refrigerant should be properly discharged into an EPA-approved recovery/recycling unit at a dealer service department or an automotive air conditioning repair facility. Always wear eye protection when disconnecting air conditioning system fittings.
Note: *The accumulator (see Section 14) should be replaced whenever the compressor is replaced.*

1   The compressor pumps refrigerant through the system and is driven by an engine drivebelt. Have the air conditioning system discharged and recovered (see **Warning** above).
2   Disconnect the cable from the negative terminal of the battery.
3   Disconnect the compressor clutch electrical connector **(see illustration)**.
4   Remove the drivebelt (see Chapter 1).
5   Disconnect the refrigerant lines from the compressor. Plug the open fittings to prevent entry of dirt and moisture.

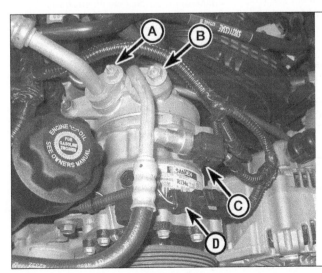

15.3 Disconnect the electrical connector from the compressor; Remove the nuts that secure the refrigerant lines at the compressor, remove the lines, then plug the openings on the compressor and the lines

A   Suction line
B   Discharge line
C   AC pressure switch connector
D   Compressor clutch connector

15.6a   Right side compressor mounting bolts - 3.7L V6 engine

15.6b   Left side compressor mounting bolt - 3.7L V6 engine

6   On 2.4L and 3.7L engnes, unbolt the compressor from the mounting brackets and lift it out of the vehicle **(see illustrations)**.

4.0L engines

7   On 4.0L models, remove the air filter housing (see Chapter 4) then remove the right engine mount through-bolt (see Chapter 2C).

8   Using an engine hoist, raise the engine enough to access the compressor mounting bolts and secure the engine in this position using a jack, then remove the air conditioning compressor mounting bolts and compressor. **Warning:** *Do not disconnect the refrigerant lines.*

All engines

9   If a new compressor is being installed, follow the directions with the compressor regarding the draining of excess oil prior to installation. **Note:** *Any replacement compressor used must be designated as compatible with refrigerant R-134a.*

10   The clutch may have to be transferred from the original to the new compressor.

11   Installation is the reverse of removal. Replace all O-rings with new ones specifically made for air conditioning system use (and compatible with refrigerant R-134a) and lubricate them with refrigerant oil. Any refrigerant oil added must also be compatible with refrigerant R-134a.

12   Have the system evacuated, recharged and leak-tested by the shop that discharged it.

16   Air conditioning system condenser - removal and installation

Refer to illustrations 16.3 and 16.6

Warning: *The air conditioning system is under high pressure. Do not loosen any hose fittings or remove any components until after the system has been discharged and recovered. Air conditioning refrigerant should be properly discharged into an EPA-approved recovery/ recycling unit at a dealer service department*

or an automotive air conditioning repair facility. Always wear eye protection when disconnecting air conditioning system fittings.

Note: *The accumulator* (see Section 14) *should be replaced whenever the condenser is replaced.*

1   The condenser is a small radiator-like component mounted in front of the radiator that cools the refrigerant. Have the air conditioning system discharged and recovered (see **Warning** above).

2   Disconnect the cable from negative terminal of the battery.

3   Unbolt and disconnect the refrigerant lines from the condenser **(see illustration)**. Plug the lines and fittings to prevent the entry of moisture and contaminants.

2007 and earlier models

4   Remove the radiator and condenser as a unit, then separate the condenser from the radiator. (see Section 6).

5   On vehicles equipped with an automatic transmission, remove the transmission cooler-to-radiator bolts and set the transmission cooler aside (see Chapter 7B).

6   Remove the mounting bolts from the condenser brackets **(see illustration)**.

7   Lift the condenser straight up and out of the condenser brackets. Plug the open fittings to prevent entry of dirt and moisture.

2008 and later models

8   On V6 models, remove the air filter housing (see Chapter 4) and fan shroud (see Section 4).

9   Remove the grille (see Chapter 11).

10   Remove the air seal around the condenser.

11   Disconnect and cap the transmission cooler line, if equipped.

12   Disconnect the nut that secures the lines to the A/C condenser and cover the lines.

13   Remove the mounting bolts from the right side of the radiator.

14   Carefully lift the condenser up and unhook or disengage the condenser mounting brackets from the radiator.

15   Remove the accumulator from the vehicle.

16   If replacing the condenser, remove the accumulator from the condenser (see Section 14).

All models

17   If the original condenser will be reinstalled, store it with the line fittings on top to prevent oil from draining out.

18   If a new condenser is being installed, pour 0.75 oz (22 ml) of refrigerant oil into it prior to installation (an oil designated as compatible with refrigerant R-134a). **Note:** *New R-134a compatible O-rings should be used in each fitting during reassembly.*

19   Reinstall the components in the reverse order of removal.

20   Have the system evacuated, recharged and leak-tested by the shop that discharged it.

17   Fixed orifice tube - removal and installation

Warning: *The air conditioning system is under high pressure. Do not loosen any hose fittings or remove any components until after the system has been discharged. Air conditioning refrigerant should be properly discharged into an EPA-approved recovery/recycling unit at a dealer service department or an automotive air conditioning repair facility. Always wear eye protection when disconnecting air conditioning system fittings.*

Note: *After operating a fully-charged air conditioner for five minutes, the liquid line should be hot near the condenser and it should be cold near the evaporator. If there isn't a significant temperature difference, the orifice tube may be plugged. If the system is checked with the appropriate gauges, and the high-pressure reads extremely high and the low-pressure reads almost a vacuum, the orifice tube is plugged. In either case, the liquid line, which contains the orifice tube, must be replaced.*

1   The fixed orifice tube is an inline metering restrictor that is located in the liquid line, between the condenser and the evaporator.

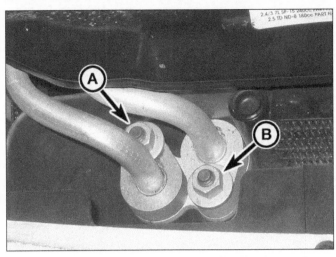

16.3 Unbolt and disconnect the refrigerant lines from the condenser

A   *Discharge line*   B   *Liquid line*

16.6 Remove the condenser mounting bolts (right side shown)

Have the air conditioning system discharged (see **Warning** above).

2   Disconnect the cable at the negative terminal of the battery.

3   Unbolt the liquid line from the condenser near the radiator **(see illustration 16.3)** and from the evaporator core at the firewall **(see illustration 14.6)**, and remove the line from the vehicle, then cap or plug the refrigerant openings to prevent any dirt or moisture from entering the system.

4   Reinstall the new line/orifice tube. **Note:** *New R-134a compatible O-rings should be used in each fitting during reassembly.*

5   Have the system evacuated, recharged and leak-tested by a dealer service department or an air conditioning repair facility.

18 Air conditioning expansion valve - removal and installation

Warning: *The air conditioning system is under high pressure. DO NOT loosen any fittings or remove any components until after the system has been discharged. Air conditioning refrigerant must be properly discharged into an EPA-approved container at a dealer service department or an automotive air conditioning repair facility. Always wear eye protection when disconnecting air conditioning system fittings.*

1   Have the air conditioning system discharged by a dealer service department or an automotive air conditioning shop before proceeding (see **Warning** above).

2   Disconnect the cable from the negative battery terminal (see Chapter 5, Section 3).

3   Remove the refrigerant lines-to-expansion valve mounting nut.

4   Separate the lines, then cap them and discard the O-rings.

5   Remove the two Allen head bolts that mount the expansion valve to the evaporator tube.

6   Remove the expansion valve and discard the O-rings. Cover or cap the openings while the expansion valve is removed.

7   Installation is the reverse of removal. Replace all the O-rings with new ones, and tighten the expansion valve bolts securely.

19 Air conditioning system evaporator - removal and installation

Warning 1: *These models are equipped with airbags. The airbag is armed and can deploy (inflate) anytime the battery is connected. To prevent accidental deployment (and possible injury), disconnect the negative battery cable whenever working near airbag components. After the battery is disconnected, wait at least 2 minutes before beginning work (the system has a back-up capacitor that must fully discharge). For more information see Chapter 12.*

Warning 2: *The air conditioning system is under high pressure. DO NOT loosen any fittings or remove any components until after the system has been discharged. Air conditioning refrigerant should be properly discharged into an EPA-approved recovery/recycling unit at a dealer service department or an automotive air conditioning repair facility. Always wear eye protection when disconnecting air condition/system fittings.*

1   The evaporator is a small radiator-like component mounted in the heater/air conditioning unit under the dash. It removes heat from the passenger compartment. Have the air conditioning system discharged (see **Warning** above).

2   Disconnect the cable at the negative terminal of the battery.

3   Remove the heater/air conditioning unit from under the dash area (see Section 12)

4   Turn the heater/air conditioning unit upside down and remove the screws that retain the two halves together.

5   Remove the evaporator from the lower housing.

6   Installation is the reverse of the removal procedure. If the evaporator was replaced with a new unit, add 1.5 oz (45 ml) of refrigerant oil (of a type designated as compatible with refrigerant R-134a) to the system. **Note:** *New R-134a-compatible O-rings should be use in each fitting during reassembly.*

7   Have the system evacuated, recharged and leak tested by a dealer service department or an air conditioning repair facility.

Notes

Chapter 4
Fuel and exhaust systems

Contents

Specifications

General

Fuel pressure (engine running at idle speed)	
2002 and 2003	47.2 to 51.2 psi (325 to 353 kPa)
2004 and later	
with JTEC PCM (three connectors)	47.2 to 51.2 psi (325 to 353 kPa)
with NGC PCM (four connectors)	53 to 63 psi (365 to 434 kPa)
Fuel injector resistance	10.8 to 13.2 ohms at 68° F (20° C)
Fuel level sending unit	
Resistance	
Empty	256 to 284 ohms
Full	19 to 21 ohms
Voltage	
Empty	8.6 volts
Full	0.6 volts

Torque specifications

Note: One foot-pound (ft-lb) of torque is equivalent to 12 inch-pounds (in-lbs) of torque. Torque values below approximately 15 foot-pounds are expressed in inch-pounds, because most foot-pound torque wrenches are not accurate at these smaller values.

	Ft-lbs (unless otherwise indicated)	Nm
Throttle body mounting bolts	105 in-lbs	12
Fuel rail mounting nuts/bolts		
2.4L four-cylinder	21	28
V6 engines	100 in-lbs	11
Fuel tank strap retaining bolts	45	61

1 General information

Refer to illustration 1.1

Warning: *Gasoline is extremely flammable, so take extra precautions when you work on any part of the fuel system. Don't smoke or allow open flames or bare light bulbs near the work area, and don't work in a garage where a gas-type appliance (such as a water heater or a clothes dryer) is present. Since gasoline is carcinogenic, wear latex gloves when there's a possibility of being exposed to fuel, and, if you spill any fuel on your skin, rinse it off immediately with soap and water. Mop up any spills immediately and do not store fuel-soaked rags where they could ignite. The fuel system is under constant pressure, so, if any fuel lines are to be disconnected, the fuel pressure in the system must be relieved first. When you perform any kind of work on the fuel system, wear safety glasses and have a Class B type fire extinguisher on hand.*

All models covered by this manual are equipped with a sequential Multi Port Fuel Injection (MPFI) system **(see Illustration)**. This system uses timed impulses to sequentially inject the fuel directly into the intake ports of each cylinder. The injectors are controlled by the Powertrain Control Module (PCM). The PCM monitors various engine parameters and delivers the exact amount of fuel, in the correct sequence, into the intake ports.

The fuel filters are extended-life parts and are not replaced during normal sched-uled maintenance. The filters will only be replaced when diagnostic testing requires it. The main fuel filter is mounted to the front of the fuel tank. An inlet fuel filter, electric fuel pump and pressure regulator are mounted inside the fuel tank. The fuel level sending unit is an integral component of the fuel pump and it must be removed from the fuel tank in the same manner. On vehicles that are not equipped with an access cover under the carpeting of the rear cargo area, it is necessary to lower the fuel tank for access to these fuel system components.

The exhaust system consists of exhaust manifold(s), a catalytic converter, an exhaust pipe and a muffler. Each of these components is replaceable. For further information regarding the catalytic converter, refer to Chapter 6.

2 Fuel pressure relief procedure

Warning: *See the* **Warning** *in Section 1.*
1   Remove the fuel filler cap (this will relieve any pressure that has built-up in the tank).
2   Remove the fuel pump relay or Totally Intergrated Power Mobile (TIPM) fuse from the power distribution center in the engine compartment (you can identify the fuses and relays by looking at the underside of the power distribution center cover).
3   Attempt to start the engine - the engine probably will not start or may start momentarily and immediately stall. Crank the engine for several seconds.

4   Turn the ignition key to the Off position. Install the fuel pump relay.
5   Disconnect the cable from the negative terminal of the battery before beginning work on the fuel system.
6   After all work on the fuel system has been completed, the CHECK ENGINE light may come on during operation because the engine was cranked with the fuel pump relay unplugged. The light will likely go out after a period of normal operation. If it does not go out, refer to Chapter 6.

3 Fuel pump/fuel pressure - check

Warning: *See the* **Warning** *in Section 1.*

Preliminary check

Refer to illustration 3.3

1   If you suspect insufficient fuel delivery, first inspect all fuel lines to ensure that the problem is not simply a leak in a line.
2   Set the parking brake and have an assistant turn the ignition switch to the ON position while you listen to the fuel pump (inside the fuel tank). You should hear a "whirring" sound, lasting for a couple of seconds indicating the fuel pump is operating. If the fuel pump is operating, proceed to the pressure check.
3   If the fuel pump does not operate, disconnect the fuel pump electrical connector from the harness connector. Connect a voltmeter to the B+ (from fuel pump relay) and to the ground terminal of the harness con-

1.1 Typical fuel system components - 3.7L V6 engine

1	*Throttle body (under resonator)*	*3*	*Fuel rail and injectors*	*5*	*Fuel pump relay (inside power distribution center)*
2	*Air intake duct/resonator*	*4*	*Air filter housing*		

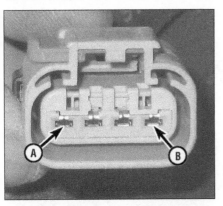

3.3 Check the battery power and ground circuits at the fuel pump connector

    A   *Battery positive*
    B   *Ground*

nector **(see illustration)**. Have an assistant attempt to start the engine - battery voltage should be indicated on the meter. If power is not available at the fuel pump connector, check the fuel pump circuit, referring to Chapter 12 and the wiring diagrams. Check the related fuses, the fuel pump relay and the related wiring to ensure power is reaching the fuel pump connector. Check the ground circuit for continuity to a good chassis ground point.

4   If the power and ground circuits are good and the fuel pump does not operate when connected, replace the fuel pump (see Section 7).

Pressure check

Refer to illustration 3.6

Note: *In order to perform the fuel pressure test, you will need a fuel pressure gauge capable of measuring high fuel pressure. The fuel gauge must be equipped with the proper fittings or adapters required to attach it to the fuel rail.*

Caution: *Do not pinch the vehicle's fuel lines shut or damage to the line may occur.*

3.6 Fuel pressure test port location - 3.7L V6 engine

5   Relieve the fuel pressure (see Section 2).
6   Remove the cap from the fuel pressure test port on the fuel rail and attach a fuel pressure gauge **(see illustration)**.
7   Start the engine and check the pressure on the gauge, comparing your reading with the pressure listed in this Chapter's Specifications.
8   If the fuel pressure is not within specifications, check the following:

    a) *If the pressure is lower than specified, check for a restriction in the fuel system.*
    b) *If the fuel pressure is higher than specified, replace the fuel pressure regulator (see Section 14).*

---

4   Fuel lines and fittings - repair and replacement

Warning: *See the* **Warning** *in Section 1.*
1   Always relieve the fuel pressure before servicing fuel lines or fittings (see Section 2).
2   The fuel feed and vapor lines extend from the fuel tank to the engine compartment. The lines are secured to the underbody with

clip and screw assemblies. These lines must be occasionally inspected for leaks, kinks and dents.
3   If evidence of dirt is found in the system or fuel filter during disassembly, the line should be disconnected and blown out. Check the fuel strainer on the fuel pump module for damage and deterioration.

Steel tubing

4   If replacement of a fuel line or emission line is called for, use tubes/hoses meeting the manufacturer's specifications.
5   Don't use copper or aluminum tubing to replace steel tubing. These materials cannot withstand normal vehicle vibration.
6   Because fuel lines used on fuel-injected vehicles are under high pressure, they require special consideration.
7   Some fuel lines have threaded fittings with O-rings. Any time the fittings are loosened to service or replace components:

    a) *Use a backup wrench while loosening and tightening the fittings.*
    b) *Check all O-rings for cuts, cracks and deterioration. Replace any that appear hardened, worn or damaged.*
    c) *If the lines are replaced, always use original equipment parts, or parts that meet the original equipment standards.*

Flexible hose

Refer to illustrations 4.11a through 4.11i

8   In the event of flexible fuel line damage it is necessary to replace the damaged lines with factory replacement parts. Others may fail from the high pressures of this system.
9   Relieve the fuel pressure before disconnecting the fitting (see Section 2).
10   Remove all fasteners attaching the lines to the vehicle body.
11   There are various methods of disconnecting the fittings, depending upon the type of quick-connect fitting installed on the fuel line **(see illustrations)**. Clean any debris from around the fitting. Disconnect the fitting and carefully remove the fuel line from the vehicle.

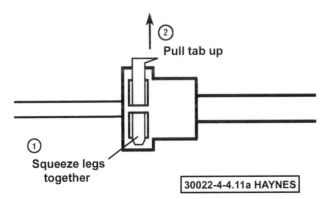

4.11a To disconnect a single-tab type fitting, squeeze the tab legs together, pull up on the tab and pull the lines apart - discard the tab and obtain a new one for reassembly

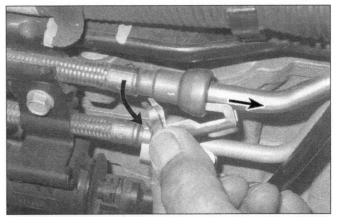

4.11b To disconnect a metal collar type fitting, pull the end of the retainer off the fuel line, then disengage the other end from the female side of the fitting . . .

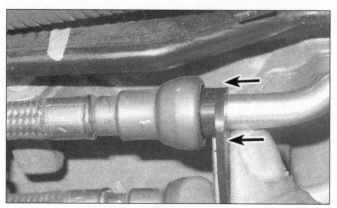

4.11c . . . insert the fuel line separator tool into the female side of the fitting, push it into the fitting until it releases the locking tabs inside the fitting . . .

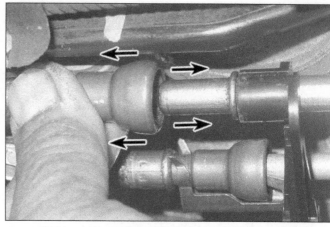

4.11d . . . then pull the two halves of the fitting apart

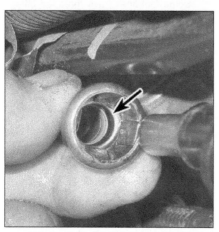

4.11e Inspect the old O-ring inside the female side of the fitting; if it's cracked, torn or deteriorated, replace it

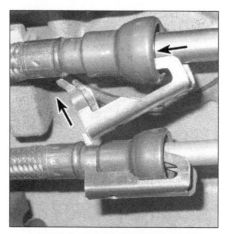

4.11f To install a retainer, insert the hooked end into the female side of the fitting, then push the clip end onto the fuel line until it snaps into place (when you're done, your retainer should look like the one already installed on the lower fitting)

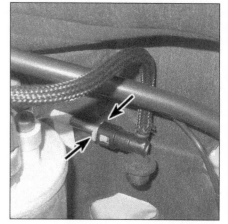

4.11g To release a plastic quick-connect fitting, squeeze the two plastic retaining tabs together . . .

Caution: *The quick-connect fittings are not serviced separately. Do not attempt to repair these types of fuel lines in the event the fitting or line becomes damaged. Replace the entire fuel line as an assembly.*

12   Don't route fuel hose within four inches of any part of the exhaust system or within ten inches of the catalytic converter. Metal lines

and rubber hoses must never be allowed to chafe against the frame. A minimum of 1/4-inch clearance must be maintained around a line or hose to prevent contact with the frame.

13   Installation is the reverse of removal with the following additions:

a) *Clean the quick-connect fittings with a lint-free cloth and apply clean engine oil to the fittings.*

b) *After connecting a quick-connect fitting, check the integrity of the connection by attempting to pull the lines apart.*

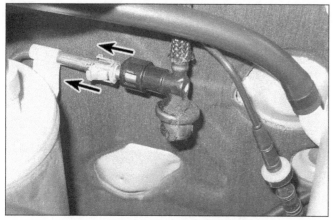

4.11h . . . and pull the line out of the fitting

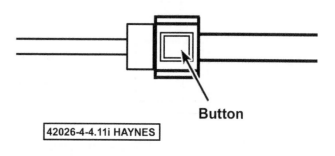

Button

42026-4-4.11i HAYNES

4.11i Some quick-connect fittings have two buttons (one on each side of the fitting) that must be depressed to release the fitting (no tools are necessary)

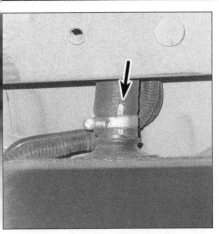

5.10 Remove the clamp that retains the fuel filler hose to the fuel tank; note the white mark on the hose (arrow) that is aligned with the notches on the fuel tank filler neck

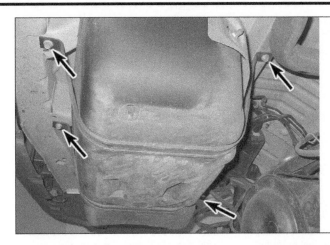

5.12 Remove the fuel tank retaining bolts

c) *Use new O-rings at the threaded fittings (if equipped).*

d) *Before starting the engine, cycle the ignition key On and Off several times and check for leaks at the fitting.*

5 Fuel tank - removal and installation

Refer to illustrations 5.10 and 5.12

Warning: *See the **Warning** in Section 1.*

1   Relieve the fuel system pressure (see Section 2).

2   Remove the cable from the negative terminal of the battery.

3   Disconnect the upper fuel pump module (see Section 8), then lift it up to gain access for siphoning the fuel out of the fuel tank. **Note:** *The fuel tank filler neck has a one-way valve which prevents siphoning from the filler neck.*

4   Using a siphoning kit (available at most auto parts stores), siphon the fuel through the fuel pump module opening, into an approved gasoline container. **Warning:** *Never start the siphoning action by mouth!* Temporarily place the fuel pump module back on the tank.

5   On 2008 and later models, raise the rear of the vehicle and support it securely on jackstands, then remove the rear driveshaft (see Chapter 8).

6   If equipped, remove the optional tow hooks, disconnect the optional tow package electrical connector and remove the trailer hitch.

7   If equipped, remove the optional fuel tank skidplate:

a) *Support the skid plate with a floor jack*

b) *Remove the fuel tank skid plate mounting bolts.*

c) *Carefully lower the skid plate from the chassis.*

8   Label and disconnect the fuel system lines attached to the fuel tank, and any retainers securing them to the chassis.

9   Disconnect the fuel pump electrical ground-strap.

10   Remove the fuel filler hose from the left side of the fuel tank **(see illustration)**.

11   Support the fuel tank with a floor jack. Position a piece of wood between the jack head and the fuel tank to protect the tank.

12   Remove the fuel tank retaining bolts **(see illustration)**.

13   Remove the tank from the vehicle. **Caution:** *Carefully lower the fuel tank several inches and disconnect the remaining electrical connector.*

14   Installation is the reverse of removal with the following additions:

a) *If the fuel tank is being replaced, remove the necessary components from the old fuel tank and install them on the new tank.*

b) *Tighten the fuel tank retaining bolts to the torque listed in this Chapter's Specifications.*

c) *Align the white-paint mark on the fuel fill hose with the notches on the fuel tank fitting and tighten the clamp securely.*

6 Fuel tank cleaning and repair - general information

1   The fuel tank installed in the vehicles covered by this manual is not repairable. If the fuel tank becomes damaged, it must be replaced.

2   Cleaning the fuel tank (due to fuel contamination) should be performed by a professional with the proper training to carry out this critical and potentially dangerous work. Even after cleaning and flushing, explosive fumes may remain inside the fuel tank.

3   If the fuel tank is removed from the vehicle, it should not be placed in an area where sparks or open flames could ignite the fumes coming out of the tank. Be especially careful inside a garage where a gas-type appliance is located.

7 Fuel filters - replacement

Warning: *See the **Warning** in Section 1.*

Note: *Fuel filter replacement is not a part of the normal maintenance on this vehicle. The main fuel filter and inlet fuel filters are extended-life parts that would only be replaced after diagnostic tests indicated that excess contamination existed in the fuel tank. The filters would be replaced after having the fuel tank cleaned (see Step 6).*

Note: *On 2005 and later models, the fuel filter is part of the pump module, and isn't replaceable separately.*

Main fuel filter

Refer to illustrations 7.1 and 7.5

1   The main fuel filter is located at the front of the fuel tank **(see illustration)**.

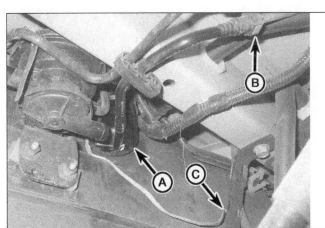

7.1 The main fuel filter is located at the front of the fuel tank

A   *Fuel filter*

B   *Third fuel line connection*

C   *Fuel filter electrical ground strap*

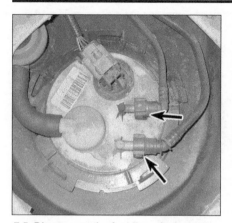

7.5 Disconnect the fuel lines from the top of the upper section of the fuel module

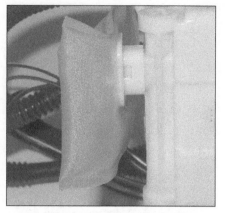

7.13 Remove the filter from the lower section of the fuel pump module by prying the two release tabs with two small screwdrivers

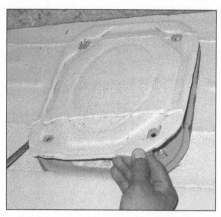

8.4 If equipped, remove the fuel pump access cover

2   Relieve the fuel system pressure (see Section 2).
3   Disconnect the cable from the negative battery terminal.
4   Check the vehicle to see if it is equipped with a fuel pump access cover, then remove it. **Note:** *If the vehicle is not equipped with a fuel pump access cover on the floorpan of the rear cargo area, the fuel tank must be lowered to gain access to the fuel pump module* (see Section 5).

a) *Drill-out the rivets on the two rear-most cargo hold-down clamps and remove the clamps.*
b) *Fold the carpeting forward.*
c) *Remove the access cover nuts* **(see illustration 8.4).**
d) *Use a heat gun to apply heat to the access cover, then carefully pry up on the cover and remove it.*

5   Clean the area surrounding the fuel lines to prevent dirt from entering the fuel system. Disconnect the fuel lines from the top of the upper section of the fuel pump module **(see illustration).**
6   Raise the rear of vehicle and support it securely on jackstands.
7   A third fuel line is connected to the bottom of the fuel filter. The connection is approximately one foot in front of the fuel filter **(see illustration 7.1).** Disconnect the third fuel line

and release it from the retention clip on the chassis.
8   Remove the fuel filter electrical ground strap from the fuel tank mounting strap **(see illustration 7.1).**
9   Remove the filter mounting nut and remove the filter.
10   Position the new filter/lines in place.
11   Installation is the reverse of removal with the following additions:

a) *If equipped, apply silicone sealant to the bottom of the fuel pump access cover.*
b) *After positioning the carpet in place, install two new rivets to secure the cargo hold-down clamps.*

Inlet fuel filter

Refer to illustration 7.13

12   The inlet fuel filter is attached to the bottom of the lower section of the fuel pump module which is located inside of the fuel tank. Remove the fuel pump module (see Section 8).
13   Remove the filter from the lower section of the fuel pump module by prying the two release tabs with two small screwdrivers **(see illustration).**
14   Clean the filter-opening on the lower section of the fuel pump module.

15   Snap the new filter into place.
16   Installation is the reverse of removal.

8   Fuel pump module - removal and installation

Refer to illustrations 8.4, 8.6a, 8.6b, 8.8 and 8.11

Warning: *See the* **Warning** *in Section 1.*
1   The fuel pump module consists of an upper section containing the fuel pressure regulator and fuel pump check valve, and a lower section consisting of the electric fuel pump and the fuel pick-up/inlet filter. The fuel pump module is replaced as a complete assembly.
2   Relieve the fuel system pressure (see Section 2).
3   Disconnect the cable from the negative battery terminal.
4   Check the vehicle to see if it is equipped with a fuel pump access cover, then remove it. **Note:** *If the vehicle is not equipped with a fuel pump access cover on the floorpan of the rear cargo area, the fuel tank must be lowered to gain access to the fuel pump module* (see Section 5).

a) *Drill-out the rivets on the two rear-most cargo hold-down clamps and remove the clamps.*
b) *Fold the carpeting forward.*
c) *Remove the access cover nuts* **(see illustration).**
d) *Use a knife blade to cut the seal or use a heat gun to apply heat to the outer edge of the access cover, then carefully pry up on the cover and remove it.*

5   Clean the area surrounding the fuel lines and upper section of the fuel module to prevent dirt from entering the fuel system. Disconnect the fuel lines and electrical connector from the upper module, as necessary **(see illustration 7.5).** Mark the orientation of the module to the fuel tank before removing the module.
6   Using a pair of large pliers to turn the lockring, or using a hammer and a brass drift, tap on the lockring to turn it counterclockwise to remove the upper section of the fuel pump module from the tank **(see Illustrations).**

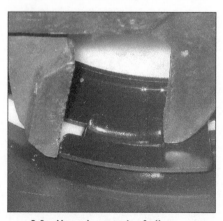

8.6a Use a large pair of pliers . . .

8.6b . . . or use a hammer and brass drift to turn the lockring securing the fuel pump module

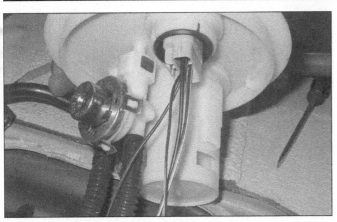

8.8 Detach the fuel lines and electrical connector from the bottom side of the upper section

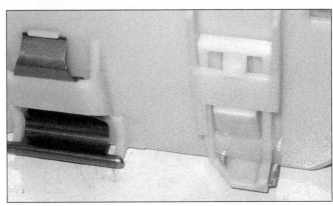

8.11 To remove the lower section of the fuel pump module, press the plastic lock, lift on the tab, slide it toward the right, out of the two guides in the tank, then lift it out of the tank

7   Carefully lift up the upper section, then secure the rubber gasket to prevent it from falling into the tank.

8   Detach and secure the fuel lines and electrical connector from the bottom side of the upper section **(see illustration)**.

9   Remove the upper section from the tank.

10   Using a siphoning kit (available at most auto parts stores), siphon the all of the fuel through the upper module opening, into an approved gasoline container. **Note:** *The fuel tank must be drained to remove the lower section of the fuel pump module.* **Warning:** *Never start the siphoning action by mouth!*

11   Reach inside the tank and press and release the plastic lock while lifting the tab **(see illustration)**. Slide the lower section to the right side of the tank to release it from the two guides in the fuel tank. Remove the lower section of the fuel pump module.

12   Installation is the reverse of removal with the following additions:

a) *Install the lower section into the fuel tank by sliding it to the left and pressing down on the locking tab.*

b) *Always replace the upper section gasket with a new one.*

c) *Position the notch on the upper section*

so that it faces the rear of the vehicle, then install it.

d) *If equipped, apply silicone sealant to the bottom of the fuel pump access cover.*

e) *After positioning the carpet in place, install two new rivets to secure the cargo hold-down clamps.*

9   Fuel level sending unit - check and replacement

Warning: *See the* **Warning** *in Section* 1.

Check

Refer to illustrations 9.3, 9.7a and 9.7b

1   With a full tank of fuel, check the voltage and resistance from the fuel level sending unit.

2   Check the vehicle to see if it is equipped with a fuel pump access cover, then remove it. **Note:** *If the vehicle is not equipped with a fuel pump access cover on the floorpan of the rear cargo area, the fuel tank must be lowered to gain access to the fuel pump module* (see Section 5).

a) *Drill-out the rivets on the two rear-most cargo hold-down clamps and remove the clamps.*

b) *Fold the carpeting forward.*

c) *Remove the access cover nuts* **(see illustration 8.4)**.

d) *Use a heat gun to apply heat to the access cover, then carefully pry up on the cover and remove it.*

3   With the ignition key in the On position, measure the voltage at the terminal coming out of the fuel pump module sending unit signal wire (dark blue/yellow) and compare it to the values listed in this Chapter's Specifications **(see illustration)**.

4   If the voltage is not within specifications, disconnect the cable from the negative terminal of the battery.

5   Disconnect the electrical connector from the upper section of the fuel pump module.

6   Connect the probes of an ohmmeter to the terminals on the upper section of the fuel pump module where the dark blue/yellow (sending unit signal) and black/light blue (ground) wires were connected. Measure the resistance and compare it to the values listed in this Chapter's Specifications.

7   If the resistance is not within specifications, remove the fuel pump module (see Section 8) and check the resistance of the sending unit out of the vehicle:

9.3 With the ignition key in the On position, measure the voltage at the dark blue/yellow wire (A) from the fuel level sending unit. B is the ground wire

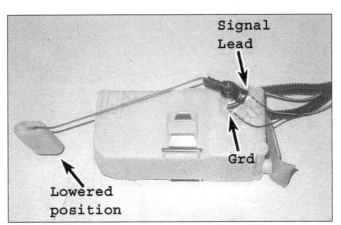

9.7a Connect an ohmmeter to the signal and ground terminals of the fuel pump module connector - measure the resistance of the fuel level sending unit with the float lowered (empty) . . .

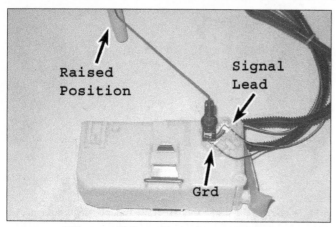

9.7b . . . and then with the float raised (full)

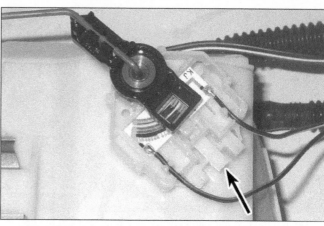

9.10 Lift the tab to remove the sending unit from the lower section of the fuel pump module

a) *Position the float in the down (Empty) position* **(see illustration)**. *Measure the resistance and compare it to the values listed in this Chapter's Specifications.*

b) *Position the float in the up (Full) position* **(see illustration)**. *Measure the resistance and compare it to the values listed in this Chapter's Specifications.*

8   If the fuel level sending unit resistance is incorrect or if the resistance does not change smoothly as the float travels from empty to full, replace the fuel level sending unit assembly.

Replacement

Refer to illustration 9.10

9   The fuel level sending unit is attached to the lower section of the fuel pump module. Remove the fuel pump module (see Section 8).

10   Lift the plastic locking tab while sliding the sending unit upward to remove it **(see illustration)**.

11   The remainder of installation is the reverse of removal.

10 Air filter housing - removal and installation

Refer to illustration 10.2

1   Remove the air filter element and air filter housing cover (see Chapter 1).

2   Loosen the hose clamps and remove the air duct **(see illustration)**.

3   On V6 models, remove the resonator box from the throttle body.

4   Remove the crankcase vent hose.

5   Pull the housing up to dislodge the locating posts from their grommets.

6   Installation is the reverse of removal.

11 Accelerator cable - replacement

Refer to illustrations 11.2, 11.3 and 11.4

Note: *2006 and later V6 engines do not have an accelerator cable, as these engines have an electronic throttle body, operated by the PCM, which receives input from the Accelerator Pedal Position sensor at the accelerator pedal (see Chapter 6).*

1   Remove the air intake duct/resonator from the throttle body (see Section 10).

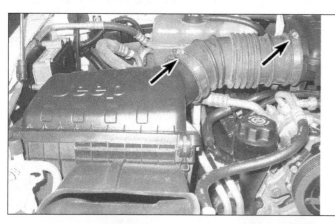

10.2 Loosen the hose clamps and remove the air duct (3.7L V6 engine shown)

11.2 Rotate the throttle lever to the wide open position and detach the cable end

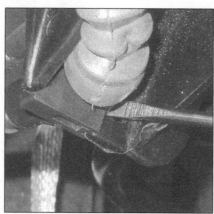

11.3 Use a small flat-bladed screwdriver to press the tab to release the cable from the bracket

11.4 Remove the plastic retainer clip to detach the cable end from the accelerator pedal

13.7 Use a stethoscope to determine if the injectors are working properly - they should make a steady clicking sound that rises and falls with engine speed changes

13.8 Measure the resistance of each injector across the two terminals of the injector

2   Rotate the throttle lever to the wide-open position and detach the cable end **(see illustration)**.

3   Using a small screwdriver, press the tab to release the accelerator cable retainer from the bracket and slide the cable out of the bracket **(see Illustration)**.

4   Working beneath the dash, remove the plastic retainer clip and detach the cable from the accelerator pedal **(see illustration)**.

5   Remove the metal clip that secures the cable to the firewall.

6   Detach the cable from the routing clip(s).

7   Push the cable through the firewall and into the engine compartment. Detach the cable from the routing clips and remove the cable from the engine compartment.

Installation is the reverse of removal.

12 Fuel injection system - general information

The sequential Multi Port Fuel Injection (MPFI) system consists of three sub-systems: air intake, emissions and engine control and fuel delivery. The system uses a Powertrain Control Module (PCM) along with the sensors (coolant temperature sensor, throttle position sensor, manifold absolute pressure sensor, oxygen sensor, etc.) to determine the proper air/fuel ratio under all operating conditions.

The fuel injection system and the emissions and engine control system are closely linked in function and design. For additional information, refer to Chapter 6.

Air intake system

The air intake system consists of the air filter, the air intake ducts, the throttle body, the idle control system, the air intake plenum and the intake manifold.

When the engine is idling, the air/fuel ratio is controlled by the idle air control system, which consists of the Powertrain Control Module (PCM) and the idle air control valve. The idle air control valve is controlled by the PCM and is opened and closed depending upon the running conditions of the engine (air conditioning system, power steering, cold and warm running etc.). This idle air

control regulates the amount of airflow past the throttle plate and into the intake manifold, thus increasing or decreasing the engine idle speed. The PCM receives information from the sensors (vehicle speed, coolant temperature, air conditioning, power steering mode etc.) and adjusts the idle according to the demands of the engine and driver. Refer to Chapter 6 for information on the idle air control valve.

Emissions and engine control system

The emissions and engine control system is described in detail in Chapter 6.

Fuel delivery system

The fuel delivery system consists of these components: The fuel pump, the fuel filter, the fuel pressure regulator, the fuel rail and the fuel injectors.

The fuel pump is an electric-type located in the fuel tank. Fuel is drawn through a screen into the pump, flows through a one-way valve, passes through the main filter and is delivered to the fuel rail and injectors.

The fuel pressure regulator supplies regulated fuel pressure to the injectors. The pressure regulator supplies fuel to the fuel rail at a constant pressure. Excessive pressure is bled off directly into the fuel tank. If the fuel pressure regulator is defective, it can be replaced.

The injectors are solenoid-actuated pintle types consisting of a solenoid, plunger, needle valve and housing. When current is applied to the solenoid coil, the needle valve raises and pressurized fuel squirts out the nozzle. The injection quantity is determined by the length of time the valve is open (the length of time during which current is supplied to the solenoid coils). The fuel injector duration is controlled by the PCM.

The Automatic Shutdown (ASD) relay and the fuel pump relay are contained within the Power Distribution Center, which is located in the engine compartment. The ASD relay connects battery voltage to the fuel injectors and the ignition coil while the fuel pump relay connects battery voltage only to the fuel pump. If the PCM senses there is NO signal

from the camshaft or crankshaft sensors (as with the engine not running or cranking), the PCM will de-energize both relays.

13 Fuel injection system - check

Refer to illustrations 13.7, 13.8 and 13.10

Note: *The following procedure is based on the assumption that the fuel pressure is adequate (see Section 3).*

1   Check all electrical connectors that are related to the system. Check the ground wire connections on the intake manifold for tightness. Loose connectors and poor grounds can cause many problems that resemble more serious malfunctions.

2   Check to see that the battery is fully charged, as the control unit and sensors depend on an accurate supply of voltage in order to properly meter the fuel.

3   Check the air filter element - a dirty or partially blocked filter will severely impede performance and economy (see Chapter 1).

4   Check the related fuses. If a blown fuse is found, replace it and see if it blows again. If it does, search for a grounded wire in the harness.

5   Check the condition of all vacuum hoses connected to the intake manifold.

6   Remove the air intake duct from the throttle body and check for dirt, carbon or other residue build-up. If it's dirty, clean it with carburetor cleaner spray, a shop towel and a toothbrush, if necessary.

7   With the engine running, place an automotive stethoscope against each injector, one at a time, and listen for a clicking sound, indicating operation **(see illustration)**. If you don't have a stethoscope, place the tip of a screwdriver against the injector and listen through the handle.

8   Disconnect the injector electrical connectors and measure the resistance of each injector **(see illustration)**. Compare the measurements with the resistance values listed in this Chapter's Specifications.

13.10 Install the "noid" light (available at most auto parts stores) into each injector electrical connector and confirm that it blinks when the engine is cranking

14.6 Throttle body mounting bolts (3.7L V6 engine shown)

9   Turn the ignition key On and check for battery voltage at the voltage supply wire terminal of one of the injector harness connectors. If battery voltage is not present, check the ASD relay and related wiring (see Chapter 12). **Note:** *The fuel injector voltage supply wire is orange/dark green. The voltage supply wire is always the same color at each injector connector while the injector control wire color is different for each injector.*

10   Install an injector test light ("noid" light) into each injector electrical connector, one at a time **(see illustration)**. Crank the engine over. Confirm that the light flashes evenly on each connector. This tests the PCM control of the injectors. If the light does not flash, have the PCM checked at a dealer service department or other properly equipped repair facility.

11   The remainder of the engine control system checks can be found in Chapter 6.

14 Throttle body - removal and installation

Refer to illustration 14.6

1   Disconnect the cable from the negative terminal of the battery.

2   Remove the air intake duct/resonator from the throttle body.

3   Disconnect the electrical connectors from the throttle position sensor and idle air control valve.

4   Label and detach all vacuum hoses from the throttle body.

5   Detach the accelerator cable (see Section 11) and, if equipped, the transmission control cable and cruise control cable. On 2006 and later V6 engines, there is one large electrical connector and no accelerator cable.

6   Remove the three mounting bolts and remove the throttle body **(see illustration)**.

7   Inspect the throttle body O-ring and replace if necessary.

8   Install the throttle body and tighten the bolts to the torque listed in this Chapter's Specifications.

9   The remainder of installation is the reverse of removal.

15 Fuel pressure regulator - replacement

Warning: *See the* **Warning** *in Section 1.*

1   The fuel pressure regulator is located on the bottom side of the upper section of the fuel pump module **(see illustration 8.8)**. Remove the upper and lower sections of the fuel pump module (see Section 8).

2   Replace the section of the fuel pump module that contains the fuel pressure regulator.

3   The remainder of installation is the reverse of removal.

16.7a To disconnect the electrical connector from the injector, pull the slider up to unlock it . . .

16 Fuel rail and injectors - removal and installation

Warning 1: *See the* **Warning** *in Section 1.*
Warning 2: *The engine must be completely cool before beginning this procedure.*
Note: *Refer to Section 13 for fuel injector diagnosis.*

Removal

Refer to illustrations 16.7a, 16.7b, 16.9a, 16.9b, 16.10a, 16.10b and 16.10c

1   Relieve the fuel pressure (see Section 2).

2   Disconnect the cable from the negative terminal of the battery.

3   Remove the air intake duct/resonator from the air filter housing and throttle body (see Section 10).

4   Using a special tool, remove the fuel line from the fuel rail (see Section 4).

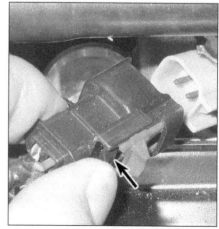

16.7b . . . then press the tab while pulling the connector off of the injector

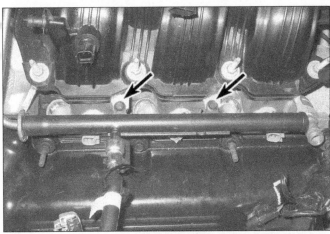

16.9 Fuel rail mounting bolts (left side shown) - 3.7L V6 engine

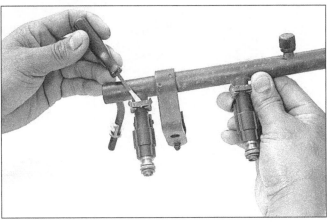

16.10a Using a screwdriver or pliers, remove the injector retaining clip . . .

5   Clearly label and remove any vacuum hoses or electrical wiring and connectors that will interfere with fuel rail removal.

6   On 2.4L four-cylinder engines, perform the following:

a) *Drain the coolant and remove the thermostat (see Chapters 1 and 5).*

b) *Remove the injector wiring from the harness brackets on the intake manifold.*

c) *Remove the PCV valve and hose (see Chapter 1).*

d) *Remove the three upper bolts and loosen the two lower bolts two turns on the intake manifold (see Chapter 2A).*

e) *On 2004 and 2005 models, disconnect and set aside the MAP sensor (see Chapter 6)*

7   Disconnect the fuel injector electrical connectors **(see illustrations)** and position the harness aside. On 2004 and later models, push in the red tab to release the injector electrical connector. **Note:** *Each connector should be numbered with the corresponding cylinder number. If the number tag is obscured or missing, renumber the connectors.*

8   On V6 engines, perform the following:

a) *Disconnect the electrical connectors from the throttle body.*

b) *Remove the ignition coils from each cylinder bank (see Chapter 5).*

c) *On 4.0L engines, remove the upper intake manifold (see Chapter 2C) and the EGR tube (see Chapter 6).*

9   Clean any debris from around the injectors. Remove the fuel rail mounting nuts/ bolts **(see illustrations)**. Gently rock the fuel rail and injectors to loosen the injectors and remove the fuel rail and fuel injectors as an assembly. **Caution:** *Do not attempt to separate the left and right fuel rails on a V6 model. Both sides are serviced together as an assembly.*

10   Remove the injectors from the fuel rail, then remove and discard the O-rings **(see illustrations)**. **Note:** *Whether you're replacing an injector or a leaking O-ring, it's a good idea to remove all the injectors from the fuel rail and replace all the O-rings.*

Installation

11   Coat the new O-rings with clean engine oil and install them on the injector(s), then insert each injector into its corresponding bore in the fuel rail. Install the injector retaining clip.

12   Clean the injector bores on the intake manifold.

13   Guide the injectors/fuel rail assembly into the injector bores on the intake manifold. Make sure the injectors are fully seated, then tighten the fuel rail mounting nuts/bolts to the torque listed in this Chapter's Specifications.

14   Connect the fuel line(s) and make sure they are securely installed. If equipped with a latch-clip type fitting, make sure the latch-clip is properly seated.

15   Connect the electrical connectors to each injector, referring to the numbered tags.

16   The remainder of installation is the reverse of removal.

17   After the injector/fuel rail assembly installation is complete, turn the ignition switch to On, but don't operate the starter (this activates the fuel pump for about two seconds, which builds up fuel pressure in the fuel lines and the fuel rail). Repeat this about two or three times, then check the fuel lines, fuel rail and injectors for fuel leakage.

17   Exhaust system servicing - general information

Warning: *Inspection and repair of exhaust system components should be done only after enough time has elapsed after driving the vehicle to allow the system components to cool completely. Also, when working under the vehicle, make sure it is securely supported on jackstands.*

1   The exhaust system consists of the exhaust manifold(s), the catalytic converter(s), the muffler, the tailpipe and all connecting pipes, brackets, hangers and clamps. The exhaust system is attached to the body with mounting brackets and rubber hangers. If any of the parts are improperly installed, excessive noise and vibration will be transmitted to the body.

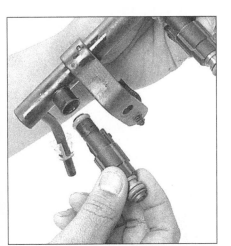

16.10b . . . and withdraw the injector from the fuel rail

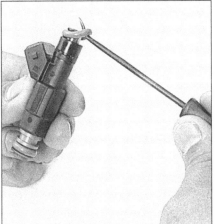

16.10c Carefully remove the O-rings from the injectors

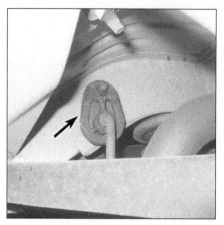

**17.2a Inspect the exhaust system
rubber hangers for damage (view from
underneath vehicle)**

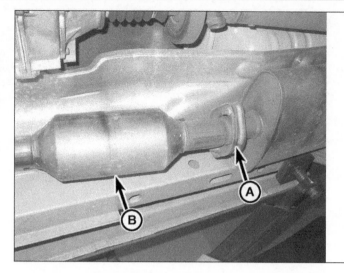

**17.2b Inspect
the exhaust pipe
connections for
leaks, and inspect
the catalytic
converter and heat
shield for damage**

A   *Exhaust pipe
connection*
B   *Catalytic
converter*

Muffler and pipes

Refer to illustrations 17.2a and 17.2b

2   Conduct regular inspections of the exhaust system to keep it safe and quiet **(see Illustrations)**. Look for any damaged or bent parts, open seams, holes, loose connections, excessive corrosion or other defects which could allow exhaust fumes to enter the vehicle. Also check the catalytic converter when you inspect the exhaust system (see below). Deteriorated exhaust system components should not be repaired; they should be replaced with new parts.

3   If the exhaust system components are extremely corroded or rusted together, welding equipment will probably be required to remove them. The convenient way to accomplish this is to have a muffler repair shop remove the corroded sections with a cutting torch. If, however, you want to save money by doing it yourself (and you don't have a welding outfit with

a cutting torch), simply cut off the old components with a hacksaw. If you have compressed air, special pneumatic cutting chisels can also be used. If you do decide to tackle the job at home, be sure to wear safety goggles to protect your eyes from metal chips and work gloves to protect your hands.

4   Here are some simple guidelines to follow when repairing the exhaust system:

a) *Work from the back to the front when removing exhaust system components.*
b) *Apply penetrating oil to the exhaust system component fasteners to make them easier to remove.*
c) *Use new gaskets, hangers and clamps when installing exhaust systems components.*
d) *Apply anti-seize compound to the threads of all exhaust system fasteners during reassembly.*
e) *Be sure to allow sufficient clearance between newly installed parts and all*

points on the underbody to avoid overheating the floor pan and possibly damaging the interior carpet and insulation. Pay particularly close attention to the catalytic converter and heat shield.

Catalytic converter

Warning: *The converter gets very hot during operation. Make sure it has cooled down before you touch it.*
Note: *See Chapter 6 for additional information on the catalytic converter.*

5   Periodically inspect the heat shield for cracks, dents and loose or missing fasteners.

6   Inspect the converter for cracks or other damage **(see Illustration 17.2b)**.

7   The catalytic converter is permanently welded into the exhaust system. If replacement is required, take the vehicle to a dealership service department or other qualified repair shop. Refer to Chapter 6 for additional information.

Chapter 5
Engine electrical systems

Contents

Specifications

General

Battery voltage	
Engine off	12.0 to 13.2 volts
Engine running	13.5 to 15 volts
Firing order	
2.4L four-cylinder engine	1-3-4-2
3.7L V6 engine	1-6-5-4-3-2
4.0L V6 engine	1-2-3-4-5-6

Ignition system

Ignition coil resistance [at 70 to 80-degrees F (21 to 27-degrees C)]	
2.4L four-cylinder engine	
Primary resistance	0.51 to 0.61 ohms
Secondary resistance	11,500 to 13,500 ohms
3.7L V6 engine	
Primary resistance	0.6 to 0.9 ohms
Secondary resistance	6,000 to 9,000 ohms
Spark plug wire resistance (2.4L four-cylinder engine)	
Minimum	250 ohms per inch
Maximum	1,000 ohms per inch

Torque specifications

	Ft-lbs	Nm
Alternator mounting bolts		
2.4L four-cylinder engine	42	57
3.7L V6 engine		
Vertical bolt	29	40
Horizontal bolts	42	57
4.0L V6 engine	42	57
Starter mounting bolts	40	54

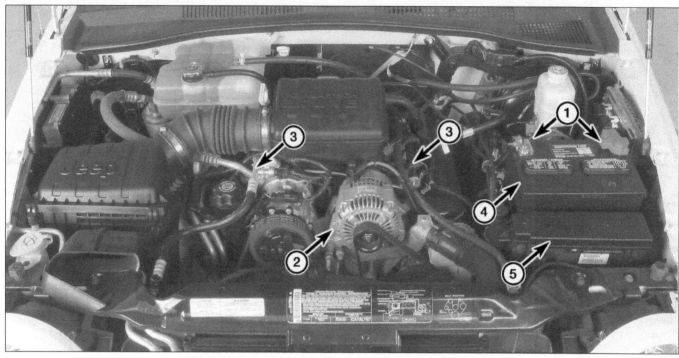

1.1 Typical engine electrical system components (3.7L V6 engine)

1	Battery cables	3	Ignition coils (on each spark plug)	5	Power distribution center
2	Alternator	4	Battery		

1   General information

Refer to illustration 1.1

The engine electrical systems include all ignition, charging and starting components **(see Illustration)**. Because of their engine-related functions, these components are discussed separately from chassis electrical devices such as the lights, the instruments, etc. (which are included in Chapter 12).

Always observe the following precautions when working on the electrical systems:

a) *Be extremely careful when servicing engine electrical components. They are easily damaged if checked, connected or handled improperly.*

b) *Never leave the ignition switch on for long periods of time with the engine off.*

c) *Don't disconnect the battery cables while the engine is running.*

d) *Maintain correct polarity when connecting a battery cable from another vehicle during jump-starting.*

e) *Always disconnect the negative cable first and hook it up last or the battery may be shorted by the tool being used to loosen the cable clamps.*

It's also a good idea to review the safety-related information regarding the engine electrical systems located in the *Safety First!* section near the front of this manual before beginning any operation included in this Chapter.

2   Battery - emergency jump starting

Refer to the *Booster battery (jump) starting* procedure at the front of this manual.

3   Battery - check and replacement

Warning: *Hydrogen gas is produced by the battery, so keep open flames and lighted cigarettes away from it at all times. Always wear eye protection when working around a battery. Rinse off spilled electrolyte immediately with large amounts of water.*
Caution: *Always disconnect the negative cable first and hook it up last or the battery may be shorted by the tool being used to loosen the cable clamps.*

Check

Refer to illustrations 3.2 and 3.3
1   Disconnect the negative battery cable, then the positive cable from the battery.
2   Check the battery state of charge. Visually inspect the indicator eye on the top of the battery. If the indicator eye is clear, charge the battery as described in Chapter 1. Next perform an open voltage circuit test using a voltmeter **(see illustration)**. **Note:** *The battery's surface charge must be removed before accurate voltage measurements can be made. Turn on the high beams for ten seconds, then turn them off and let the vehicle stand for two minutes.* With the engine and all accessories off, touch the negative probe of the voltmeter to the negative terminal of the battery and the positive probe to the positive terminal of the battery. The battery voltage should be 12.4 volts or more. If the battery is less than the specified voltage, charge the battery before proceeding to the next test. Do not proceed with the battery load test unless the battery charge is correct.
3   Perform a battery load test. An accurate check of the battery condition can only be performed with a battery load tester (available at most auto parts stores). This test evaluates the ability of the battery to operate the starter and other accessories during periods of heavy amperage draw (load). The tool utilizes a carbon pile to increase the load demand (amperage draw) on the battery. Install a special battery load-testing tool onto the terminals **(see illustration)**. Load test the battery according to the tool manufacturer's instructions. Typically a load of 50-percent of the cold cranking amperage rating is applied during the test. The cold cranking amperage rating can usually be found on the battery label. Maintain the load on the battery for a maximum of 15 seconds. The battery voltage should not drop below 9.6 volts during the test. If the battery condition is weak or defective, the tool will indicate this condition immediately. **Note:** *Cold temperatures will cause the voltage readings to drop slightly. Follow the chart given in the tool manufacturer's instructions to compensate for cold climates. Minimum load voltage for freezing temperatures (32-degrees F) should be approximately 9.1 volts.*

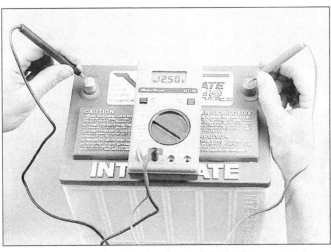

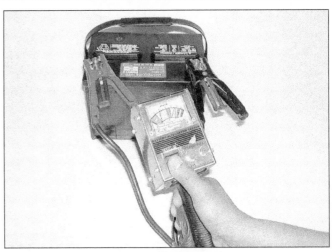

3.2 To test the open circuit voltage of the battery, connect a voltmeter to the battery - a fully charged battery should measure at least 12.4 volts (depending on outside air temperature)

3.3 Connect a battery load tester to the battery and check the battery condition under load following the tool manufacturer's instructions

Replacement

Refer to illustrations 3.5 and 3.8

4   Disconnect the cable from the negative terminal of the battery, then disconnect the cable from the positive terminal of the battery.
5   Remove the battery hold-down clamp **(see illustration)**.
6   Lift out the battery. Be careful - it's heavy. **Note:** *Battery straps and handlers are available at most auto parts stores for a reasonable price. They make it easier to remove and carry the battery.*
7   While the battery tray is out, inspect the area underneath the tray for corrosion.
8   If corrosion has leaked down past the battery tray, remove the battery tray:

a) *Detach the power distribution center from the battery tray.*
b) *Disconnect the electrical connector from the battery temperature sensor.*
c) *Remove the battery tray mounting bolts/ nuts and remove the battery tray* **(see illustration).**

9   Clean the battery tray, then use baking soda to clean any deposits from the metal underneath to prevent further oxidation. Spray the area with a rust-inhibiting paint.
10   If you are replacing the battery, make sure you get one that's identical, with the same dimensions, amperage rating, cold cranking rating, etc.
11   Installation is the reverse of removal.

4   Battery cables - check and replacement

Refer to illustration 4.4

1   Periodically inspect the entire length of each battery cable for damage, cracked or burned insulation and corrosion. Poor battery cable connections can cause starting problems and decreased engine performance.
2   Check the cable-to-terminal connections at the ends of the cables for cracks, loose wire strands and corrosion. The presence of white, fluffy deposits under the insu-

lation at the cable terminal connection is a sign that the cable is corroded and should be replaced. Check the terminals for distortion, missing mounting bolts and corrosion.
3   When removing the cables, always disconnect the negative cable first and hook it up last or the battery may be shorted by the tool used to loosen the cable clamps. Even if only the positive cable is being replaced, be sure to disconnect the negative cable from the battery first (see Chapter 1 for further information regarding battery cable removal).
4   Disconnect the old cables from the battery, then trace each of them to their opposite ends and detach them from the starter solenoid, power distribution center and ground terminals, as necessary. Note the routing of each cable to ensure correct installation. Cut the electrical tape and remove the protective sheathing from the battery cables **(see illustration)**.
5   If you are replacing either or both of the battery cables, take them with you when buying new cables. It is vitally important that you

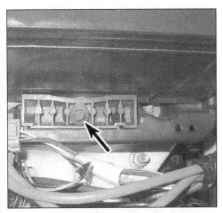

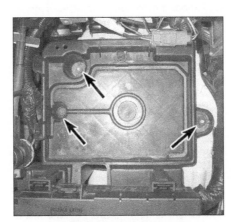

3.5 Remove the bolt and detach the hold-down clamp from the battery tray

3.8 Remove the battery tray mounting bolts/nuts

4.4 The battery cables are wrapped in a protective sheathing - if necessary, cut the electrical tape and remove the sheathing

6.4 To use a calibrated ignition tester, simply disconnect a spark plug wire (2.4L four-cylinder engines) or remove a coil (3.7L V6 engines). Connect the spark plug wire or coil to the tester, clip the tester to a convenient ground and crank the engine over - if there's enough power to fire the plug, bright blue sparks will be visible between the electrode tip and the tester body (weak sparks or intermittent sparks are the same as no sparks) (3.7L V6 shown)

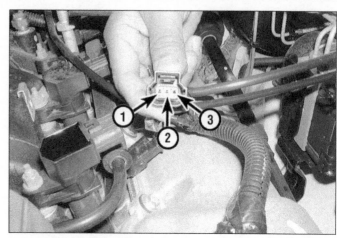

6.7a Disconnect the ignition coil electrical connector and check for battery voltage at the battery positive terminal with the ignition On; if battery voltage is available to the coil(s), check the coil control terminal(s) for a blinking voltage while cranking the engine (3.7L V6 shown)

1   *Coil driver*
2   *B+ (from auto shut-down relay)*
3   *Not used*

replace the cables with identical parts. Cables have characteristics that make them easy to identify: positive cables are usually red and larger in cross-section; ground cables are usually black and smaller in cross-section.

6   Clean the threads of the solenoid or ground connection with a wire brush to remove rust and corrosion. Apply a light coat of battery terminal corrosion inhibitor or petroleum jelly to the threads to prevent future corrosion.

7   Attach the cable to the solenoid or ground connection and tighten the mounting nut/bolt securely.

8   Before connecting a new cable to the battery, make sure that it reaches the battery post without having to be stretched.

9   Connect the positive cable first, followed by the negative cable.

5   Ignition system - general information

All models are equipped with a distributorless ignition system. The ignition system consists of the battery, ignition coil(s), spark plugs, camshaft position sensor, crankshaft position sensor and the Powertrain Control Module (PCM). The PCM controls the ignition timing and spark advance characteristics for the engine. The ignition timing is not adjustable.

The 2.4L four-cylinder ignition system utilizes two ignition coils combined into one unit. Each coil-half fires two cylinders in the firing order simultaneously (1-4, 2-3) each time one cylinder is under compression and its companion cylinder is on the exhaust stroke. Since the cylinder on the exhaust stroke requires very little of the available voltage to fire its plug, most of the voltage is

used to fire the plug of the cylinder on the compression stroke.

The 3.7L V6 uses a "coil-over-plug" system. The system utilizes six individual coils located above each spark plug and connected directly to the spark plug. The PCM controls the operation of the ignition coils, firing each coil in sequence.

The PCM controls the ignition system by opening and closing the ignition coil ground circuit. The computerized ignition system provides complete control of the ignition timing by determining the optimum timing in response to engine speed, coolant temperature, throttle position and vacuum pressure in the intake manifold. These parameters are relayed to the PCM by the camshaft position sensor, crankshaft position sensor, throttle position sensor, coolant temperature sensor and manifold absolute pressure sensor. The PCM and the crankshaft position sensor are very important components of the ignition system. The ignition system will not operate and the engine will not start if the PCM or the crankshaft position sensor are defective. Refer to Chapter 6 for additional information on the various sensors.

6   Ignition system - check

Refer to illustrations 6.4, 6.7a and 6.7b
Warning 1: *Because of the high voltage generated by the ignition system, extreme care should be taken whenever an operation is performed involving ignition components. This not only includes the ignition coil, but also related components and test equipment.*
Warning 2: *The following procedure requires the engine to be cranked during testing. Make sure the meter leads, loose clothing, long hair, etc. are away from the moving parts of the*

engine (drivebelt, cooling fan, etc.) before cranking the engine.

1   Before proceeding with the ignition system, check the following items:

a) *Make sure the battery cable clamps, where they connect to the battery, are clean and tight.*

b) *Test the condition of the battery (see Section 3). If it does not pass all the tests, replace it with a new battery.*

c) *Check the ignition system wiring and connections for tightness, damage, corrosion or any other signs of a bad connection.*

d) *Check the related fuses inside the power distribution center (see Chapter 12). If they're burned, determine the cause and repair the circuit.*

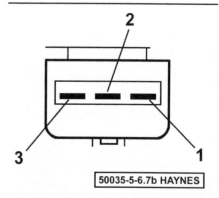

50035-5-6.7b HAYNES

6.7b Ignition coil electrical connector terminal guide (harness side) - four cylinder engine

1   *Coil driver (no. 2)*
2   *B+ (from auto shut-down relay)*
3   *Coil driver (no. 1)*

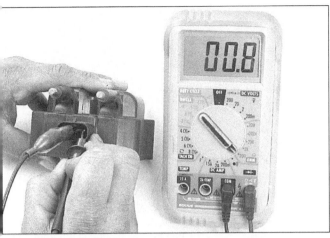

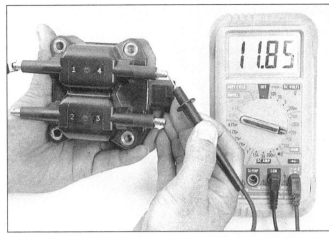

7.3 Test each coil's primary resistance by connecting one probe of the ohmmeter to the center terminal and the other probe to each of the end terminals. The primary resistance of each coil should be within specifications

7.4 Checking the secondary resistance of coil "1 and 4" spark plug terminals - repeat the test for coil "2 and 3." The secondary resistance of each coil should be within specifications

2   If the engine turns over but won't start or has a severe misfire, perform the following steps using a calibrated ignition tester to make sure there is sufficient secondary ignition voltage to fire the spark plugs.

3   Disable the fuel system by removing the fuel pump relay from the power distribution center (see Chapter 12).

4   On 2.4L four-cylinder engines, remove a spark plug wire. On 3.7L V6 engines, remove an ignition coil (see Section 7). Attach a calibrated ignition system tester (available at most auto parts stores) to the spark plug boot (on 3.7L V6 engines, be sure to reconnect the electrical connector to the coil). Connect the clip on the tester to a bolt or metal bracket on the engine **(see illustration)**. Crank the engine and watch the end of the tester to see if a bright blue, well-defined spark occurs (weak spark or intermittent spark is the same as no spark).

5   If spark occurs, sufficient voltage is reaching the plug to fire it (repeat the check at the remaining spark plug terminals to verify that the ignition coil(s) are good). If the ignition system is operating properly the problem lies elsewhere; such as a mechanical or fuel system problem. However, the spark plugs may be fouled, so remove and check them as described in Chapter 1.

6   If no spark occurs at one or more coil, remove the coil and check the spring inside the spark plug boot for damage. Using an ohmmeter, measure the primary and secondary resistance of the coil (see Section 7). Replace the coil if defective.

7   If no spark occurs at any of the coils, check for battery voltage to the ignition coils. Attach a 12-volt test light or voltmeter to a good engine ground point. Disconnect the coil electrical connector(s) and check for battery voltage at the power wire terminal **(see illustrations)**. Battery voltage should be available with the ignition key On. If there is no battery voltage present, check the auto shut-down relay, and the circuits from the

power distribution center to the ignition coil(s) and auto shut-down relay (see Chapter 12).

8   If battery voltage is available to the ignition coil, attach a test light to the coil control terminal(s) on the ignition coil connector **(see illustrations 6.7a and 6.7b)**. Crank the engine. The test light should blink with the engine cranking if a trigger signal is present. If a trigger signal is present at the coil, the Powertrain Control Module, the camshaft position sensor and the crankshaft position sensor are functioning properly.

9   If a trigger signal is not present at the ignition coil, refer to Chapter 6 and check the camshaft position sensor and crankshaft position sensor. If the camshaft position sensor and crankshaft position sensor are good, have the PCM checked by a dealer service department or other qualified repair shop.

7   Ignition coil - check and replacement

2.4L four-cylinder engines
Check
Refer to illustrations 7.3 and 7.4

1   Remove the ignition coil unit (see Steps 6 through 9).

2   Clean the outer case and check it for cracks and other damage. Clean the coil primary terminals and check the coil tower terminal for corrosion. Clean it with a wire brush if any corrosion is found. It is essential for proper operation of the ignition system that all coil terminals and wires be kept clean and dry.

3   Check the coil primary resistance by attaching the leads of an ohmmeter to one of the outer terminals and the center terminal. Repeat this step with the other outer terminal and the center terminal. **(see illustration)**.

4   Check the coil secondary resistance by connecting one of the ohmmeter leads to the

no. 1 terminal and the other ohmmeter lead to the no. 4 terminal **(see illustration)**. Repeat this step for the no. 2 and no. 3 coil terminals.

5   Compare your measurements to the resistance values listed in this Chapter's Specifications. If either measured resistance value is not as specified, replace the ignition coil unit.

Replacement
Refer to illustration 7.8

6   Remove the air filter housing and outlet hose (see Chapter 4).

7   Mark the spark plug wires to ensure the proper firing order during installation, then carefully remove them from the ignition coil with a twisting motion.

8   Disconnect the electrical connector from the coil **(see illustration)**.

9   Remove the four coil mounting bolts and remove the coil from the engine compartment **(see illustration 7.8)**.

10   Installation is the reverse of removal.

7.8 After carefully disconnecting the spark plug wires and electrical connector, remove the coil pack mounting bolts and remove the coil from the engine compartment

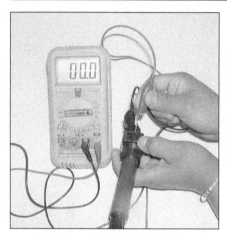

7.13 On 3.7L V6 engines, measure the primary resistance of the ignition coil across the two terminals of the coil connector

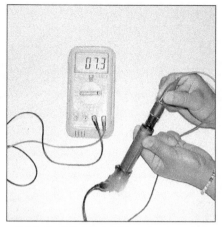

7.14 On 3.7L V6 engines, measure the secondary resistance across one of the primary terminals and the spark plug terminal

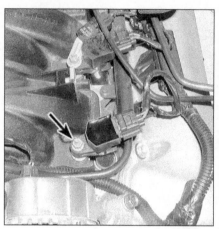

7.19 Ignition coil mounting stud/nut and electrical connector - 3.7L V6 engines

3.7L V6 engines

Check

Refer to illustrations 7.13 and 7.14

11   Remove the ignition coil(s) (see Steps 16 through 19).

12   Clean the outer case and check it for cracks and other damage. Clean the coil primary terminals and check the coil tower terminal for corrosion. Clean it with a wire brush if any corrosion is found. It is essential for proper operation of the ignition system that all coil terminals and wires be kept clean and dry.

13   Check the coil primary resistance by attaching the leads of an ohmmeter to the two primary coil terminals **(see illustration)**. Repeat this step for each coil that is being checked.

14   Check the coil secondary resistance by connecting one of the ohmmeter leads to a coil primary terminal and the other ohmmeter lead to the spark plug terminal **(see illustration)**. Repeat this step for each coil that is being checked.

15   Compare your measurements to the resistance values listed in this Chapter's Specifications. If either measured resistance value is not as specified, replace the ignition coil(s).

Replacement

Refer to illustration 7.19

16   Depending on which ignition coil is being replaced, remove the air inlet duct/resonator if necessary.

17   Disconnect the electrical connector from the coil.

18   Clean the area surrounding the ignition coil(s) with compressed air to prevent debris from falling into the engine cylinder.

19   Remove the nut from the ignition coil mounting stud **(see illustration)**. Carefully pull the coil up with a twisting motion and remove the coil from the engine compartment.

20   Installation is the reverse of removal.

8   Charging system - general information and precautions

The charging system includes the alternator, a charge indicator light, the battery, the Powertrain Control Module (PCM), a fusible link and the wiring between all the components. The charging system supplies electrical power for the ignition system, the lights, the radio, etc. The alternator is driven by a serpentine drivebelt at the front of the engine.

The alternator control system within the PCM regulates the DC current output of the alternator in accordance with driving conditions. Depending upon electric load, vehicle speed, battery temperature and accessories (air conditioning system, radio, cruise control etc.), the system will adjust the amount of DC current generated, creating less load on the engine.

The purpose of the voltage regulator is to limit the alternator voltage output to a preset value. This prevents power surges, circuit overloads, etc., during peak voltage output. The voltage regulator is contained within the PCM and in the event of failure, the PCM must be replaced.

The charging system doesn't ordinarily require periodic maintenance. However, the drivebelt, battery, wiring harness and connections should be inspected at the intervals outlined in Chapter 1.

The dashboard warning light should come ON when the ignition key is turned to ON, but it should go off immediately after the engine is started. If it remains ON, there is a malfunction in the charging system. Some vehicles are also equipped with a voltmeter. If the voltmeter indicates abnormally high or low voltage, check the charging system (see Section 9).

Be very careful when making electrical circuit connections to a vehicle equipped with an alternator and note the following:

a) *When reconnecting wires to the alternator from the battery, be sure to note the polarity.*

b) *Before using arc-welding equipment to repair any part of the vehicle, disconnect the wires from the alternator and the battery terminals.*

c) *Never start the engine with a battery charger connected.*

d) *Always disconnect both battery cables before using a battery charger.*

e) *The alternator is turned by an engine drivebelt which could cause serious injury if your hands, hair or clothes become entangled in it with the engine running.*

f) *Because the alternator is connected directly to the battery, it could arc or cause a fire if overloaded or shorted out.*

g) *Wrap a plastic bag over the alternator and secure it with rubber bands before steam-cleaning the engine.*

9   Charging system - check

Note: *These vehicles are equipped with an On-Board Diagnostic (OBD) system that is useful for detecting charging system problems. Refer to Chapter 6 for the list of diagnostic codes and procedures for obtaining the codes.*

1   If a malfunction occurs in the charging circuit, do not immediately assume that the alternator is causing the problem. First check the following items:

a) *The battery cables where they connect to the battery. Make sure the connections are clean and tight.*

b) *The battery electrolyte specific gravity (by observing the charge indicator on the battery). If it is low, charge the battery.*

c) *Check the external alternator wiring and connections.*

d) *Check the drivebelt condition and tension (see Chapter 1).*

e) *Check the alternator mounting bolts for tightness.*

f) *Run the engine and check the alternator for abnormal noise.*

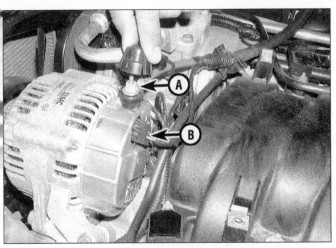

10.2 Disconnect the alternator electrical connections

A   B+ terminal (output)   *B   Field terminals*

10.4 Alternator mounting bolts (3.7L V6 engine shown, four-cylinder engine similar)

2   Using a voltmeter, check the battery voltage with the engine off. It should be approximately 12.4 to 12.6 volts with a fully charged battery.

3   Start the engine and check the battery voltage again. It should now be greater than the voltage recorded in Step 2, but not more than 15 volts.

4   If the indicated voltage reading is less or more than the specified charging voltage, have the charging system checked at a dealer service department or other properly equipped repair facility. The voltage regulator on these models is contained within the PCM and it cannot be adjusted, removed or tampered with in any way.

10   Alternator - removal and installation

Refer to illustrations 10.2 and 10.4

1   Disconnect the cable from the negative terminal of the battery.

2   Disconnect the output wire and the field terminals from the alternator **(see illustration)**.

3   Loosen the tension on the drivebelt, then remove the serpentine drivebelt (see Chapter 1).

4   Remove the mounting nuts/bolts and separate the alternator from the engine **(see illustration)**.

5   If you are replacing the alternator, take the old one with you when purchasing a replacement unit. Make sure the new/rebuilt unit looks identical to the old alternator. Look at the terminals - they should be the same in number, size and location as the terminals on the old alternator. Finally, look at the identification numbers - they will be stamped into the housing or printed on a tag attached to the housing. Make sure the numbers are the same on both alternators.

6   Many new/rebuilt alternators do not have a pulley installed, so you may have to

switch the pulley from the old unit to the new/rebuilt one. When buying an alternator, find out the shop's policy regarding pulleys; some shops will perform this service free of charge.

7   Installation is the reverse of removal. Tighten the mounting bolts to the torque listed in this Chapter's Specifications.

8   Install the drivebelt (see Chapter 1).

9   Check the charging voltage to verify proper operation of the alternator (see Section 9).

11   Starting system - general information and precautions

The starter motor assembly installed on these engines uses a planetary gear reduction drive. This type starter motor assembly is serviced as a complete unit. If any component of the starter motor fails, including the solenoid, the entire assembly must be replaced.

The starting system consists of the battery, starter relay, starter motor assembly and the wiring connecting the components.

The starter motor assembly is installed on the lower part of the engine, next to the transmission bellhousing.

When the ignition key is turned to the START position, the starter solenoid is actuated through the starter control circuit which includes a starter relay located in the power distribution center. The starter solenoid then connects the battery to the starter motor. The battery supplies the electrical energy to the starter motor, which does the actual work of cranking the engine.

Always observe the following precautions when working on the starting system:

a)   *Excessive cranking of the starter motor can overheat it and cause serious damage. Never operate the starter motor for more than 15 seconds at a time without pausing to allow it to cool for at least two minutes.*

b)   *The starter is connected directly to the battery and could arc or cause a fire if mishandled, overloaded or shorted.*

c)   *Always detach the cable from the negative terminal of the battery before working on the starting system.*

12   Starter motor and circuit - check

1   If a malfunction occurs in the starting circuit, do not immediately assume that the starter is causing the problem. First, check the following items:

a)   *Make sure the battery cable clamps, where they connect to the battery, are clean and tight.*

b)   *Check the condition of the battery cables (see Section 4). Replace any defective battery cables with new parts.*

c)   *Test the condition of the battery (see Section 3). If it does not pass all the tests, replace it with a new battery.*

d)   *Check the starter motor wiring and connections.*

e)   *Check the starter motor mounting bolts for tightness.*

f)   *Check the related fuses in the engine compartment fuse box (see Chapter 12). If they're blown, determine the cause and repair the circuit.*

g)   *Check the ignition switch circuit for correct operation (see Chapter 12).*

h)   *Check the starter relay (see Chapter 12).*

i)   *Check the operation of the clutch safety switch (see Chapter 8) or the Park/Neutral position switch (see Chapter 7B). These systems must operate correctly to provide battery voltage to the starter solenoid.*

2   If the starter does not activate when the ignition switch is turned to the start position, check for battery voltage to the starter solenoid. This will determine if the solenoid is receiving the correct voltage from the ignition switch. Connect a 12-volt test light or a volt-

13.3 On 3.7L V6 models with 4WD, unbolt the left side exhaust down-pipe from the crossover pipe, then lower the crossover pipe to allow clearance for the removal of the front driveshaft

13.6 Remove the battery cable mounting nut and disconnect the electrical connector from the starter motor solenoid (arrows)

meter to the starter solenoid positive terminal. While an assistant turns the ignition switch to the start position, observe the test light or voltmeter. The test light should shine brightly or battery voltage should be indicated on the voltmeter. If voltage is not available to the starter solenoid, refer to the wiring diagrams in Chapter 12 and check the fuses and starter relay in series with the starting system. If voltage is available but there is no movement from the starter motor, remove the starter from the engine (see Section 13) and bench test the starter (see Step 4).

3   If the starter turns over slowly, check the starter cranking voltage and the current draw from the battery. This test must be performed with the starter assembly on the engine. Crank the engine over (for 10 seconds or less) and observe the battery voltage. It should not drop below 9.6 volts. Also, observe the current draw using an amp meter. Typically a starter should not exceed 160 amps. If the starter motor amperage draw is excessive, have it tested by a dealer service department or other qualified repair shop. There are several conditions that may affect the starter cranking potential. The battery must be in good condition and the battery cold-cranking rat-

ing must not be under-rated for the particular application. Be sure to check the battery specifications carefully. The battery terminals and cables must be clean and not corroded. Also, in cases of extreme cold temperatures, make sure the battery and/or engine block is warmed before performing the tests.

4   If the starter is receiving voltage but does not activate, remove and check the starter motor assembly on the bench. Most likely the solenoid is defective. In some rare cases, the engine may be seized so be sure to try and rotate the crankshaft pulley (see Chapter 2A, 2B or 2C) before proceeding. With the starter assembly mounted in a vise on the bench, install one jumper cable from the positive terminal of a test battery to the B+ terminal on the starter. Install another jumper cable from the negative terminal of the battery to the body of the starter. Install a starter switch and apply battery voltage to the solenoid S terminal (for 10 seconds or less) and observe the solenoid plunger, shift lever and overrunning clutch extend and rotate the pinion drive. If the pinion drive extends but does not rotate, the solenoid is operating but the starter motor is defective. If there is no movement but the solenoid clicks, the solenoid and/or the starter

motor is defective. If the solenoid plunger extends and rotates the pinion drive, the starter assembly is operating properly.

13   Starter motor - removal and installation

1   Detach the cable from the negative terminal of the battery.
2   Raise the vehicle and support it securely on jackstands.

V6 engines

Refer to illustration 13.3

3   On models equipped with 4WD, detach the left side exhaust down-pipe from the crossover pipe and lower the crossover pipe to allow clearance for removal of the front driveshaft **(see illustration)**.
4   If equipped, remove the front driveshaft (see Chapter 8).
5   Remove the heat shield from the starter motor.

All models

Refer to illustrations 13.6 and 13.7

6   Disconnect the battery cable and wires from the terminals on the starter motor solenoid **(see illustration)**. Disconnect any clips securing the wiring to the starter. **Note:** *It may be easier to disconnect the battery cable from the starter after removing the mounting bolts and lowering the starter. Be sure to support the starter, do not allow it to hang by the wiring.*
7   Remove the two starter mounting bolts and remove the starter from the engine **(see illustration)**. **Note:** *On 3.7L V6 engines, position the starter so that the front of the starter faces the rear of the vehicle and the solenoid faces downward, then remove the starter by passing it between the exhaust pipe and transmission bellhousing.*
8   Installation is the reverse of removal.

13.7 Remove the starter motor mounting bolts (arrows)

Chapter 6
Emissions and engine control systems

Contents

1 General information

Refer to illustrations 1.1 and 1.6

To prevent pollution of the atmosphere from incompletely burned and evaporating gases, and to maintain good driveability and fuel economy, a number of emission control systems are incorporated **(see illustration)**. They include the:

Catalytic converter
Crankcase ventilation system
Electronic engine control system
Evaporative emissions control system
Exhaust gas recirculation (2007 and later)

All of these systems are linked, directly or indirectly, to the emission control system.

The Sections in this Chapter include general descriptions, checking procedures within the scope of the home mechanic and component replacement procedures (when possible) for each of the systems listed above. Before assuming that an emissions control system is malfunctioning, check the fuel and ignition systems carefully. The diagnosis of some emission control devices requires specialized tools, equipment and training. If checking and servicing become too difficult or if a procedure is beyond your ability, consult a dealer service department or other qualified repair shop. Remember, the most frequent cause of emissions problems is simply a loose or broken vacuum hose or wire, so always check the hose and wiring connections first.

This doesn't mean, however, that emission control systems are particularly difficult to maintain and repair. You can quickly and easily perform many checks and do most of the regular maintenance at home with common tune-up and hand tools. **Note:** *Because of a federally mandated warranty which covers the emission control system components, check with your dealer about warranty coverage before working on any emissions-related systems. Once the warranty has expired, you*

may wish to perform some of the component checks and/or replacement procedures in this Chapter to save money.

Pay close attention to any special precautions outlined in this Chapter. It should be noted that the illustrations of the various systems may not exactly match the system installed on the vehicle you're working on because of changes made by the manufacturer during production or from year-to-year.

A Vehicle Emissions Control Information (VECI) label is located in the engine compartment **(see illustration)**. This label contains important emissions specifications and adjustment information, as well as a vacuum hose schematic with emissions components identified. When servicing the engine or emissions systems, the VECI label in your particular vehicle should always be checked for up-to-date information.

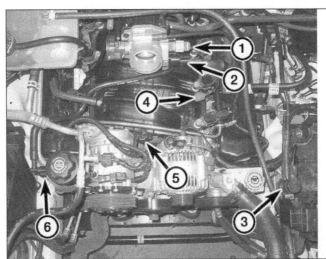

1.1 Typical emission and engine control system components - 3.7L V6 engines

1 *Throttle Position Sensor (TPS)*
2 *Idle Air Control (IAC) valve*
3 *EVAP purge solenoid*
4 *Intake Air Temperature (IAT) sensor*
5 *Manifold Absolute Pressure (MAP) sensor*
6 *PCV valve*

1.6 The Vehicle Emission Control Information (VECI) label, located in the engine compartment, contains information on the emission devices installed on your vehicle and a vacuum hose routing schematic

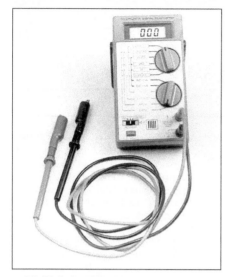

2.1 Digital multimeters can be used for testing all types of circuits; because of their high impedance, they are much more accurate than analog meters for measuring low-voltage computer circuits

2.2 Hand-held scan tools like these can extract computer codes and also perform diagnostics

---

2   On Board Diagnosis system and trouble codes

Diagnostic tool information

Refer to illustrations 2.1 and 2.2

1   A digital multimeter is necessary for checking fuel injection and emission related components **(see illustration)**. A digital voltohmmeter is preferred over the older style analog multimeter for several reasons. The analog multimeter cannot display the volts-ohms or amps measurement in hundredths and thousandths increments. When working with electronic circuits which are often very low voltage, this accurate reading is most important. Another good reason for the digital multimeter is the high impedance circuit. The digital multimeter is equipped with a high resistance internal circuitry (10 million ohms). Because a voltmeter is hooked up in parallel with the circuit when testing, it is vital that none of the voltage being measured be allowed to travel the parallel path set up by the meter itself. This dilemma does not show itself when measuring larger amounts of voltage (9 to 12 volt circuits) but if you are measuring a low voltage circuit such as the oxygen sensor signal voltage, a fraction of a volt may be a significant amount when diagnosing a problem. However, there are several exceptions where using an analog voltmeter may be necessary to test certain sensors.

2   Hand-held scanners are the most powerful and versatile tools for analyzing engine management systems used on later model vehicles **(see illustration)**. Each brand scan tool must be examined carefully to match the year, make and model of the vehicle you are working on. Often interchangeable cartridges are available to access the particular manu-

facturer (Ford, GM, Chrysler, etc.). Some manufacturers will specify by continent (Asia, Europe, USA, etc.).

3   With the arrival of the federally mandated emission control system (OBD-II), a specially designed scanner has been developed. Several tool manufacturers have released OBD-II scan tools for the home mechanic. Ask the parts salesman at a local auto parts store for additional information.

4   Other types of code readers may be available for some models at parts stores. These tools simplify the procedure for extracting codes from the engine management computer by simply "plugging in" to the diagnostic connector on the vehicle wiring harness.

On Board Diagnostic system general description

5   All models described in this manual are equipped with an On Board Diagnostic system. The systems consist of an onboard computer, known as the Powertrain Control Module (PCM)/Transmission Control Module (TCM), and information sensors, which monitor various functions of the engine and send data to the PCM. All models covered by this manual are equipped with the second generation On Board Diagnostic system (OBD-II).

6   Based on the data and the information programmed into the computer's memory, the PCM generates output signals to control various engine functions via control relays, solenoids and other output actuators. The PCM is specifically calibrated to optimize the emissions, fuel economy and driveability of the vehicle.

7   Because of a federally mandated warranty which covers the emissions system components and because any owner-induced damage to the PCM, the sensors and/or the control devices may void the warranty, it isn't a good idea to attempt diagnosis or replacement of the PCM at home while the vehicle is under warranty. Take the vehicle to a dealer service department if the PCM or a system component malfunctions.

Information sensors

8   **Battery temperature sensor** - The battery temperature sensor senses the temperature of the battery. The PCM uses this infor-

mation in controlling the voltage output of the alternator.

9   **Camshaft position (CMP) sensor** - The camshaft position sensor provides information on camshaft position. The PCM uses this information, along with the crankshaft position sensor information, to control ignition timing and fuel injection synchronization.

10   **Crankshaft position (CKP) sensor** - The crankshaft position sensor senses crankshaft position (TDC) during each engine revolution. The PCM uses this information to control ignition timing and fuel injection synchronization.

11   **Engine coolant temperature (ECT) sensor** - The engine coolant temperature sensor senses engine coolant temperature. The PCM uses this information to control the fuel delivery, ignition timing and temperature light/gauge.

12   **Intake air temperature (IAT) sensor** - The intake air temperature senses the temperature of the air entering the intake manifold. The PCM uses this information to control fuel delivery.

13   **Knock sensors (3.7L V6 engines)** -There are two knock sensors located underneath the intake manifold - one on the left side and one on the right side of the engine block. The sensors send voltage signals to the PCM based on the intensity of the vibrations inside the cylinders. When cylinder knock ("pinging") is detected, the PCM will respond by retarding the ignition timing for all of the cylinders.

14   **Manifold absolute pressure (MAP) sensor** - The manifold absolute pressure monitors intake manifold pressure and ambient barometric pressure. The PCM uses this input signal to determine engine load and adjust fuel delivery accordingly.

15   **Oxygen (O2) sensor** - The oxygen sensors generate a voltage signal that varies with the difference between the oxygen content of the exhaust and the oxygen in the surrounding air. The PCM uses this information to determine if the fuel system is running rich or lean.

16   **Throttle position sensor (TPS)** - The throttle position sensor senses throttle movement and position. This signal enables the

PCM to determine when the throttle is closed, in a cruise position, or wide open. The PCM uses this information to control fuel delivery and ignition timing.

17 **Transmission pressure and temperature sensors** - The transmission pressure and temperature sensor signals are used by the PCM/TCM to control shift operation.

18 **Miscellaneous PCM inputs** - In addition to the various sensors, the PCM monitors various switches and circuits to determine vehicle operating conditions. The switches and circuits include:

a) *Air conditioning system*
b) *Auto shutdown relay*
c) *Battery voltage*
d) *Brake On/Off switch*
e) *Cruise control system*
f) *Engine oil pressure*
g) *EVAP leak detection pump operation*
h) *Fuel level*
i) *Ignition switch*
j) *Overdrive switch*
k) *Park/neutral position switch*
l) *Sensor signal and ground circuits*

Output actuators

19 **Air conditioning clutch relay** - The PCM controls the operation of the air conditioning compressor clutch with the air conditioning clutch relay.

20 **Automatic shutdown (ASD) relay** - The automatic shutdown relay supplies battery power to the fuel injectors, ignition coil and oxygen sensor heaters. The PCM controls the operation of the automatic shutdown relay.

21 **Check Engine light** - The PCM will illuminate the Check Engine light if a malfunction in the emissions system or electronic engine control system occurs.

22 **Cruise control vacuum and vent solenoids** - The cruise control system operation is controlled by the PCM.

23 **Engine cooling fan relay** - The auxiliary electric engine cooling fan, on 1999 and 2000 models, is controlled by the PCM according to information received from the engine coolant temperature sensor.

24 **EVAP canister purge solenoid** - The evaporative emission canister purge solenoid is a solenoid valve, operated by the PCM to purge the fuel vapor canister and route fuel vapor to the intake manifold for combustion.

25 **EVAP leak detection pump** - Some models are equipped with a self-diagnostic EVAP leak detection system. The system checks the integrity of the EVAP system when the engine is started cold. The PCM controls the operation of the EVAP leak detection pump.

26 **Fuel injectors** - The PCM opens the fuel injectors individually in firing order sequence. The PCM also controls the time the injector is held open (pulse width). The pulse width of the injector (measured in milliseconds) determines the amount of fuel delivered. For more information on the fuel delivery system and the fuel injectors, including injector replacement, refer to Chapter 4.

27 **Fuel pump relay** - The fuel pump relay is activated by the PCM with the ignition switch in the Start or Run position. When the ignition switch is turned on, the relay is activated to supply initial line pressure to the system. Refer to Chapter 12 or your owner's manual for more information on relay location. For more information on fuel pump check and replacement, refer to Chapter 4.

28 **Idle air control (IAC) valve** - The idle air control valve controls the amount of air allowed to bypass the throttle plate when the throttle valve is closed or at idle position. The more air allowed to bypass the throttle plate, the higher the idle speed. The idle air control valve opening and the resulting idle speed is controlled by the PCM.

29 **Ignition coil(s)** - The PCM controls the ignition coil(s) by grounding the primary circuit to the ignition coil which generates high voltage in the secondary circuit, thus sending spark from the ignition coil to the spark plugs. The PCM controls ignition timing depending on engine operation conditions. Refer to Chapter 5 for more information on the ignition coil.

30 **Tachometer** - The PCM operates the tachometer from the information received from the crankshaft position sensor.

31 **Transmission shift control solenoids** - The PCM receives input signals from various sensors and switches such as the speed sensors, transmission temperature sensor, throttle position sensor and manifold absolute pressure sensor to determine shifting points, required line pressure and torque converter lock-up operations of the transmission.

32 **Voltage regulator** - The PCM controls the charging system voltage by grounding the alternator field driver circuit depending on electrical load requirements. The voltage regulator circuitry is fully contained within the PCM. Any failure in the voltage regulator circuitry requires replacement of the PCM.

Obtaining diagnostic trouble codes

Refer to illustration 2.34

Note: *The diagnostic trouble codes on all models can be extracted from the Powertrain Control Module (PCM) using a specialized scan tool. Have the vehicle diagnosed by a dealer service department or other qualified automotive repair facility if the proper scan tool is not available.*

33 The PCM will illuminate the CHECK ENGINE light (also known as the Malfunction Indicator Lamp) on the dash if it recognizes a fault in the system. The light will remain illuminated until the problem is repaired and the code is cleared or the PCM does not detect any malfunction for several consecutive drive cycles.

34 The preferred code extraction method requires a special scan tool that is programmed to interface with the OBD system by plugging into the diagnostic connector

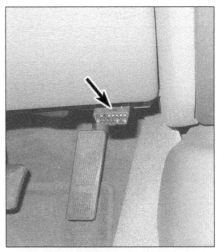

2.34 The diagnostic connector is located under the instrument panel

(see illustration). When used, the scan tool has the ability to diagnose in-depth driveability problems and it allows freeze frame data to be retrieved from the PCM stored memory. Freeze frame data is an OBD II PCM feature that records all related sensor and actuator activity on the PCM data stream whenever an engine control or emissions fault is detected and a trouble code is set. This ability to look at the circuit conditions and values when the malfunction occurs provides a valuable tool when trying to diagnose intermittent driveability problems. If the tool is not available and intermittent driveability problems exist, have the vehicle checked at a dealer service department or other qualified repair shop.

Clearing diagnostic trouble codes

35 After the system has been repaired, the codes must be cleared from the PCM memory using a scan tool. **Caution:** *Do not disconnect the battery from the vehicle in an attempt to clear the codes. If necessary, have the codes cleared by a dealer service department or other qualified repair facility.*

36 Always clear the codes from the PCM before starting the engine after a new electronic emission control component is installed. The PCM stores the operating parameters of each sensor. The PCM may set a trouble code if a new sensor is allowed to operate before the parameters from the old sensor have been erased.

Diagnostic trouble code identification

37 The accompanying list of diagnostic trouble codes is a compilation of all the codes that may be encountered. Not all codes pertain to all models and not all codes will illuminate the Check Engine light when set. All models require a scan tool to access the diagnostic trouble codes.

Diagnostic trouble code chart

Code	Possible cause
P0016	Crankshaft position/camshaft position, bank 1, sensor A - correlation
P0031	HO2S heater control circuit low (bank 1, sensor 1)
P0032	HO2S heater control circuit high (bank 1, sensor 1)
P0037	HO2S heater control circuit low (bank 1, sensor 2)
P0038	HO2S heater control circuit high (bank 1, sensor 2)
P0051	HO heater 2S control circuit low (bank 2, sensor 1)
P0052	HO2S heater control circuit high (bank 2, sensor 1)
P0057	HO2S heater control circuit low (bank 2, sensor 2)
P0058	HO2S heater control circuit high (bank 2, sensor 2)
P0068	Throttle position (TP) sensor inconsistent with mass air flow (MAF) sensor
P0070	Ambient temperature sensor, stuck
P0071	Ambient air temperature sensor range/performance problem
P0072	Ambient air temperature sensor circuit low input
P0073	Ambient air temperature sensor circuit high input
P0107	Manifold absolute pressure or barometric pressure circuit, low input
P0108	Manifold absolute pressure or barometric pressure circuit, high input
P0110	Intake Air Temperature (IAT) sensor, stuck
P0111	Intake air temperature circuit, range or performance problem
P0112	Intake air temperature circuit, low input
P0113	Intake air temperature circuit, high input
P0116	Engine coolant temperature circuit range/performance problem
P0117	Engine coolant temperature circuit, low input
P0118	Engine coolant temperature circuit, high input

Code	Possible cause
P0121	Throttle position or pedal position sensor/switch circuit, range or performance problem
P0122	Throttle position or pedal position sensor/switch circuit, low input
P0123	Throttle position or pedal position sensor/switch circuit, high input
P0124	Throttle position or pedal position sensor/switch circuit, intermittent
P0125	Insufficient coolant temperature for closed loop fuel control
P0128	Coolant thermostat (coolant temperature below thermostat regulating temperature)
P0129	Barometric pressure - too low
P013A	O2 sensor slow response, rich to lean (bank 1, sensor 2)
P013C	O2 sensor slow response, rich to lean (bank 2, sensor 2)
P0131	O2 sensor circuit, low voltage (bank 1, sensor 1)
P0132	O2 sensor circuit, high voltage (bank 1, sensor 1)
P0133	O2 sensor circuit, slow response (bank 1, sensor 1)
P0134	Upstream oxygen sensor (cylinder bank no. 1), sensor remains at center (not switching)
P0135	O2 sensor heater circuit malfunction (bank 1, sensor 1)
P0137	O2 sensor circuit, low voltage (bank 1, sensor 2)
P0138	O2 sensor circuit, high voltage (bank 1, sensor 2)
P0139	O2 sensor circuit, slow response (bank 1, sensor 2)
P0140	Downstream oxygen sensor (cylinder bank no. 1), sensor remains at center (not switching)
P0141	O2 sensor heater circuit malfunction (bank 1, sensor 2)
P0151	O2 sensor circuit, low voltage (bank 2, sensor 1)
P0152	O2 sensor circuit, high voltage (bank 2, sensor 1)
P0153	O2 sensor circuit, slow response (bank 2, sensor 1)

Code	Possible cause
P0155	O2 sensor heater circuit malfunction (bank 2, sensor 1)
P0157	O2 sensor circuit, low voltage (bank 2, sensor 2)
P0158	O2 sensor circuit, high voltage (bank 2, sensor 2)
P0159	O2 sensor circuit, slow response (bank 2, sensor 2)
P0161	O2 sensor heater circuit malfunction (bank 2, sensor 2)
P0171	System too lean (bank 1)
P0172	System too rich (bank 1)
P0174	System too lean (bank 2)
P0175	System too rich (bank 2)
P0201	Injector circuit malfunction - cylinder no. 1
P0202	Injector circuit malfunction - cylinder no. 2
P0203	Injector circuit malfunction - cylinder no. 3
P0204	Injector circuit malfunction - cylinder no. 4
P0205	Injector circuit malfunction - cylinder no. 5
P0206	Injector circuit malfunction - cylinder no. 6
P0218	Transmission overheating condition
P0221	Throttle position or pedal position sensor/switch B, range or performance problem
P0222	Throttle position or pedal position sensor/switch B circuit, low input
P0223	Throttle position or pedal position sensor/switch B circuit, high input
P0300	Random/multiple cylinder misfire detected
P0301	Cylinder no. 1 misfire detected
P0302	Cylinder no. 2 misfire detected
P0303	Cylinder no. 3 misfire detected
P0304	Cylinder no. 4 misfire detected
P0305	Cylinder no. 5 misfire detected
P0306	Cylinder no. 6 misfire detected
P0315	Crankshaft position system - variation not learned

Code	Possible cause
P0320	No crankshaft reference signal at Powertrain Control Module (PCM)
P0325	Knock sensor no. 1 circuit malfunction (bank 1 or single sensor)
P0330	Knock sensor no. 2 circuit malfunction (bank 2)
P0335	Crankshaft position sensor "A" - circuit malfunction
P0339	Crankshaft position sensor "A" - circuit intermittent
P0340	Camshaft position sensor "A" - circuit malfunction (bank 1)
P0344	Camshaft position sensor "A" - circuit intermittent (bank 1)
P0401	Exhaust gas recirculation - insufficient flow detected
P0402	Exhaust gas recirculation - excessive flow detected
P0403	Exhaust gas recirculation - circuit malfunction
P0404	Exhaust gas recirculation - range or performance problem
P0405	Exhaust gas recirculation valve position sensor A - circuit low
P0406	Exhaust gas recirculation valve position sensor A - circuit high
P0420	Catalyst system efficiency below threshold (bank 1)
P0430	Catalyst system efficiency below threshold (bank 2)
P0432	Catalyst system efficiency below threshold (cylinder bank no. 2)
P0440	Evaporative emission control system malfunction
P0441	Evaporative emission control system, incorrect purge flow
P0442	Evaporative emission control system, small leak detected
P0443	Evaporative emission control system, purge control valve circuit malfunction
P0452	Evaporative emission control system, pressure sensor low input
P0453	Evaporative emission control system, pressure sensor high input
P0455	Evaporative emission (EVAP) control system leak detected (no purge flow or large leak)
P0456	Evaporative emission (EVAP) control system leak detected (very small leak)

Code	Possible cause
P0457	Evaporative emission control system leak detected (fuel cap loose/off)
P0460	Fuel level sending unit, no change as vehicle is operated
P0461	Fuel level sensor circuit, range or performance problem
P0462	Fuel level sensor circuit, low input
P0463	Fuel level sensor circuit, high input
P0480	Cooling fan no. 1, control circuit malfunction
P0481	Cooling fan no. 2, control circuit malfunction
P0498	Evaporative emission system, vent control - circuit low
P0499	Evaporative emission system, vent control - circuit high
P050B	Cold start ignition timing performance
P050D	Cold start rough idle
P0501	Vehicle speed sensor, range or performance problem
P0503	Vehicle speed sensor circuit, intermittent, erratic or high input
P0506	Idle control system, rpm lower than expected
P0507	Idle control system, rpm higher than expected
P0508	Idle control system circuit low
P0509	Idle control system circuit high
P0513	Incorrect immobilizer key
P0516	Battery temperature sensor circuit low
P0517	Battery temperature sensor circuit high
P0519	Idle speed performance
P0522	Engine oil pressure sensor/switch circuit, low voltage
P0532	A/C refrigerant pressure sensor, low input
P0533	A/C refrigerant pressure sensor, high input
P0551	Power steering pressure sensor circuit, range or performance problem
P0562	System voltage low
P0563	System voltage high

Code	Possible cause
P0571	Cruise control/brake switch A, circuit malfunction
P0572	Cruise control/brake switch A, circuit low
P0573	Cruise control/brake switch A, circuit high
P0579	Cruise control system, multi-function input "A" - circuit range/performance problem
P0580	Cruise control system, multi-function input "A" - circuit low
P0581	Cruise control system, multi-function input "A" - circuit high
P0582	Cruise control system, vacuum control - circuit open
P0585	Cruise control system, multi-function input A/B - correlation
P0586	Cruise control system, vent control - circuit open
P0591	Cruise control system, multi-function input B - circuit range/performance problem
P0592	Cruise control system, multi-function, input B - circuit low
P0593	Cruise control system, multi-function input B - circuit high
P0594	Cruise control system, servo control - circuit open
P0600	Serial communication link malfunction
P0601	Internal control module, memory check sum error
P0602	Control module, programming error
P0604	Internal control module, random access memory (RAM) error
P0605	Internal control module, read only memory (ROM) error
P0606	PCM processor fault
P060B	Electronic Throttle Control (ETC) A to D ground performance
P060D	Electronic Throttle Control (ETC) level 2 Accelerator Pedal Position Sensor (APPS) performance
P060E	Electronic Throttle Control (ETC) level 2 Throttle Position Sensor (TPS) performance
P060F	Electronic Throttle Control (ETC) level 2 Engine coolant Temperature (ECT) performance

Diagnostic trouble code chart (continued)

Code	Possible cause	Code	Possible cause	Code	Possible cause
P061A	Electronic Throttle Control (ETC) level 2 torque performance	P0716	Input/turbine speed sensor circuit, range or performance problem	P0868	Transmission fluid pressure (TFP) sensor - low
P061C	Electronic Throttle Control (ETC) level 2 RPM performance	P0720	Output speed sensor malfunction	P0869	Transmission fluid pressure (TFP) sensor- high
P0613	Transmission control module (TCM) processor error	P0721	Output speed sensor circuit, range or performance problem	P0870	Transmission fluid pressure (TFP) circuit malfunction - sensor/switch C
P0627	Fuel pump control - circuit open	P0725	Engine speed input circuit malfunction	P0871	Transmission fluid pressure (TFP) sensor C - range/performance problem
P062C	Electronic Throttle Control (ETC) level 2 MPH performance	P0726	Engine speed input circuit, range or performance problem		
P0622	Alternator field control circuit malfunction or field not switching correctly	P0731	Incorrect gear ratio, first gear	P0875	Transmission fluid pressure (TFP) sensor D - circuit malfunction
		P0732	Incorrect gear ratio, second gear	P0876	Transmission fluid pressure (TFP) sensor D - range/performance problem
P0630	VIN not programmed or mismatch - ECM/PCM	P0733	Incorrect gear ratio, third gear		
P0632	Odometer not programmed - ECM	P0734	Incorrect gear ratio, fourth gear	P0882	Transmission control module (TCM) power input signal low
P0633	Immobilizer key not programmed - ECM	P0735	Incorrect gear ratio, fifth gear		
		P0736	Incorrect gear ratio, reverse gear	P0883	Transmission control module (TCM) power input signal high
P0642	Engine control module (ECM), knock control - defective	P0740	Torque converter clutch, circuit malfunction	P0884	Transmission control module (TCM) power input signal intermittent malfunction
P0643	Sensor reference voltage A - circuit high	P0750	Shift solenoid A malfunction		
P0645	A/C clutch relay control circuit	P0755	Shift solenoid B malfunction	P0888	Transmission control module (TCM) power relay sense circuit malfunction
P0652	Sensor reference voltage B - circuit low	P0760	Shift solenoid C malfunction		
		P0765	Shift solenoid D malfunction	P0890	Transmission control module (TCM) power relay - sense circuit low
P0653	Sensor reference voltage B - circuit high	P0770	Shift solenoid E malfunction		
P0685	EGM power relay, control - circuit open	P0833	Clutch released switch circuit	P0891	Transmission control module (TCM) power relay - sense circuit high
		P0836	Four wheel drive (4WD) switch circuit		
P0688	Engine, control relay - short to positive	P0836	Four wheel drive (4WD) switch circuit	P0897	Transmission fluid deteriorated
P0700	Transmission control system malfunction	P0837	Four wheel drive (4WD) switch circuit range/performance problem	P0932	Hydraulic pressure sensor - circuit malfunction
P0703	Torque converter/brake switch B, circuit malfunction	P0838	Four wheel drive (4WD) switch circuit low	P0933	Hydraulic pressure sensor - range/performance problem
P0706	Transmission range sensor circuit, range or performance problem	P0839	Four wheel drive (4WD) switch circuit high	P0934	Hydraulic pressure sensor - circuit low input
P0711	Transmission fluid temperature sensor circuit, range or performance problem	P0841	Transmission fluid pressure sensor/switch "A" circuit range/performance problem	P0935	Hydraulic pressure sensor - circuit high input
P0712	Transmission fluid temperature sensor circuit, low input	P0845	Transmission fluid pressure sensor/switch "B" circuit malfunction	P0944	Hydraulic pressure unit - loss of pressure
P0713	Transmission fluid temperature sensor circuit, high input	P0846	Transmission fluid pressure sensor/switch "B" circuit range/performance problem	P0987	4C Hydraulic pressure test failure
P0714	Transmission fluid temperature sensor circuit, intermittent input	P0850	Park/neutral position (PNP) switch - input circuit malfunction	P0988	Transmission fluid pressure (TFP) sensor E - circuit range/performance problem
P0715	Input/turbine speed sensor circuit malfunction	P0856	Traction control torque request circuit	P0992	Transmission fluid pressure (TFP) sensor F - circuit malfunction

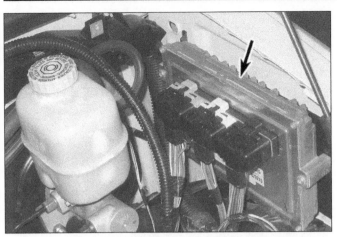

3.1a The Powertrain Control Module (PCM) is located in the engine compartment, behind the battery on early models

3.1b On 2008 and later Liberty models and all Nitro models, the PCM is located on the right inner fender panel

3 Powertrain Control Module (PCM) - removal and installation

Refer to illustrations 3.1a and 3.1b

Caution: *Avoid static electricity damage to the Powertrain Control Module (PCM) by grounding yourself to the body of the vehicle before touching the PCM and using a special anti-static pad to store the PCM on, once it is removed.*

Note 1: *Anytime the PCM is replaced with a new unit, the PCM must be reprogrammed with a scan tool by a dealership service department or other qualified repair shop.*

Note 2: *Anytime the battery is disconnected, stored operating parameters may be lost from the PCM causing the engine to run rough for a period of time while the PCM relearns the information.*

1   On 2007 and earlier Liberty models, the PCM is located in the engine compartment, on the left inner fender panel, behind the battery **(see illustration)**. On all Nitro and 2008 and later Liberty models, the PCM is located in the engine compartment, on the right inner fender panel, just in front of the cowl **(see illustration)**. On 2004 and later models, the PCM incorporates a Transmission Control Module (TCM) as an integral part of the PCM. The PCM/TCM are replaced as a unit.

2   Disconnect the cable from the negative terminal of the battery.

3   Unsnap and remove the cover over the electrical connectors. On later models, the "NGC" modules have four 58-way connectors, while the JTEC modules have three 32-way connectors.

4   Carefully disconnect the PCM electrical connectors.

5   Remove the PCM mounting bolts/nuts.

6   Installation is the reverse of removal.

4 Throttle Position Sensor (TPS) - replacement

Refer to illustrations 4.1 and 4.5

1   The throttle position sensor is located

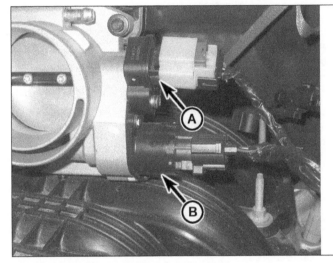

4.1 Throttle position sensor and idle air control valve locations (V6 model shown, four-cylinder similar)

a) *Throttle position sensor*

b) *Idle air control valve*

on the throttle body at the end of the throttle shaft **(see illustration)**. By monitoring the output voltage from the throttle position sensor, the PCM can determine fuel delivery based on throttle valve angle (driver demand). A broken or loose throttle position sensor can cause intermittent bursts of fuel from the injectors and an unstable idle because the PCM thinks the throttle is moving.

2   Remove the air intake duct/resonator.

3   Disconnect the electrical connector from the throttle position sensor. On 2007 and later models, the throttle body is electronic and controlled by the PCM. There is only one large electrical connector at the throttle body, and no accelerator cable.

4   Remove the throttle position sensor mounting screws and remove the throttle position sensor from the throttle body.

5   When installing the throttle position sensor, be sure to align the socket locating tangs on the throttle position sensor with the throttle shaft in the throttle body **(see illustration)**. Install the throttle position sensor and slightly rotate it against the throttle shaft to apply a slight amount of tension on the throttle position sensor. If necessary remove the sensor

and install it with the socket locating tangs on the other side of the throttle shaft.

6   The remainder of installation is the reverse of removal.

4.5 When installing the throttle position sensor, align the socket locating tangs on the throttle position sensor with the throttle shaft

5.2 Manifold absolute pressure (MAP) sensor location and intake air temperature sensor (IAT) - 3.7L V6 models shown, 4.0L similar

a) *MAP sensor*   b) *IAT sensor*

6.6 IAT sensor location (Nitro and 2008 and later Liberty models)

5   Manifold Absolute Pressure (MAP) sensor - replacement

Refer to illustration 5.2

1   The manifold absolute pressure sensor monitors the intake manifold pressure changes resulting from changes in engine load and speed and converts the information into a voltage output. The PCM uses the manifold absolute pressure sensor to control fuel delivery and ignition timing. The PCM will receive information as a voltage signal that will vary from 0.5 volts at closed throttle (high vacuum) and 4.5 volts at wide-open throttle (low vacuum).

2   On 2.4L four-cylinder engines, the manifold absolute pressure sensor is located on the rear of the intake manifold. On V6 engines, the manifold absolute pressure sensor is located at the front of the intake manifold **(see illustration)**.

3   Disconnect the electrical connector from the manifold absolute pressure sensor and remove the screws retaining the manifold absolute pressure sensor to the intake manifold and withdraw the sensor from the intake manifold. Inspect the O-ring for damage and replace it, if necessary.

4   Installation is the reverse of removal.

6   Intake Air Temperature (IAT) sensor - replacement

1   The intake air temperature sensor is a thermistor (a resistor which varies the value of its resistance in accordance with temperature changes). The change in the resistance values will directly affect the voltage signal from the sensor to the PCM. As the sensor temperature INCREASES, the resistance values will DECREASE. As the sensor temperature DECREASES, the resistance values will INCREASE.

2   On 2.4L four-cylinder engines, the intake air temperature sensor is located on the rear lower part of the intake manifold. On 2007

and earlier 3.7L V6 engines, the intake air temperature sensor is located on the left side of the intake manifold **(see illustration 5.2)**.

2007 and earlier Liberty models

3   Disconnect the electrical connector from the intake air temperature sensor.

4   Remove the sensor mounting bolts from the intake manifold and remove the intake air temperature sensor.

5   Installation is the reverse of removal.

All Nitro and 2008 and later Liberty models

Refer to illustration 6.6

6   The intake manifold air temperature (IAT) sensor is installed into the rubber air intake tube near the throttle body or beginning of the air filter housing **(see illustration)**. Disconnect electrical connector from IAT sensor.

7   Remove the sensor by turning it 1/4-turn counterclockwise and pulling it out of the air filter housing or air intake resonator.

8   Check condition of sensor O-ring and replace as necessary.

9   Installation is the reverse of removal.

7   Engine coolant temperature (ECT) sensor - replacement

Refer to illustration 7.2
Warning: *Wait until the engine is completely cool before beginning this procedure.*

1   The engine coolant temperature sensor is a thermistor (a resistor which varies the value of its resistance in accordance with temperature changes) which is mounted into an engine water jacket. The change in the resistance values will directly affect the voltage signal from the sensor to the PCM. As the sensor temperature INCREASES, the resistance values will DECREASE. As the sensor temperature DECREASES, the resistance values will INCREASE.

2   On 2.4L four-cylinder engines, the engine coolant temperature sensor is located on the left side of the front of the cylinder

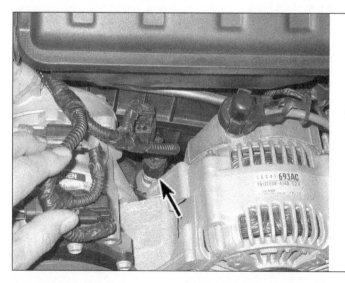

7.2 Engine coolant temperature (ECT) sensor location - 3.7L V6 engines

8.2 Crankshaft position (CKP) sensor location - 3.7L V6 engines

9.2 Camshaft position sensor location and mounting bolt - 3.7L V6 engines (seen from below)

head. On 3.7L V6 engines, the engine coolant temperature sensor is located on the front of the intake manifold, behind the alternator **(see illustration)**.

3   Partially drain the cooling system (see Chapter 1).

4   Disconnect the electrical connector from the sensor and carefully unscrew the sensor.

5   Before installing the new sensor, wrap the threads with Teflon sealing tape to prevent leakage and thread corrosion.

6   Installation is the reverse of removal.

7   Refill the cooling system (see Chapter 1).

8   Crankshaft position (CKP) sensor - replacement

Refer to illustration 8.2

1   The crankshaft position sensor determines the timing for the fuel injection and ignition on each cylinder. It also detects engine RPM. The crankshaft position sensor is a Hall-Effect device triggered by a crankshaft-mounted reluctor wheel. The engine will not operate if the PCM does not receive a crankshaft position sensor input.

2   On 2.4L four-cylinder engines, the sensor is mounted on the right side of the engine block, toward the front, near the engine mount. On 3.7L V6 engines, the sensor is mounted on the right side of the engine block bedplate, toward the rear **(see illustration)**. On 4.0L engine models, the sensor is mounted on the right side of the transmission bellhousing.

3   Disconnect the electrical connector from the crankshaft position sensor.

4   Remove the mounting bolt and carefully twist the sensor and remove it from the engine or transmission.

5   Inspect the O-ring and replace it if necessary.

6   Apply a small amount of engine oil on the O-ring and press the sensor in the engine block or transmission until fully seated. Install

the mounting bolt and tighten it to 21 ft-lbs (28 Nm) on four-cylinder and 3.7L V6 engines or 105 in-lbs (11.8 Nm) on 4.0L V6 engines.

9   Camshaft position (CMP) sensor - replacement

Refer to illustration 9.2

1   The camshaft position sensor, in conjunction with the crankshaft position sensor, determines the timing for the fuel injection on each cylinder. The camshaft position sensor is a Hall-Effect device that is triggered by the reluctor wheel on the camshaft.

2   On 2.4L four-cylinder engines, the sensor is located on right side of the cylinder head, toward the front. On 3.7L V6 engines, the sensor is located at the front of the right-bank cylinder head **(see illustration)**. On 4.0L models, the sensor is located at the front-top of the timing belt cover, below the alternator.

Removal

3   Disconnect the electrical connector from the camshaft position sensor.

2.4L four-cylinder engines

4   Disconnect the electrical connector from the engine coolant temperature sensor on the front of the cylinder head **(see illustration 7.2)**.

5   Remove the upper timing belt cover.

6   Rotate the crankshaft clockwise until the camshaft sprocket and camshaft position sensor trigger wheel are aligned in front of the face of the camshaft position sensor.

7   Remove the sensor mounting bolts.

8   Slide the sensor toward the rear, then remove it from the cylinder head.

3.7L V6 engines

9   Remove the sensor mounting bolt **(see illustration 9.2)**.

10   Carefully twist the sensor and remove it from the cylinder head.

11   Inspect the O-ring and replace it if necessary.

4.0L engines

Warning: *Wait for the engine to cool completely before performing this procedure.*

12   Remove the alternator (see Chapter 5).

13   Disconnect the electrical connector from the sensor.

14   Unscrew the sensor mounting bolt and carefully twist the sensor out from the timing belt cover.

Installation

2.4L four-cylinder engines

15   Insert the sensor into the cylinder head and install the mounting bolts finger-tight. **Caution:** *The depth-adjustment on the camshaft position sensor must be set before starting the engine. Otherwise, the sensor and/or trigger wheel could be damaged.*

16   Using a .030-inch (0.762 mm) L-shaped wire gauge, adjust the air gap between the face of the sensor and the backside of the trigger wheel by gently pushing forward on the sensor until contact is made with the wire gauge. Tighten the sensor mounting bolts.

17   The remainder of the installation is the reverse of removal.

3.7L V6 engines

18   Clean the machined camshaft position sensor mounting bore on the cylinder head.

19   Apply a small amount of engine oil on the O-ring and carefully twist and rock the sensor into the cylinder head until it is fully seated to the cylinder head.

20   Install the mounting bolt and tighten it to 106 in-lbs (12 Nm).

21   Connect the electrical connector to the sensor.

22   Apply a film of clean engine oil to the O-ring, then install the sensor and mounting bolt, tightening the bolt securely.

23   The remainder of installation is the reverse of removal.

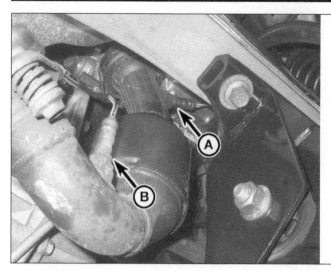

10.3 **3.7L V6 engines are equipped with four oxygen sensors - two sensors for each cylinder bank (right side shown)**

a) *Pre-converter oxygen sensor*
b) *Post-converter oxygen sensor*

10.10 **A special slotted socket, allowing clearance for the wiring harness, may be required for oxygen sensor removal (the tool is available at most auto parts stores) (typical)**

10 Oxygen (O2) sensor - replacement

General description

Refer to illustration 10.3

1   The oxygen in the exhaust reacts with the elements inside the oxygen sensor to produce a voltage output that varies from 0.1 volt (high oxygen, lean mixture) to 0.9 volt (low oxygen, rich mixture). The pre-converter oxygen sensor (mounted in the exhaust system before the catalytic converter) provides a feedback signal to the PCM that indicates the amount of left-over oxygen in the exhaust. The PCM monitors this variable voltage continuously to determine the required fuel injector pulse width and to control the engine air/fuel ratio. A mixture ratio of 14.7 parts air to 1 part fuel is the ideal ratio for minimum exhaust emissions, as well as the best combination of fuel economy and engine performance. Based on oxygen sensor signals, the PCM tries to maintain this air/fuel ratio of 14.7:1 at all times.

2   The post-converter oxygen sensor (mounted in the exhaust system after the catalytic converter) has no effect on PCM control of the air/fuel ratio. However, the post-converter sensor is identical to the pre-converter sensor and operates in the same way. The PCM uses the post-converter signal to monitor the efficiency of the catalytic converter. A post-converter oxygen sensor will produce a slower fluctuating voltage signal that reflects the lower oxygen content in the post-catalyst exhaust.

3   Oxygen sensor configuration varies depending on model, and is as follows **(see illustration)**:

a) *2.4L four-cylinder engines are equipped with one pre-converter oxygen sensor (referred to as 1/1) and one post-converter oxygen sensor (referred to as 1/2).*
b) *V6 engines are equipped with four oxygen sensors; two pre-converter oxygen sensors (referred to as 1/1 on the left side and 2/1 on the right side) and two post-converter oxygen sensors (referred to as 1/2 on the left side and 2/2 on the right side).*

4   An oxygen sensor produces no voltage when it is below its normal operating temperature of about 600-degrees F. During this warm-up period, the PCM operates in an open-loop fuel control mode. It does not use the oxygen sensor signal as a feedback indication of residual oxygen in the exhaust. Instead, the PCM controls fuel metering based on the inputs of other sensors and its own programs.

5   Proper operation of an oxygen sensor depends on four conditions:

a) *Electrical - The low voltages generated by the sensor require good, clean connections which should be checked whenever a sensor problem is suspected or indicated.*
b) *Outside air supply - The sensor needs air circulation to the internal portion of the sensor. Whenever the sensor is installed, make sure the air passages are not restricted.*
c) *Proper operating temperature - The PCM will not react to the sensor signal until the sensor reaches approximately 600-degrees F. This factor must be considered when evaluating the performance of the sensor.*
d) *Unleaded fuel - Unleaded fuel is essential for proper operation of the sensor.*

6   The PCM can detect several different oxygen sensor problems and set diagnostic trouble codes to indicate the specific fault (see Section 2). When an oxygen sensor fault occurs, the PCM will disregard the oxygen sensor signal voltage and revert to open-loop fuel control as described previously.

Replacement

Refer to illustration 10.10

7   Observe the following guidelines when replacing an oxygen sensor:

a) *The sensor has a permanently attached pigtail and electrical connector which should not be removed from the sensor. Damage or removal of the pigtail or electrical connector can harm operation of the sensor.*
b) *Keep grease, dirt and other contaminants away from the electrical connector and the louvered end of the sensor.*
c) *Do not use cleaning solvents of any kind on the oxygen sensor.*
d) *Do not drop or roughly handle the sensor.*

8   Raise the vehicle and place it securely on jackstands.

9   Disconnect the electrical connector from the sensor.

10   Using a suitable wrench or specialized oxygen sensor socket, unscrew the sensor from the exhaust pipe **(see illustration)**.

11   Anti-seize compound must be used on the threads of the sensor to aid future removal. The threads of most new sensors will be coated with this compound. If not, be sure to apply anti-seize compound before installing the sensor.

12   Install the sensor and tighten it to 22 ft-lbs (30 Nm).

13   Reconnect the electrical connector to the sensor and lower the vehicle.

11 Battery temperature sensor - replacement

Refer to illustration 11.4

1   The battery temperature sensor provides the PCM with information on battery temperature. The PCM uses this information to regulate the charging system voltage. The PCM increases the charging system voltage at colder temperatures and decreases the charging system voltage at warmer temperatures.

2   The battery temperature sensor is mounted in the battery tray, beneath the battery.

3   Remove the battery (see Chapter 5).

4   Pry the sensor from the battery tray and pull the harness and connector through the hole **(see illustration)**.

5   Disconnect the electrical connector from the sensor harness.

6   Installation is the reverse of removal.

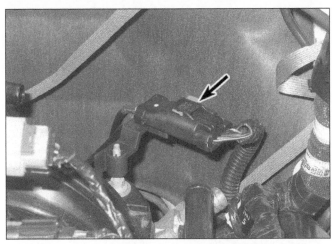

11.4 The battery temperature sensor is mounted in the battery tray, beneath the battery (view from underneath tray)

12.2 Disconnect the knock sensor electrical connector located near the rear of the left valve cover

12 Knock sensors - replacement (V6 engine only)

3.7L engines

Refer to illustration 12.2

1   There are two knock sensors located underneath the intake manifold - one on the left side and one on the right side of the engine block. The sensors send voltage signals to the PCM based on the intensity of the vibrations inside the cylinders. When cylinder knock ("pinging") is detected, the PCM will respond by retarding the ignition timing for all of the cylinders. **Note:** *The left and right knock sensors are different (the left sensor has an identification tag). Don't mix them up.*
2   Disconnect the knock sensor electrical connector, which is located near the left valve cover, toward the rear **(see illustration)**.
3   Remove the intake manifold (see Chapter 2B).
4   Remove the knock sensor mounting bolts and remove the sensors.
5   Thoroughly clean the bolts and bolt holes.
6   Install the knock sensors and tighten the mounting bolts to 15 ft-lbs (20 Nm). **Note:** *Do not apply sealant to the bolts. It is important to properly torque the mounting bolts to ensure proper knock sensor operation and accurate ignition timing control.*
7   Installation is the reverse of removal.

4.0L engines

Warning: *Wait until the engine is completely cool before beginning this procedure.*
Note: *The single knock sensor is located in the valley between the cylinder heads.*
8   Relieve the fuel system pressure (see Chapter 4) and drain the cooling system (see Chapter 1).
9   Disconnect the cable from the negative terminal of the battery (see Chapter 5).
10   Remove the upper intake manifold and lower intake manifold (see Chapter 2C).
11   Disconnect the electrical connector from the knock sensor.

12   Remove the mounting bolt and detach the sensor from the engine block.
13   Install the sensor and tighten the knock sensor bolt to 84 in-lbs.
14   The remainder of installation is the reverse of removal.
15   Refill the cooling system.

13 Idle Air Control (IAC) valve - replacement

Note: *This procedure applies to 2006 and earlier models only.*
1   The idle speed is controlled by the idle air control valve, located on the throttle body **(see illustration 4.1)**. The idle air control valve regulates the air bypassing the throttle plate by moving the pintle in or out of the air passage. The idle air control valve is controlled by the PCM, adjusting the idle speed depending upon the running conditions of the engine (air conditioning system, power steering, cold and warm running etc.).
2   Disconnect the electrical connector from the idle air control valve.
3   Remove the two mounting screws and withdraw the idle air control valve from the throttle body.
4   Installation is the reverse of removal. Be sure to install a new O-ring.

14 Crankcase ventilation (PCV) system

1   The crankcase ventilation system reduces hydrocarbon emissions by scavenging crankcase vapors. It does this by circulating fresh air from the air cleaner through the crankcase, where it mixes with blow-by gases. The gases are drawn by intake manifold vacuum into the intake manifold, where they mix with the incoming air/fuel mixture and are consumed during normal combustion. These models utilize a Positive Crank-

case Ventilation (PCV) system. The main component of this system is the PCV valve. Fresh air flows from the air intake duct through a vent tube into the crankcase. Crankcase vapors are drawn from the crankcase by the PCV valve. To maintain idle quality and good driveability, the PCV valve restricts the flow when the intake manifold vacuum is high. When intake manifold vacuum is lower, maximum vapor flow is allowed through the valve.
2   Checking and replacement of the PCV valve is covered in Chapter 1.

15 Evaporative emissions control (EVAP) system

General description

1   The fuel evaporative emissions control (EVAP) system absorbs fuel vapors from the fuel tank and, during engine operation, releases them into the engine intake system where they mix with the incoming air/fuel mixture.
2   The evaporative emissions control system employs a canister filled with activated charcoal to absorb fuel vapors.
3   The fuel tank filler cap is fitted with a two-way valve as a safety device. The valve vents fuel vapors to the atmosphere if the evaporative control system fails. After passing through the check valve and flow management valve, fuel vapor is carried by vent hoses to the charcoal canister. The activated charcoal in the canister absorbs and stores these vapors.
4   The check valve, mounted on top of the fuel tank, regulates fuel vapor flow from the fuel tank to the charcoal canister, based on the pressure or vacuum caused by temperature changes. The check valve prevents fuel flow to the canister when the fuel tank is full of fuel or in the event the vehicle rolls over in an accident.
5   The system is an electronic system controlled by the PCM.
6   When the engine is running and warmed to a pre-set temperature, a purge control

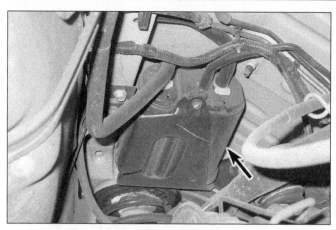

15.8 The EVAP canister is mounted underneath the vehicle, at the left front of the fuel tank - 2007 and earlier models

15.14 The purge control solenoid is located in the engine compartment, next to the battery (battery removed for clarity)

solenoid allows a purge control diaphragm valve in the charcoal canister to be opened by intake manifold vacuum. Fuel vapors from the canister are then drawn through the purge control diaphragm valve by intake manifold vacuum. The duty cycle of the EVAP purge control solenoid regulates the rate of flow of the fuel vapors from the canister to the throttle body. The PCM controls the purge control solenoid. During cold running conditions and hot start time delay, the PCM does not energize the solenoid. After the engine has warmed up to the correct operating temperatures, the PCM purges the vapors into the throttle body according to the running conditions of the engine. The PCM will cycle (On then Off) the purge control solenoid about 5 to 10 times per second. The flow rate will be controlled by the pulse width or length of time the solenoid is allowed to be energized.

7   A leak detection monitor system is a self-diagnostic system designed to detect a leak in the EVAP system. Each time the engine is started cold, the PCM energizes the leak detection pump. The pump pressurizes the EVAP system then shuts off. The PCM is able to detect a leak if the pump continues to

run, unable to pressurize the system. If a leak is detected, the PCM will trigger a diagnostic trouble code (see Section 2).

Component replacement

EVAP canister
Refer to illustration 15.8

8   The EVAP canister is located underneath the vehicle. On 2007 and earlier models it is at the left front of the fuel tank **(see illustration)**. On 2008 and later models it is at the left rear wheel inner splash shield.
9   Raise the vehicle and place it securely on jackstands.
10   Disconnect the hoses from the canister.
11   Remove the canister mounting bracket bolts
12   Remove the canister-to-bracket mounting bolt and slide the canister mounting pins out of the bracket.
13   Installation is the reverse of removal.

Purge control solenoid
Refer to illustration 15.14

14   The purge control solenoid is mounted in the engine compartment, next to the battery **(see illustration)**.
15   Disconnect the electrical connector. Label and remove the hoses from the purge control solenoid.
16   Pull straight up to remove the purge control solenoid and bracket.
17   Installation is the reverse of removal.

Leak detection pump
Refer to illustration 15.18

18   The leak detection pump is located on the front of the fuel tank **(see illustration)**.
19   Raise the vehicle and place it securely on jackstands.
20   Carefully label and remove the hoses from the leak detection pump.
21   To gain access to the leak detection pump electrical connector, lower the front of the fuel tank about 1/2-inch:

a) *Support the tank with a floor jack with a piece of wood between the jack and the tank (or skid plate, if equipped).*

b) *Loosen the front fuel tank strap nuts about 10 turns.*
c) *Lower the tank.*

22   Remove the mounting nuts and slightly lower the leak detection pump.
23   Disconnect the electrical connector from the leak detection pump.
24   Remove the leak detection pump.
25   Installation is the reverse of removal. **Note:** *The vapor and vacuum lines must be securely connected to the leak detection pump or a diagnostic trouble code may be set.*

---

16   Catalytic converter(s)

Refer to illustrations 16.3a and 16.3b
Note: *Because of a federally mandated extended warranty which covers emissions-related components such as the catalytic converter, check with a dealer service department before replacing the converter at your own expense.*

1   The catalytic converter is an emission control device added to the exhaust system to reduce pollutants from the exhaust gas stream. A monolithic converter design is used with a three-way (reduction) catalyst. The catalytic coating on the three-way catalyst contains platinum, palladium and rhodium, which lowers the levels of oxides of nitrogen (NOx) as well as hydrocarbons (HC) and carbon monoxide (CO).
2   The test equipment for a catalytic converter is expensive and highly sophisticated. If you suspect that the converter on your vehicle is malfunctioning, take it to a dealer or authorized emissions inspection facility for diagnosis and repair.
3   Whenever the vehicle is raised for servicing of underbody components, check the converter for leaks, corrosion, dents and other damage **(see illustrations)**. Check the welds/flange bolts that attach the front and rear ends of the converter to the exhaust system. If damage is discovered, the converter should be replaced.
4   Although catalytic converters don't break

15.18 The leak detection pump is mounted to the front of the fuel tank

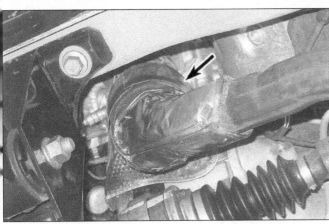

16.3a On 3.7L V6 models, a mini-catalytic converter is located near each exhaust manifold (left side shown)...

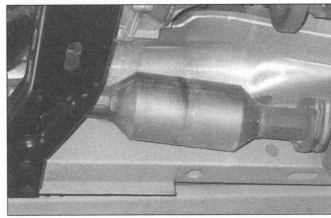

16.3b ... and another catalytic converter is located ahead of the muffler

too often, they can become plugged. The easiest way to check for a restricted converter is to use a vacuum gauge to diagnose the effect of a blocked exhaust on intake vacuum.

a) Connect a vacuum gauge to an intake manifold vacuum source.

b) Warm the engine to operating temperature, place the transmission in park (automatic) or neutral (manual) and apply the parking brake.

c) Note and record the vacuum reading at idle.

d) Open the throttle until the engine speed is about 2000 rpm.

e) Release the throttle quickly and record the vacuum reading.

f) Perform the test three more times, recording the reading after each test.

g) If the reading after the fourth test is more than one in-Hg lower than the reading recorded at idle, the catalytic converter, muffler or exhaust pipes may be plugged or restricted.

5   If the catalytic converter requires replacement, take the vehicle to a dealership service department or other qualified repair shop. Refer to the exhaust system servicing section in Chapter 4 for additional information.

17 Exhaust Gas Recirculation (EGR) - description and replacement

1   Under certain engine conditions, such as pulling a heavy load or climbing a steep hill, an engine can produce too much NOx emissions. Controlled by the PCM, the EGR valve recirculates some exhaust gasses back to the intake manifold to reduce combustion temperatures under these circumstances.

2   On four-cylinder engines, the EGR valve and cooler are mounted underneath the intake manifold, at the left rear of the engine. On 3.7L engines, the EGR valve is located at the rear of the left cylinder head. On 4.0L engines, the electronic EGR valve and solenoid assembly is attached to the rear of the right cylinder head. Disconnect the electrical connector, remove the EGR tube, and unbolt the EGR valve.

Four-cylinder engine

Warning: Wait until the engine is completely cool before beginning this procedure.

3   Disconnect the cable from the negative terminal of the battery (see Chapter 5).

4   Drain the cooling system (see Chapter 1).

5   Remove the engine cover and disconnect the EGR valve electrical connector.

6   Disconnect the hoses from the EGR valve cooler, then remove the bolts and detach the EGR pipe from the housing.

7   Remove the EGR valve housing mounting bolts, then remove the housing, valve and cooler assembly from the engine block.

8   Remove the bolts and detach the EGR valve from the housing.

9   Remove all traces of gasket material from the valve and housing.

10   Install the valve to the housing, using a new gasket. Tighten the bolts to 95 inch-lbs (10.8 Nm).

11   Position the housing/valve/cooler assembly against the engine block, install the bolts, then tighten them to 18 ft-lbs (24.5 Nm).

12   Connect the EGR pipe to the housing, using a new gasket, and tighten the bolts to 24 ft-lbs (32.4 Nm).

13   Reconnect the electrical connector and install the engine cover.

14   Refill the cooling system (see Chapter 1).

V6 engines

15   Installation is the reverse of removal. When replacing the EGR valve, use a new gasket. When replacing the EGR pipe, replace the gasket at the EGR valve end, and a seal at the intake manifold end (3.7L V6 engine) or a new gasket at the flange (4.0L V6 engine).

18 Accelerator Pedal Position (APP) sensor - description and replacement

Refer to illustration 18.4

1   On 2007 and later models, the throttle system is electronically controlled. The throttle body incorporates a throttle body motor that opens or closes the throttle. There is no accelerator cable attached to the throttle body. It is controlled by the PCM with information from various input sensors and the Accelerator Pedal Position (APP) sensor, which is a component of the accelerator pedal assembly. When the driver steps down on the accelerator pedal, the PCM is informed just how much and opens the throttle body accordingly, and vice-versa when the driver wishes to slow down.

2   If diagnostic trouble codes indicate the APP sensor needs to be replaced, the APP sensor cannot be replaced separately, but only as an assembly with the accelerator pedal.

3   Remove the knee bolster trim panel and the knee bolster (see Dashboard trim panels - removal and installation in Chapter 11).

4   Disconnect the electrical connector from the upper end of the APP sensor assembly (see illustration).

5   Remove the accelerator pedal/APP sensor assembly mounting nuts and remove the assembly.

6   Installation is the reverse of removal.

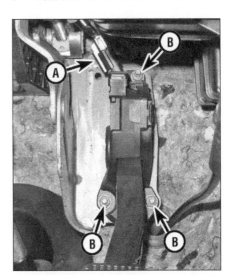

18.4 APP sensor electrical connector (A) and mounting fasteners (B)

Notes

Chapter 7 Part A
Manual transmission

Contents

Specifications

General

Transmission lubricant type	See Chapter 1

Torque specifications   Ft-lbs (unless otherwise indicated) Nm

Note: *One foot-pound (ft-lb) of torque is equivalent to 12 inch-pounds (in-lbs) of torque. Torque values below approximately 15 ft-lbs are expressed in inch-pounds, because most foot-pound torque wrenches are not accurate at these smaller values.*

	Ft-lbs	Nm
Shift tower-to-transmission bolts	120 in-lbs	14
Transmission mount-to-transmission bolts	34	46
Transmission-to-engine bolts		
2002 through 2004 models	50	68
2005 through 2008 models with NSG370		
Upper transmission-to-engine mounting bolts	30	41
Lower transmission-to-engine mounting bolts	40	54
Engine-to-transmission bolts	50	67

1 General information

The vehicles covered by this manual are equipped with a manual or an automatic transmission. Information on the manual transmission is included in this Part of Chapter 7. Information on the automatic transmission can be found in Part B of this Chapter. Information on the transfer case is in Part C of this Chapter.

2004 and earlier models equipped with a manual transmission use either an NV1500 or an NV3550. Both units are top-loader five-speeds with internal shift mechanisms. Models from 2005 to 2008 were offered with the NSG370, six-speed, manual transmission. 2009 and later models are only available with an automatic transmission.

Depending on the cost of having a transmission overhauled, it might be a better idea to replace it with a used or rebuilt unit. Your local dealer or transmission shop should be able to supply information concerning cost, availability and exchange policy. Regardless of how you decide to remedy a transmission problem, you can still save a lot of money by removing and installing the unit yourself.

2 Extension housing oil seal - replacement

Refer to illustrations 2.4 and 2.6
Note: *This procedure also applies to the transfer case extension housing seal on 4WD models.*

1 Oil leaks frequently occur due to wear of the extension housing oil seal. Replacement of this seal is relatively easy, since it can be performed without removing the transmission from the vehicle.

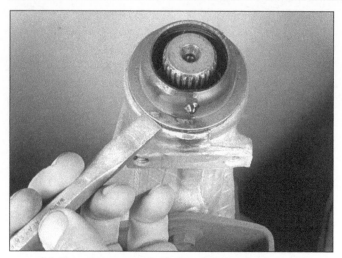

2.4 Use a hammer and chisel to dislodge the rear seal

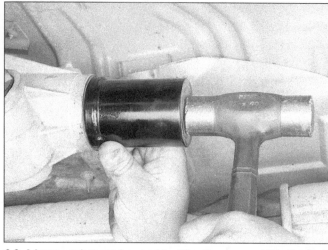

2.6 A large socket can be used to drive the new seal into the bore

2   The extension housing oil seal is located at the extreme rear of the transmission, where the driveshaft is attached. If leakage at the seal is suspected, raise the vehicle and support it securely on jackstands. If the seal is leaking, transmission lubricant will be built up on the front of the driveshaft and may be dripping from the rear of the transmission.
3   Remove the driveshaft (see Chapter 8).
4   Using a chisel and hammer, carefully pry the oil seal out of the rear of the transmission **(see illustration)**. Do not damage the splines on the transmission output shaft.
5   If the oil seal cannot be removed with a chisel, a special oil seal removal tool (available at most auto parts stores) will be required.
6   Using a large section of pipe or a very large deep socket as a drift, install the new oil seal **(see illustration)**. Drive it into the bore squarely and make sure it's completely seated.
7   Lubricate the splines of the transmission output shaft and the outside of the driveshaft yoke with lightweight grease, then install the driveshaft (see Chapter 8). Be careful not to damage the lip of the new seal.

3   Shift lever - removal and installation

1   Place the shift lever in Neutral.
2   Unscrew the shifter extension, then refer to Chapter 11 for the console removal procedure.
3   Remove the bolts attaching the shift tower and shift lever assembly to the transmission.
4   Remove the shift tower and lever as an assembly.
5   If necessary, disassemble the shift lever assembly, clean the parts in solvent, blow everything dry with compressed air, coat all friction surfaces with clean multi-purpose grease and reassemble.

6   Installation is the reverse of removal. Be sure to tighten the shift tower bolts to the torque listed in this Chapter's Specifications.

4   Back-up light switch - check and replacement

1   The back-up light switch is located on the left side of the transmission case.

Check

2   Turn the ignition key to the On position and move the shift lever to the Reverse position; the back-up lights should go on.
3   If the lights don't go on, check the back-up light fuse first (see Chapter 12). If the fuse is blown, trace the back-up light circuit for a short-circuit condition.
4   If the fuse is okay, raise the vehicle and support it securely on jackstands. Place the shifter in Reverse.
5   Working under the vehicle, unplug the back-up light switch electrical connector. Using an ohmmeter, check for continuity across the terminals of the switch. Continuity should exist. If not, replace the switch.
6   If the switch has continuity, check for voltage at the electrical connector; one of the two terminals should have battery voltage present with the ignition key in the On position. If no voltage is present, trace the circuit between the fuse block and the electrical connector for an open circuit condition.
7   If voltage is present, trace the back-up light circuit between the electrical connector and the back-up light bulbs for an open circuit condition. **Note:** *Although not very likely, the back-up light bulbs could both be burned out; don't rule out this possibility.*

Replacement

8   Raise the vehicle and support it securely on jackstands, if not already done.
9   Unplug the back-up light switch electri-

cal connector.
10   Unscrew the back-up light switch from the transmission case.
11   Apply RTV sealant or Teflon tape to the threads of the new switch to prevent leakage. Install the switch in the transmission case and tighten it securely. Plug in the electrical connector.
12   Lower the vehicle and check the operation of the back-up lights.

5   Transmission - removal and installation

Removal

1   Disconnect the negative cable from the battery.
2   Shift the transmission into Neutral.
3   Raise the vehicle sufficiently to provide clearance to easily remove the transmission. Support the vehicle securely on jackstands.
4   Remove the skid plate, if equipped.
5   Disconnect the electrical connectors from the transmission.
6   If the transmission is going to be torn down, drain the lubricant (see Chapter 1).
7   Remove the driveshaft (see Chapter 8). Use a plastic bag to cover the end of the transmission to prevent fluid loss and contamination.
8   On 4WD models, remove the shift cable and vent hose from the transfer case, then remove the transfer case (see Chapter 7C).
9   Disconnect the clutch hydraulic release cylinder from the clutch housing. Move the cylinder aside for clearance.
10   Remove the starter motor (see Chapter 5).
11   Support the engine from above with an engine hoist, or place a jack (with a block of wood as an insulator) under the engine oil pan. The engine must remain supported at all times while the transmission is out of the vehicle.

12 Support the transmission with a jack - preferably a special jack made for this purpose. **Note:** *These jacks can be obtained at most equipment rental yards.* Safety chains will help steady the transmission on the jack.

13 Remove the exhaust crossover pipe.

14 Remove the exhaust pipe hanger from the transmission crossmember.

15 Remove the engine-to-transmission structural collar and the transmission dust shield.

16 Raise the engine slightly and remove the transmission mount.

17 Remove the transmission crossmember.

18 Lower the jacks supporting the transmission and engine assembly.

19 Remove the shift tower and lever assembly (see Section 3).

20 Remove the nuts and bolts attaching the transmission to the engine.

21 Make a final check for any wiring or hoses connected to the transmission, then move the transmission and jack toward the rear of the vehicle until the transmission input shaft clears the splined-hub in the clutch disc. Keep the transmission level as this is done.

22 Once the input shaft is clear, lower the transmission slightly and remove it from under the vehicle.

23 While the transmission is removed, be sure to remove and inspect all clutch components (see Chapter 8). In most cases, new clutch components should be routinely installed if the transmission is removed.

Installation

24 Insert a small amount of multi-purpose grease into the pilot bearing in the crankshaft and lubricate the inner surface of the bearing. Also apply a light film of grease on the input shaft splines, input shaft bearing retainer and the release lever/bearing contact points (see Chapter 8).

25 Install the clutch components (see Chapter 8).

26 With the transmission secured to the jack as on removal, raise the transmission into position and carefully slide it forward, engaging the input shaft with the clutch plate hub. Do not use excessive force to install the transmission - if the input shaft does not slide into place, readjust the angle of the transmission so it is level and/or turn the input shaft so the splines engage properly with the clutch.

27 Install the transmission-to-engine bolts and tighten them to the torque listed in this Chapter's Specifications. **Caution:** *Don't use the bolts to draw the transmission to the engine. If the transmission doesn't slide forward easily and mate with the engine block, find out why before proceeding.*

28 Install the shift tower and lever assembly.

29 Raise the transmission into place, install the crossmember and attach it to the frame rails.

30 Install the transmission mount between the extension housing and the crossmember. Carefully lower the transmission extension housing onto the mount and the crossmember. When everything is properly aligned, tighten all nuts and bolts securely.

31 Remove the jacks supporting the transmission and the engine.

32 On 4WD models, install the transfer case and shift cable (see Chapter 7C).

33 Install the various items removed previously, referring to Chapter 8 for the installation of the driveshaft and clutch release cylinder, Chapter 5 for the starter motor, and Chapter 4 for the exhaust system components.

34 Connect all wiring that was disconnected during removal.

35 Remove the jackstands and lower the vehicle.

36 Fill the transmission with the specified lubricant to the proper level (see Chapter 1).

37 Connect the negative battery cable.

38 Road test the vehicle for proper operation and check for leakage.

6 Transmission overhaul - general information

Overhauling a manual transmission is a difficult job for the do-it-yourselfer. It involves the disassembly and reassembly of many small parts. Numerous clearances must be precisely measured and, if necessary, changed with select fit spacers and snaprings. As a result, if transmission problems arise, it can be removed and installed by a competent do-it-yourselfer, but overhaul should be left to a transmission repair shop. Rebuilt transmissions may be available - check with your dealer parts department and auto parts stores. At any rate, the time and money involved in an overhaul is almost sure to exceed the cost of a rebuilt unit.

Nevertheless, it's not impossible for an inexperienced mechanic to rebuild a transmission if the special tools are available and the job is done in a deliberate step-by-step manner so nothing is overlooked.

The tools necessary for an overhaul include internal and external snap-ring pliers, a bearing puller, a slide hammer, a set of pin punches, a dial indicator and possibly a hydraulic press. In addition, a large, sturdy workbench and a vise or transmission stand will be required.

During disassembly of the transmission, make careful notes of how each piece comes off, where it fits in relation to other pieces and what holds it in place. If you note how each part is installed before removing it, getting the transmission back together again will be much easier.

Before taking the transmission apart for repair, it will help if you have some idea what area of the transmission is malfunctioning. Certain problems can be closely tied to specific areas in the transmission, which can make component examination and replacement easier. Refer to the *Troubleshooting* Section at the front of this manual for information regarding possible sources of trouble.

Notes

Chapter 7 Part B
Automatic transmission

Contents

Specifications

Torque specifications	Ft-lbs	Nm
Driveplate-to-torque converter bolts		
45RFE	23	31
42RLE	65	88
NAG1 (4.0L V6 engine)	31	42
Engine support collar	40	54
Transmission-to-engine bolts		
All except 4.0L V6 engine	50	68
4.0L V6 engine	29	39
Transmission crossmember-to-frame fasteners	50	68
Transmission mount-to-transmission bolts	39	47
Transmission mount through-bolt/nut	50	68
Transmission crossmember-to-support bracket fasteners	39	47

1 General information and identification

General information

All vehicles covered in this manual come equipped with a five-speed manual transmission, six-speed manual transmission, or an automatic transmission. All information on the automatic transmission is included in this Part of Chapter 7. Information on the manual transmission can be found in Part A of this Chapter. Vehicles with four-wheel-drive also use a transfer case. Information on the transfer case is in Part C of this Chapter.

Due to the complexity of the automatic transmissions covered in this manual and the need for specialized equipment to perform most service operations, this Chapter contains only general diagnosis, routine maintenance, adjustment and removal and installation procedures.

If the transmission requires major repair work, it should be left to a dealer service department or an automotive or transmission repair shop. You can, however, remove and install the transmission yourself and save the expense, even if the repair work is done by a transmission shop.

Identification

Two different electronically controlled four-speed automatic transmissions are used in the vehicles covered by this manual.

The 45RFE is used on 2002 and 2003 models. It can be identified by the stamped numbers on the left side, above the transmission pan.

The 42RLE is used on 2003 and later models. It can be identified by the label on the left side, on the upper area of the bellhousing.

2 Diagnosis - general

Note: *Automatic transmission malfunctions may be caused by five general conditions: poor engine performance, improper adjustments, hydraulic malfunctions, mechanical malfunctions or malfunctions in the computer or its signal network. Diagnosis of these prob-*

lems should always begin with a check of the easily repaired items: fluid level and condition (see Chapter 1) and shift linkage adjustment. Next, perform a road test to determine if the problem has been corrected or if more diagnosis is necessary. If the problem persists after the preliminary tests and corrections are completed, additional diagnosis should be done by a dealer service department or transmission repair shop. Refer to the Troubleshooting Section at the front of this manual for information on symptoms of transmission problems.

Preliminary checks

1   Drive the vehicle to warm the transmission to normal operating temperature.

2   Check the fluid level as described in Chapter 1:

  a) *If the fluid level is unusually low, add enough fluid to bring the level within the designated area of the dipstick, then check for external leaks (see below).*

  b) *If the fluid level is abnormally high, drain off the excess, then check the drained fluid for contamination by coolant. The presence of engine coolant in the automatic transmission fluid indicates that a failure has occurred in the internal radiator walls that separate the coolant from the transmission fluid (see Chapter 3).*

  c) *If the fluid is foaming, drain it and refill the transmission, then check for coolant in the fluid or a high fluid level.*

3   Check the engine idle speed. **Note:** *If the engine is malfunctioning, do not proceed with the preliminary checks until it has been repaired and runs normally.*

4   Inspect the shift control linkage (see Section 4). Make sure it's properly adjusted and the linkage operates smoothly.

Fluid leak diagnosis

5   Most fluid leaks are easy to locate visually. Repair usually consists of replacing a seal or gasket. If a leak is difficult to find, the following procedure may help.

6   Identify the fluid. Make sure it's transmission fluid and not engine oil or brake fluid (automatic transmission fluid is a deep red color).

7   Try to pinpoint the source of the leak. Drive the vehicle several miles, then park it over a large sheet of cardboard. After a minute or two, you should be able to locate the leak by determining the source of the fluid dripping onto the cardboard.

8   Make a careful visual inspection of the suspected component and the area immediately around it. Pay particular attention to gasket mating surfaces. A mirror is often helpful for finding leaks in areas that are hard to see.

9   If the leak still cannot be found, clean the suspected area thoroughly with a degreaser or solvent, then dry it.

10   Drive the vehicle several miles at normal operating temperature and varying speeds. After driving the vehicle, visually inspect the

suspected component again.

11   Once the leak has been located, the cause must be determined before it can be properly repaired. If a gasket is replaced but the sealing flange is bent, the new gasket will not stop the leak. The bent flange must be straightened.

12   Before attempting to repair a leak, check to make sure the following conditions are corrected or they may cause another leak. **Note:** *Some of the following conditions cannot be fixed without highly specialized tools and expertise. Such problems must be referred to a transmission repair shop or a dealer service department.*

Gasket leaks

13   Check the pan periodically. Make sure the bolts are tight, no bolts are missing, the gasket is in good condition and the pan is flat (dents in the pan may indicate damage to the valve body inside).

14   If the pan gasket is leaking, the fluid level or the fluid pressure may be too high, or the vent may be plugged, or the pan bolts may be too tight, or the pan sealing flange may be warped, or the sealing surface of the transmission housing may be damaged, or the gasket may be damaged, or the transmission casting may be cracked or porous. If sealant instead of gasket material has been used to form a seal between the pan and the transmission housing, it may be the wrong sealant.

Seal leaks

15   If a transmission seal is leaking, the fluid level or pressure may be too high, or the vent may be plugged, or the seal bore may be damaged, or the seal itself may be damaged or improperly installed, or the surface of the shaft protruding through the seal may be damaged, or a loose bearing may be causing excessive shaft movement.

16   Make sure the dipstick tube seal is in good condition and the tube is properly seated. Periodically check the area around the speedometer gear or sensor for leakage. If transmission fluid is evident, check the O-ring for damage.

Case leaks

17   If the case itself appears to be leaking,

the casting is porous and will have to be repaired or replaced.

18   Make sure the oil cooler hose fittings are tight and in good condition.

Fluid comes out the vent pipe or fill tube

19   If this condition occurs, the transmission is overfilled, there is coolant in the fluid, the case is porous, the dipstick is incorrect, the vent is plugged or the drain back holes are plugged.

---

3   Shift lever - removal and installation

Refer to illustration 3.6

Warning: *These models are equipped with airbags. Always disable the airbag system before working in the vicinity of any airbag system component to avoid the possibility of accidental deployment of the airbag(s), which could cause personal injury (see Chapter 12).*

1   Remove the center console (see Chapter 11).

2   Place the shift lever in Park.

3   Detach the shift cable from the pin on the shift lever and from the shifter base (see Section 4).

4   Detach the Park/lock cable from the shift lever and from the shifter base (see Section 7).

5   Disconnect the electrical connector(s) at from the shifter assembly.

6   Unscrew the mounting nuts and remove the shifter assembly **(see illustration)**.

7   Installation is the reverse of removal. If necessary, adjust the shift cable (see Section 4) and the Park interlock cable (see Section 7).

---

4   Shift cable - check, replacement and adjustment

Check

1   Firmly apply the parking brake and try to momentarily operate the starter in each shift lever position. The starter should operate in Park and Neutral only. If the starter operates

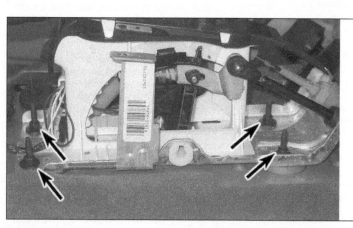

3.6 Remove the shift lever mounting nuts (arrows)

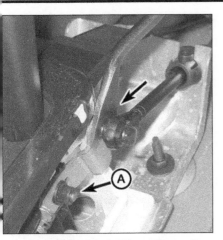

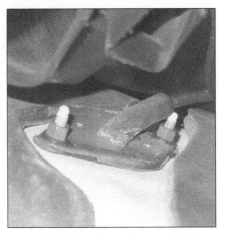

4.5 Using a prying tool (arrow), pop the shift cable off the pin at the shift lever; (A) indicates the cable adjuster nut

4.6 Squeeze the tangs on the shift cable retainer and detach the cable from the shifter base

4.7 Remove the cable seal plate mounting nuts

in any position other than Park or Neutral, adjust the shift linkage (see below).

Replacement
Refer to illustrations 4.5, 4.6, 4.7, 4.9 and 4.10
Warning: *These models are equipped with airbags. Always disable the airbag system before working in the vicinity of any airbag system component to avoid the possibility of accidental deployment of the airbag(s), which could cause personal injury (see Chapter 12).*
2   Place the shift lever in Park.
3   Remove the center console (see Chapter 11).
4   Unbolt the shifter assembly from the floor pan (see Section 3).
5   Pry the cable end off the pin on the shift lever **(see illustration)**.
6   Detach the cable from the shifter base **(see illustration)**.
7   Remove the cable seal plate from the floor pan **(see illustration)**.

8   Raise the vehicle and support it securely on jackstands.
9   Detach the cable end from the shift lever on the transmission **(see illustration)**.
10   Detach the cable from the bracket at the transmission **(see illustration)**.
11   Push the cable through the floor pan and remove it from the vehicle.
12   Installation is the reverse of removal. Adjust the cable (see the next Step).

Adjustment
13   Place the shift lever in Park.
14   Remove the center console (see Chapter 11).
15   Loosen the shifter cable adjustment screw at the shift lever **(see illustration 4.5)**.
16   Raise the vehicle and support it securely on jackstands.
17   Detach the cable end from the shift lever on the transmission **(see illustration 4.9)**.
18   Detach the cable from the bracket at the transmission **(see illustration 4.10)**.

19   Be sure the shift lever on the transmission is all the way to the rear, in the last detent. This is the Park position.
20   Make sure the park lock is engaged by trying to rotate the driveshaft. The driveshaft will not rotate if the park lock is functioning properly.
21   Recheck to make sure the gearshift lever inside the vehicle is still in the Park position.
22   Attach the cable end to the shift lever on the transmission.
23   Attach the cable to the bracket at the transmission
24   Tighten the shifter cable adjustment screw at the shifter to 65 in-lbs (7 Nm) **(see illustration 4.5)**.
25   Lower the vehicle.
26   With the parking brake firmly applied, make sure the engine starts only in Park and Neutral.
27   The remainder of installation is the reverse of removal.

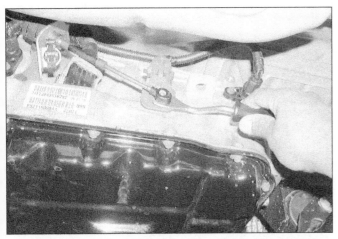

4.9 The shift cable is connected to the transmission shift lever with a ball-and-socket type connector; to disconnect it, pry the cable end loose from the small pin on the lever (typical)

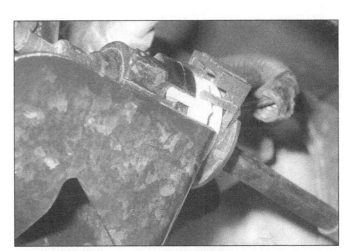

4.10 To detach the shift cable from the bracket at the transmission, pinch the housing tabs together with a pair of pliers, then pull the cable and housing through the bracket (typical)

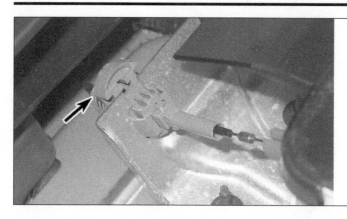

5.5a Pull the cable lock button up (arrow) to release the cable

5.5b Pull the cable forward, then release it and press down the cable lock button until it snaps into place

5   Park/Lock system - description, adjustment and component replacement

Warning: *These models are equipped with airbags. Always disable the airbag system before working in the vicinity of any airbag system component to avoid the possibility of accidental deployment of the airbag(s), which could cause personal injury* (see Chapter 12).

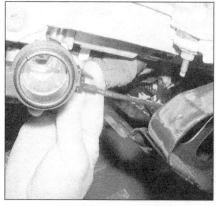

5.10 To detach the Park/Lock cable from the lock cylinder housing, depress the plastic locking tang and pull the cable end out of the housing (typical)

Description

1   The Park/Lock system prevents the shift lever from being moved out of Park unless the brake pedal is depressed simultaneously. It also prevents the Ignition key from being removed from the ignition switch unless the shift lever is in the Park position. When the ignition key is turned to the Accessory or Run position, a solenoid in-line with the interlock cable is energized, locking the shift lever in Park; when the brake pedal is depressed, the solenoid is de-energized. unlocking the shift lever so that it can be moved into some other gear.

Park/Lock cable
Check and adjustment
Refer to illustrations 5.5a and 5.5b

2   With the ignition key turned to the Run position and the brake pedal depressed, ver-ify that the shift lever moves through all gear positions.
3   While moving the shift lever through all gear positions, verify that the ignition key can't be turned to the Lock position.
4   Verify that the ignition key can be removed when it's in the Lock position and the shift lever is in the Park position.
5   If the Park/Lock cable fails any of the above tests, adjust the gear shift cable (see Section 4), then adjust the Park/Lock cable as follows.

a) *Put the shift lever in Park.*
b) *Turn the ignition key to Lock.*
c) *Remove the console* (see Chapter 11).
d) *Pry up the locking tab on the cable adjuster, then pull the cable forward and release it* **(see illustrations)**. *The adjuster is spring loaded, so it will automatically remove the slack from the cable.*
e) *Push in on the locking tab until it clicks back into place.*

6   Following Steps 2 through 4, check the cable operation.

Replacement
Refer to illustrations 5.10, 5.11 and 5.12

7   Disconnect the cable from the negative terminal of the battery.
8   Remove the console, the left-side under-dash panel and steering column knee bolster panel (see Chapter 11). On later models, lower the steering column (see Chapter 10).
9   Put the shift lever in the Park position.
10   Depress the locking tang and slide the end of the interlock cable out of the lock cyl-inder housing **(see illustration)**.
11   Detach the shifter end of the cable from the Park/Lock lever **(see illustration)**.
12   Depress the tabs on the sides of the cable adjuster **(see illustration)** and disen-gage the adjuster from the cable bracket.

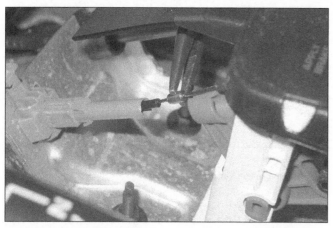

5.11 Detach the Park/Lock cable from the shifter lever

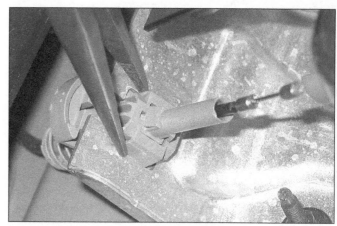

5.12 Squeeze the tabs on the cable housing to detach the cable from the shifter base

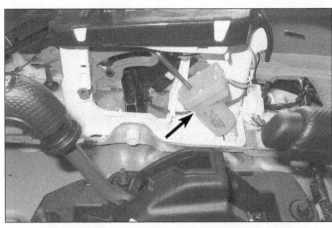

5.16 The BTSI solenoid (arrow) is mounted inside the shifter assembly

6.3 To check the transmission mount, insert a large screwdriver or prybar between the transmission and crossmember and pry the transmission up, looking for cracks in the rubber (typical)

13   Remove any cable clips and remove the Park/Lock cable.
14   Installation is the reverse of removal. Be sure to adjust the cable as described earlier in this Section.
15   Check the operation of the Park/Lock cable.

Brake/Transmission Shift Interlock (BTSI) solenoid Check

Refer to illustration 5.16

16   Turn the ignition key to the Run position and, without depressing the brake pedal, attempt to move the shift lever out of Park; you shouldn't be able to depress the button on the shift lever. If you can depress the button, either the solenoid is not receiving voltage, is not grounded, or it is defective. The solenoid is located inside the shifter assembly.

a) *Check the fuses.*
b) *Check the adjustment of the shift cable.*
c) *Remove the console (See Chapter 11) and check for power to the BTSI solenoid (see illustration). If voltage is not available, repair the circuit between the solenoid and the fuse block. If voltage is available, check the ground circuit for continuity. If continuity is present, replace the solenoid.*

17   Now depress the brake pedal - you should be able to push the button on the shifter in and move the lever out of Park. If you can't, the solenoid isn't de-activating.

a) *Make sure the shift cable is properly adjusted.*
b) *Unplug the electrical connector from the solenoid and check for continuity to ground on the harness terminal that does not have power to it. There should be continuity when the brake pedal is at rest and no continuity (or high resistance) when the pedal is depressed at least 1/2-inch.*
c) *If the ground path from the solenoid is cut when the brake pedal is depressed, but the solenoid still doesn't release, the solenoid is stuck - replace it.*

Replacement

Note: *On 2007 and later models, the BTSI solenoid is an integral component of the shift lever assembly and must be replaced as a unit (see Section 3).*

18   The solenoid is located inside the shifter assembly.
19   Disconnect the cable from the negative terminal of the battery.
20   Remove the console (see Chapter 11).
21   Disconnect the electrical connector from the solenoid.
22   Remove the solenoid mounting bolts **(see illustration 5.16)**.
23   Installation is the reverse of removal.

6   Transmission mount - check and replacement

Refer to illustrations 6.3, 6.5, 6.6 and 6.7

1   The transmission mount is located on the bottom side of the transmission case, toward the rear of the transmission.
2   Raise the vehicle and support it securely on jackstands.
3   Working underneath the vehicle, insert a large screwdriver or prybar into the space between the transmission and the crossmember **(see illustration)**. Try to pry the transmission or transfer case up slightly. There should be very little movement between the center of the mount and the rubber insulator. If any cracks are noticed, replace the mount.
4   To replace the mount, slightly raise the transmission with a jack.
5   Remove the transmission crossmember-to-frame bolts from the left and right side frame rails **(see illustration)**.
6   Remove the through-bolt that attaches the transmission support bracket to the mount **(see illustration)**. Support the crossmember during removal to prevent it from falling as the bolt is removed.
7   Remove the transmission mount bolts

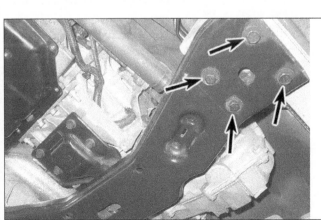

6.5 To remove the transmission mount, slightly raise the transmission and remove the crossmember-to-frame bolts (arrows) (left side shown) from the left and right sides of the vehicle

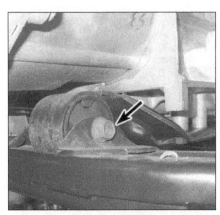

6.6 Remove the transmission mount through-bolt (arrow) . . .

6.7 . . . and the mount-to-transmission bolts

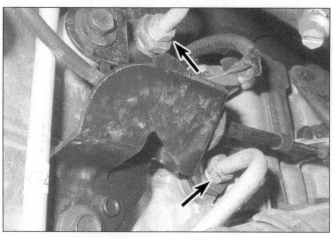

7.8 Disconnect the transmission cooler lines (arrows) from the fittings on the transmission

from the transmission **(see illustration)**.

8   Installation is the reverse of the removal procedure. Tighten the transmission mount and crossmember fasteners to the torque values listed in this Chapter's Specifications.

7   Automatic transmission - removal and installation

Removal

Refer to illustrations 7.8, 7.16, 7.18, and 7.19

Note 1: *A faulty transmission should not be removed before the vehicle has been assessed by a knowledgeable technician equipped with the proper tools, as trouble-shooting must be performed with the transmission installed in the vehicle.*

Note 2: *After rebuilding or replacing the 42RLE or NAG1 transmission, a DRB scan tool must be used to perform a "Quick Learn" procedure, which recalibrates the transmission and Transmission Control Module (TCM) to work in the most efficient manner. This is a procedure which must be performed by a dealer service department or other repair shop equipped with the required special tool.*

1   Disconnect the cable from the negative terminal of the battery.

2   Raise the vehicle and support it securely on jackstands.

3   Remove the skid plate, if equipped.

4   Unplug all electrical connectors, unclip the wiring harnesses from their retaining clips on the transmission and/or transfer case, and set the harnesses aside. It's a good idea to label all connectors with colored electrical tape to simplify reassembly.

5   Remove the exhaust system crossover pipe (see Chapter 4).

6   Drain the transmission fluid (see Chapter 1).

7   Reinstall the transmission pan.

8   Using a flare-nut wrench and back-up wrench, disconnect the transmission cooler lines **(see illustration)**. Plug the lines to prevent leakage and to protect the transmission from contamination.

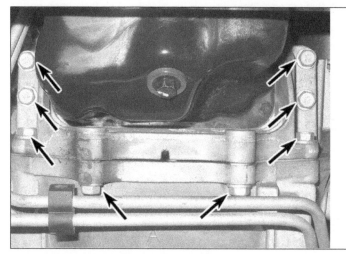

7.16 Remove the engine support collar mounting bolts from the engine bedplate and the transmission torque converter housing (arrows)

9   Disconnect the shift cable from the transmission shift lever and bracket (see Section 4), and secure it aside.

10   On 4WD models, disconnect the transfer case shift cable and the transfer case vent hose from the transfer case (see Chapter 7 Part C).

11   Support the transmission with a floor jack and, if available, a transmission jack adapter. Safety chains will help steady the transmission on the jack.

12   Remove the transmission crossmember and the transmission mount (see Section 6).

13   Mark the relationship of the drive shaft to the flange with white paint so they can be installed in the same position and remove the rear drive shaft (see Chapter 8).

14   On 4WD models, remove the transfer case (see Chapter 7 Part C). **Note:** *If you are not planning to replace the transmission, but are removing it in order to gain access to other components such as the torque converter, it's not really necessary to remove the transfer case. However, the front drive-shaft will need to be removed (see Chapter 8). If the transfer case is not removed, the transmission and transfer case will be awk-ward and heavy when removed and installed*

as a single assembly; they're much easier to maneuver off and on as separate units. **Warning:** *If you decide to leave the transfer case attached to the transmission, be sure to use safety chains to help stabilize the transmission and transfer case assembly and to prevent it from falling off the jack head, which could cause serious damage to the transmission and/or transfer case and serious bodily injury to you.*

15   Remove the starter motor (see Chapter 5).

16   Remove the engine support collar **(see illustration)**.

17   If equipped, remove the torque converter access cover from the transmission.

18   Mark the relationship of the torque converter to the driveplate with white paint so they can be installed in the same position **(see illustration)**.

19   Remove the torque converter-to-drive-plate bolts **(see illustration)**. **Caution:** *Always turn the crankshaft in a clockwise direction only (as viewed from the front) to gain access to each torque converter bolt. Otherwise, engine damage may occur.*

20   Support the engine with a floor jack or an engine hoist (see Chapter 2).

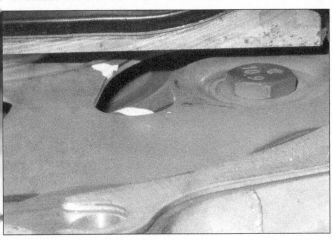

7.18 Mark the relationship of the torque converter to the driveplate with white paint so they can be correctly aligned on reassembly - this is critical for engine balancing! (typical)

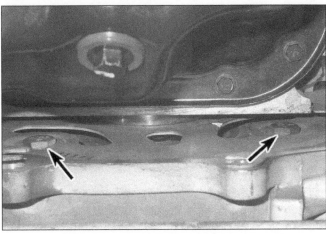

7.19 To remove the torque converter-to-driveplate bolts (arrows), turn the crankshaft for access to each bolt; turn the crankshaft in a clockwise direction only (as viewed from the front)

21   Remove the lower transmission-to-engine bolts.

22   Lower the transmission slightly and remove the upper transmission-to-engine bolts.

23   Remove the transmission dipstick tube. Remove and discard the dipstick tube O-ring.

24   Move the transmission to the rear to disengage it from the engine block dowel pins and make sure the torque converter is detached from the driveplate. Attach a C-clamp to the transmission housing to temporarily secure the torque converter so it won't fall out during transmission removal.

25   Lower the transmission and move it out from under the vehicle.

Installation

Note: *If you removed the torque converter, make sure the O-ring on the converter hub is in good condition.*

26   Prior to installing the torque converter, make sure the torque converter hub is securely engaged in the transmission front pump. The distance from the bellhousing mating surface to the converter bolt-lugs should be at least 1/2-inch (13 mm). If it isn't, rotate the converter while pushing in until the converter "clunks" into place (then recheck the measurement). Attach a C-clamp to the transmission housing to temporarily secure the torque converter so it won't fall out during transmission installation.

27   Apply a light coat of high-temperature grease to the torque converter hub pocket on the back of the engine crankshaft.

28   With the transmission (and, on 4WD models, the transfer case, if you left it

attached) secured to the jack, raise it into position. If a C-clamp is not attached to the transmission housing, be sure to keep the transmission level so the torque converter does not slide forward. Connect the transmission fluid cooler lines.

29   Remove the C-clamp and turn the torque converter to line up its bolt holes with the holes in the driveplate. The white paint mark on the torque converter and the driveplate made in Step 18 must line up.

30   Move the transmission (and on 4WD models, the transfer case, if already attached) forward carefully until the dowel pins and the torque converter are engaged.

31   Install the transmission housing-to-engine bolts. Tighten the bolts to the torque values listed in this Chapter's Specifications.

32   Install the torque converter-to-driveplate bolts. Tighten the bolts to the torque listed in this Chapter's Specifications. **Note:** *Install all of the bolts before tightening any of them.*

33   The remainder of installation is the reverse of removal.

34   Fill the transmission with the specified fluid (see Chapter 1), run the engine and check for fluid leaks

35   Test-drive the vehicle to make sure everything is working properly.

8   Automatic transmission overhaul - general information

In the event of a fault occurring, it will be necessary to establish whether the fault is electrical, mechanical or hydraulic in nature, before repair work can be contemplated. Diagnosis requires detailed knowl-

edge of the transmission's operation and construction, as well as access to specialized test equipment, and so is deemed to be beyond the scope of this manual. It is therefore essential that problems with the automatic transmission are referred to a dealer service department or other qualified repair facility for assessment.

9   Transmission control module (TCM) - general information

Based on the information programmed into the computer's memory and shift lever position, throttle position, engine load and fluid temperature data, the Transmission Control Module (TCM) generates output signals to control the electronic transmission functions via control relays, solenoids and other output actuators. The TCM is specifically calibrated to optimize the operation and driveability of the vehicle by selecting the transmission shift schedule and managing the transmission shifting points.

The TCM is an integral component of the Powertrain Control Module (PCM). Refer to Chapter 6 for the PCM replacement procedure. **Note:** *On models equipped with the 42RLE or NAG1 transmission, anytime the PCM/TCM is replaced, a DRB scan tool must be used to perform a "Quick Learn" procedure, which recalibrates the transmission and Transmission Control Module (TCM) to work in the most efficient manner. This is a procedure which must be performed by a dealer service department or other repair shop equipped with the required special tool.*

Notes

Chapter 7 Part C Transfer case

Contents

Specifications

Torque specifications

	Ft-lbs (unless otherwise indicated)	Nm

Note: *One foot-pound (ft-lb) of torque is equivalent to 12 inch-pounds (in-lbs) of torque. Torque values below approximately 15 ft-lbs are expressed in inch-pounds, since most foot-pound torque wrenches are not accurate at these smaller values.*

	Ft-lbs	Nm
Transfer case-to-transmission nuts	26 to 35	35 to 47
Transfer case shift motor bolts		
MP1522 models	18.5 to 20.5	25 to 28
MP3022 models	108 inch-lbs	12

1 General information

Installed on 4WD models, the transfer case is a device which transmits power from the transmission to the front and rear driveshafts. The models covered by this manual may be equipped with any one of the following transfer cases:

a) The NV231 is a part-time transfer case with a low-range reduction gear system.

b) The NV241 Gen II is a part-time transfer case with a low-range gear system (2.72:1).

c) The NV242 is a full- and part-time transfer case with four operating ranges. It provides both 2WD and full-time 4WD, and has a locking mechanism for undifferentiated 4WD in high and low ranges. The low-range reduction system provides increased low-speed torque capability.

d) The MP1522 is a two-speed, part-time four-wheel drive transfer case with an electric shift console-mounted switch.

e) The MP3022 is a two-speed, on-demand, active four-wheel drive transfer case that provides maximum traction for various conditions.

f) The MP 140 single-speed transfer case is used only with automatic transmissions in the Nitro model Full-Time Four-Wheel Drive System. It provides full-time four-wheel drive operation and functions working with the Traction Control and ESP (Electronic Stability Program) systems to provide maximum traction and stability automatically.

g) The MP143 single-speed, part-time transfer case used with both manual and automatic transmissions is a part-time transfer case in the Nitro model Part-Time Four-Wheel Drive System. Two-wheel drive and four-wheel drive modes are electronically controlled via a switch in the center console. Either mode can be selected at any time and works with the Traction Control and ESP (Electronic Stability Program) systems to provide maximum traction and stability.

2.3a Transfer case shift cable installation details - at transfer case shift lever

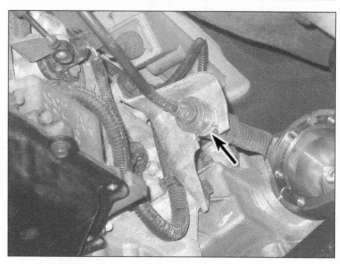

2.3b Transfer case shift cable installation details - at transfer case bracket

2   Transfer case shift cable (2007 and earlier models) - removal and installation

Refer to illustrations 2.3a, 2.3b and 2.5
Warning: *These models are equipped with airbags. Always disable the airbag system before working in the vicinity of any airbag system component to avoid the possibility of accidental deployment of the airbag(s), which could cause personal injury (see Chapter 12).*
1   Apply the parking brake, shift the transmission into the Park position and shift the transfer case into the 4L position.
2   Raise the vehicle and support it securely on jackstands.
3   Disconnect the shift cable from the shift lever at the transfer case. Remove the retaining clip and detach the cable from the support bracket at the transfer case **(see illustrations)**.
4   Remove the center console (see Chapter 11).
5   Remove the shifter mounting nuts from the floor pan studs and pull the shift cable through the floor pan and into the passenger compartment **(see illustration)**.
6   Disconnect the cable from the shift lever. Remove the clip and detach the cable from the shift lever base.
7   Installation is the reverse of removal. Check the transfer case operation in all operating ranges.

3   Transfer case shift motor (MP1522 and MP3022 models) - removal and installation

Note: *All new shift motor assemblies are shipped in the "NEUTRAL" position. If a new shift motor is installed, the transfer case must be shifted into the "NEUTRAL" position before shift motor installation.*

1   Raise the vehicle and support it securely on jackstands.

MP1522 models

Removal
2   Disconnect the electrical connector from the shift motor.
3   Remove the shift motor mounting bolts.
4   Carefully pull the shift motor from the transfer case and selector shaft, then discard the three-tab sealing-ring.

Installation
5   Install a new shift motor three-tab sealing-ring into the groove of the shift motor.
6   Align the "missing wide tooth" of the shift motor drive spline with the "wide tooth" of the transfer case shaft.
7   Install the shift motor onto the transfer case, making sure the teeth are splined together.
8   Install the mounting bolts and tighten them to the torque listed in this Chapter's Specifications.
9   The remainder of installation is the reverse of removal.

MP3022 models
Removal
10   Disconnect the electrical connector from the shift motor.
11   Remove the shift motor mounting bolts.
12   Carefully pull the shift motor from the transfer case and discard the O-ring.

Installation
13   Install a new shift motor O-ring into the groove of the shift motor.
14   Align the shift motor slotted shaft with the slot in the transfer case worm shaft.
15   Install the shift motor onto the transfer case.
16   Install the mounting bolts and tighten them to the torque listed in this Chapter's Specifications.

17   The remainder of installation is the reverse of removal.

4   Transfer case shift mode sensor (MP3022 models) - removal and installation

1   Raise the vehicle and support it securely on jackstands.

Removal
2   Disconnect the wiring harness electrical connector from the shift mode sensor.
3   Remove the fasteners attaching the shift mode sensor to the transfer case.
4   Carefully pull the shift mode sensor from the transfer case and discard the O-ring.

Installation
5   Install a new shift mode sensor O-ring into the grove of the sensor.
6   Align the shift mode sensor slotted shaft with the slot in the shift shaft.
7   Install the shift mode sensor and fasteners, tightening them securely.
8   The remainder of installation is the reverse of removal.

5   Transfer case - removal and installation

Removal
Refer to illustration 5.10
1   Raise the vehicle and support it securely on jackstands.
2   Shift the transfer case into Neutral.
3   Remove the skid plate, if equipped.
4   Drain the transfer case lubricant (see Chapter 1).
5   Disconnect the electrical connectors from the transfer case position sensor, shift motor and shift mode sensor (if equipped).

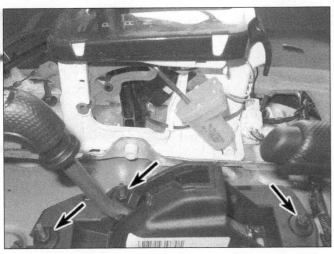

2.5 Remove the transfer case shifter mounting nuts (arrows) from the studs

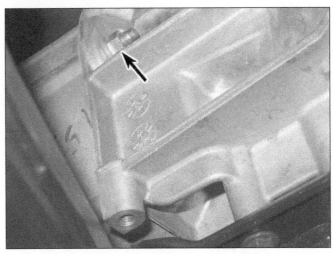

5.10 Remove the nuts (arrow) (typical) from the transfer case mounting studs

6   Disconnect the transfer case vent hose from the transfer case.

7   Disconnect the shift cable from the transfer case shift lever and bracket (see Section 2).

8   Remove the driveshafts (see Chapter 8).

9   Support the transfer case with a jack - preferably a special jack made for this purpose. Safety chains will help steady the transfer case on the jack.

10   Remove the transfer case-to-transmission nuts from the mounting studs **(see illustration)**.

11   Make a final check that all wires and hoses have been disconnected from the transfer case and then move the transfer case and jack toward the rear of the vehicle until the transfer case is clear of the transmission. Keep the transfer case level as this is done.

12   Once the input shaft is clear, lower the transfer case and remove it from under the vehicle.

Installation

13   With the transfer case secured to the jack as on removal, raise it into position behind the transmission and then carefully slide it forward, engaging the input shaft with the transmission output shaft. Do not use excessive force to install the transfer case - if the input shaft does not slide into place, readjust the angle so it is level and/or turn the input shaft so the splines engage properly with the transmission.

14   Install the transfer case-to-transmission

nuts. Tighten the nuts to the torque listed in this Chapter's Specifications.

15   Remove the safety chains and remove the jack supporting the transfer case.

16   Install the skid plate (if equipped).

17   Install the driveshafts (see Chapter 8).

18   Reattach the transfer case vent hose.

19   Reconnect the shift cable to the transfer case shift lever.

20   Reconnect the electrical connector.

21   Check the lubricant level in the transfer case and refill as necessary (see Chapter 1). If the vehicle is equipped with a manual transmission, check the lubricant level for the manual transmission, as well (see Chapter 1).

22   Remove the jackstands and lower the vehicle.

23   Road test the vehicle for proper operation and check for leakage.

---

6   Transfer case overhaul - general information

Overhauling a transfer case is a difficult job for the do-it-yourselfer. It involves the disassembly and reassembly of many small parts. Numerous clearances must be precisely measured and, if necessary, changed with select fit spacers and snap-rings. As a result, if transfer case problems arise, it can be removed and installed by a competent do-it-yourselfer, but overhaul should be left to a transmission repair shop. Rebuilt transfer cases may be avail-

able - check with your dealer parts department and auto parts stores. At any rate, the time and money involved in an overhaul is almost sure to exceed the cost of a rebuilt unit.

Nevertheless, it's not impossible for an inexperienced mechanic to rebuild a transfer case if the special tools are available and the job is done in a deliberate step-by-step manner so nothing is overlooked.

The tools necessary for an overhaul include internal and external snap-ring pliers, a bearing puller, a slide hammer, a set of pin punches, a dial indicator and possibly a hydraulic press. In addition, a large, sturdy workbench and a vise or transmission stand will be required.

During disassembly of the transfer case, make careful notes of how each piece comes off, where it fits in relation to other pieces and what holds it in place. Noting how they are installed when you remove the parts will make it much easier to get the transfer case back together.

Before taking the transfer case apart for repair, it will help if you have some idea what area of the transfer case is malfunctioning. Certain problems can be closely tied to specific areas in the transfer case, which can make component examination and replacement easier. Refer to the *Troubleshooting* section at the front of this manual for information regarding possible sources of trouble.

Notes

Chapter 8
Clutch and driveline

Contents

Specifications

Fluid type	See Chapter 1

Torque specifications

Ft-lbs (unless otherwise indicated) **Nm**

Note: *One foot-pound (ft-lb) of torque is equivalent to 12 inch-pounds (in-lbs) of torque. Torque values below approximately 15 foot-pounds are expressed in inch-pounds, because most foot-pound torque wrenches are not accurate at these smaller values.*

Clutch

	Ft-lbs	Nm
Clutch master cylinder mounting nuts	28	38
Clutch slave cylinder mounting nuts	17	23
Flywheel-to-crankshaft bolts	60	81
Pressure plate-to-flywheel bolts		
2.4L four-cylinder	23	31
3.7L V6	37	50

Drivetrain

	Ft-lbs	Nm
Driveshaft		
Front driveaxle CV joint bolts	22	30
Front driveshaft bolts (4WD only)	22	30
Rear driveshaft fasteners		
2007 and earlier models		
With straps	156 in-lbs	17.5
With companion flange	85	115
2008 and later models	80	108
Front axle assembly (4WD models only)		
Bracket-to-frame mount fasteners	65	88
Differential cover bolts	14 to 19	19 to 26
Front and rear bracket-to-differential housing bolts	45	61
Right-side bracket-to-differential housing bolts	65	88
Front driveaxle hub nut	100	136
Front hub and bearing assembly-to-steering knuckle bolts	96	130
Pinion mate shaft lock bolt, 8-1/4 rear differentials		
2007 and earlier models	96 in-lbs	11
2008 and later models	19	26
Rear axle retaining plate nuts, 198RBI rear differential	35 to 55	47 to 75

1   General information

The information in this Chapter deals with the components from the rear of the engine to the drive wheels, except for the transmission and transfer case, which are dealt with in the previous Chapter. For the purposes of this Chapter, these components are grouped into three categories: clutch, driveshaft and axles. Separate Sections within this Chapter offer general descriptions and checking procedures for components in each of the three groups.

Since nearly all the procedures covered in this Chapter involve working under the vehicle, make sure it's securely supported on sturdy jackstands or on a hoist where the vehicle can be easily raised and lowered.

2   Clutch - description and check

1   All vehicles with a manual transmission use a single dry plate, diaphragm spring type clutch. The clutch disc has a splined hub which allows it to slide along the splines of the transmission input shaft. The clutch and pressure plate are held in contact by spring pressure exerted by the diaphragm in the pressure plate.

2   The clutch release system is operated by hydraulic pressure. The hydraulic release system consists of the clutch pedal, a master cylinder and fluid reservoir, the hydraulic line, a slave cylinder, and a release lever mounted inside the clutch housing.

3   When pressure is applied to the clutch pedal to release the clutch, hydraulic pressure is exerted against the release bearing through the release lever. The bearing pushes against the fingers of the diaphragm spring of the pressure plate assembly, which in turn releases the clutch plate.

4   Terminology can be a problem when discussing the clutch components because common names are in some cases different from those used by the manufacturer. For example, the driven plate is also called the clutch plate or disc. The clutch release bearing is sometimes called a throwout bearing. The release cylinder is sometimes called the operating or slave cylinder.

5   Other than replacement of components with obvious damage, some preliminary checks should be performed to diagnose clutch problems.

a) *The first check should be of the fluid level in the clutch master cylinder. If the fluid level is low, add fluid as necessary and inspect the hydraulic system for leaks.*

b) *To check "clutch spin down time," run the engine at normal idle speed with the transmission in Neutral (clutch pedal up - engaged). Disengage the clutch (pedal down), wait several seconds and shift the transmission into Reverse. No grinding noise should be heard. A grinding noise would most likely indicate a prob-*

lem in the pressure plate or the clutch disc (assuming the transmission is in good condition).

c) *To check for complete clutch release, run the engine (with the parking brake applied to prevent movement) and hold the clutch pedal approximately 1/2-inch from the floor. Shift the transmission between 1st gear and Reverse several times. If the shift is rough, component failure is indicated. Check the release cylinder pushrod travel. With the clutch pedal depressed completely, the release cylinder pushrod should extend substantially. If it doesn't, check the fluid level in the clutch master cylinder.*

d) *Visually inspect the pivot bushing at the top of the clutch pedal to make sure there is no binding or excessive play.*

3   Clutch hydraulic release system - removal and installation

Note: *The clutch hydraulic release system is serviced as an assembly. If the system is leaking or has air in it, replace the entire assembly.*

1   Raise the vehicle and support it securely on jackstands.

2   Remove the nuts attaching the slave cylinder to the clutch housing.

3   Disengage the clutch hydraulic fluid line from the retaining clips on the body.

4   Working from the driver's compartment, remove the retainer bushing that attaches the clutch master cylinder pushrod to the clutch pedal, then slide the clutch master cylinder pushrod off of the clutch pedal pin. If the bushing is worn, replace it.

5   To avoid spillage, verify that the cap on the master cylinder reservoir is tight. Unplug the electrical connector for the clutch pedal position switch. Remove the master cylinder mounting nuts that attach the master cylinder to the firewall.

6   Remove the entire assembly through the engine compartment. Do not attempt to disconnect the clutch fluid hydraulic line from the master cylinder or the slave cylinder.

7   Installation is the reverse of removal. Be sure to tighten all fasteners to the torque values listed in this Chapter's Specifications.

4   Clutch components - removal, inspection and installation

Warning: *Dust produced by clutch wear and deposited on clutch components is hazardous to your health. DO NOT blow it out with compressed air and DO NOT inhale it. DO NOT use gasoline or petroleum-based solvents to remove the dust. Brake system cleaner should be used to flush the dust into a drain pan. After the clutch components are wiped clean with a rag, dispose of the contaminated rags and cleaner in a covered, marked container.*

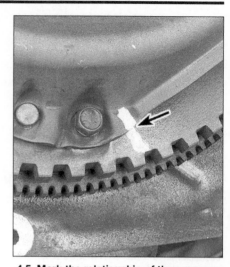

4.5 Mark the relationship of the pressure plate to the flywheel (in case you are going to re-use the same pressure plate) (typical)

Removal

Refer to illustration 4.5

1   Access to the clutch components is normally accomplished by removing the transmission, leaving the engine in the vehicle. If, of course, the engine is being removed for major overhaul, then check the clutch for wear and replace worn components as necessary. However, the relatively low cost of the clutch components compared to the time and trouble spent gaining access to them warrants their replacement anytime the engine or transmission is removed, unless they are new or in near perfect condition. The following procedures are based on the assumption the engine will stay in place.

2   Disconnect the clutch slave cylinder from the clutch housing (see Section 3). **Caution:** *Do not disconnect the clutch hydraulic fluid line from the slave cylinder. The hydraulic release system cannot be bled if air is allowed to enter the system.*

3   Remove the transmission from the vehicle (see Chapter 7, Part A). Support the engine while the transmission is out. Preferably, an engine hoist should be used to support it from above. However, if a jack is used underneath the engine, make sure a piece of wood is positioned between the jack and oil pan to spread the load. **Caution:** *The pickup for the oil pump is very close to the bottom of the oil pan. If the pan is bent or distorted in any way, engine oil starvation could occur.*

4   To support the clutch disc during removal, install a clutch alignment tool through the clutch disc hub.

5   Carefully inspect the flywheel and pressure plate for indexing marks. The marks are usually an X, an O or a white letter. If they cannot be found, scribe marks yourself so the pressure plate and the flywheel will be in the same alignment during installation **(see illustration).**

6   Turning each bolt a little at a time, loosen the pressure plate-to-flywheel bolts.

4.8 Check the surface of the flywheel for cracks, hot spots (dark colored areas) and other obvious defects; resurfacing by a machine shop will correct *minor* defects

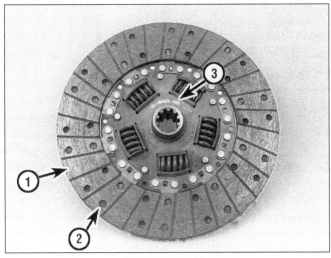

4.9 The clutch disc (typical)

1)   ***Lining*** *- will wear down in use*
2)   ***Rivets*** *- secure the lining and will damage the pressure plate or flywheel surface if allowed to contact it*
3)   ***Hub*** *- "Flywheel side" or other marks (flat-side shown)*

Work in a criss-cross pattern until all spring pressure is relieved. Then hold the pressure plate securely and completely remove the bolts, followed by the pressure plate and clutch disc.

Inspection

Refer to illustrations 4.8, 4.9, 4.10a and 4.10b
7   Ordinarily, when a problem occurs in the clutch, it can be attributed to wear of the clutch driven plate assembly (clutch disc). However, all components should be inspected at this time.
8   Inspect the flywheel for cracks, heat checking, grooves and other obvious defects **(see illustration)**. If the imperfections are slight, a machine shop can machine the surface flat and smooth, which is highly recommended regardless of the surface appearance. Refer to Chapter 2 for the flywheel removal and installation procedure.
9   Inspect the lining on the clutch disc. There should be at least 1/16-inch of lining above the rivet heads. Check for loose rivets, distortion, cracks, broken springs and other

4.10a Inspect the surface of the pressure plate for cracks, dark-colored areas (signs of overheating) and other obvious defects

obvious damage **(see illustration)**. As mentioned above, ordinarily the clutch disc is routinely replaced, so if in doubt about the condition, replace it with a new one.
10   Check the machined surfaces and the diaphragm spring fingers of the pressure

plate **(see illustrations)**. If the surface is grooved or otherwise damaged, replace the pressure plate. Also check for obvious damage, distortion, cracking, etc. Light glazing can be removed with sandpaper or emery cloth. If a new pressure plate is required, new

NORMAL FINGER WEAR

EXCESSIVE WEAR

EXCESSIVE FINGER WEAR

BROKEN OR BENT FINGERS

4.10b Replace the pressure plate if excessive wear or damage is noted on the diaphragm fingers

and factory-rebuilt units are available.

11   Check the condition of the release bearing (see Section 5).

12   Inspect the pilot bearing (see Section 6).

Installation

Refer to illustration 4.14

13   Before installation, clean the flywheel and pressure plate machined surfaces with brake system cleaner, lacquer thinner or acetone. It's important that no oil or grease is on these surfaces or the lining of the clutch disc. Handle the parts only with clean hands.

14   Position the clutch disc and pressure plate against the flywheel with the clutch held in place with an alignment tool **(see illustration)**. Make sure it's installed properly (most replacement clutch discs will be marked "flywheel side" or something similar. **Caution:** *If the disc hub is not marked, install it as follows:*

   a)   *On 2.4L four-cylinder engines, install the flat-side of the disc hub toward the transmission.*

   b)   *On 3.7L V6 engines, install the flat-side of the disc hub facing toward the flywheel.*

15   Tighten the pressure plate-to-flywheel bolts only finger tight, working around the pressure plate.

16   Center the clutch disc by ensuring the alignment tool extends through the splined hub and into the pilot bearing in the crankshaft. Wiggle the tool up, down or side-to-side as needed to bottom the tool in the pilot bearing. Tighten the pressure plate-to-flywheel bolts a little at a time, working in a criss-cross pattern to prevent distorting the cover. After all of the bolts are snug, tighten them to the torque listed in this Chapter's Specifications. Remove the alignment tool.

17   Using high-temperature grease, lubricate the inner groove of the release bearing (see Section 5). Also place grease on the release lever contact areas and the transmission input shaft bearing retainer.

18   Install the clutch release bearing (see Section 5).

19   Apply a very light coat of grease to the transmission input shaft splines. **Caution:** *Applying too much grease to the splines could cause the clutch to become contaminated, causing premature failure.*

20   Install the transmission (see Chapter 7, Part A).

21   Install the slave cylinder (see Section 3).

22   Remove the jackstands and lower the vehicle.

23   Test drive the vehicle to verify that the transmission and clutch assembly are working properly.

---

5   Clutch release bearing - removal, inspection and installation

Warning: *Dust produced by clutch wear and deposited on clutch components is hazard-*

4.14 Center the clutch disc in the pressure plate with an alignment tool before the bolts are tightened

ous to your health. DO NOT blow it out with compressed air and DO NOT inhale it. DO NOT use gasoline or petroleum-based solvents to remove the dust. Brake system cleaner should be used to flush the dust into a drain pan. After the clutch components are wiped clean with a rag, dispose of the contaminated rags and cleaner in a covered, marked container.

Removal

1   Unbolt the clutch slave cylinder from the clutch housing (see Section 3) and pull it out of its recess. Hang it out of the way with a piece of wire but be careful not to kink the clutch hydraulic fluid line. **Caution:** *Do not disconnect the clutch hydraulic fluid line. The clutch release system is a one-piece assembly and cannot be bled.*

2   Remove the transmission (see Chapter 7, Part A).

3   Disengage the release lever from the return spring and the pivot stud, slide the bearing off the transmission input shaft bearing retainer and separate the bearing from the lever.

Inspection

Refer to illustration 5.4

4   Hold the center portion of the bearing stationary and rotate the outer portion while applying pressure. If the bearing doesn't turn smoothly or if it's noisy, replace it with a new one **(see illustration)**. Wipe the bearing with a clean rag and inspect it for damage and wear. Don't immerse the bearing in solvent - it is sealed for life and immersion would ruin it. Inspect the bearing for fluid leakage. If any leakage is evident, replace the entire assembly.

5   Inspect the release lever and the pivot stud. Make sure the pivot stud is tight and in good condition. Make sure the lever isn't distorted or worn. Inspect the release lever return spring. If it's bent or damaged, replace it.

5.4 To check the clutch release bearing, hold the hub (center) of the bearing and rotate the outer portion while applying pressure; if the bearing doesn't turn smoothly or if it's noisy, replace it

Installation

6   Installation is basically the reverse of removal. Be sure to lightly lubricate the crankshaft pilot bearing, the transmission input shaft splines, the transmission bearing retainer sliding surfaces, the release lever and the release lever pivot stud with high-temperature grease. **Caution:** *Applying too much grease to the input shaft splines could cause the clutch to become contaminated, causing premature failure.*

7   Install the transmission (see Chapter 7).

8   Reattach the clutch slave cylinder (see Section 3).

---

6   Pilot bearing - inspection and replacement

Refer to illustrations 6.5a, 6.5b and 6.6

1   The clutch pilot bearing is pressed into the rear of the crankshaft. It is greased at the factory and does not require additional lubrication. Its primary purpose is to support the front of the transmission input shaft. The pilot bearing should be inspected whenever the clutch components are removed from the engine. Due to its inaccessibility, if you are in doubt as to its condition, replace it with a new one. **Note:** *If the engine has been removed from the vehicle, disregard the following steps which do not apply.*

2   Remove the transmission (refer to Chapter 7 Part A).

3   Remove the clutch components (see Section 4).

4   Inspect for any excessive wear, scoring, lack of grease, dryness or obvious damage. If any of these conditions are noted, the bearing should be replaced. A flashlight will be helpful to direct light into the recess.

5   Removal can be accomplished with a pilot bearing puller or a slide hammer fitted

6.5a Use a pilot bearing puller . . .

6.5b . . . or a small slide-hammer and puller attachment to remove the pilot bearing from the crankshaft

with a puller attachment (see illustrations), which are available at most auto parts stores or equipment rental yards.

6   To install the new bearing, lightly lubricate the outside surface with multi-purpose grease, then drive it into the recess with a hammer and bearing/bushing driver (see illustration). Make sure the bearing seal faces toward the transmission. If you don't have a bearing driver, carefully tap it into place with a hammer and a socket. Caution: *Be careful not to let the bearing become cocked in the bore.*

7   Install the clutch components, transmission and all other components removed previously, tightening all fasteners properly.

7   Clutch pedal position switch - check and replacement

Check

1   The clutch pedal position switch, which is part of the starter relay circuit, closes the starter relay circuit only when the clutch pedal is fully depressed.

2   To test the switch, verify that the engine will not crank over when the clutch pedal is in the released position, and that it does crank over with the pedal depressed. Warning: *Make sure the transmission is in the Neutral position before performing this check.*

3   If the engine starts without depressing the clutch pedal, replace the switch. The continuity of the switch can also be checked with an ohmmeter; there should be no continuity through the switch with the pedal at rest, and continuity with the pedal depressed.

Replacement

4   The switch is an integral part of the clutch master cylinder pushrod and can't be serviced separately. If the switch must be replaced, so must the clutch hydraulic release system (see Section 3).

8   Driveshafts, differentials and axles - general information

Driveshafts

The driveshaft transmits power from the transmission (2WD models) or the transfer case (4WD models) to the rear axle. On 4WD models, this driveshaft is referred to as the rear driveshaft. On 4WD models, another driveshaft also transmits power from the transfer case to the front axle; this driveshaft is referred to as the front driveshaft.

As the vehicle travels over irregular surfaces, the axles "float" up and down, so the driveshaft must be able to operate while constantly changing its angle relative to the transmission/transfer case and the axle. This is made possible by universal joints (U-joints) at each end of the rear driveshaft, or constant velocity joints (CV joints) at each end of the front driveshaft (on 4WD models). As the angle between the transmission/transfer case and the axle changes, so does the *length* between them, so the driveshaft must also be able to change its length. The necessary change in length is made possible by a sliding yoke, which is also referred to as a sleeve yoke or slip joint.

The U-joints installed by the factory require no lubrication.

On 4WD models, the front driveshaft CV-joints are not lubricated or repairable. If the joint is worn or damaged, the driveshaft must be replaced. A splined shaft on the CV joint allows the overall driveshaft length to be adjusted for optimal joint travel.

The driveshafts are finely balanced during production. Whenever they are removed or disassembled, they must be reassembled and installed in the exact manner and positions they were originally installed, to avoid excessive vibration.

Differentials

The differential is located inside a cast-iron or aluminum housing which is an inte-

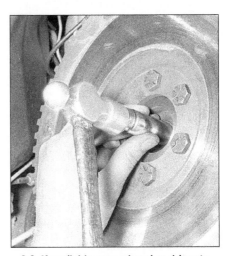

6.6 If available, use a bearing driver to install the pilot bearing; note that the seal in the bearing (if equipped with one) must face the transmission

gral part of the axle assembly. A stamped steel cover bolted to the open rear side of the housing allows inspection and servicing of the differential. The differential itself is a hypoid-gear design with the centerline of the pinion below the centerline of the ring gear. Besides the standard differential, an optional limited-slip unit is also available. If one wheel spins and loses traction, the differential transfers torque from that wheel to the other wheel.

Differentials are often designated by the diameter of the ring gear in the differential. The 2002 models have the 198RBI (198 mm) or 8-1/4 (8-1/4 inch) rear axle. The 2003 and later models have the 8-1/4 rear axle. 4WD models have the 186FIA (186 mm) aluminum front axle.

Because of the special tools and skills needed to adjust and/or overhaul the differential, this work should be left to a dealer service department or qualified repair shop.

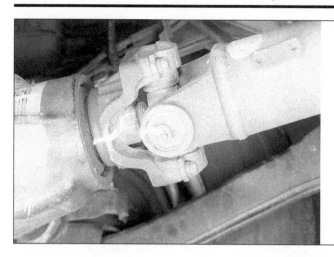

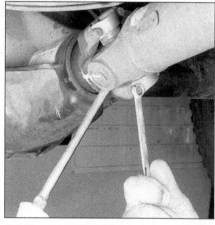

9.4 Before removing the bolts and straps that attach the rear end of the rear driveshaft to the rear differential, be sure to mark the relationship of the two yokes to one another

9.5 To prevent the driveshaft from turning when loosening the flange bolts, insert a screwdriver through the yoke as shown (typical)

Axles

Axle bearings and CV-joints are serviced as an assembly, i.e. they cannot be rebuilt.

On 4WD models, both ends of the front driveaxles are equipped with constant velocity (CV) joints to allow the front wheels to turn left and right.

On some models, the axle is equipped with an anti-lock brake system (ABS) as optional equipment.

The rear axleshafts are housed inside the rear axle assembly. On 198RBI differentials, the axleshafts are retained by nuts at the hub bearings. On 8-1/4 differentials, the axleshafts are retained by C-locks inside the differential housing.

The part number and gear ratio are sometimes listed on a tag attached to the differential housing cover and the build-date identification codes stamped on the backside of the axleshaft tube. Refer to these numbers when replacing the axle.

---

9   Driveshafts - removal and installation

---

1   Raise the vehicle and support it securely on jackstands. Chock the wheels at the opposite end of the vehicle being raised to prevent the vehicle from rolling.

2   Place the transmission in Neutral with the parking brake off.

Rear driveshaft

Removal

Refer to illustrations 9.4 and 9.5

3   If equipped, pry open the dust boot clamp on the driveshaft yoke at the front of the driveshaft. **Note:** *A special tool will be required to crimp-tighten the dust boot clamp during installation.*

4   Mark the relationship of the driveshaft to the differential yoke to ensure that the driveshaft is reinstalled in the same position to preserve the balance **(see illustration)**.

5   Remove the rear universal joint bolts and straps **(see illustration)**. Turn the driveshaft (or tires) as necessary to bring the bolts into

the most accessible position. **Note:** *When necessary, wrap a piece of tape from the top of one U-joint bearing cap to the tops of the other U-joint bearing caps. to prevent the caps from coming off during removal.*

6   Lower the rear of the driveshaft and then slide the front out of the transmission or transfer case.

7   To prevent loss of transmission fluid and protect against contamination while the driveshaft is out, insert a plastic extension housing cap, or wrap a plastic bag over the transmission or transfer case housing and hold it in place with a rubber band.

Installation

8   Remove the plastic bag from the transmission or transfer case and wipe the area clean. Inspect the oil seal carefully. If it's damaged or worn, replace it (see Chapter 7, Part A).

9   Slide the front of the driveshaft into the transmission (2WD models) or transfer case (4WD models).

10   Raise the rear of the driveshaft into position, checking to be sure the marks are in alignment. If not, turn the rear wheels to match the pinion flange and the driveshaft.

11   Remove the tape securing the bearing caps and install new straps and bolts. Tighten the bolts to the torque listed in this Chapter's Specifications.

12   If the driveshaft is equipped with a dust boot around the yoke at the front of the drive-

shaft, use a special tool to tighten the dust boot clamp.

13   Remove the jackstands and lower the vehicle.

Front driveshaft (4WD models only)

Refer to illustration 9.14

14   Using white paint, chalk or a scribe, mark the relationship of the CV joints to the companion flanges at the transfer case and front differential **(see illustration)**.

15   Unbolt the front and rear driveshaft CV-joints from the companion flanges at the transfer case and front differential **(see illustration 9.14)**.

16   Push the driveshaft forward and detach the rear CV-joint from the transfer case companion flange.

17   Detach the front CV-joint from the differential companion flange, then tilt the front of the driveshaft downward. Pull the driveshaft forward and remove it from the vehicle.

18   Installation is the reverse of removal. Clean the bolt threads and apply Loctite®242 to the threads during installation. Be sure to tighten all fasteners to the torque listed in this Chapter's Specifications.

9.14 On 4WD models, mark the CV-joints and companion flanges on both ends of the front driveshaft before removing the CV-joint mounting bolts

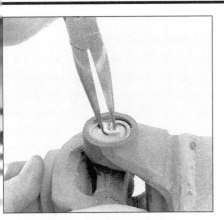

10.2 A pair of needle-nose pliers can be used to remove the universal joint snap-rings

10.4 To press the universal joint out of the driveshaft yoke, set it up in a vise with the small socket pushing the joint and bearing cap into the large socket

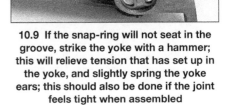

10.9 If the snap-ring will not seat in the groove, strike the yoke with a hammer; this will relieve tension that has set up in the yoke, and slightly spring the yoke ears; this should also be done if the joint feels tight when assembled

10 Universal joints - replacement

Refer to illustrations 10.2, 10.4 and 10.9
Note 1: *A press or large vise will be required for this procedure. It may be a good idea to take the driveshaft to a repair or machine shop where the universal joints can be replaced for you, normally at a reasonable charge.*
Note 2: *On 4WD models, the CV-joints on the front driveshaft cannot be repaired or replaced. If either of the joints is worn or damaged, replace the front driveshaft assembly (see Section 9).*

1   Remove the driveshaft (see Section 9).

U-joint (rear drive shaft)

2   Using a small pair of pliers, remove the snap-rings from the spider **(see illustration)**.
Note: *If it is difficult to remove the snap ring, gently tap the snap-ring/bearing cap with a drift or socket to loosen the snap-ring.*
3   Supporting the driveshaft, place it in position on either an arbor press or on a workbench equipped with a vise.
4   Place a piece of pipe or a large socket with the same inside diameter over one of the bearing caps. Position a socket which is of slightly smaller diameter than the cap on the

opposite bearing cap **(see illustration)** and use the vise or press to force the cap out (inside the pipe or large socket), stopping just before it comes completely out of the yoke. Use the vise or large pliers to work the cap the rest of the way out.
5   Transfer the sockets to the other side and press the opposite bearing cap out in the same manner.
6   Pack the new universal joint bearings with grease. Ordinarily, specific instructions for lubrication will be included with the universal joint servicing kit and should be followed carefully.
7   Position the spider in the yoke and partially install one bearing cap in the yoke. If the replacement spider is equipped with a grease fitting, be sure it's offset in the proper direction (toward the driveshaft).
8   Start the spider into the bearing cap and then partially install the other cap. Align the spider and press the bearing caps into position, being careful not to damage the dust seals.
9   Install the snap-rings. If difficulty is encountered in seating the snap-rings, strike the driveshaft yoke sharply with a hammer. This will spring the yoke ears slightly and allow the snap-rings to seat in the groove **(see illustration)**.

10   Install the grease fitting and fill the joint with grease. Be careful not to overfill the joint, as this could blow out the grease seals.
11   Install the driveshaft (see Section 9).

11 Pinion oil seal - replacement

Refer to illustrations 11.6a and 11.6b
1   A pinion shaft oil seal failure results in the leakage of differential gear lubricant past the seal and onto the driveshaft yoke or flange. The seal is replaceable without removing or disassembling the differential.
2   Loosen the wheel lug nuts, raise the vehicle and support it securely on jackstands. Chock the wheels at the opposite end to prevent the vehicle from rolling.
3   Remove the wheels and brake drums or brake calipers and discs (see Chapter 9).
4   Remove the driveshaft (see Section 9).
5   Rotate the yoke at least ten times, then, using an inch-pound torque wrench, measure and record the torque required to turn the pinion.
6   Mark the relationship of the pinion shaft yoke to the pinion shaft **(see illustration)**. Remove the pinion shaft nut while preventing

11.6a Mark the relationship of the pinion shaft to the pinion shaft yoke

11.6b Hold the pinion shaft yoke stationary while removing the nut

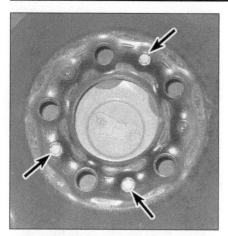

12.1 When necessary, remove the center cap screws from the back side of the wheel, then reinstall the wheel (4WD models only)

12.2 On 4WD models, before raising the vehicle, loosen the driveaxle/hub nut

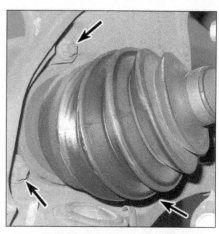

12.9a Unbolt the hub assembly from the steering knuckle, then carefully tap the hub assembly out of the steering knuckle

the yoke from turning, using a special tool (available at most auto parts stores) or a chain wrench **(see illustration)**.

7   Remove the yoke from the shaft, using a puller if necessary.

8   Place an oil drain pan under the pinion seal area. Using a large screwdriver or seal removal tool, carefully pry the pinion oil seal out of the differential housing. Be careful not to damage the splines of the pinion shaft.

9   Lubricate the new seal lip with multi-purpose grease or differential lubricant and carefully place it in position in the differential housing. Using a hammer and a seal driver or a large socket with a diameter slightly smaller than that of the new seal, drive the seal into place.

10   Clean the sealing lip contact surface of the pinion shaft yoke. Apply a thin coat of multi-purpose grease to the seal contact surface and the shaft splines and, aligning the matchmarks made prior to removal, install the yoke onto the shaft. If necessary, use the old nut to pull the yoke into place. **Caution:** *The manufacturer recommends replacing the nut with a new one whenever it has been removed.*

11   Install the washer and a new pinion shaft nut. **Caution:** *Tighten the nut just enough to eliminate all endplay in the pinion shaft. The following are the pinion nut MINIMUM-torque values:*

   a) *200 ft-lbs (271 Nm) on the 198RBI rear differential*
   b) *210 ft-lbs (285 Nm) on the 8-1/4 rear differential*
   c) *160 ft-lbs (216 Nm) on the 186FIA front differential*

12   Turn the pinion shaft yoke several times to seat the bearings. Using an inch-pound torque wrench, measure the torque required to turn the pinion shaft. Tighten the nut in 5 ft-lb (6.8 Nm) increments until the torque recorded in Step 5 plus an additional 5 in-lbs (0.56 Nm) is reached. **Caution:** *Be very careful not to exceed the torque reading from this Step, which applies the proper preload to the*

pinion bearings. The following are the pinion nut MAXIMUM-torque values:

   a) *350 ft-lbs (474 Nm) on the 198RBI rear differential*
   b) *260 ft-lbs (352 Nm) on the 186FIA front differential*

If the preload or the maximum torque is exceeded, the differential will have to be disassembled and a new collapsible spacer installed. DO NOT loosen the pinion nut once the tightening process has begun.

13   Check the differential lubricant level (see Chapter 1).

14   Install the driveshaft (see Section 9).

15   Install the brake drums or brake discs and calipers (see Chapter 9). Install the wheels and lug nuts. Lower the vehicle and tighten the lug nuts to the torque listed in the Chapter 1 Specifications.

16   Test drive the vehicle and check around the pinion shaft yoke for evidence of leakage.

---

12   Hub and bearing assembly (front) - removal and installation

---

Note: *The hub and bearing assembly cannot be disassembled. If either the hub or bearing is damaged or worn, the entire assembly must be replaced.*

Removal

Refer to illustrations 12.1, 12.2, 12.9a and 12.9b

1   If you're working on a 4WD model, remove the wheel center cap/hubcap (if equipped). **Note:** *When necessary, remove the wheel to access the center cap screws, then reinstall the wheel* **(see illustration)**.

2   On 4WD models, loosen the driveaxle/hub nut 1/4-turn **(see illustration)**.

3   Loosen the front wheel lug nuts.

4   Raise the front of the vehicle and support it securely on jackstands.

5   Remove the wheel.

6   Remove the disc brake caliper, caliper mounting bracket and brake disc (see Chapter 9).

12.9b If the driveaxle sticks in the hub splines, push it out of the hub with a puller

7   Remove the ABS wheel speed sensor and the sensor wire bracket.

8   On 4WD models, remove the axle hub nut.

9   Unbolt the hub assembly from the steering knuckle and tap it out of the knuckle bore **(see illustration)**. **Note:** *On 4WD models, if the driveaxle sticks in the hub splines, push it out of the hub with a puller* **(see illustration)**.

Installation

10   Using sandpaper or emery cloth, clean the opening in the steering knuckle to remove any rust or dirt that may be present.

11   Install the hub assembly onto the steering knuckle and tighten the hub-to-knuckle bolts to the torque listed in this Chapter's Specifications.

12   On 4WD models, install the hub nut and tighten it securely.

13   The remainder of installation is the reverse of removal.

14   Lower the vehicle, then tighten the hub nut to the torque listed in this Chapter's Specifications.

15   Tighten the lug nuts to the torque listed in the Chapter 1 Specifications.

13.4 Disconnect the driveaxle from the steering knuckle by swinging the steering knuckle outward and pushing the driveaxle inward

13.5 Pry the driveaxle from the axle assembly with a large screwdriver or prybar (arrow)

13.8 After expanding the snap-ring, the outer joint assembly can be removed

13 Driveaxle - removal, boot replacement and installation (front) (4WD models only)

Removal

Refer to illustrations 13.4 and 13.5

1 Remove the driveaxle/hub nut **(see illustrations 12.1 and 12.2)**.
2 Remove the stabilizer bar link (see Chapter 10).
3 Detach the lower ball joint from the control arm and the tie-rod end from the steering knuckle (see Chapter 10).
4 Swing the steering knuckle outward and push the end of the driveaxle out away of the knuckle **(see illustration)**.
5 Using a large screwdriver or prybar, pry the driveaxle from the differential **(see illustration)**. **Note:** *On the right side, an axleshaft will remain inside of the axle housing tube.*

Boot replacement

Note: *The constant velocity joints are not repairable - if one becomes worn, noisy or breaks, the entire joint must be replaced. The joint can and should, however, be disassem-*

bled, cleaned, inspected and repacked with grease in the event of a boot failure.
6 Mount the driveaxle in a soft-jaw vise and support the CV-joint. The jaws of the vise must be lined with wood or rags to avoid damage to the driveaxle.
7 Cut both of the CV-joint dust boot retaining clamps and discard them. Slide the boot back onto the driveaxle. **Note:** *To install the new clamps, a special pair of clamp-tightening pliers will be needed (most auto parts stores carry them)* **(see illustration 13.23)**.

Outer CV-joint

Refer to illustrations 13.8, 13.9, 13.10a, 13.10b, 13.11, 13.12, 13.13a, 13.13b, 13.16, 13.17, 13.18, 13.22 and 13.23

8 Clean the grease off of the CV-joint and spread the snap-ring located in the inner bearing race, **(see illustration)** then slide the CV-joint off of the driveaxle. Discard the dust boot.
9 Remove the driveaxle from the vise. Place the joint assembly in the vise, with the stub axle pointing down. Again, wood or rags must line the vise jaws to prevent marring the stub axle surface. Mark the outer bearing race, bearing cage and inner race with paint or a marker **(see illustration)**.

13.9 After removing the grease, mark the bearing housing, cage and inner race

10 Tap the inner race with the brass punch to angle it far enough to allow a ball bearing to be removed **(see illustration)**. Repeat this until all the balls are removed. If the balls are stuck, a screwdriver can be used to pry them from the cage **(see illustration)**.
11 With all of the balls removed, tilt the inner race and cage assembly 90-degrees, align the cage windows with the outer race (joint housing) lands and remove the assembly from the outer race **(see illustration)**.

13.10a Tilt the inner race far enough to allow ball removal - a brass punch can be used if the inner race is difficult to move

13.10b If necessary, pry the ball bearings out with a screwdriver

13.11 Tilt the inner race and cage 90-degrees, then align the windows in the cage with the lands and rotate the inner race up and out of the outer race

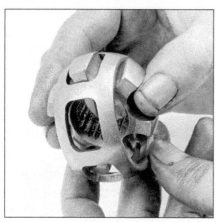

13.12 Align the inner race lands with the cage windows and rotate the inner race out of the cage

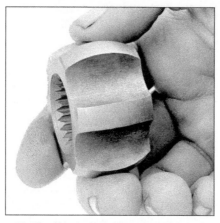

13.13a Check the inner race lands and grooves for pitting and score marks

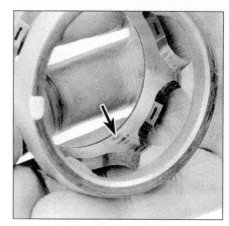

13.13b Check the cage for cracks, pitting and score marks (shiny spots are normal and don't affect operation)

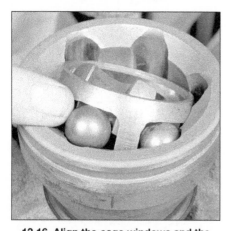

13.16 Align the cage windows and the inner and outer race grooves, then tilt the cage and inner race to insert the balls

13.17 Apply grease through the splined hole, then insert a wooden dowel into the hole and push down - the dowel will force grease into the joint

13.18 Wrap the driveaxle splines with electrical tape to prevent damaging the boot as it is slid onto the shaft

12   Align the inner race lands with the cage windows and rotate the inner race out of the cage **(see illustration)**.
13   Wash all of the components in solvent. Inspect the cage and races for pitting, score marks, cracks and other signs of wear and damage **(see illustrations)**. Shiny, polished spots are normal and don't affect CV joint operation.
14   Install the inner race in the cage by reversing the technique described in Step 12.
15   Install the inner race and cage assembly in the outer race by reversing the removal method used in Step 11. The small diameter of the cage must face out and the stopping groove in the inner race must face in.
16   Tilt the cage and inner race, then press the balls into the cage windows **(see illustration)**.
17   Pack the CV joint assembly with half of the lubricant supplied in the boot/joint kit through the inner splined hole. Force the grease into the bearing by inserting a wooden dowel through the splined hole and pushing it to the bottom of the joint **(see illustration)**. Repeat this procedure until the joint is thoroughly packed (the other half of the grease is

to be placed in the boot). **Caution:** *Use CV joint grease only.*
18   Wrap the driveaxle splines with tape to avoid damaging the boot **(see illustration)**. Place the small boot clamp in the groove on the small end of the boot, then slide the boot onto the driveaxle. Remove the tape and smear the remainder of the grease into the boot.
19   Install the snap-ring in the inner race.
20   Slide the CV-joint onto the driveaxle until the snap-ring seats into the driveaxle groove.
21   Install the large boot clamp in its groove in the boot and slide the boot over the outer race (joint housing).
22   Insert a small screwdriver between the boot and the housing to equalize the pressure inside the boot **(see illustration)**.
23   Tighten both boot clamps **(see illustration)**.
24   Installation is the reverse of removal.

Inner CV-joint
Refer to illustration 13.25
25   Clean the grease off of the CV-joint and remove the snap-ring from the inside the outer housing **(see illustration)**.

26   Slide the outer housing off of the drive-axle, then remove the balls from the cage.
27   With all of the balls removed, tilt the cage 30-degrees, then slide the cage off of the inner race and down onto the driveaxle.

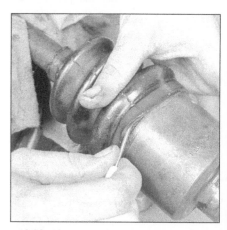

13.22 After installing the boot, equalize the pressure inside the boot by inserting a small, dull screwdriver between the boot and the CV-joint housing

13.23 A tool like this, available at most auto parts stores, is required for tightening the boot clamps

13.25 After cleaning the grease from the CV-joint, remove the snap-ring from inside the housing

14.3 If you are just replacing the oil seal, pry the seal out of the axle housing with a seal removal tool or a prybar

28 Spread the snap-ring located in the inner race and slide the inner race off of the drive-axle.

29 Slide the dust boot off of the driveaxle and discard the dust boot.

30 Wash all of the components in solvent. Inspect the cage and races for pitting, score marks, cracks and other signs of wear and damage **(see illustrations 13.13a and 13.13b)**. Shiny, polished spots are normal and don't affect CV joint operation.

31 Wrap the driveaxle splines with tape to avoid damaging the boot **(see illustration 13.18)**. Place the small boot clamp in the groove on the small end of the boot, then slide the boot onto the driveaxle. Remove the tape.

32 Slide the cage onto the driveaxle. **Note:** *The end of the cage with the smaller diameter should face toward the dust boot.*

33 Slide the inner race onto the driveaxle until the snap-ring seats into the driveaxle groove.

34 Apply the lubricant supplied in the boot/joint kit onto the races, cage and balls. **Caution:** *Use CV-joint grease only.*

35 Tilt the cage 30-degrees, then slide the cage onto the inner race.

36 Insert the balls.

37 Place the large boot clamp in the groove on the large end of the boot.

38 Slide the housing/outer race onto the cage/bearing assembly, then install the hous-ing/outer race snap-ring. **Caution:** *Make sure the snap-ring is properly seated into the groove in the outer race.*

39 Pack grease into the bearing assembly. Repeat this procedure until the joint is thoroughly packed. Apply the remaining grease into the dust boot. **Caution:** *Use CV-joint grease only.*

40 Slide the boot over the housing.

41 Move the joint in and out, then position it midway through its travel. Insert a small screwdriver between the boot and the housing to equalize the pressure inside the boot **(see illustration 13.22)**.

42 Tighten both boot clamps **(see illustration 13.23)**.

43 Installation is the reverse of removal.

---

14 Axleshaft, seal and bearing (front) (4WD models only) - removal and installation

Note 1: *If the bearing is being replaced, the seal must also be replaced.*

Note 2: *The axleshaft inside the right-side axle housing must be removed to access the seal and/or bearing on that side. To remove the right-side axleshaft, it is necessary to use a slide hammer and a special axleshaft puller, or a similar tool that clamps over the end of the axleshaft. If a tool cannot be purchased or borrowed, the work for the right-side should be done by a dealer service department or other qualified repair shop.*

Removal

Refer to illustrations 14.3, 14.4 and 14.5

1 Remove the driveaxle (see Section 13).

2 On the right side, remove the axleshaft:

a) *Remove the snap-ring from the end of the axleshaft.*

b) *Install the special axleshaft puller tool over the end of the axleshaft.*

c) *Attach a slide hammer onto the axleshaft puller tool, then remove the axleshaft.*

3 If you are just replacing the oil seal, pry the seal out of the axle housing with a seal removal tool or a prybar **(see illustration)**. If you are replacing the bearing, skip to the next step - the seal will come out as you remove the bearing.

4 A bearing puller which grips the bearing from behind **(see illustration)** will be required for bearing removal. Attach a slide hammer to the puller and extract the bearing from the axle housing.

5 Clean out the bearing recess and drive in the new bearing with a bearing installer or a large socket positioned against the outer bearing race **(see illustration)**. Make sure the bearing is tapped in to the full depth of the recess

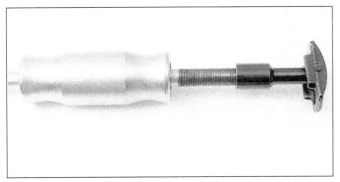

14.4 Using the slide hammer and puller attachment, remove the axleshaft bearing

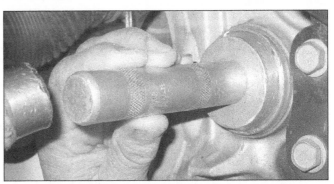

14.5 Install the axleshaft seal and bearing with a bearing installer or large socket

15.10a After supporting the axle assembly with a floor jack, remove the bracket-to-frame bolts/nuts from the right side of the axle assembly . . .

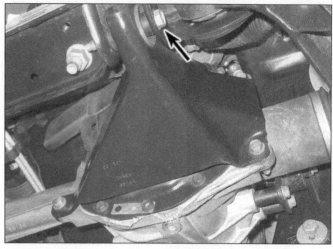

15.10b . . . the front of the axle assembly . . .

and the numbers on the bearing are visible from the outer end of the axle housing.

6   Apply high-temperature grease to the oil seal recess and tap the new seal evenly into place with a hammer and seal installation tool or a large socket so the lips are facing in and the metal face is visible from the end of the axle housing. When correctly installed, the face of the oil seal should be flush with the end of the axle housing.

7   Lubricate the seal lip and the bore in the axle bearing with clean differential lubricant.

8   If it was removed, install the right-side axleshaft.

9   The remainder of installation is the reverse of removal.

10   Lower the vehicle and tighten the lug nuts to the torque listed in the Chapter 1 Specifications.

15   Axle assembly (front) (4WD models only) - removal and installation

Removal

Refer to illustrations 15.10a, 15.10b and 15.10c

1   Raise the vehicle and support it securely on jackstands.

2   On 2007 and earlier models, remove the lower control arms (see Chapter 10).

3   On 2008 and later models, remove the stabilizer bar links, the shock absorber-to-control arm bolts and the steering knuckles (see Chapter 10).

4   Remove the driveaxles (see Section 13).

5   If equipped, remove the front axle skid plate.

6   Drain the differential lubricant (see Chapter 1).

7   Detach the vent hose from the differential cover.

8   Remove the front driveshaft from the differential companion flange (see Section 9).

9   Support the front axle assembly with a floor jack placed underneath the differential.

10   Remove the bracket-to-frame bolts/nuts

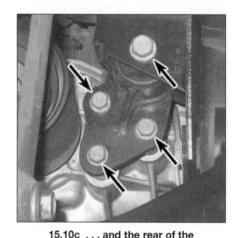

15.10c . . . and the rear of the axle assembly

on the right side, front and rear of the axle housing **(see illustrations)**.

11   Slowly lower the axle assembly to the ground and remove the assembly out from under the vehicle.

12   If necessary, remove the brackets from the axle housing.

Installation

13   Installation is the reverse of removal.

14   Tighten the axle assembly fasteners to the torque listed in this Chapter's Specifications.

15   Refill the differential with lubricant (see Chapter 1).

16   Lower the vehicle and tighten the lug nuts to the torque listed in the Chapter 1 Specifications.

17   Check the vehicle wheel alignment.

16   Axleshaft, seal and bearing (rear) - removal and installation

Note: *2002 models are equipped with either the 198RBI or the 8-1/4 rear differential. 2003 and later models are equipped with the 8-1/4 rear differential. The 198RBI differential can be*

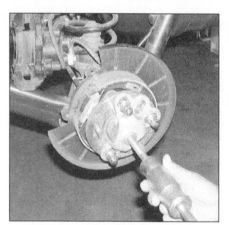

16.4 After removing the axle retaining plate fasteners, pull the axle from the housing with a slide hammer and adapter connected to the axle flange (typical)

identified by the four bolts and axle plate on each side, which retain the axleshaft. On the 8-1/4 differential, the differential cover must be removed to allow removal of the axle(s).

198RBI differential models

Removal

Refer to illustration 16.4

1   Loosen the wheel lug nuts, raise the rear of the vehicle and support it securely on jackstands. Chock the front wheels to prevent the vehicle from rolling. Remove the wheel.

2   Remove the brake drum (see Chapter 9).

3   Remove the nuts, located behind the axle flange, which secures the axle retaining plate to the axle housing. **Note:** *There is a circular access hole(s) in the axle flange. Rotate the axleshaft as necessary to access each nut.*

4   Connect a slide hammer and adapter to the axle flange and pull the axle from the housing **(see Illustration)**.

5   If the bearing or seal must be replaced, take the assembly to an automotive machine shop to have the old bearing and seal removed and new ones installed.

16.15 On the 8-1/4 differential, remove the lock bolt securing the pinion mate shaft (typical)

16.16 Carefully remove the pinion mate shaft from the differential case (don't turn the differential case after the shaft has been removed, or the spider gears may fall out)

Installation

6   Wipe the bearing bore in the axle housing clean. Pack the bearing with wheel bearing grease and apply a thin coat of grease to the outer surface of the bearing.

7   Guide the axleshaft straight into the axle housing. Rotate the shaft slightly to engage the splines on the shaft with the splines in the differential side gear.

8   Install the axle retaining plate studs and nuts and tighten them to the torque listed in this Chapter's Specifications.

9   The remainder of installation is the reverse of removal.

10   Check the differential lubricant level and add some, if necessary.

11   Install the wheel and lug nuts. Lower the vehicle and tighten the lug nuts to the torque listed in the Chapter 1 Specifications.

8-1/4 differential models

Removal

Refer to illustrations 16.15, 16.16, 16.17 and 16.19

12   Loosen the wheel lug nuts, raise the rear of the vehicle and support it securely on jack-stands. Chock the front wheels to prevent the vehicle from rolling. Remove the wheel.

13   Remove the brake drum or brake disc (see Chapter 9). Don't disconnect the brake hose from the caliper. Hang the caliper with a piece of wire - don't let it hang by the hose. If equipped, remove the ABS wheel speed sensor (so it doesn't get damaged).

14   Remove the cover from the differential carrier and allow the oil to drain into a container (see Chapter 1).

15   Remove the lock bolt that secures the differential pinion mate shaft **(see illustration)**.

16   Remove the pinion mate shaft **(see illustration)**. **Caution:** *Do not rotate the differential with the shaft removed or the spider gears may be dislodged.*

17   Push the outer (flanged) end of the axleshaft in and remove the C-lock from the inner end of the shaft **(see illustration)**. Withdraw the axleshaft.

18   If you are just replacing the oil seal, pry the seal out of the axle housing with a seal removal tool or a prybar (if you are replacing the bearing, skip to the next step - the seal will come out as you remove the bearing).

19   A bearing puller which grips the bearing from behind will be required for bearing removal **(see illustration 14.4a)**. Attach a slide hammer to the puller and extract the bearing from the axle housing **(see illustration)**.

Installation

20   Clean out the bearing recess and drive in the new bearing with a bearing installer or a large socket positioned against the outer bearing race. Make sure the bearing is tapped in to the full depth of the recess and the numbers on the bearing are visible from the outer end of the axle housing.

21   Apply high-temperature grease to the oil seal recess and tap the new seal evenly into place with a hammer and seal installation tool or a large socket so the lips are facing in and the metal face is visible from the end of the axle housing. When correctly installed, the face of the oil seal should be flush with the end of the axle housing.

22   Lubricate the seal lip and the bore in the axle bearing with clean differential lubricant.

23   Install the axleshaft and C-lock, then insert the pinion mate shaft. Apply a non-hardening thread-locking compound to the threads of the lock bolt, then tighten the lock

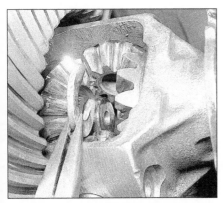

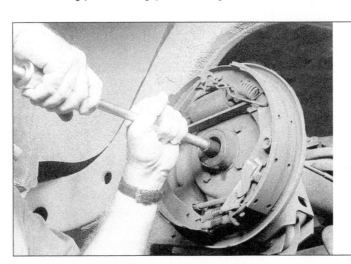

16.19 Using the slide hammer and puller attachment to remove the axleshaft bearing

16.17 Push the axle flange in, then remove the C-lock from the inner end of the axleshaft

bolt to the torque listed in this Chapter's Specifications.

24 Clean off all traces of sealant or old gasket material from the differential cover and rear axle housing, then apply a bead of RTV sealant to the cover. Install the cover and bolts, tightening the bolts in a criss-cross pattern to the torque listed in the Chapter 1 Specifications.

25 Refill the differential with the correct grade of lubricant (see Chapter 1).

26 The remainder of installation is the reverse of removal.

27 Install the wheel and lug nuts, lower the vehicle and tighten the lug nuts to the torque listed in the Chapter 1 Specifications.

17 Axle assembly (rear) - removal and installation

1 Loosen the rear wheel lug nuts, raise the vehicle and support it securely on jackstands placed underneath the frame.

2 Remove the wheels.

3 Support the rear axle assembly with a floor jack placed underneath the differential. If you have two floor jacks, use both of them; one placed under each axle tube.

4 Detach the shock absorbers from the axle assembly brackets and compress the shocks to get them out of the way (see Chapter 10).

5 Mark the driveshaft and disconnect it from the differential pinion shaft yoke and hang the rear of the driveshaft from the underbody with a piece of wire (see Section 9).

6 Unbolt the stabilizer bar from the lower control arms (see Chapter 10).

7 On 2008 and later models, disconnect the track bar from the axle housing (see Chapter 10).

8 Disconnect the brake sensor, if equipped, then remove the brake caliper, brackets and rotor or brake (see Chapter 9). **Note:** *The parking brake is self-adjusting, and the cables must be locked out before removing them* (see Chapter 9, **illustration 13.4**).

9 Disconnect the parking brake cables from the equalizer (see Chapter 9).

10 Disconnect the flexible brake hose from the fitting where it meets the rigid brake line.

Plug the end of the line and hose or wrap plastic bags tightly around the open fittings to prevent excessive fluid loss and contamination.

11 Remove the upper control arm ball joint pinch bolt (see Chapter 10).

12 Detach the control arms from the axle assembly brackets (see Chapter 10).

13 Detach the vent hose from the axle tube.

14 Lower the axle assembly enough to remove the rear coil springs (see Chapter 10).

15 Lower the jack(s) and move the axle assembly out from under the vehicle.

16 Installation is the reverse of removal. Tighten all fasteners to their proper torque values. **Note:** *Don't tighten the control arm-to-axle housing fasteners until the vehicle is sitting at normal ride height (this can be simulated by supporting the vehicle with jackstands placed under the rear axle tubes).*

17 Bleed the brake system (see Chapter 9). (It will probably only be necessary to bleed the rear brakes if you plugged the brake line right after it was disconnected.)

18 Tighten the wheel lug nuts to the torque listed in the Chapter 1 Specifications.

Notes

Notes

Chapter 9   Brakes

Contents

Specifications

General
Brake fluid type .... See Chapter 1

Disc brakes

Disc minimum thickness	Stamped or cast into disc
Maximum allowable disc thickness variation	
Front	
2002	No more than 0.0005 inch point to point
2003 through 2007	No more than 0.0006 inch point to point
2008 and later	No more than 0.00035 inch point to point
Rear	
2007 and earlier	No more than 0.0007 inch point to point
2008 and later	No more than 0.00035 inch point to point
Disc runout (maximum)	
Front	
2002	0.005 inch
2003 through 2007	0.004 inch
2008 and later	0.0008 inch
Rear	
2007 and earlier	0.004 inch
2008 and later	0.0008 inch
Brake pad minimum thickness	See Chapter 1

Drum brakes

Drum maximum diameter	Stamped or cast into drum
Minimum brake lining thickness	See Chapter 1

Torque specifications

	Ft-lbs (unless otherwise indicated)	Nm

Note: *One foot-pound (ft-lb) of torque is equivalent to 12 inch-pounds (in-lbs) of torque. Torque values below approximately 15 ft-lbs are expressed in inch-pounds, because most foot-pound torque wrenches are not accurate at these smaller values.*

	Ft-lbs	Nm
Brake hose-to-caliper banjo bolt (front and rear)	23	31
Caliper adapter (mounting bracket) bolts		
Front	100	136
Rear		
2007 and earlier	100	136
2008 and later	75	102
Caliper-to-adapter mounting bolts		
Front		
2007 and earlier	132 in-lbs	15
2008 and later	28	38
Rear		
2007 and earlier	18	24
2008 and later	28	38
Caliper mounting bolts		
Front	132 in-lbs	15
Rear	18	24
Master cylinder mounting nuts	18	24
Power brake booster mounting nuts		
2002	29	39
2003 and later	18	24
Wheel cylinder mounting bolts	84 in-lbs	9.5
Wheel lug nuts	See Chapter 1	
Wheel speed sensor mounting bolt		
Front		
2002	132 in-lbs	15
2003 through 2007	120 in-lbs	13.5
2008 and later	71 in-lbs	8
Rear	80 in-lbs	9

1 General information

The vehicles covered by this manual are equipped with hydraulically operated front and rear brake systems. The front brakes are ventilated discs; the rear brakes are either drums (2002 models) or solid discs (2003 and later models). Both the front and rear brakes are self-adjusting. The front and rear disc brakes automatically compensate for pad wear, while the rear drum brakes incorporate an adjustment mechanism, which is activated as the brakes are applied when the vehicle is stopped in reverse. Anti-lock brakes are optional on most models.

Hydraulic circuits

The hydraulic system consists of two separate circuits; if a leak or failure occurs in one hydraulic circuit, the other circuit will remain operative. The master cylinder has separate reservoirs for each circuit. A visual warning of circuit failure or air in the system is given by a warning light activated by displacement of the piston in the pressure differential switch portion of the combination valve (on non-ABS systems) from its normal "in balance" position.

Master cylinder/reservoir assembly

The master cylinder used on all models is an aluminum type with a nylon reservoir. It is not rebuildable. If found to be defective, it must be replaced. The reservoir grommets are replaceable, however.

Combination valve

The combination valve, used on non-ABS models, performs two functions. A proportioning valve controls front-to-rear brake action during rapid stops.

A brake pressure differential switch continuously compares the front and rear brake pressures. If a failure occurs somewhere in the system, a red warning light on the instrument cluster comes on. Once a failure has occurred, the warning light remains on until the system has been repaired.

On ABS-equipped vehicles, the ABS unit performs the functions of the combination valve.

Power brake booster

A dual-diaphragm, vacuum-operated power brake booster, mounted between the firewall and the master cylinder, utilizes engine manifold vacuum and atmospheric pressure to provide assistance to the hydraulic brake system.

Parking brake system

The parking brake system consists of a lever between the front seats, the cables connecting this lever to the rear brakes, an equalizer for balancing the tension of the two rear cables, and the parking brake shoes. On all 2002 models with rear drums, the parking brake system utilizes the brake shoes themselves. On 2003 and later models with rear discs, small parking brake shoes are housed inside the rear disc assemblies.

Service

After completing any operation involving disassembly of any part of the brake system, always test drive the vehicle to check for proper braking performance before resuming normal driving. When testing the brakes, perform the tests on a clean, dry flat surface. Conditions other than these can lead to inaccurate test results.

Test the brakes at various speeds with both light and heavy pedal pressure. The vehicle should stop evenly without pulling to one side or the other. Avoid locking the brakes because this slides the tires and diminishes braking efficiency and control of the vehicle.

Tires, vehicle load and front-end alignment are factors which also affect braking performance.

2 Anti-lock Brake System (ABS) - general information

The optional Anti-lock Brake System (ABS) is designed to maintain steering control, directional stability and optimal deceleration under severe braking conditions, regardless of the road surface. It does so by monitoring the rotational speed of the wheels and controlling the brake line pressure to each wheel during braking. This prevents the wheels from locking-up, which means the vehicle is easier to control during panic stops.

Components

Control Unit (CAB)

The Control Unit (CAB), which is attached to the hydraulic control unit, is the "brain" of the ABS system. Both units are located in the left side of the engine compartment. The CAB has a microprocessor that receives sensor signals that are processed and compared. A self-check program illuminates an amber warning light on the instrument cluster when a system fault is detected. Fault codes are stored in a diagnostic program memory, but cannot be accessed without a DRB scan tool. This tool is usually only available at dealer service departments.

Hydraulic control unit (HCU)

The hydraulic control unit (HCU) consists of a valve body and a pump/motor assembly. The valve body contains three electrically-operated solenoid valves - one for each of the front wheels and one for both rear wheels - that modulate brake hydraulic pressure when the brakes are applied. During braking, these solenoid valves are cycled on and off rapidly and continuously by the CAB, modulating brake hydraulic pressure as necessary to control wheel lock-up.

The pump/motor assembly supplies the extra fluid needed during anti-lock braking. The pump is connected to the master cylinder reservoir by supply and return hoses, and is operated by an integral DC electric motor which is controlled by the CAB.

Wheel speed sensors

Wheel speed sensors generate small electrical pulsations when the wheels are turning, sending a signal to the CAB, indicating wheel rotational speed.

The front wheel speed sensors are mounted at the steering knuckles in close relationship to the brake discs.

On 2005 and earlier models, the rear sensor is installed in the top center of the rear axle housing. It measures the speed of the rear wheels, but not individually. On 2006 and later models two rear wheel speed sensors are used, mounted at the end of each axle tube.

Diagnosis and repair

If a dashboard warning light comes on and stays on while the vehicle is in operation, the ABS system requires attention. Although a DRB scan tool is necessary to properly diagnose the ABS system (and is required to access any trouble codes that may have been stored), you can perform a few preliminary checks before taking the vehicle to a dealer who is equipped with this tester.

a) *Check the brake fluid level in the master cylinder reservoir.*
b) *Verify that the electrical connector for the CAB is plugged in.*
c) *Check the electrical connector at the pump motor.*
d) *Check the ABS fuses (see Chapter 12).*
e) *Follow the wiring harness to each wheel and check that all connections are secure and that the wiring is not damaged.*

If the above preliminary checks do not rectify the problem, the vehicle should be diagnosed by a dealer service department or other repair facility equipped with the necessary scan tool. Due to the rather complex nature of this system and the high operating pressures involved, all actual repair work must be done by a dealer service department or other qualified repair shop.

3 Disc brake pads - replacement

Refer to illustrations 3.5a through 3.5m and 3.6a through 3.6n

Warning: *Disc brake pads must be replaced on both front wheels or both rear wheels at the same time - never replace the pads on only one wheel. Also, the dust created by the brake system is harmful to your health. Never blow it out with compressed air and don't inhale any of it. An approved filtering mask should be worn when working on the brakes. Do not, under any circumstances, use petro-*

leum-based solvents to clean brake parts. Use brake system cleaner only!

Note: *This procedure applies to both the front and the rear disc brakes.*

1   Remove the cap from the brake fluid reservoir.

2   Loosen the wheel lug nuts, raise the front of the vehicle and support it securely on jackstands.

3   Remove the front or rear wheels. Work on one brake assembly at a time, using the assembled brake for reference if necessary.

4   Inspect the brake disc carefully as outlined in Section 5. If machining is necessary, follow the information in that Section to remove the disc.

5   If you're replacing the front brake pads, follow the photo sequence beginning with **illustration 3.5a**. Be sure to stay in order and read the caption under each illustration.

3.5a Before removing the front caliper, wash off all traces of brake dust with brake system cleaner

3.5b Insert a screwdriver through the caliper window and pry the caliper outward to depress the piston - make sure the fluid in the master cylinder reservoir doesn't overflow

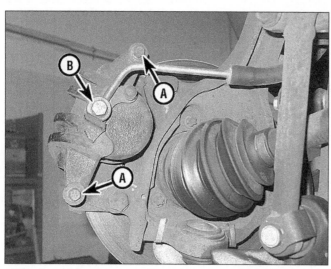

3.5c To remove the caliper, remove the two mounting bolts (A); don't remove the brake hose-to-caliper banjo bolt (B) unless you intend to replace the caliper or hose

3.5d Remove the front caliper; this is a good time to check for fluid leakage around the caliper piston boot, which would indicate the need for replacement

3.5e Whenever you have to let go of the caliper, hang it from the control arm (for front brakes) or the spring (for rear brakes) with a piece of wire - DON'T let it hang by the brake hose!

3.5f Remove the brake pads from the caliper adapter

3.5g Pull out both pins and inspect them for corrosion and wear; if either pin is damaged, replace it

6   If you're replacing the rear brake pads, follow the photo sequence beginning with **illustration 3.6a**. Be sure to stay in order and read the caption under each illustration.

7   When reinstalling the caliper, be sure to tighten the mounting bolts to the torque listed in this Chapter's Specifications. Tighten the wheel lug nuts to the torque in the Chapter 1 Specifications. After the job has been completed, firmly depress the brake pedal a few times to bring the pads into contact with the disc.

8   Check for fluid leakage and make sure the brakes operate normally before driving in traffic.

3.5h If either rubber dust boot for the pins is cracked, torn or dried out, replace it

3.5i Coat the pins with high-temperature grease . . .

3.5j . . . then install them in the caliper adapter

3.5k Before installing the new brake pads, depress the piston all the way into the caliper to facilitate installation of the new pads over the disc (front brake shown); again, make sure the fluid in the master cylinder reservoir doesn't overflow

3.5l Install both new front pads into the caliper adapter

3.5m Install the caliper over the disc and the new pads, install the caliper bolts and tighten them to the torque listed in this Chapter's Specifications

3.6a On the rear brakes, start by washing the caliper with brake system cleaner, then using a large C-clamp to compress the piston into its bore - make sure the fluid in the master cylinder reservoir doesn't overflow

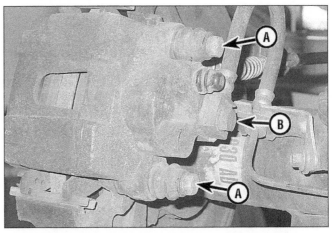

3.6b Remove the caliper mounting bolts (A). Don't remove the brake hose-to-caliper banjo bolt (B) unless you intend to replace the caliper or hose

3.6c Swing the caliper upwards to remove it from the disc -
on later models, remove the anti-rattle springs at
each end of the caliper

3.6d Wire the caliper to the rear coil spring to avoid
stressing the brake hose

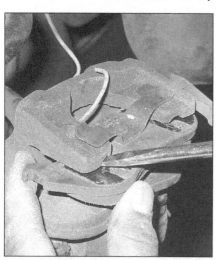

3.6e Pry the outer pad from the rear
caliper with a screwdriver

3.6f The inner pads have prongs that clip
into the piston

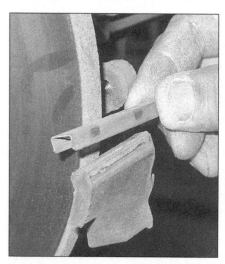

3.6g Remove the anti-rattle clips from the
caliper mounting bosses

3.6h Remove the bushings
from the caliper

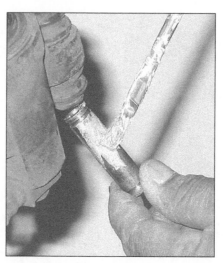

3.6i Apply anti-seize to the bushings after
cleaning them (also clean and lubricate
inside the bushings)

3.6j After careful cleaning, install the
anti-rattle clips

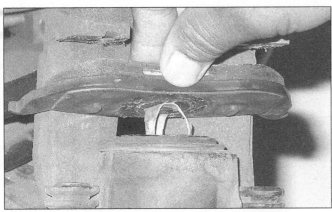

3.6k Install the inner pad, clipping it into the piston

3.6l Align the protrusions on the outer brake pad with the slots in the caliper frame, then push the pad into the caliper until it seats properly

3.6m Engage the upper part of the brake pads with the upper anti-rattle clip (A) and swing the rear of the caliper down (B)

3.6n Install the rear caliper bolts and tighten them to the torque listed in this Chapter's Specifications

4   Disc brake caliper - removal and installation

Warning: *Dust created by the brake system is harmful to your health. Never blow it out with compressed air and don't inhale any of it. An approved filtering mask should be worn when working on the brakes. Do not, under any circumstances, use petroleum-based solvents to clean brake parts. Use brake system cleaner only!*

Note: *If replacement is indicated (usually because of fluid leakage) explore all options before beginning the job. New and factory rebuilt calipers are available on an exchange basis, which makes this job quite easy. Always replace the calipers in pairs (on the front or rear) - never replace just one of them.*

Removal

1   Remove as much fluid as possible from the reservoir with a suction gun, large syringe or a poultry baster. **Warning:** *If a poultry baster is used, never again use it for the preparation of food. Brake fluid is poisonous - never siphon it by mouth.* Place rags under the brake line fittings and prepare caps or plastic bags to cover the ends of the lines once they're disconnected. **Caution:** *Brake fluid will damage*

paint. Cover all painted surfaces and avoid spilling fluid during this procedure.
2   Loosen the wheel lug nuts, raise the vehicle and support it securely on jackstands. Remove the wheels.
3   Push the piston back into the caliper bore **(see illustration 3.5b or 3.6a)**.
4   To disconnect the brake hose from the caliper, remove the brake hose-to-caliper banjo bolt. **Note:** *Do not disconnect the brake hose from the caliper if you are only removing (not replacing) the caliper.* Have a rag handy to catch spilled fluid and immediately wrap a plastic bag tightly around the end of the hose to prevent fluid loss and contamination.
5   Unscrew the two caliper mounting bolts and detach the caliper **(see illustration 3.5c or 3.6b)**. If you're removing the caliper for access to other components, hang it with a piece of wire from the control arm or the coil spring - DON'T let it hang by the brake hose **(see illustration 3.5e or 3.6d)**.

Installation

6   Inspect the caliper mounting bolts for excessive corrosion, replacing them if necessary.
7   Clean and lubricate the caliper mounting bolts.
8   Install the caliper, tightening the mount-

ing bolts to the torque listed in this Chapter's Specifications.
9   Using new sealing washers, install the brake hose and banjo bolt. Tighten the bolt to the torque listed in this Chapter's Specifications.
10   Bleed the brakes (see Section 10).
11   Install the wheels and lower the vehicle. Tighten the wheel lug nuts to the torque listed in the Chapter 1 Specifications.
12   After the job has been completed, firmly depress the brake pedal a few times to bring the pads into contact with the disc.
13   Check brake operation before driving the vehicle in traffic.

5   Brake disc - inspection, removal and installation

Inspection

Refer to illustrations 5.4a, 5.4b, 5.5a and 5.5b
1   Loosen the wheel lug nuts, raise the vehicle and support it securely on jackstands. Remove the wheel and install the lug nuts to hold the disc in place. **Note:** *If the lug nuts don't contact the disc when screwed on all the way, install washers under them.* If you're working on a rear disc brake, release the parking brake.

5.4a Use a dial indicator to check disc runout - if the reading exceeds the specified allowable runout limit, the disc will have to be machined or replaced

5.4b Using a swirling motion, remove the glaze from the disc surface with sandpaper or emery cloth

4   To check disc runout, place a dial indicator at a point about 1/2-inch from the outer edge of the disc **(see illustration)**. Set the indicator to zero and turn the disc. The indicator reading should not exceed the specified allowable runout limit. If it does, the disc should be refinished by an automotive machine shop. **Note:** *Professionals recommend resurfacing of brake discs regardless of the dial indicator reading (to produce a smooth, flat surface that will eliminate brake pedal pulsations and other undesirable symptoms related to questionable discs). At the very least, if you elect not to have the discs resurfaced, deglaze them with sandpaper or emery cloth (use a swirling motion to ensure a non-directional finish)* **(see illustration)**.

5   The disc must not be machined to a thickness less than the specified minimum refinish thickness. The minimum (or discard) thickness is cast into the disc **(see illustration)**. The disc thickness can be checked with a micrometer **(see illustration)**.

Removal

Refer to illustrations 5.7a and 5.7b

6   Remove the lug nuts you installed to hold the disc in place.

7   On front discs, remove the wheel stud clips (if present) and the caliper adapter **(see illustration)**. On rear discs, remove the assembly-line wheel stud clips and discard them **(see illustration)**.

8   Slide the disc off the hub.

Installation

9   Place the disc over the wheel studs on the hub. On front discs, install the caliper adapter, tightening the bolts to the torque listed in this Chapter's Specifications.

10   Install the brake pads and caliper (see Section 3). Tighten the caliper mounting bolts to the torque listed in this Chapter's Specifications.

11   Install the wheel, then lower the vehicle to the ground. Tighten the wheel lug nuts to the torque listed in the Chapter 1 Specifications.

5.5a The minimum thickness is cast into the inside of the disc on most models (typical)

2   Remove the brake caliper (see Section 4). It isn't necessary to disconnect the brake hose. After removing the caliper mounting bolts, suspend the caliper out of the way with a piece of wire. Don't let the caliper hang by the hose and don't stretch or twist the hose.

3   Visually check the disc surface for score marks and other damage. Light scratches

and shallow grooves are normal after use and may not always be detrimental to brake operation, but deep score marks require disc removal and refinishing by an automotive machine shop. Be sure to check both sides of the disc. If pulsating has been noticed during application of the brakes, suspect disc runout. Be sure to check the wheel bearings to make sure they're not worn.

5.5b Use a micrometer to measure disc thickness at several points

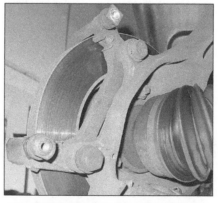

5.7a On front brakes, the caliper adapter (mounting bracket) will have to be removed to allow removal of the disc - here are the front caliper mounting bracket-to-steering knuckle bolts

5.7b Discs are often held on with assembly-line clips. It is not necessary to reinstall them

12   Depress the brake pedal a few times to bring the brake pads into contact with the disc. Bleeding of the system will not be necessary unless the brake hose was disconnected from the caliper. Check the operation of the brakes carefully before placing the vehicle into normal service.

6   Drum brake shoes - replacement

Refer to illustrations 6.4a through 6.4w and 6.5

Warning: *Drum brake shoes must be replaced on both wheels at the same time - never replace the shoes on only one wheel. Also, the dust created by the brake system is harmful to your health. Never blow it out with compressed air and don't inhale any of it. An approved filtering mask should be worn when working on the brakes. Do not, under any circumstances, use petroleum-based solvents to clean brake parts. Use brake system cleaner only!*

Caution: *Whenever the brake shoes are replaced, the retracting and hold-down springs should also be replaced. Due to the*

6.4a Before removing any internal drum brake components, wash them off with brake cleaner and allow them to dry - position a drain pan under the brake to catch the residue - DO NOT USE COMPRESSED AIR TO BLOW THE BRAKE DUST FROM THE PARTS!

continuous heating/cooling cycles to which the springs are subjected, they lose their tension over a period of time and may allow the shoes to drag on the drum and wear at a much faster rate than normal.

1   Loosen the wheel lug nuts, raise the rear of the vehicle and support it securely on jackstands. Block the front wheels to keep the vehicle from rolling.

2   Release the parking brake.

3   Remove the wheel and brake drum (see Chapter 1). **Note:** *All four rear brake shoes must be replaced at the same time, but to avoid mixing up parts, work on only one brake assembly at a time.*

4   Follow the accompanying illustrations **(6.4a through 6.4w)** for the replacement of the brake shoes. Be sure to stay in order and

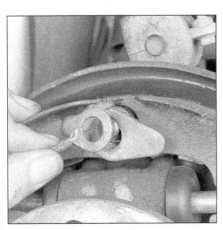

6.4b Pry out the adjuster lever and turn the star wheel to retract the brake shoes

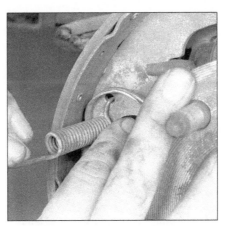

6.4c Unhook the adjuster cable from the lever

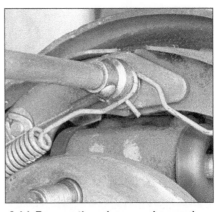

6.4d Remove the primary and secondary shoe retracting springs - the spring removal tool shown here can be purchased at most auto parts stores and greatly simplifies this step

6.4e Remove the self-adjuster cable and anchor pin plate from the anchor pin

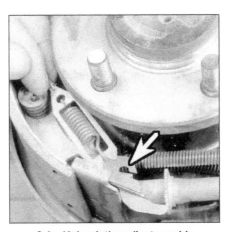

6.4f Remove the secondary shoe retracting spring and cable guide from the secondary shoe

6.4g Remove the secondary shoe hold-down spring and pin . . .

6.4h ... then lift the secondary shoe and adjusting screw from the backing plate

6.4i Remove the parking brake link

read the caption under each illustration. **Note:** *If the brake drum cannot be easily pulled off, make sure the parking brake is completely released, then apply some penetrating oil at the hub-to-drum joint. Allow the oil to soak*

in and try to pull the drum off. If the drum still cannot be pulled off, the brake shoes will have to be retracted. This is accomplished by first removing the plug from the backing plate. With the plug removed, push

the lever off the adjusting star wheel with one narrow screwdriver while turning the adjusting wheel with another narrow screwdriver, moving the shoes away from the drum. The drum should now come off.

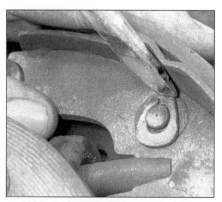

6.4j Remove the primary shoe hold-down spring and pin then lift the shoe from the backing plate - pry the parking brake lever retaining clip off of the pivot pin . . .

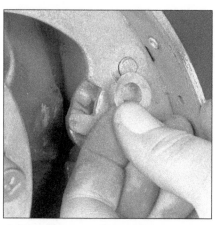

6.4k ... then separate the lever from the primary shoe - be careful not to lose the spring washer

6.4l Lubricate the brake shoe contact areas with high-temperature grease

6.4m Attach the new shoe to the parking brake lever, install the spring washer and retaining clip on the pivot pin, then crimp the clip closed with a pair of pliers

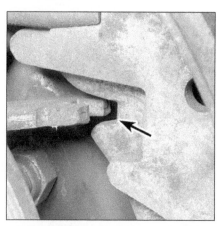

6.4n Install the primary shoe and hold-down spring to the backing plate, then position the end of the parking brake link into the notch (arrow)

6.4o Lubricate the adjusting screw threads and socket with high-temperature grease

6.4p Place the secondary shoe against the backing plate, then install the hold-down spring - make sure the parking brake strut and wheel cylinder pushrods engage in the brake shoe slots (arrows)

6.4q Install the anchor pin plate . . .

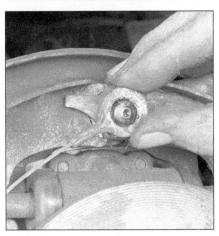

6.4r . . . and the self-adjuster cable

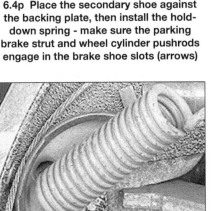

6.4s Hook the end of the secondary shoe retractor spring through the cable guide and into the hole in the shoe, then stretch the spring over the anchor pin

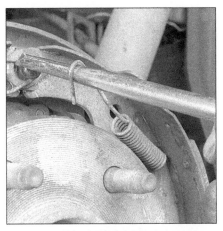

6.4t Install the primary shoe retractor spring - the tool shown here is available at most auto parts stores and makes this step much easier and safer

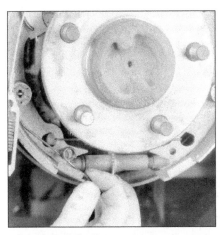

6.4u Install the lower return spring and the adjuster lever spring

5   Before reinstalling the drum it should be checked for cracks, score marks, deep scratches and hard spots, which will appear as small discolored areas. If the hard spots

cannot be removed with fine emery cloth or if any of the other conditions listed above exist, the drum must be taken to an automotive machine shop to have it resurfaced. **Note:** *Professionals recommend resurfacing*

the drums whenever a brake job is done. Resurfacing will eliminate the possibility of out-of-round drums. If the drums are worn so much that they can't be resurfaced without exceeding the maximum allowable diam-

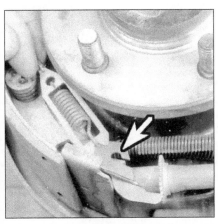

6.4v Install the adjuster lever and connect the cable to it

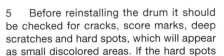

6.4w Wiggle the assembly to ensure the shoes are centered on the backing plate

6.5 The maximum allowable drum diameter is cast into the drum (typical)

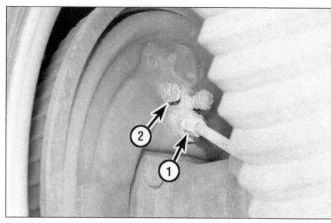

7.3 Completely loosen the brake line fitting (1), then remove the two wheel cylinder mounting bolts (2) (the front bolt isn't visible in this photo)

eter (cast into the drum) **(see illustration),** then new ones will be required. At the very least, if you elect not to have the drums resurfaced, remove the glazing from the surface with sandpaper or emery cloth using a swirling motion.

6   Once the new shoes are in place, install the drums on the axle flanges. Remove the rubber plugs from the brake backing plates.

7   Insert a narrow screwdriver or brake adjusting tool through the adjustment hole and turn the star wheel until the brakes drag slightly as the drum is turned.

8   Turn the star wheel in the opposite direction until the shoes don't drag and the drum turns freely.

9   Repeat the adjustment on the opposite wheel and install the backing plate plugs.

10   Mount the wheel, install the lug nuts, then lower the vehicle. Tighten the lug nuts to the torque listed in the Chapter 1 Specifications.

11   Make a number of alternating forward and reverse stops to allow the brakes to further adjust themselves.

12   Check brake operation before driving the vehicle in traffic.

7   Wheel cylinder - removal and installation

Note: If replacement is indicated (usually because of fluid leakage or sticky operation), it is recommended that new wheel cylinders be installed. Never replace only one wheel cylinder - always replace both of them at the same time.

Removal

Refer to illustration 7.3

1   Loosen the rear lug nuts. Raise the rear of the vehicle and support it securely on jackstands. Block the front wheels to keep the vehicle from rolling. Remove the rear wheels

2   Remove all dirt and foreign material from around the wheel cylinder.

3   Completely loosen the brake line fitting using a flare wrench if possible **(see illustration).** Don't pull the brake line away from the wheel cylinder.

4   Remove the wheel cylinder mounting bolts.

5   Remove the brake drum. Disconnect the brake return springs and disengage the shoes from the wheel cylinder pushrods.

6   Detach the wheel cylinder from the brake backing plate. Immediately plug the brake line to prevent fluid loss and contamination. **Note:** If the brake shoe linings are contaminated with brake fluid, install new brake shoes.

Installation

7   Place the wheel cylinder in position and loosely connect the brake line to it, being careful not to cross-thread the fitting.

8   Install the bolts and tighten them to the torque listed in this Chapter's Specifications. Tighten the brake line fitting securely. Install the brake shoe springs and engage the shoes with the wheel cylinder pushrods.

9   Install the brake drum and wheel.

10   Bleed the brakes (see Section 10).

11   Check brake operation before driving the vehicle in traffic.

8   Master cylinder - removal, installation and reservoir/ grommet replacement

Removal

Refer to illustration 8.3

1   Remove as much fluid as possible from the reservoir with a suction gun, large syringe or a poultry baster. **Warning:** If a poultry baster is used, never again use it for the preparation of food. Place rags under the brake line fittings and prepare caps or plastic bags to cover the ends of the lines once they're disconnected. **Caution:** Brake fluid will damage paint. Cover all painted surfaces and avoid spilling fluid during this

procedure. Brake fluid can be siphoned out of the reservoir using a squeeze bulb, but wear safety goggles.

2   Loosen the tube nuts at the ends of the brake lines where they enter the master cylinder. To prevent rounding off the flats on these nuts, a flare-nut wrench, which wraps around the nut, should be used. Pull the brake lines away from the master cylinder slightly and plug the ends to prevent contamination.

3   Disconnect the electrical connector for the fluid level switch **(see illustration).**

4   Remove the two master cylinder mounting nuts. Remove the master cylinder from the booster, taking care not to kink the hydraulic lines.

5   Remove the reservoir cap, then discard any fluid remaining in the reservoir.

Installation

Refer to illustrations 8.9 and 8.15

Warning: Non-ABS systems can be bled in the conventional manner. ABS systems may require the HCU to be bled. To bleed air from the hydraulic control unit, the manufacturer states that the system must be bled using the

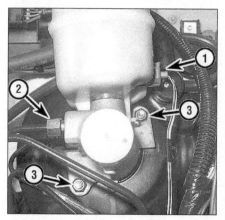

8.3 Master cylinder mounting details

1   Electrical connector for fluid level sensor

2   Brake line fitting

3   Mounting nuts

8.9 The best way to bleed air from the master cylinder before installing it on the vehicle is with a pair of bleeder tubes that direct brake fluid into the reservoir during bleeding

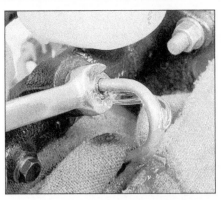

8.15 Have an assistant depress the brake pedal and hold it down, then loosen the fitting nut, allowing air and brake fluid to escape; repeat this procedure on both fittings until the fluid is clear of air bubbles

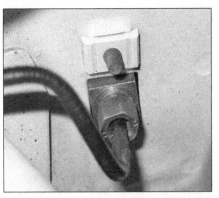

9.2 Working in the engine compartment, unscrew the brake line from the hose fitting, using a flare-nut wrench . . .

conventional method (as described in Section 10), then with the use of a DRB scan tool to purge any air from the HCU. After that has been done the system must be bled using the conventional method once again. This tool is very expensive and generally available only to dealership service departments. If, after bleeding the system you do not have a firm brake pedal, or if the ABS light on the instrument panel does not go off, or if you have any doubts whatsoever about the effectiveness of the brake system, have it towed to a dealer service department or other repair shop equipped with the necessary tools for bleeding the system.

6   Bench bleed the new master cylinder before installing it. Mount the master cylinder in a vise, with the jaws of the vise clamping on the mounting flange.

7   Attach a pair of master cylinder bleeder tubes to the outlet ports of the master cylinder.

8   Fill the reservoir with brake fluid of the recommended type (see Chapter 1).

9   Slowly push the pistons into the master cylinder (a large Phillips screwdriver can be used for this) - air will be expelled from the pressure chambers and into the reservoir **(see illustration)**. Because the tubes are submerged in fluid, air can't be drawn back into the master cylinder when you release the pistons.

10   Repeat the procedure until no more air bubbles are present.

11   Remove the bleed tubes, one at a time, and install plugs in the open ports to prevent fluid leakage and air from entering. Install the reservoir cap.

12   Install the master cylinder over the studs on the power brake booster and tighten the attaching nuts only finger tight at this time.

13   Thread the brake line fittings into the master cylinder. Since the master cylinder is still a bit loose, it can be moved slightly in order for the fittings to thread in easily. Do not strip the threads as the fittings are tightened.

14   Tighten the mounting nuts to the torque listed in this Chapter's Specifications, then the brake line fittings.

15   Fill the master cylinder reservoir with fluid, then bleed the master cylinder. **Caution:** *Have plenty of rags on hand to catch the fluid - brake fluid will ruin painted surfaces. After the bleeding procedure is completed, rinse the area under the master cylinder with clean water.* To bleed the cylinder on the vehicle, have an assistant depress the brake pedal and hold the pedal to the floor. Loosen the fitting nut to allow air and fluid to escape **(see illustration)**. Repeat this procedure on both fittings at the master cylinder until the fluid is clear of air bubbles, then bleed the combination valve (on non-ABS models).

16   Test the operation of the brake system carefully before placing the vehicle into normal service. **Warning:** *Do not operate the vehicle if you are in doubt about the effectiveness of the brake system.*

Reservoir/grommet replacement

Note: *The brake fluid reservoir can be replaced separately from the master cylinder body if it becomes damaged. If there is leakage between the reservoir and the master cylinder body, the grommets in the master cylinder can be replaced.*

17   Remove as much fluid as possible from the reservoir with a suction gun, large syringe or a poultry baster. **Warning:** *If a poultry baster is used, never again use it for the preparation of food.*

18   Place rags under the master cylinder to absorb any fluid that may spill out once the reservoir is detached from the master cylinder. **Caution:** *Brake fluid will damage paint. Cover all body parts and be careful not to spill fluid during this procedure.*

19   Disconnect the electrical wiring from the reservoir.

20   Remove the reservoir mounting bolt.

21   Pull the reservoir out of the master cylinder body.

22   If you are simply replacing the grommets, remove them from the master cylinder and install new ones. **Caution:** *Use only your*

fingers to press the grommets into place. A tool of any kind may damage the grommets.

23   Lubricate the reservoir O-rings with clean brake fluid, then insert the reservoir into place on the master cylinder body. Press it down while wiggling it back and forth.

24   Install the mounting bolt.

25   Connect the wiring to the reservoir.

26   Refill the reservoir with the recommended brake fluid (see Chapter 1) and check for leaks.

27   If air has entered the system, bleed the master cylinder **(see illustration 8.15)**.

9   Brake hoses and lines - inspection and replacement

Inspection

1   About every six months, with the vehicle raised and supported securely on jackstands, the flexible hoses that connect the steel brake lines with the front and rear brake assemblies should be inspected for cracks, chafing of the outer cover, leaks, blisters and other damage. These are important and vulnerable parts of the brake system and inspection should be complete. A light and mirror will be helpful for a thorough check. If a hose exhibits any of the above conditions, replace it with a new one.

Replacement

2007 and earlier Liberty models

Front brake hose

Refer to illustrations 9.2 and 9.4

2   Working in the engine compartment, disconnect the brake line from the hose fitting, being careful not to bend the brake line **(see illustration)**. Use a flare nut wrench, if available.

3   Raise the front of the vehicle and support it securely on jackstands. Remove the banjo bolt from the brake caliper **(see illustration 3.6b)** and separate the hose from the caliper. Discard the sealing washers.

9.4 . . . then, working in the fenderwell, unbolt the brake line from the inner fender panel

9.11 To disconnect a rear brake hose from its metal line, unscrew the line and then remove the bolt from the fitting

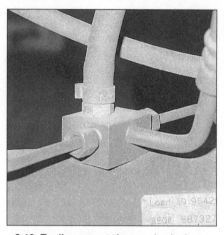

9.13 To disconnect the rear brake hose from the axle, unscrew the two brake line fittings from the junction block, then unscrew the vent hose bolt

4   Remove the bolt that attaches the hose to the vehicle **(see illustration)**. Remove the hose.

5   To install the hose, first attach it to the caliper, using new sealing washers on both sides of the fitting. Tighten the banjo fitting bolt to the torque listed in this Chapter's Specifications.

6   Without twisting the hose, connect the brake line to the hose fitting, but don't tighten it yet.

7   Install the bolt retaining the hose to the frame, tightening it securely.

8   Tighten the brake line fitting securely.

9   When the brake hose installation is complete, there should be no kinks in the hose. Make sure the hose doesn't contact any part of the suspension. Check this by turning the wheels to the extreme left and right positions. If the hose makes contact, remove it and correct the installation as necessary. Bleed the system (see Section 10).

Rear brake hose

Refer to illustrations 9.11 and 9.13

10   Raise the vehicle and support it securely on jackstands.

11   Disconnect the steel brake line from the hose **(see illustration)**.

12   Unbolt the hose bracket from the body.

13   At the rear of the hose, disconnect the vent tube from the hose end **(see illustration)**. This will expose the mounting bolt.

14   Disconnect the two steel brake lines from the hose end, using a flare wrench if possible.

15   Unbolt the hose from the rear axle housing and remove it.

16   Installation is the reverse of removal. Before final tightening of the fittings, check that the hose is not twisted or kinked. Tighten the steel brake lines securely.

17   Fill the master cylinder reservoir and bleed the system (see Section 10).

Metal brake lines

18   When replacing brake lines be sure to use the correct parts. Don't use copper tubing for any brake system components. Purchase steel brake lines from a dealer or auto parts store.

19   Prefabricated brake line, with the tube ends already flared and fittings installed, is available at auto parts stores and dealers. These lines can be bent to the proper shapes using a tubing bender.

20   When installing the new line, make sure it's securely supported in the brackets and has plenty of clearance between moving or hot components.

21   After installation, check the master cylinder fluid level and add fluid as necessary. Bleed the brake system as outlined in Section 10 and test the brakes carefully before driving the vehicle in traffic.

2008 and later Liberty models/all Nitro models

Refer to illustrations 9.30, 9.32, 9.46 and 9.47

Note: *On these models the flexible brake lines are crimped to the rigid metal lines and are not replaceable separately - the entire line must be replaced.*

22   Depress the brake pedal a couple of inches and support it in that position with a prop rod.

Left front brake hose/line

23   Using a flare-nut wrench, disconnect the left front brake line from the ABS Hydraulic Control Unit (HCU).

24   Loosen the left front wheel lug nuts, raise the front of the vehicle and support it securely on jackstands, then remove the wheel.

25   Detach the ABS wheel speed sensor harness from the brake hose, then remove the clip securing the hose/line to the bracket on the chassis **(see illustration 9.32)**.

26   Unscrew the brake hose banjo bolt and detach the hose from the caliper. Obtain new sealing washers for installation.

27   Guide the brake line/hose out from the fenderwell.

28   Installation is the reverse of removal. Be sure to use new sealing washers when connecting the hose to the caliper, and tighten the banjo bolt to the torque listed in this Chapter's Specifications.

29   Bleed the hydraulic system (see Section 10).

Right front brake hose/line

30   Using a flare-nut wrench, disconnect the right front brake line from the ABS Hydraulic Control Unit (HCU), then free the line retainers from the threaded studs on the firewall **(see illustration)**.

31   Loosen the right front wheel lug nuts, raise the front of the vehicle and support it securely on jackstands, then remove the wheel.

32   Detach the ABS wheel speed sensor harness from the brake hose, then remove the clip securing the hose/line to the bracket on the chassis **(see illustration)**.

33   Unscrew the brake hose banjo bolt and detach the hose from the caliper. Obtain new sealing washers for installation.

34   Guide the brake line/hose out from the fenderwell and into the engine compartment.

35   Installation is the reverse of removal. Be sure to use new sealing washers when connecting the hose to the caliper, and tighten the banjo bolt to the torque listed in this Chapter's Specifications.

36   Bleed the hydraulic system (see Section 10).

Left rear brake hose/line

37   Relieve the fuel system pressure (see Chapter 4), then disconnect the cable from the negative terminal of the battery (see Chapter 5).

38   Using a flare-nut wrench, disconnect the left front brake line from the ABS Hydraulic Control Unit (HCU).

39   Raise the rear of the vehicle and support it securely on jackstands.

40   Remove the exhaust system and the fuel tank (see Chapter 4).

41   Unscrew the brake hose banjo bolt and detach the hose from the caliper. Obtain new sealing washers for installation.

9.30 Right front brake line details in the engine compartment

1 Line fitting at the ABS HCU

2 Retainers

9.32 Front brake hose details

1 Banjo bolt

2 ABS harness-to-hose retainers

3 Hose/line retaining clip

42 Free the brake line retaining clips from the studs on the body and guide the front of the line down and out of the engine compartment.

43 Installation is the reverse of removal. Be sure to use new sealing washers when connecting the hose to the caliper, and tighten the banjo bolt to the torque listed in this Chapter's Specifications.

44 Bleed the hydraulic system (see Section 10).

Right rear brake hose/line

45 Raise the rear of the vehicle and support it securely on jackstands.

46 Unscrew the brake hose banjo bolt and detach the hose from the caliper **(see illustration)**. Obtain new sealing washers for installation.

47 Using a pair of Ifare-nut wrenches, unscrew the fitting connecting the hose/line to the line leading to the engine compartment **(see illustration)**.

48 Unbolt the retaining bracket from the body.

49 Installation is the reverse of removal. Be sure to use new sealing washers when connecting the hose to the caliper, and tighten the banjo bolt to the torque listed in this Chapter's Specifications.

50 Bleed the hydraulic system (see Section 10).

10 Brake system bleeding

Refer to illustration 10.10

Warning 1: *The following procedure is a manual bleeding procedure. This is the only bleeding procedure that can be performed at home without special tools. However, on ABS-equipped vehicles if air has found its way into the hydraulic control unit, the entire system must be bled manually, then with a DRB scan tool, then manually a second time. If the brake pedal feels "spongy" even after bleeding the brakes, or the ABS light on the instrument panel does not go off, or if you*

have any doubts whatsoever about the effectiveness of the brake system, have the vehicle towed to a dealer service department or other repair shop equipped with the necessary tools for bleeding the system.

Warning 2: *Wear eye protection when bleeding the brake system. If the fluid comes in contact with your eyes, immediately rinse them with water and seek medical attention.*

Note: *Bleeding the hydraulic system is necessary to remove any air that manages to find its way into the system when it's been opened during removal and installation of a hydraulic component.*

1 It will be necessary to bleed the complete system if air has entered the system due to low fluid level, or if the brake lines have been disconnected at the master cylinder.

2 If a brake line was disconnected only at a wheel, then only that caliper or wheel cylinder must be bled.

3 If a brake line is disconnected at a fitting located between the master cylinder and any

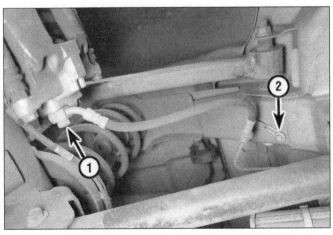

9.46 Right rear brake hose/line details

1 Banjo bolt

2 Retaining bracket

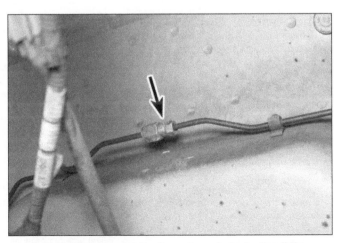

9.47 Right rear brake line/hose-to-main brake line fitting

10.10 When bleeding the brakes, a hose is connected to the bleeder valve at the caliper or wheel cylinder and then submerged in brake fluid. Air will be seen as bubbles in the tube and container. All air must be expelled before moving to the next wheel

11.10 Pry off the clip retaining the brake booster pushrod to the brake pedal

of the brakes, that part of the system served by the disconnected line must be bled. The following procedure describes bleeding the entire system, however.

4   Remove any residual vacuum from the brake power booster by applying the brake several times with the engine off.

5   Remove the cap from the master cylinder reservoir and fill the reservoir with brake fluid. Reinstall the cap(s). **Note:** *Check the fluid level often during the bleeding operation and add fluid as necessary to prevent the fluid level from falling low enough to allow air bubbles into the master cylinder.*

6   Have an assistant on hand, as well as a supply of new brake fluid, a clear container partially filled with clean brake fluid, a length of clear tubing to fit over the bleeder valve and a wrench to open and close the bleeder valve.

7   Begin the bleeding process by bleeding the master cylinder **(see illustration 8.15)**.

8   Using the same technique, bleed the lines at the combination valve (non-ABS models only) and where the lines enter the ABS HCU (ABS models only).

9   Moving to the right rear wheel, loosen the bleeder valve slightly, then tighten it to a point where it is snug but can still be loosened quickly and easily.

10   Place one end of the hose over the bleeder valve and submerge the other end in the brake fluid in the container **(see illustration)**.

11   Have the assistant push the brake pedal slowly to the floor, then hold the pedal firmly depressed.

12   While the pedal is held depressed, open the bleeder valve just enough to allow a flow of fluid to leave the valve. Watch for air bubbles to exit the submerged end of the tube. When the fluid flow slows after a couple of seconds, close the valve and have your assistant release the pedal.

13   Repeat Steps 11 and 12 until no more air is seen leaving the tube, then tighten the bleeder valve and proceed to the left rear

wheel, the right front wheel and the left front wheel, in that order, and perform the same procedure. Be sure to check the fluid in the master cylinder reservoir frequently.

14   Never use old brake fluid. It contains moisture, which can boil, rendering the brakes inoperative.

15   Refill the master cylinder with fluid at the end of the operation.

16   Check the operation of the brakes. The pedal should feel solid when depressed, with no sponginess. If necessary, repeat the entire process. **Warning:** *If, after bleeding the system you do not have a firm brake pedal, or if the ABS light on the instrument panel does not go off, or if you have any doubts whatsoever about the effectiveness of the brake system, have the vehicle towed to a dealer service department or other repair shop equipped with the necessary tools for bleeding the system.*

---

11   Power brake booster - check, removal and installation

Operating check

1   Depress the pedal and start the engine. If the pedal goes down slightly, operation is normal.

2   Depress the brake pedal several times with the engine running and make sure that there is no change in the pedal reserve distance.

Airtightness check

3   Start the engine and turn it off after one or two minutes. Depress the brake pedal several times slowly. If the pedal goes down farther the first time but gradually rises after the second or third depression, the booster is airtight.

4   Depress the brake pedal while the engine is running, then stop the engine with

the pedal depressed. If there is no change in the pedal reserve travel after holding the pedal for 30 seconds, the booster is airtight.

Removal

Refer to illustration 11.10

5   Refer to Section 8 and remove the master cylinder.

6   Disconnect the vacuum hose from the booster.

7   Remove any interfering steel brake lines.

8   On ABS-equipped vehicles, remove the ABS hydraulic control unit from its mount and move it aside for clearance.

9   Working inside the vehicle, refer to Chapter 11 and remove the steering column lower cover.

10   Working on the brake pedal arm, remove the brake pushrod clip and slip the pushrod off of the pin **(see illustration)**. Remove and discard the brake light switch (see Section 12).

11   Remove the four nuts holding the brake booster to the inside of the firewall. You may need a light to see these, as they are up under the dash area.

12   Slide the booster straight out from the firewall until the studs clear the holes and pull the booster, brackets and gaskets from the engine compartment area.

Installation

13   Installation is the reverse of removal. Be sure to tighten the booster mounting nuts to the torque listed in this Chapter's Specifications.

---

12   Brake light switch - check, replacement and adjustment

Refer to illustration 12.1

Warning: *The models covered by this manual are equipped with a Supplemental Restraint System (SRS), more commonly known as air-*

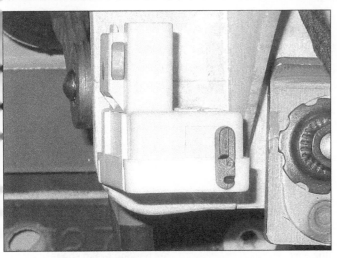

12.1 The brake light switch one-time adjustment lever is located on the end opposite the plunger

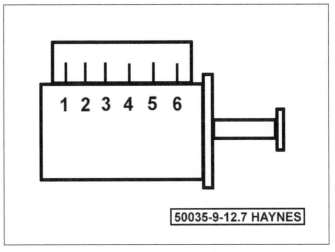

50035-9-12.7 HAYNES

12.7 Brake light switch terminal guide

bags. *Always disarm the airbag system before working in the vicinity of any airbag system component to avoid the possibility of accidental deployment of the airbag, which could cause personal injury (see Chapter 12).*

1   The brake light switch is mounted on the steering column bracket at the brake pedal **(see illustration)**. The switch adjusts itself automatically when it is installed. It cannot be repaired or adjusted - it must be replaced if it is defective. If it has been removed for any reason it must also be replaced.

2   The brake light switch also incorporates circuits that supply information about the brake pedal position to the Powertrain Control Module as well as the cruise control module.

Check

Refer to illustration 12.7

3   Verify that the brake lights come on when the brake pedal is depressed.

4   If the brake lights do not come on when the pedal is depressed, check the fuse and the brake light bulbs (see Chapter 12).

5   To check the switch, refer to Chapter 11 and remove the lower trim panel from the dash.

6   Disconnect the wiring from the brake light switch.

7   Use an ohmmeter to check the continuity of the terminals of the switch. With the plunger extended, there should be continuity between terminals 1 and 2. With the plunger in, there should be continuity between terminal 3 and 4 and also between terminals 5 and 6 **(see illustration)**. If the switch fails any of these tests it must be replaced.

8   If the switch is operating properly but the brake lights are inoperative, locate the open circuit condition between the switch and the brake lights.

Replacement

9   Remove the driver's knee bolster from the dash (see Chapter 11).

10   Disconnect the cable from the negative terminal of the battery (see Chapter 5).

11   Unplug the switch electrical connector.

12   Rotate the switch counterclockwise so that its tabs index with the slots on the mounting bracket. Pull the switch to the rear and remove it.

13   Throw the switch away as it cannot be reused.

14   Push down on the brake pedal and hold it until the new switch is locked in place.

15   Insert the new switch, making sure it is in place against the bracket.

16   Twist the switch clockwise 30-degrees until it locks into position. Keep the brake pedal depressed until the plunger adjustment is made.

17   Locate the adjustment release lever on the switch **(see illustration 12.1)**. Turn it in a clockwise direction until it locks. It should be parallel to the wiring connector. The adjustment procedure is now complete. **Note:** *This is a one-time adjustment procedure. Once done, it cannot be readjusted. It must be replaced.*

18   The remainder of the installation is the reverse of removal.

19   Depress the brake pedal and verify that the switch is operating properly.

13   Parking brake shoes (2003 and later models) - replacement

Refer to illustrations 13.4 and 13.5a through 13.5o

Warning: *Dust created by the brake system is harmful to your health. Never blow it out with compressed air and don't inhale any of it. An approved filtering mask should be worn when working on the brakes. Do not, under any circumstances, use petroleum-based solvents to clean brake parts. Use brake system cleaner only!*

Note: *On 2007 and later models, the brake*

shoe assembly has been rotated 90-degrees from the illustrations shown here.

1   Loosen the wheel lug nuts, release the parking brake, raise the rear of the vehicle and support it securely on jackstands. Block the front wheels to keep the vehicle from rolling. Remove the rear wheels.

2   Refer to Section 4 and remove the rear brake calipers (don't disconnect the hoses). Hang them with a length of wire.

3   Refer to Section 5 and remove the rear brake discs. **Note:** *If the brake disc cannot be easily pulled off the axle and shoe assembly, make sure that the parking brake is completely released, then apply some penetrating oil at the hub-to-disc joint. Allow the oil to soak in and try to pull the disc off.*

4   Remove the center console (see Chapter 11). Pull the parking brake cable up **(see illustration)**. Move the spring until the two tabs pass each other and then insert a screwdriver into the side hole. Release the tension on the spring.

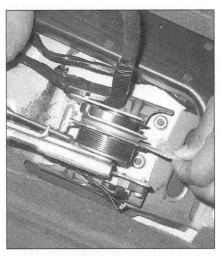

13.4 Pull the cable up as shown, then place a punch or screwdriver in the hole

13.5a Pushing on the pin from the backing plate side with your finger, pry the front hold-down clip loose, then pull the pin out

13.5b Remove the rear hold-down clip and pin the same way

13.5c Disengage the lower spring from the parking brake shoes

13.5d Remove the adjuster

13.5e Disengage the upper spring from the rear parking brake shoe

13.5f Remove both parking brake shoes

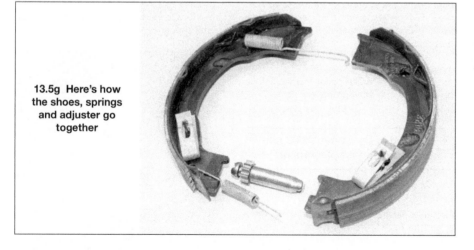

13.5g Here's how the shoes, springs and adjuster go together

13.5h Holding them together with the upper spring, install the new shoes

5   Clean the parking brake shoe assembly with brake system cleaner then follow the accompanying illustrations **(13.5a through 13.5o)** for the parking brake shoe replacement procedure. Be sure to stay in order and read the caption under each illustration. **Note:** *All four parking brake shoes must be replaced at the same time, but to avoid mixing up parts, work on only one brake assembly at a time.*

6   Before installing the disc, check the parking brake surfaces of the disc for cracks, score marks, deep scratches and hard spots, which will appear as small discolored areas. If any of the above listed conditions cannot

13.5i Spread the upper ends of the shoes apart and engage them with the anchor as shown

13.5j Install the adjuster

13.5k Make sure the adjuster is engaged with the shoes as shown

13.5l Install the lower spring

13.5m Make sure the coils of the lower spring are facing down, away from the adjuster

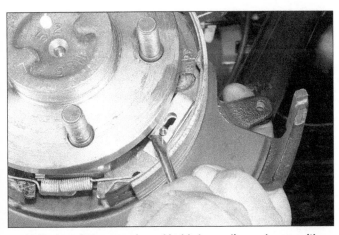

13.5n Install the rear pin and hold-down clip as shown, with the head of the pin firmly seated into the lower, smaller part of the hole in the clip

13.5o Install the front pin and hold-down clip the same way

be removed with sandpaper or emery cloth, the disc must be replaced.

7   The remainder of installation is the reverse of removal.

8   Check brake operation before driving in traffic.

9   To increase the effectiveness of the new parking brake shoes, make one stop from about 25 mph using only light force on the parking brake. Make sure that you are not in traffic while you do this and that it is otherwise safe.

Notes

Chapter 10
Suspension and steering systems

Contents

Specifications

Torque specifications

Ft-lbs (unless otherwise indicated)   **Nm**

Note: *One foot-pound (ft-lb) of torque is equivalent to 12 inch-pounds (in-lbs) of torque. Torque values below approximately 15 foot-pounds are expressed in inch-pounds, because most foot-pound torque wrenches are not accurate at these smaller values.*

Front suspension

	Ft-lbs	Nm
Balljoints		
2007 and earlier models	60	81
2008 and later models		
Step 1	30	40
Step 2	Tighten an additional 90-degrees	
Hub/bearing bolts	See Chapter 8	
Driveaxle/hub nut	See Chapter 8	
Shock absorber		
Clevis bracket-to-shock bolt	100	136
Clevis bracket-to-lower control arm bolt	110	149
Shock absorber/coil spring-to-body nuts (4)	80	108
Shock absorber damper shaft nut	30	40
Stabilizer bar		
Bushing clamp nuts	110	149
Link-to-bar nut	100	136
Link-to-lower control arm nut	85	115
Suspension arm pivot bolt nuts		
Lower	125	169
Upper	90	122

Torque specifications (continued)

Ft-lbs (unless otherwise indicated)   **Nm**

Note: *One foot-pound (ft-lb) of torque is equivalent to 12 inch-pounds (in-lbs) of torque. Torque values below approximately 15 foot-pounds are expressed in inch-pounds, because most foot-pound torque wrenches are not accurate at these smaller values.*

Rear suspension

	Ft-lbs	Nm
Lower suspension arm-to-axle nut		
2007 and earlier models	120	163
2008 and later models	150	203
Lower suspension arm-to-body nut	120	163
Shock absorber		
Upper nut/bolt	80	108
Lower nut	85	115
Track bar-to-axle bolt	130	176
Track bar-to-body bracket nut *	135	183
Stabilizer bar bracket bolts		
2007 and earlier models	73	99
2008 and later models	35	47
Stabilizer bar-to-stabilizer link nut	65	88
Stabilizer link-to-body bolt	75	102
Upper center balljoint bracket bolts	100	136
Upper suspension arm-to-axle nut/bolt		
2007 and earlier models	74	100
2008 and later models	70	95
Upper suspension arm-to-body nut/bolt		
2007 and earlier models	74	100
2008 and later models	70	95
Upper suspension bar center balljoint nut	70	95

* Tighten at normal ride height.

Steering

	Ft-lbs	Nm
Airbag module screws	90 in-lbs	10
Power steering bracket-to-engine bolts	35	47
Power steering lines	23	31
Power steering pump-to-bracket	21	28
Steering column coupler bolt		
2007 and earlier models	36	49
2008 and later models	39	53
Steering column intermediate shaft support bearing nuts		
2007 and earlier models	125 in-lbs	14
2008 and later models	106 in-lbs	12
Steering column mounting bolts	150 in-lbs	17
Steering gear intermediate shaft bolt		
2007 and earlier models	36	49
2008 and later models	39	53
Steering gear mounting bolts		
2007 and earlier models	120	163
2008 and later models	130	176
Steering wheel-to-steering shaft bolt		
2007 and earlier models	40	54
2008 and later models	32	43
Tie-rod end adjustment jam nut	55	75
Tie-rod end nut	80	108
Wheel lug nuts	See Chapter 1	

1 General information and precautions

Front suspension

Refer to illustration 1.1

The front suspension **(see illustration)** is fully independent. Each wheel is connected to the frame by a steering knuckle, upper and lower balljoints and upper and lower control arms. Coil spring-over-shock absorbers are used. A stabilizer bar connected to the frame and to the two lower control arms reduces body roll during cornering.

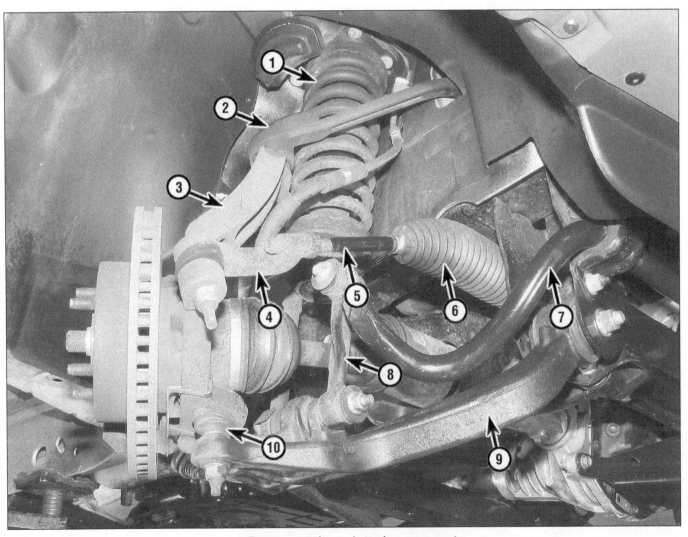

1.1 Front suspension and steering components

1	Shock absorber/coil spring assembly	5	Tie-rod	9	Lower control arm
2	Upper control arm	6	Steering gear boot	10	Lower balljoint
3	Steering knuckle	7	Stabilizer bar		
4	Tie-rod end	8	Stabilizer bar link		

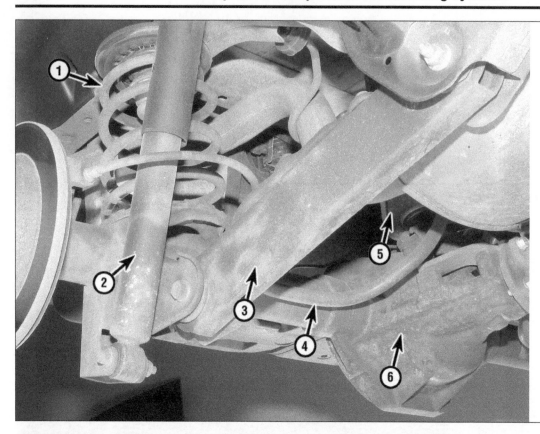

1.2 Rear suspension components (2007 and earlier Liberty models)

1   Coil spring
2   Shock absorber
3   Lower suspension arm
4   Stabilizer bar
5   Upper suspension arm
6   Rear axle assembly

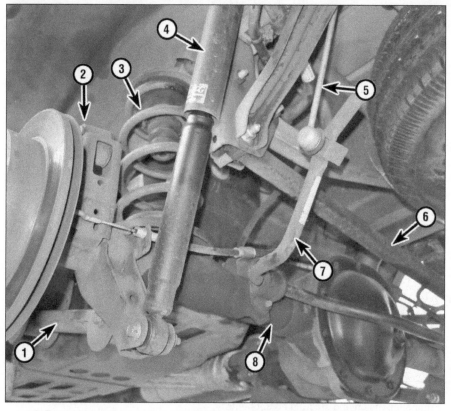

1.3 Rear suspension components (2008 and later Liberty models/all Nitro models)

1	Lower control arm	5	Stabilizer bar link
2	Upper control arm	6	Track bar
3	Coil spring	7	Stabilizer bar
4	Shock absorber	8	Rear axle

Rear suspension

Refer to illustration 1.2 and 1.3

The solid rear axle **(see illustration)** consists of a center housing for the differential with a pair of steel tubes for the axleshafts welded or pressed to each side (for more information on the components of the rear axle, see Chapter 8). The rear axle is suspended by two shock absorbers and two coil springs. Two lower control arms allow the axle to move vertically. On 2007 and earlier models, lateral movement and rotation are controlled by a single V-shape upper control arm. On 2008 and later Liberty and all Nitro models, two upper control arms are used, and lateral movement is controlled by a track bar. A stabilizer bar controls body roll.

Steering

The steering system consists of a rack-and-pinion steering gear and two adjustable tie-rods. Power assist is standard. The steering shaft is connected to the pinion of the steering gear. When the shaft is rotated, it forces the rack to move from side to side, thereby operating the tie-rods, which are attached to the steering knuckles. When the knuckles pivot, the front wheels turn.

Precautions

Frequently, when working on the suspension or steering system components, you may come across fasteners which seem impossible to loosen. These fasteners on the underside of the vehicle are continually subjected to water, road grime, mud, etc., and can become rusted or "frozen," making them extremely difficult to remove. In order to unscrew these stubborn

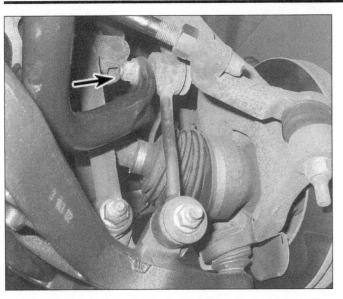

2.2 To disconnect the stabilizer bar link from the bar, remove the upper link nut; it isn't necessary to remove the lower link nut and bolt unless you're planning to replace the link itself

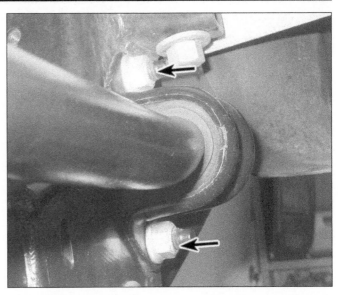

2.3 To disconnect the stabilizer bar from the frame, remove the bushing clamp nuts; inspect the rubber bushings and replace them if they're hard, cracked or otherwise deformed

fasteners without damaging them (or other components), be sure to use lots of penetrating oil and allow it to soak in for a while. Using a wire brush to clean exposed threads will also ease removal of the nut or bolt and prevent damage to the threads. Sometimes a sharp blow with a hammer and punch is effective in breaking the bond between a nut and bolt threads, but care must be taken to prevent the punch from slipping off the fastener and ruining the threads. Heating the stuck fastener and surrounding area with a torch sometimes helps too, but isn't recommended because of the obvious dangers associated with fire. Long breaker bars and extension, or "cheater," pipes will increase leverage, but never use an extension pipe on a ratchet - the ratcheting mechanism could be damaged. Sometimes, turning the nut or bolt in the tightening (clockwise) direction first will help to break it loose. Fasteners that require drastic measures to unscrew should always be replaced with new ones.

Since most of the procedures that are dealt with in this chapter involve jacking up the vehicle and working underneath it, a good pair of jackstands will be needed. A hydraulic floor jack is the preferred type of jack to lift the vehicle, and it can also be used to support certain components during various operations. **Warning:** *Never, under any circumstances, rely on a jack to support the vehicle while working on it.*

Whenever any of the suspension or steering fasteners are loosened or removed they must be inspected and, if necessary, be replaced with new ones of the same part number or of original equipment quality and design. Torque specifications must be followed for proper reassembly and component retention. Never attempt to heat or straighten any suspension or steering components. Instead, replace any bent or damaged part with a new one.

2 Stabilizer bar and bushings (front) - removal and installation

Removal

Refer to illustrations 2.2 and 2.3

1   Apply the parking brake. Loosen the front lug nuts. Raise the front of the vehicle and support it securely on jackstands. Remove the front wheels.
2   Detach the upper ends of the links from the stabilizer bar **(see illustration)**. If it is necessary to remove the links, simply unbolt them from the axle brackets.
3   Remove the nuts from the stabilizer bar bushing clamps **(see illustration)** and detach the bar from the vehicle.
4   Pull the bushings off the stabilizer bar and inspect them for cracks, hardness and other signs of deterioration. If the bushings are damaged, replace them. Inspect the bushings in the lower ends of the links, replacing them if necessary.

Installation

5   Position the stabilizer bar bushings on the bar. Don't use any grease or oil on the bushings.
6   Push the brackets over the bushings and raise the bar up to the frame. Install the bracket nuts but don't tighten them completely at this time.
7   Connect the link to the stabilizer bar and the axle. Tighten the fasteners to the torques listed in this Chapter's Specifications. Tighten the wheel lug nuts to the torque listed in the Chapter 1 Specifications.
8   Tighten the bracket bolts to the torque listed in this Chapter's Specifications.

3 Coil spring and shock absorber (front) - removal, component replacement and installation

Refer to illustrations 3.4 and 3.6

Warning 1: *Before undertaking the following procedure, be aware that disassembling the shock absorber/coil spring assemblies is a potentially dangerous job. Careless or unsafe work can cause injury. Use only a high-quality spring compressor and be sure to follow the spring compressor manufacturer's instructions. After removing the compressed spring, set it aside in an isolated area.*

Warning 2: *Always replace the shock absorbers or coil springs in pairs - never replace just one shock absorber or coil spring, as handling abnormalities may result.*

Note: *If the shock absorber/coil spring assemblies must be replaced, you can save time by simply installing new complete shock/coil assemblies. Therefore, before deciding which way you want to go, find out the cost of each option. You may find that the cost for two assembled complete shock/coil assemblies is only slightly higher than the cost for individual components.*

Removal

1   If you're working on the left side, remove the battery and battery tray (see Chapter 5). Release the clips securing the underhood fuse/relay box and set it aside without disconnecting its electrical connectors.
2   If you're working on the right side, refer to Chapter 4 and remove the air filter housing. On later models, you may have to remove the nuts securing the cruise control servo and set it aside for access to all the shock upper nuts.

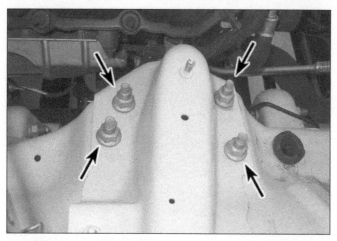

3.4 The top of the front coil spring/shock absorber assembly is secured with these four nuts

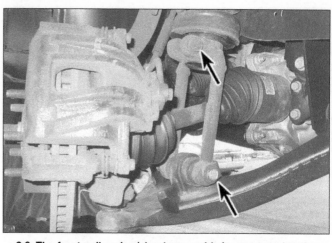

3.6 The front coil spring/shock assembly is connected to the lower control arm by this clevis. Disconnect it at the lower end first, then disconnect it at the top

3   If you're working on the right side, remove the cruise control servo mounting nuts.

4   Remove the upper mounting nuts for the shock absorber/coil spring assembly **(see illustration)**.

Note: *On 2008 and later models, to access the four nuts on the driver's side, the battery and battery tray must be removed (see Chapter 5).*

5   Apply the parking brake. Loosen the front wheel lug nuts, raise the vehicle and support it securely on jackstands, then remove the front wheels.

6   Remove the clevis bracket-to-lower control arm bolt **(see illustration)**.

7   Refer to Section 2 and detach the stabilizer bar link from the lower control arm.

8   Disconnect and then separate the lower balljoint from the control arm (see Section 4). Pull the lower control arm down to provide working room.

9   Disconnect the clevis bracket from the shock absorber **(see illustration 3.6)**. Remove the spring/shock absorber from the wheelwell.

Component replacement

10   Secure the shock/coil assembly in a bench vise. If you're planning to reuse the old components, line the jaws of the vise with wood or shop rags to protect them. Don't tighten the jaws any more than necessary.

11   Install a spring compressor according to the tool manufacturer's instructions. You can buy a spring compressor at most auto parts stores or rent one from most equipment yards on a daily basis. Compress the spring enough to relieve all force on the spring seat; when you can wiggle the spring it's compressed enough to disassemble the shock/spring assembly. **Warning:** *Don't compress the spring any more than necessary.*

12   Remove the damper shaft nut from the shock absorber upper mount, then remove the mount, coil spring and lower insulator. Place the spring in an isolated area.

13   Reassemble the unit by reversing the disassembly procedure. **Warning:** *Be careful*

4.3 Hold the balljoint stud with one wrench while loosening the nut with another

to avoid bumping the compressed spring or the compressor. Make sure that end of the bottom of the spring seats against the recess in the bottom spring retainer plate.

14   Install the damper shaft nut, tightening it to the torque listed in this Chapter's Specifications.

15   Slowly and carefully release the spring compressor.

Installation

16   Installation is the reverse of removal. Tighten all fasteners to the torques listed in this Chapter's Specifications. Tighten the wheel lug nuts to the torque listed in the Chapter 1 Specifications.

---

4   Suspension arms (front) - removal and installation

---

1   Loosen the wheel lug nuts. Raise the front of the vehicle and support it securely on jackstands placed under the frame.

2   Remove the wheel.

4.4 This type of balljoint separation tool will prevent damage to the balljoint seal

Upper suspension arm

Refer to illustrations 4.3, 4.4 and 4.7

3   Loosen the nut from the upper balljoint a few turns **(see illustration)**.

4   Using an appropriate balljoint separator, separate the upper balljoint from the steering knuckle **(see illustration)**. Remove the nut.

5   If you're working on the left control arm, remove the battery and the battery tray (refer to Chapter 5). Unclip the power center from its support and shift it aside.

6   If you're working on the right control arm, remove the air filter assembly (refer to Chapter 4). Also remove the nuts that secure the cruise control servo.

7   Remove the nuts from the pivot bolts and lift the control arm from the vehicle **(see illustration)**. **Note:** *To remove the left rear bolt, it will be necessary to use a long extension with a ratchet run under the steering shaft and near the power steering reservoir.*

8   Installation is the reverse of removal. Tighten the pivot bolt nuts to the torque listed in this Chapter's Specifications. Tighten the

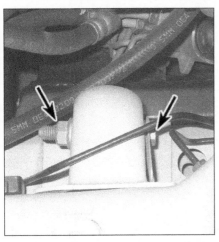

4.7 The upper control arm pivot bolts/ nuts require the use of two wrenches for removal

4.15 Mark the position of each lower control arm bolt to the crossmember

4.16 Front lower control arm bolts

6.4 Use a large pair of pliers to check for play in the balljoint

wheel lug nuts to the torque listed in the Chapter 1 Specifications.

9   Have the front wheel alignment checked and, if necessary, adjusted.

Lower suspension arm

Refer to illustrations 4.15 and 4.16

10   Loosen the front wheel lug nuts. Raise the vehicle and support it securely on jackstands.

11   Remove the wheel.

12   Unbolt the shock absorber clevis bracket from the lower control arm **(see illustration 3.6)**.

13   Remove the bolt that attaches the stabilizer bar link to the lower control arm.

14   Loosen the lower balljoint nut. Using an appropriate balljoint separator, disconnect the balljoint from the steering knuckle (see Step 4), then remove the nut.

15   Mark the positions of the front and rear cam bolts to the crossmember **(see illustration)**.

16   Remove the nuts from the bolts, then drive the bolts out with a hammer and punch **(see illustration)**.

17   Remove the control arm.

18   Installation is the reverse of removal. Be

sure to align the marks you made in Step 15 before tightening the nuts, then tighten the nuts to the torque listed in this Chapter's Specifications. Tighten the wheel lug nuts to the torque listed in the Chapter 1 Specifications.

19   Have the front wheel alignment checked and, if necessary, adjusted.

5   Steering knuckle - removal and installation

Removal

1   Loosen the wheel lug nuts, raise the vehicle and support it securely on jackstands. Remove the wheel.

2   Refer to Chapter 9 and remove the brake disc.

3   Remove the ABS wheel speed sensor.

4   On four-wheel drive models, refer to Chapter 8 and remove the driveaxle/hub nut.

5   Remove the three hub/bearing assembly bolts (refer to Chapter 8). Set the hub and bearing assembly aside.

6   Refer to Section 12 and disconnect the tie-rod end from the steering knuckle.

7   Refer to Section 6 and disconnect both balljoints from the knuckle. Remove the knuckle from the vehicle.

Installation

8   Position the knuckle onto both balljoints. Install the balljoint nuts, tightening them to the torque listed in this Chapter's Specifications.

9   The remainder of installation is the reverse of removal. Tighten the wheel lug nuts to the torque listed in the Chapter 1 Specifications.

10   Have the front wheel alignment checked and, if necessary, adjusted.

6   Balljoints - check and replacement

Refer to illustrations 6.4 and 6.6

Check

1   Inspect the control arm balljoints for looseness anytime either of them is separated from the steering knuckle. See if you can turn the ballstud in its socket with your fingers. If the balljoint is loose or if the ballstud can be turned, replace the balljoint. You can also check the balljoints with the suspension assembled as follows:

Upper balljoints

2   Raise the vehicle and support it securely on jackstands placed under the frame. Put a floor jack under the lower control arm and raise it slightly.

3   Attach a dial indicator in a vertical position to one of the upper control arms with the plunger touching the steering knuckle.

4   Squeeze the assembly with a large pair of pliers **(see illustration)**. The dial indicator should show no more than 0.060-inch deflection. If the indicated reading exceeds this figure it is considered bad.

Lower balljoints

5   Raise the vehicle and support it securely on jackstands placed under the frame. Put a floor jack under the lower control arm and raise it slightly.

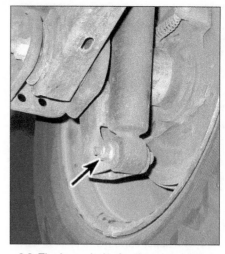

6.6 Attach a dial indicator with a magnetic base to the control arm and zero it. Pry between the lower control arm and the steering knuckle and read the dial indicator

7.2 These bolts secure each end of the rear stabilizer bar

8.3 The lower bolts for the rear shocks are readily accessible

6   Attach a dial indicator to the lower control arm and position the plunger against the steering knuckle **(see illustration)**.

7   Use a prybar to separate the balljoint as much as possible. The dial indicator should show no more that 0.060-inch deflection. If the indicated reading exceeds this figure, replace the balljoint.

Replacement

Upper balljoints

8   Upper balljoints are not replaceable separately. Refer to Section 4 and replace the upper control arm assembly.

Lower balljoints

9   Refer to Section 4 and remove the lower control arm.

10   Take the control arm assembly to an automotive shop that is competent in this type of repair. Replacement of the balljoint involves pressing out the old unit and pressing in a new one. On-the-car balljoint presses may be available for rental at a local auto supply store or rental yard.

11   Installation is the reverse of the removal procedure. Be sure to lubricate the new balljoint before installation if it is equipped with a grease fitting.

---

7   Stabilizer bar and bushings (rear) - removal and installation

2007 and earlier models

Removal

Refer to illustration 7.2

1   Raise the rear of the vehicle and support it securely on jackstands.

2   Unbolt the stabilizer bar from the lower suspension arms **(see illustration)**.

3   Slide the stabilizer bar out from above

the axle housing. Inspect it for cracks and other signs of fatigue.

Installation

4   Guide the stabilizer bar into place, making sure it's even on both sides.

5   Install the mounting bolts and tighten them to the torque listed in this Chapter's Specifications.

6   Install the wheels and lug nuts, then lower the vehicle. Tighten the lug nuts to the torque listed in the Chapter 1 Specifications.

2008 and later models

Removal

7   Loosen the rear wheel lug nuts. Raise the front of the vehicle and support it securely on jackstands. Block the front wheels to keep the vehicle from rolling off the stands. Remove the rear wheels.

8   Detach the stabilizer bar link nuts from the stabilizer bar. If the ballstud turns with the nut, use an Allen wrench to hold the stud.

9   Unbolt the stabilizer bar bushing clamps and remove the stabilizer bar.

10   While the stabilizer bar is off the vehicle, slide off the retainer bushings and inspect them. If they're cracked, worn or deteriorated, replace them.

11   Clean the bushing area of the stabilizer bar with a stiff wire brush to remove any rust or dirt.

Installation

12   Lubricate the inside and outside of the new bushing with vegetable oil (used in cooking) to simplify reassembly. **Caution:** *Don't use petroleum or mineral-based lubricants or brake fluid - they will lead to deterioration of the bushings.* The slits of the bushings must face the rear of the vehicle.

13   Installation is the reverse of removal. Lower the vehicle and tighten the fasteners to the torque listed in this Chapter's Specifications.

---

8   Shock absorber (rear) - removal and installation

Refer to illustrations 8.3 and 8.4

Warning: *Always replace the shock absorbers in pairs - never replace just one of them.*

1   Raise the rear of the vehicle and support it securely on jackstands.

2   Support the rear axle assembly with a floor jack placed under the differential. Raise the jack just enough to take the spring force off the shock absorbers (the shock absorbers limit downward travel of the suspension).

3   Remove the lower mounting bolt and nut **(see illustration)**.

4   Remove the upper mounting bolt and nut **(see illustration)**.

5   Installation is the reverse of the removal procedure. Install the shock mounting bolts hand-tight, then lower the vehicle before tightening the fasteners to the torque listed in this Chapter's Specifications.

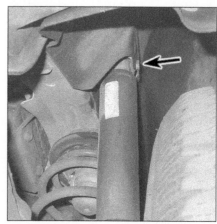

8.4 Upper rear shock absorber bolt location

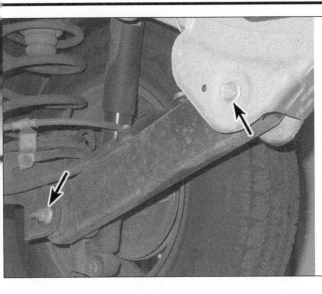

9.3 Lower suspension arm mounting fasteners. Remove only one of the rear lower suspension arms at a time

9 Suspension arms (rear) - removal and installation

1   Raise the rear of the vehicle and support it securely on jackstands placed under the frame rails. Support the rear axle housing with a jack.

Lower suspension arm

Refer to illustration 9.3

2   Refer to Section 7 and disconnect the stabilizer bar from the lower arm. Also remove the shock absorber lower mounting bolt and swing the shock absorber forward, out of the way (see Section 8).

3   Unbolt the suspension arm at the front and rear mounts **(see illustration)**. **Warning:** *If both lower control arms are removed at the same time, the axle assembly will have a tendency to rotate. Replace only one at a time.* **Note:** *In order to remove the right front mounting bolt, it will be necessary to pry the exhaust pipe over slightly using a long flat prybar.*
Note: *On 2008 and later models, in order to*

remove the left side lower suspension arm bolt, the fuel tank must be lowered (see Chapter 4).
4   Installation is the reverse of the removal procedure. Tighten all fasteners to the torques listed in this Chapter's Specifications.

Upper suspension arm
2007 and earlier models
Refer to illustrations 9.5 and 9.7
5   Remove the center balljoint pinch bolt **(see illustration)**. Pull the upper control arm up from the rear axle housing.
6   Remove the exhaust heat shield near the right pivot bolt for adequate bolt-removal clearance.
7   Remove the bolts from each of the pivots at the front of the control arm **(see illustration)**.
8   Remove the control arm from the vehicle.
9   Installation is the reverse of the removal procedure. Tighten all fasteners to the torques listed in this Chapter's specifications.

2008 and later models
10   Lower the fuel tank to access the left

side upper arm bolts (see Chapter 4).
11   Remove the arm-to-axle bracket nut/bolt.
12   Remove the arm-to-frame rail nut/bolt and remove the upper arm.
13   Installation is the reverse of removal. Lower the vehicle and tighten the fasteners to the torque listed in this Chapter's Specifications.

Track bar
14   Support the rear axle with a floor jack.
15   Remove the track bar-to-frame nut and slide the special "flag or ear" type bolt out.
16   Remove the track bar-to-axle nut/bolt and remove the track bar.
17   Installation is the reverse of removal. Lower the vehicle and tighten the fasteners to the torque listed in this Chapter's Specifications.

10 Coil spring (rear) - removal and installation

Refer to illustration 10.4
1   Raise the rear of the vehicle and support it securely on jackstands placed underneath the frame rails.
2   Support the axle with a floor jack placed under the center of the axle housing.
3   Remove the lower mounting bolt from each rear shock absorber (see Section 8).
4   Slowly lower the axle until the coil springs are fully extended. Pull the axle down and tilt it to release the coil springs **(see illustration)**.
5   Check the springs for deep nicks and corrosion, which could cause premature failure of the spring. Replace the springs as a set if these or any other questionable conditions are evident.
6   Also check the spring isolators for damage, replacing them as necessary.
7   Installation is the reverse of removal. Tighten the shock absorber lower mounting bolts/nuts to the torque listed in this Chapter's Specifications.

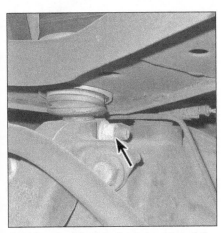

9.5 The pinch bolt in the top center of the rear axle housing secures the V-shaped upper control arm

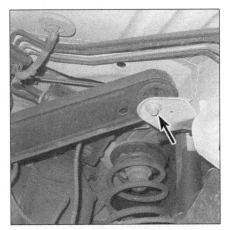

9.7 The upper suspension arm left attachment point

10.4 When the rear axle housing is pulled down sufficiently, the spring can be removed

11.2 Two screws must be removed from the rear of the steering wheel to release the airbag

11.4 First disconnect the small horn wire (arrow) . . .

11.5 . . . then disconnect the two larger airbag clockspring wiring connectors. There are latches on each side of the connectors that must be depressed

11 Steering wheel - removal and installation

Refer to illustrations 11.2, 11.4, 11.5, 11.8a, 11.8b and 11.12

Warning: *The models covered by this manual are equipped with a Supplemental Restraint System (SRS), more commonly known as airbags. Always disarm the airbag system before working in the vicinity of any airbag system component to avoid the possibility of accidental deployment of the airbag, which could cause personal injury (see Chapter 12). Do not use a memory-saving device to preserve the PCM's memory when working on or near airbag system components.*

1   Park the vehicle with the wheels pointing straight ahead. Disconnect the cable from the negative terminal of the battery (see Chapter 5).

2   On 2007 and earlier models, remove the two airbag module screws from the back of the steering wheel **(see illustration)**. **Note:** *On 2008 and later models, there are three airbag retainer access holes back side of the steering wheel, located at the two o'clock, six o'clock and ten o'clock positions. The steering wheel must be rotated to bring each of these access holes up to the twelve o'clock*

position of the steering column one at a time for removal access. On 2008 and later models, insert the short blade screwdriver or plastic trim prying tool into either of the upper access holes of the steering wheel cover and use the tool to move the retainer wire loop off of the latch.

3   Remove the airbag away from the steering wheel far enough to reach the wiring on the rear of the airbag.

4   Disconnect the horn wiring connector **(see illustration)**.

5   To disconnect the two airbag electrical connectors, press down on the latches on each side of the connectors where they attach to the airbag **(see illustration)**. Pull each connector straight away from the socket.

6   Remove the airbag from the vehicle. **Warning:** *Whenever handling the airbag module, always keep the airbag opening (the trim side) pointed away from your body. Never set the airbag down with the airbag opening facing the surface. Always place the airbag module in a safe location with the airbag opening facing up.*

7   Loosen the steering wheel retaining bolt but don't remove it completely.

8   Use a puller to loosen the steering wheel from the shaft **(see illustrations)**. Don't hammer on the shaft to dislodge the steering

wheel. **Caution:** *Don't allow the steering shaft to turn while the steering wheel is removed.*

9   Remove the bolt and discard it. It must be replaced with a new one. Remove the steering wheel. **Caution:** *Don't allow the steering shaft to turn with the steering wheel removed. If the shaft turns, the airbag clockspring will become uncentered, which may cause the wire inside to break when the vehicle is returned to service.*

10   If it's necessary to remove the clockspring from the steering column, remove the steering column covers (see Chapter 11), unclip the latches, disconnect the electrical connector and remove the clockspring **(see illustration)**.

Installation

11   If it was removed, reinstall the clockspring:

a) *Install the clockspring on the steering column, making sure the latch hooks engage properly. Also make sure the electrical connectors are plugged in completely.*

b) *If a new clockspring is being installed, it will already be in the centered position.*

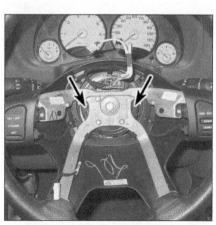

11.8a There are two pockets on the steering wheel. The puller arms must fit into these pockets

11.8b A two-jaw puller can be used to remove the steering wheel

11.12 Clockspring alignment details

A   *Release button*
B   *Alignment arrows*
C   *Inspection window*

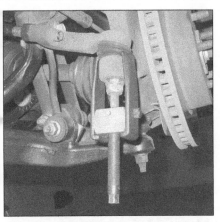

12.3 Use a two-jaw puller to separate the tie-rod ends from the steering knuckles

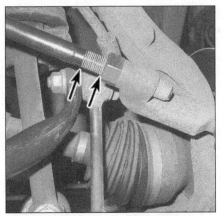

12.4 Count the threads exposed behind the tie rod jam nut and jot it down. The new tie rod end should be installed in the same position as the old one

13.8 The correct position of the intermediate shaft coupler must be marked with paint prior to disassembly

12   If you aren't sure if the clockspring is centered or not, depress the release button and turn the hub clockwise until it stops (don't apply too much force). Now, rotate the hub in the other direction about 2-1/2 turns and let go of the release button. The wires should now be positioned at the top, the arrows on the hub and clockspring body should be aligned, and the blue roller should be visible in the inspection window **(see illustration)**. Install the clockspring, making sure the locking tabs engage securely. Install the steering column covers.

13   Install the steering wheel onto the shaft. Install the new bolt and tighten it to the torque listed in this Chapter's Specifications.

14   Install the airbag module and tighten the fasteners to the torque listed in this Chapter's Specifications. **Caution:** *The two airbag wiring connectors must snap into their sockets with a distinct clicking sound.*

15   The remainder of installation is the reverse of removal.

12 Tie-rod ends - removal and installation

Refer to illustrations 12.3 and 12.4

1   Loosen the wheel lug nuts, then raise the vehicle and support it securely on jackstands. Apply the parking brake. Remove the wheel.

2   Loosen the nut on the balljoint end of the tie-rod but do not remove it. If the ballstud turns while loosening the nut, hold it with a wrench.

3   Using a puller, separate the tie-rod end from the steering knuckle **(see illustration)**. **Caution:** *The use of a picklefork-type balljoint separator will most likely damage the tie-rod boot.* Remove the nut.

4   Count the number of threads exposed on the steering gear rod **(see illustration)**. Jot this number down. Loosen the tie-rod jam nut a quarter turn. Unscrew the tie-rod from the steering gear.

5   Screw the replacement tie-rod end onto the steering gear. Screw it up to the jam nut and then back it off a quarter turn. Don't tighten the jam nut yet. If the jam nut has been moved, position it so that the same

number of threads is showing as before.

6   Insert the tie-rod end into the knuckle. Install the nut and tighten it to the torque listed in this Chapter's Specifications. If the ballstud spins in the steering knuckle when attempting to tighten the nut, hold it with a wrench. Tighten the jam nut.

7   Install the wheel, tightening the lug nuts to the torque listed in the Specifications in Chapter 1.

8   Have the wheel alignment checked and, if necessary, adjusted.

13 Steering gear - removal and installation

Warning: *Make sure the steering shaft is not turned while the steering gear is removed or you could damage the airbag system clockspring. One way to prevent the shaft from turning is to place the ignition key in the LOCK position or run the seat belt through the steering wheel and clip the seat belt into place.*

Removal

1   Loosen the front wheel lug nuts. Raise the front of the vehicle and support it securely on jackstands. Apply the parking brake.

2   Remove the front wheels.

3   Lock the steering wheel in position (see the **Warning** above).

4   Remove the skid plate.

4WD models

5   Refer to Section 4 and remove the lower control arms.

Note: *On 2008 and later models, only the left front control arm needs to be lowered to remove the front axle bracket mounting bolts and bracket. The oil filter must also be removed to allow the bracket to clear the engine and be removed from the vehicle.*

6   On 2007 and earlier models, remove the driveaxles and front axle assembly (see Chapter 8).

All models

Refer to illustrations 13.8 and 13.11

7   Refer to Section 12 and separate the tie-rod ends from the steering knuckles.

8   Use a dab of paint to mark the position of the intermediate shaft coupler to the steering gear **(see illustration)**. Remove the bolt from the coupling. Pull the shaft off of the steering gear. **Note:** *On 2008 and later models, to access the coupler bolt, the battery and battery tray must be removed (see Chapter 5).*

9   Remove the power steering pressure hose mounting bracket.

10   Disconnect both power steering hoses from the steering gear. Plug the ends to prevent contamination.

11   Unbolt the steering gear from the bracket and remove it **(see illustration)**.

Installation

12   If necessary, transfer the tie-rod ends to the new steering gear. Before removing the tie-rods from the old steering gear, count the number of threads visible next to the jam nut. When installing the tie-rod ends on the new gear, adjust them so that the same number of threads is exposed. Do not tighten the jam nuts until the steering gear has been bolted in place.

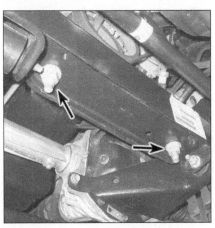

13.11 Two bolts secure the steering gear

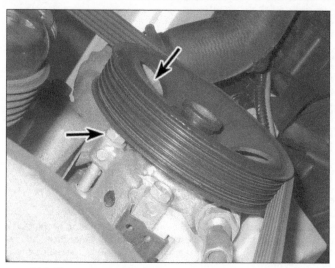

14.9 You must use the holes in the power steering pump pulley to get at the mounting bolts

14.10 This special tool, designed for removing power steering pump pulleys, is available for rent or sale at most auto parts stores

13   Raise the steering gear into position and connect the intermediate shaft, aligning the marks.

14   Install the mounting bolts and washers and tighten them to the torque listed in this Chapter's Specifications.

15   Install the intermediate shaft lower pinch bolt and tighten it to the torque listed in this Chapter's Specifications.

16   Connect the power steering hoses/lines to the steering gear and fill the power steering pump reservoir with the recommended fluid (see Chapter 1).

17   The remainder of installation is the reverse of removal.

18   Lower the vehicle and bleed the steering system (see Section 15). Be sure to tighten the wheel lug nuts to the torque listed in the Chapter 1 Specifications.

19   Have the wheel alignment checked and, if necessary, adjusted.

---

14   Power steering pump - removal and installation

Refer to illustrations 14.9, 14.10, 14.11a and 14.11b

Caution: *The plastic pulley used on the pump of the 2.4L engine is not reusable. Once it is removed it cannot be reinstalled on a new pump.*

Note: *If you're replacing the pump with a new one, you may have to swap the pump pulley to the new unit. A special puller/installer is required for this. They are available at most automotive parts stores and they can be rented at some equipment rental yards.*

1   Using a poultry baster or a suction gun, remove as much fluid as possible from the pump reservoir. **Warning:** *Do not attempt to siphon by mouth. If using a poultry baster, do not use it for cooking again.*

3.7L models

2   Remove the radiator crossmember. **Note:** *On 2005 and earlier models, remove the grille, hood latch and the radiator upper crossmember (see Chapter 11). To remove the crossmember, some rivets must be drilled out.*

3   Refer to Chapter 3 and remove the cooling fan and the fan shroud.

4.0L models

4   Remove the air filter housing (see Chapter 4).

5   Remove the coolant reservoir and windshield washer reservoir (see Chapter 3).

6   Remove the fan shroud (see Chapter 3).

All models

7   Refer to Chapter 1 and remove the accessory drivebelt.

8   Disconnect both hoses from the power

14.11a The pulley must be pressed back onto the pump with a special tool . . .

14.11b . . . or a long bolt with the same threads as the internal threads of the shaft, a nut, washer and a socket

10-13

steering pump. Plug the ends to avoid contamination.
9   Unbolt the pump from its bracket.
Access the bolts through the holes in the pulley **(see illustration)**.
10   If it is necessary to remove the pulley
from the pump, use a special puller designed
for this purpose **(see illustration)**. If the pulley is made of plastic, it cannot be reused. A
new one must be purchased and installed.
Note the position of the old plastic pulley on
the shaft before removing it.
11   A special pulley installation tool is also
available (usually supplied with the removal
kit) **(see illustration)**. If it is not available, an
alternate tool can be made from a long bolt,
nut, washer and a socket of the same diameter as the pulley hub **(see illustration)**. Press
the steel pulley on until it is flush with the end
of the shaft. A plastic pulley must be installed
so that it is in the same position on the shaft
as the original pulley.
12   Installation of the pump is the reverse
of the removal procedure. Be sure to tighten
the fasteners securely and bleed the power
steering system following the procedure in
Section 15.

15   Power steering system - bleeding

1   Following any operation in which the
power steering fluid lines have been disconnected, the power steering system must be
bled to remove all air and obtain proper steering performance.
2   With the front wheels in the straight
ahead position, check the power steering
fluid level and, if low, add fluid until it reaches
the Cold (C) mark on the dipstick.
3   Start the engine and allow it to run at
fast idle. Recheck the fluid level and add
more if necessary to reach the Cold (C) mark
on the dipstick.
4   Bleed the system by turning the wheels
from side-to-side, without hitting the stops.
This will work the air out of the system. Keep
the reservoir full of fluid as this is done.
5   When the air is worked out of the system, return the wheels to the straight ahead
position and leave the vehicle running for
several more minutes before shutting it off.
6   Road test the vehicle to be sure the
steering system is functioning normally and
noise-free.
7   Recheck the fluid level to be sure it is up
to the Hot (H) mark on the dipstick while the
engine is at normal operating temperature.
Add fluid if necessary (see Chapter 1).

16   Wheel studs - replacement

Refer to illustration 16.3
Note: *This procedure applies to both the front
and rear wheel studs.*
1   Loosen the wheel lug nuts, raise the
vehicle and support it securely on jackstands.
Remove the wheel.

**16.3 Use a press tool
to remove the stud
from the flange**

2   Remove the brake disc or drum (see
Chapter 9).
3   Install a lug nut part way onto the stud
being replaced. Push the stud out of the hub
flange with a press tool **(see illustration)**.
4   Insert the new stud into the hub flange
from the rear and install some flat washers
and a lug nut on the stud.
5   Tighten the lug nut until the stud is
seated in the flange.
6   Reinstall the brake drum or disc. Install
the wheel and lug nuts. Lower the vehicle and
tighten the lug nuts to the torque listed in the
Chapter 1 Specifications.

17   Wheels and tires -
general information

Refer to illustration 17.1
The vehicles covered by this manual
are equipped with metric-sized fiberglass or
steel-belted radial tires **(see illustration)**.
Use of other size or type of tires may affect
the ride and handling of the vehicle. Don't
mix different types of tires, such as radials
and bias belted, on the same vehicle as
handling may be seriously affected. It's recommended that tires be replaced in pairs on

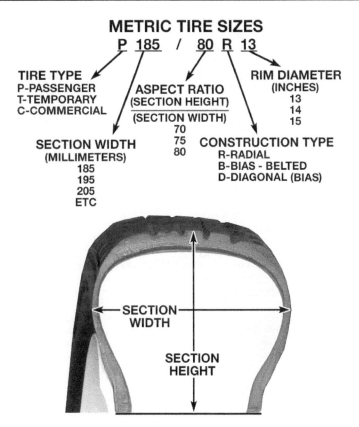

17.1 Metric tire size code

the same axle, but if only one tire is being replaced, be sure it's the same size, structure and tread design as the other.

Because tire pressure has a substantial effect on handling and wear, the pressure on all tires should be checked at least once a month or before any extended trips (see Chapter 1).

Wheels must be replaced if they are bent, dented, leak air, have elongated bolt holes, are heavily rusted, out of vertical symmetry or if the lug nuts won't stay tight. Wheel repairs that use welding or peening are not recommended.

Tire and wheel balance is important to the overall handling, braking and performance of the vehicle. Unbalanced wheels can adversely affect handling and ride characteristics as well as tire life. Whenever a tire is installed on a wheel, the tire and wheel should be balanced by a shop with the proper equipment.

18   Front end alignment - general information

Refer to illustration 18.1

A front end alignment refers to the adjustments made to the front wheels so they are in proper angular relationship to the suspension and the ground. Front wheels that are out of proper alignment not only affect steering control, but also increase tire wear.

Getting the proper front wheel alignment is a very exacting process, one in which complicated and expensive machines are necessary to perform the job properly. Because of this, you should have a technician with the proper equipment perform these tasks. We will, however, use this space to give you a basic idea of what is involved with front end alignment so you can better understand the process and deal intelligently with the shop that does the work.

Toe-in is the turning in of the front wheels **(see illustration)**. The purpose of a toe specification is to ensure parallel rolling of the front wheels. In a vehicle with zero toe-in, the distance between the front edges of the wheels will be the same as the distance between the rear edges of the wheels. The actual amount of toe-in is normally only a fraction of an inch. Toe-in adjustment is controlled by the tie-rod end position on the tie-rod. Incorrect toe-in will cause the tires to wear improperly by making them scrub against the road surface.

Caster is the tilting of the steering axis from vertical **(see illustration)**. A tilt toward the rear is positive caster and a tilt toward

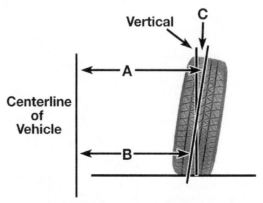

CAMBER ANGLE (FRONT VIEW)

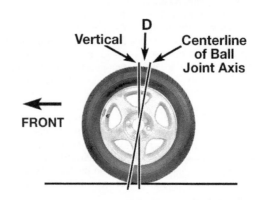

CASTER ANGLE (SIDE VIEW)

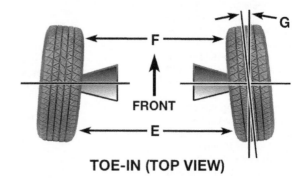

TOE-IN (TOP VIEW)

18.1 Front end alignment details

A minus B = C (degrees camber)   E minus F = toe-in (measured in inches)
D = degrees caster   G = toe-in (expressed in degrees)

the front is negative caster.

Camber is the tilting of the front wheels from vertical when viewed from the front of the vehicle. Both camber and caster are

adjusted by changing the positions of the lower control arms. The control arms are attached with bolts that use eccentric cams to adjust them.

Chapter 11   Body

Contents

---

1   General information

Warning: *The models covered by this manual are equipped with a Supplemental Restraint System (SRS), more commonly known as airbags. Always disable the airbag system before working in the vicinity of any airbag system components to avoid the possibility of accidental deployment of the airbags, which could cause personal injury (see Chapter 12).*

Certain body components are particularly vulnerable to accident damage and can be unbolted and repaired or replaced. Among these parts are the hood, doors, tailgate, liftgate, bumpers and front fenders.

Only general body maintenance practices and body panel repair procedures within the scope of the do-it-yourselfer are included in this Chapter.

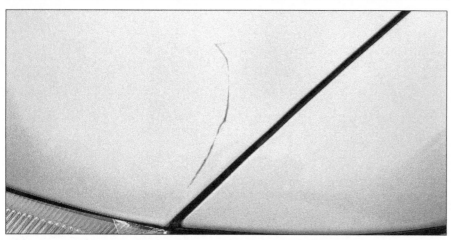

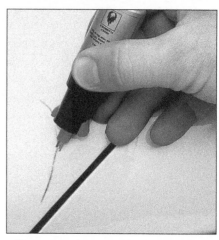

Make sure the damaged area is perfectly clean and rust free. If the touch-up kit has a wire brush, use it to clean the scratch or chip. Or use fine steel wool wrapped around the end of a pencil. Clean the scratched or chipped surface only, not the good paint surrounding it. Rinse the area with water and allow it to dry thoroughly

Thoroughly mix the paint, then apply a small amount with the touch-up kit brush or a very fine artist's brush. Brush in one direction as you fill the scratch area. Do not build up the paint higher than the surrounding paint

2   Repair minor paint scratches

No matter how hard you try to keep your vehicle looking like new, it will inevitably be scratched, chipped or dented at some point. If the metal is actually dented, seek the advice of a professional. But you can fix minor scratches and chips yourself. Buy a touch-up paint kit from a dealer parts department or an auto parts store. To ensure that you get the right color, you'll need to have the specific make, model and year of your vehicle and, ideally, the paint code, which is located on a special metal plate under the hood or in the door jamb.

3   Body repair - minor damage

Plastic body panels

The following repair procedures are for minor scratches and gouges. Repair of more serious damage should be left to a dealer service department or qualified auto body shop. Below is a list of the equipment and materials necessary to perform the following repair procedures on plastic body panels.

Wax, grease and silicone removing solvent
Cloth-backed body tape
Sanding discs
Drill motor with three-inch disc holder
Hand sanding block
Rubber squeegees
Sandpaper
Non-porous mixing palette
Wood paddle or putty knife
Curved-tooth body file
Flexible parts repair material

Flexible panels (bumper trim)

1   Remove the damaged panel, if necessary or desirable. In most cases, repairs can be car-

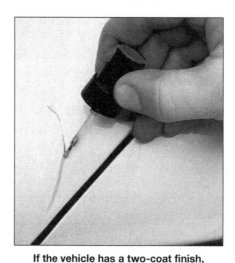

If the vehicle has a two-coat finish, apply the clear coat after the color coat has dried

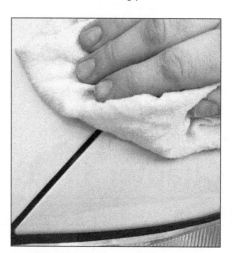

Wait a few days for the paint to dry thoroughly, then rub out the repainted area with a polishing compound to blend the new paint with the surrounding area. When you're happy with your work, wash and polish the area

ried out with the panel installed.
2   Clean the area(s) to be repaired with a wax, grease and silicone removing solvent applied with a water-dampened cloth.
3   If the damage is structural, that is, if it extends through the panel, clean the backside of the panel area to be repaired as well. Wipe dry.
4   Sand the rear surface about 1-1/2 inches beyond the break.
5   Cut two pieces of fiberglass cloth large enough to overlap the break by about 1-1/2 inches. Cut only to the required length.
6   Mix the adhesive from the repair kit according to the instructions included with the kit, and apply a layer of the mixture approximately 1/8-inch thick on the backside of the panel. Overlap the break by at least 1-1/2 inches.
7   Apply one piece of fiberglass cloth to the adhesive and cover the cloth with additional adhesive. Apply a second piece of fiberglass

cloth to the adhesive and immediately cover the cloth with additional adhesive in sufficient quantity to fill the weave.
8   Allow the repair to cure for 20 to 30 minutes at 60-degrees to 80-degrees F.
9   If necessary, trim the excess repair material at the edge.
10   Remove all of the paint film over and around the area(s) to be repaired. The repair material should not overlap the painted surface.
11   With a drill motor and a sanding disc (or a rotary file), cut a "V" along the break line approximately 1/2-inch wide. Remove all dust and loose particles from the repair area.
12   Mix and apply the repair material. Apply a light coat first over the damaged area; then continue applying material until it reaches a level slightly higher than the surrounding finish.
13   Cure the mixture for 20 to 30 minutes

at 60-degrees to 80-degrees F.

14   Roughly establish the contour of the area being repaired with a body file. If low areas or pits remain, mix and apply additional adhesive.

15   Block sand the damaged area with sandpaper to establish the actual contour of the surrounding surface.

16   If desired, the repaired area can be temporarily protected with several light coats of primer. Because of the special paints and techniques required for flexible body panels, it is recommended that the vehicle be taken to a paint shop for completion of the body repair.

Steel body panels

See photo sequence

Repair of dents

17   When repairing dents, the first job is to pull the dent out until the affected area is as close as possible to its original shape. There is no point in trying to restore the original shape completely as the metal in the damaged area will have stretched on impact and cannot be restored to its original contours. It is better to bring the level of the dent up to a point that is about 1/8-inch below the level of the surrounding metal. In cases where the dent is very shallow, it is not worth trying to pull it out at all.

18   If the backside of the dent is accessible, it can be hammered out gently from behind using a soft-face hammer. While doing this, hold a block of wood firmly against the opposite side of the metal to absorb the hammer blows and prevent the metal from being stretched.

19   If the dent is in a section of the body which has double layers, or some other factor makes it inaccessible from behind, a different technique is required. Drill several small holes through the metal inside the damaged area, particularly in the deeper sections. Screw long, self-tapping screws into the holes just enough for them to get a good grip in the metal. Now pulling on the protruding heads of the screws with locking pliers can pull out the dent.

20   The next stage of repair is the removal of paint from the damaged area and from an inch or so of the surrounding metal. This is easily done with a wire brush or sanding disk in a drill motor, although it can be done just as effectively by hand with sandpaper. To complete the preparation for filling, score the surface of the bare metal with a screwdriver or the tang of a file or drill small holes in the affected area. This will provide a good grip for the filler material. To complete the repair, see the Section on filling and painting.

Repair of rust holes or gashes

21   Remove all paint from the affected area and from an inch or so of the surrounding metal using a sanding disk or wire brush mounted in a drill motor. If these are not available, a few sheets of sandpaper will do the job just as effectively.

22   With the paint removed, you will be able to determine the severity of the corrosion and decide whether to replace the whole panel, if possible, or repair the affected area. New body panels are not as expensive as most people think and it is often quicker to install a new panel than to repair large areas of rust.

23   Remove all trim pieces from the affected area except those which will act as a guide to the original shape of the damaged body, such as headlight shells, etc. Using metal snips or a hacksaw blade, remove all loose metal and any other metal that is badly affected by rust. Hammer the edges of the hole in to create a slight depression for the filler material.

24   Wire-brush the affected area to remove the powdery rust from the surface of the metal. If the back of the rusted area is accessible, treat it with rust inhibiting paint.

25   Before filling is done, block the hole in some way. This can be done with sheet metal riveted or screwed into place, or by stuffing the hole with wire mesh.

26   Once the hole is blocked off, the affected area can be filled and painted. See the following subsection on filling and painting.

Filling and painting

27   Many types of body fillers are available, but generally speaking, body repair kits which contain filler paste and a tube of resin hardener are best for this type of repair work. A wide, flexible plastic or nylon applicator will be necessary for imparting a smooth and contoured finish to the surface of the filler material. Mix up a small amount of filler on a clean piece of wood or cardboard (use the hardener sparingly). Follow the manufacturer's instructions on the package, otherwise the filler will set incorrectly.

28   Using the applicator, apply the filler paste to the prepared area. Draw the applicator across the surface of the filler to achieve the desired contour and to level the filler surface. As soon as a contour that approximates the original one is achieved, stop working the paste. If you continue, the paste will begin to stick to the applicator. Continue to add thin layers of paste at 20-minute intervals until the level of the filler is just above the surrounding metal.

29   Once the filler has hardened, the excess can be removed with a body file. From then on, progressively finer grades of sandpaper should be used, starting with a 180-grit paper and finishing with 600-grit wet-or-dry paper. Always wrap the sandpaper around a flat rubber or wooden block, otherwise the surface of the filler will not be completely flat. During the sanding of the filler surface, the wet-or-dry paper should be periodically rinsed in water. This will ensure that a very smooth finish is produced in the final stage.

30   At this point, the repair area should be surrounded by a ring of bare metal, which in turn should be encircled by the finely feathered edge of good paint. Rinse the repair area with clean water until all of the dust produced by the sanding operation is gone.

31   Spray the entire area with a light coat of primer. This will reveal any imperfections in the surface of the filler. Repair the imperfections with fresh filler paste or glaze filler and once more smooth the surface with sandpaper. Repeat this spray-and-repair procedure until you are satisfied that the surface of the filler and the feathered edge of the paint are perfect. Rinse the area with clean water and allow it to dry completely.

32   The repair area is now ready for painting. Spray painting must be carried out in a warm, dry, windless and dust free atmosphere. These conditions can be created if you have access to a large indoor work area, but if you are forced to work in the open, you will have to pick the day very carefully. If you are working indoors, dousing the floor in the work area with water will help settle the dust that would otherwise be in the air. If the repair area is confined to one body panel, mask off the surrounding panels. This will help minimize the effects of a slight mismatch in paint color. Trim pieces such as chrome strips, door handles, etc., will also need to be masked off or removed. Use masking tape and several thickness of newspaper for the masking operations.

33   Before spraying, shake the paint can thoroughly, then spray a test area until the spray painting technique is mastered. Cover the repair area with a thick coat of primer. The thickness should be built up using several thin layers of primer rather than one thick one. Using 600-grit wet-or-dry sandpaper, rub down the surface of the primer until it is very smooth. While doing this, the work area should be thoroughly rinsed with water and the wet-or-dry sandpaper periodically rinsed as well. Allow the primer to dry before spraying additional coats.

34   Spray on the top coat, again building up the thickness by using several thin layers of paint. Begin spraying in the center of the repair area and then, using a circular motion, work out until the whole repair area and about two inches of the surrounding original paint is covered. Remove all masking material 10 to 15 minutes after spraying on the final coat of paint. Allow the new paint at least two weeks to harden, then use a very fine rubbing compound to blend the edges of the new paint into the existing paint. Finally, apply a coat of wax

4   Body repair - major damage

1   Major damage must be repaired by an auto body shop specifically equipped to perform body and frame repairs. These shops have the specialized equipment required to do the job properly.

2   If the damage is extensive, the frame must be checked for proper alignment or the vehicle's handling characteristics may be adversely affected and other components may wear at an accelerated rate.

3   Due to the fact that all of the major body components (hood, fenders, etc.) are separate and replaceable units, any seriously damaged components should be replaced rather than repaired. Sometimes the components can be found in a wrecking yard that specializes in used vehicle components, often at considerable savings over the cost of new parts.

These photos illustrate a method of repairing simple dents. They are intended to supplement *Body repair - minor damage* in this Chapter and should not be used as the sole instructions for body repair on these vehicles.

1 If you can't access the backside of the body panel to hammer out the dent, pull it out with a slide-hammer-type dent puller. Tap with a hammer near the edge of the dent to help 'pop' the metal back to its original shape, about 1/8-inch below the surface of the surrounding metal

2 Using coarse-grit sandpaper, remove the paint down to the bare metal. Clean the repair area with wax/silicone remover.

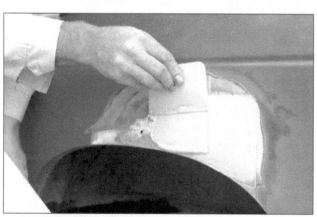

3 Following label instructions, mix up a batch of plastic filler and hardener, then quickly press it into the metal with a plastic applicator. Work the filler until it matches the original contour and is slightly above the surrounding metal

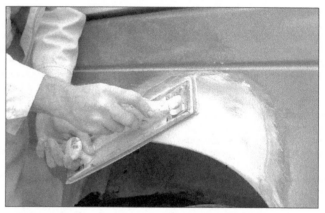

4 Let the filler harden until you can just dent it with your fingernail. File, then sand the filler down until it's smooth and even. Work down to finer grits of sandpaper - always using a board or block - ending up with 360 or 400 grit

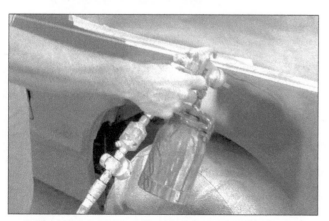

5 When the area is smooth to the touch, clean the area and mask around it. Apply several layers of primer to the area. A professional-type spray gun is being used here, but aerosol spray primer works fine

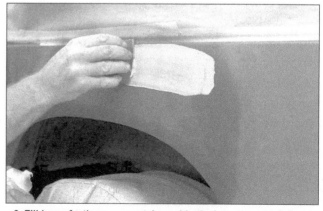

6 Fill imperfections or scratches with glazing compound. Sand with 360 or 400-grit and re-spray. Finish sand the primer with 600 grit, clean thoroughly, then apply the finish coat. Don't attempt to rub out or wax the repair area until the paint has dried completely (at least two weeks)

5 Upholstery, carpets and vinyl trim - maintenance

Upholstery and carpets

1   Every three months remove the floor-mats and clean the interior of the vehicle (more frequently if necessary). Use a stiff whiskbroom to brush the carpeting and loosen dirt and dust, then vacuum the upholstery and carpets thoroughly, especially along seams and crevices.

2   Dirt and stains can be removed from carpeting with basic household or automotive carpet shampoos available in spray cans. Follow the directions and vacuum again, then use a stiff brush to bring back the "nap" of the carpet.

3   Most interiors have cloth or vinyl upholstery, either of which can be cleaned and maintained with a number of material-specific cleaners or shampoos available in auto supply stores. Follow the directions on the product for usage, and always spot-test any upholstery cleaner on an inconspicuous area (bottom edge of a backseat cushion) to ensure that it doesn't cause a color shift in the material.

4   After cleaning, vinyl upholstery should be treated with a protectant. **Note:** *Make sure the protectant container indicates the product can be used on seats - some products may make a seat too slippery.* **Caution:** *Do not use protectant on vinyl-covered steering wheels.*

5   Leather upholstery requires special care. It should be cleaned regularly with saddlesoap or leather cleaner. Never use alcohol, gasoline, nail polish remover or thinner to clean leather upholstery.

6   After cleaning, regularly treat leather upholstery with a leather conditioner, rubbed in with a soft cotton cloth. Never use car wax on leather upholstery.

7   In areas where the interior of the vehicle is subject to bright sunlight, cover leather seating areas of the seats with a sheet if the vehicle is to be left out for any length of time.

Vinyl trim

8   Don't clean vinyl trim with detergents, caustic soap or petroleum-based cleaners. Plain soap and water works just fine, with a soft brush to clean dirt that may be ingrained. Wash the vinyl as frequently as the rest of the vehicle.

9   After cleaning, application of a high-quality rubber and vinyl protectant will help prevent oxidation and cracks. The protectant can also be applied to weather-stripping, vacuum lines and rubber hoses, which often fail as a result of chemical degradation, and to the tires.

6 Fastener and trim removal

Refer to illustration 6.4

1   There is a variety of plastic fasteners used to hold trim panels, splash shields and other parts in place in addition to typical screws, nuts and bolts. Once you are familiar with them, they can usually be removed without too much difficulty.

2   The proper tools and approach can prevent added time and expense to a project by minimizing the number of broken fasteners and/or parts.

3   The following illustration shows various types of fasteners that are typically used on most vehicles and how to remove and install them **(see illustration)**. Replacement fasteners are commonly found at most auto parts stores, if necessary.

Fasteners

This tool is designed to remove special fasteners. A small pry tool used for removing nails will also work well in place of this tool

A Phillips head screwdriver can be used to release the center portion, but light pressure must be used because the plastic is easily damaged. Once the center is up, the fastener can easily be pried from its hole

Here is a view with the center portion fully released. Install the fastener as shown, then press the center in to set it

This fastener is used for exterior panels and shields. The center portion must be pried up to release the fastener. Install the fastener with the center up, then press the center in to set it

This type of fastener is used commonly for interior panels. Use a small blunt tool to press the small pin at the center in to release it . . .

. . . the pin will stay with the fastener in the released position

Reset the fastener for installation by moving the pin out. Install the fastener, then press the pin flush with the fastener to set it

This fastener is used for exterior and interior panels. It has no moving parts. Simply pry the fastener from its hole like the claw of a hammer removes a nail. Without a tool that can get under the top of the fastener, it can be very difficult to remove

6.4 These small plastic pry tools are ideal for prying off trim panels

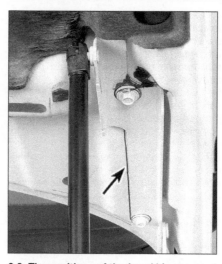

9.2 The positions of the hood hinges must be marked prior to removal of the hood

9.10 Twist the bumper stop to make fine adjustments to the hood closed height

4   Trim panels are typically made of plastic and their flexibility can help during removal. The key to their removal is to use a tool to pry the panel near its retainers to release it without damaging surrounding areas or breaking-off any retainers. The retainers will usually snap out of their designated slot or hole after force is applied to them. Stiff plastic tools designed for prying on trim panels are available at most auto parts stores **(see illustration)**. Tools that are tapered and wrapped in protective tape, such as a screwdriver or small pry tool, are also very effective when used with care.

7   Hinges and locks - maintenance

Once every 3000 miles, or every three months, the hinges and latch assemblies on the doors, hood and trunk should be given a few drops of light oil or lock lubricant. The door latch strikers should also be lubricated with a thin coat of grease to reduce wear and ensure free movement. Lubricate the door and trunk locks with spray-on graphite lubricant.

8   Windshield and fixed glass - replacement

Replacement of the windshield and fixed glass requires the use of special fast-setting urethane adhesive materials and specialized tools and techniques. These operations should be left to a shop specializing in glass work.

9   Hood - removal, installation and adjustment

Note: *The hood is heavy and somewhat awkward to remove and install - at least two people should perform this procedure.*

Removal and installation
Refer to illustration 9.2
1   Use blankets or pads to cover the cowl area of the body and the fenders. This will protect the body and paint as the hood is lifted off.
2   Scribe alignment marks around the hinge plates to insure proper alignment dur-

ing installation (paint or a permanent-type felt-tip marker also will work for this) **(see illustration)**.
3   Have an assistant support one side of the hood while you support the other. Simultaneously remove the hinge-to-hood bolts.
4   Lift off the hood.
5   Installation is the reverse of removal.

Adjustment
Refer to illustration 9.10
6   Fore-and-aft and side-to-side adjustment of the hood is done by moving the hood in relation to the hinge plates after loosening the bolts. Scribe or trace a line around the entire hinge plate so you can judge the amount of movement.
7   Loosen the bolts and move the hood into correct alignment. Move it only a little at a time. Tighten the hinge bolts and carefully lower the hood to check the alignment.
8   If the rear of the hood is too low in relation to the fenders, install washers between the hinges and the hood.
9   If necessary after installation, the entire hood latch assembly can be adjusted up-and-down as well as from side-to-side on the radiator support so the hood closes securely and flush with the fenders. To make the adjustment, scribe a line or mark around the hood latch to provide a reference point, then loosen them and reposition the latch assembly, as necessary **(see illustration 10.7)**. Following adjustment, retighten the mounting nuts.
10   Finally, adjust the hood bumpers on the radiator support so the hood, when closed, is flush with the fenders **(see illustration)**.
11   The hood latch assembly, as well as the hinges, should be periodically lubricated with white lithium-base grease to prevent sticking and wear.

10.7 Remove the two nuts from the latch . . .

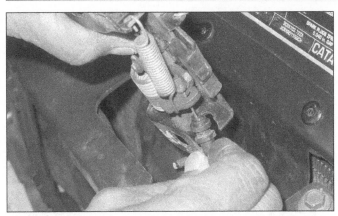

10.8 . . . then remove the latch and disconnect the latch cable

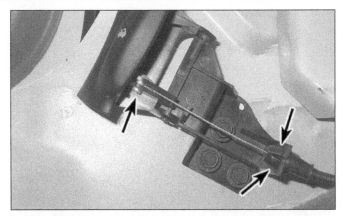

10.13 The hood latch cable is disconnected from the handle by collapsing the plastic retainer ears, then unhooking the cable end

10 Hood latch and cable - removal and installation

Refer to illustrations 10.7, 10.8 and 10.13

Latch

2002 models

1   Refer to Section 11 and remove the grille.
2   Remove the bolts from the center vertical latch bracket.
3   Scribe or mark a line around the latch, then remove the two nuts from the latch.
4   Disconnect the cable and remove the latch.
5   Installation is the reverse of the removal procedure.

2003 and later models

6   On 2007 and earlier models, remove the three fasteners at the upper part of the grille support. Pull the grille support forward to access the latch assembly.
7   Scribe or mark a line around the latch, then remove the two nuts from the latch **(see illustration)**.
8   Disconnect the cable and remove the latch **(see illustration)**.
9   Installation is the reverse of the removal procedure.

Cable

10   Refer to Chapter 5 and remove the battery.
11   Refer to Step 1 and remove the latch.
12   Refer to Chapter 6 and remove the Powertrain Control Module.
13   Remove the left side scuff plate and kick panel. Unbolt the hood release handle, then detach the cable from it **(see illustration)**.
14   Working in the engine compartment, detach the cable from its retaining clips, then push the cable grommet through the firewall. From inside the vehicle, pull the cable through the firewall.
15   Installation is the reverse of the removal procedure. Make sure the cable grommet seats completely in the firewall.

11 Radiator grille - removal and installation

2007 and earlier models

Refer to illustrations 11.1 and 11.2
1   Remove the top mounting screws or clips **(see illustration)**.
2   Pull the grille forward and release the hooks at the bottom **(see illustration)**.
3   Installation is the reverse of removal.

2008 and later models

4   Remove the hood latch (see Section 10).
5   Remove the plastic push pins securing the top of the grille to the radiator support.
6   Remove the bumper stops (see Section 9).
7   Using a plastic trim tool, disengage the hooks from the upper corners of the grille, just above the headlights.
8   Using a plastic trim tool, disengage the six hooks along the bottom edge of the grille, and remove the grille.
9   On Nitro models, pull the grille forward enough to disconnect the electrical connector to the lamps.
Note: *If you're replacing the grille, remove he front lamp assembly fasteners and remove the lamp assemblies from the grille.*
10   Installation is the reverse of removal.

12 Front fender - removal and installation

Refer to illustrations 12.2a, 12.2b, 12.8, 12.9, 12.11a, 12.11b and 12.11c
1   Loosen the front wheel lug nuts. Raise the vehicle and support it securely on jackstands.

11.1 There are four screws that secure the top of the grille

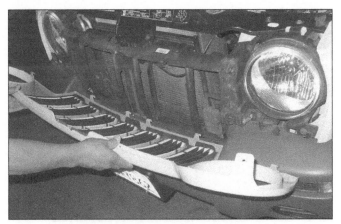

11.2 After swinging the grille forward and down, disengage the hooks at the bottom

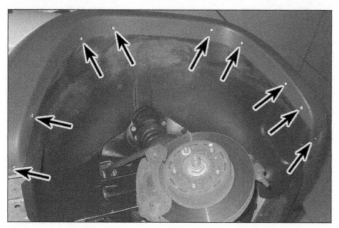

12.2a Front fender flare rivet locations

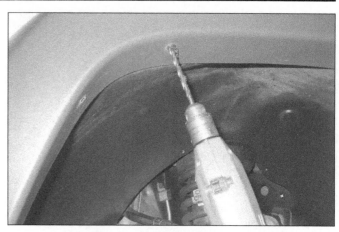

12.2b The rivets must be carefully drilled out

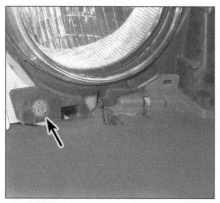

12.8 The front fender bracket bolts are accessible after the fascia is pulled aside

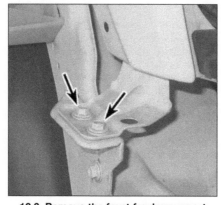

12.9 Remove the front fender support bracket bolts

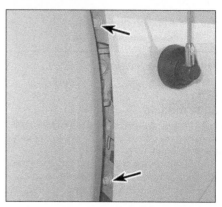

12.11a The two upper rear mounting bolts can be removed when the door is partly open

Remove the front wheel. Refer to Section 11 and remove the grille.

2   On 2007 and earlier models, remove the rivets securing the flare molding to the fender and the fascia **(see illustrations)**.

3   Carefully pry the fender flare away from the front fascia to release the clip. **Note:** *On 2008 and later models, once the flare is pried out, remove the mounting screws.*

4   Remove the pushpin securing the inner splash shield to the air dam.

5   Remove the remainder of the pushpins securing the splash shield, then remove the shield from the vehicle.

6   On 2005 and later models, remove the front bumper fascia to access the front fender bolts (see Section 13).

7   Release the clips located between the lights in the fascia, if equipped.

8   Pull the fascia to one side to gain access to the front fender bracket bolts **(see illustration)**. **Note:** *On 2008 and later models, remove*

the fascia support bracket bolts and bracket.

9   Remove the bolts from the front of the fender at the lower support bracket **(see illustration)**.

10   Remove the three bolts from the top of the fender.

11   Remove the bolts from the rear of the fender **(see illustrations)**.

12   The antenna body must be removed if you are working on the right fender. Unscrew the antenna mast.

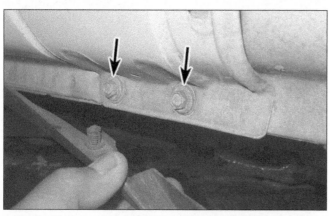

12.11b The lower pinchweld trim must be pulled away to allow access to the lower bolts

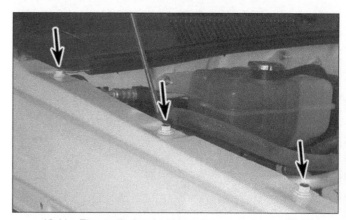

12.11c The top bolts should be the last to be removed

13.6 The six pushpins that secure the upper fascia are highlighted here

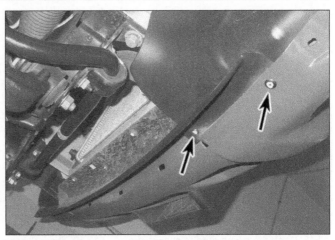

13.7 Remove the fasteners from the air dam (two shown here)

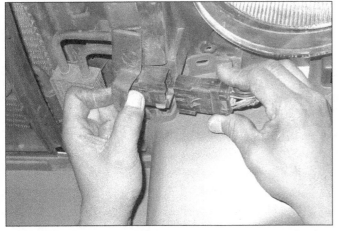

13.10 Disconnect the fascia wiring before removing the fascia

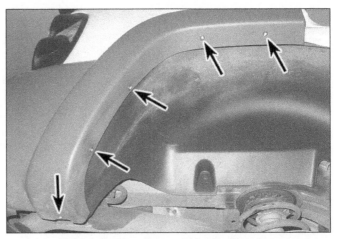

13.15 Locations of the rear fender flare rivets

13   Pull outward on the upper rear of the fender. Remove the antenna base cover, the nut beneath it and the bezel adapter.

14   Carefully lift the fender from the vehicle. It's a good idea to have an assistant support the fender while it's being moved away from the vehicle to prevent damage to the surrounding body panels.

15   Installation is the reverse of removal.

13   Bumper fascia - removal and installation

Note: *The number and exact location of the bumper fascia fasteners and clips may vary slightly between model years.*

Front

Refer to illustrations 13.6, 13.7 and 13.10

1   The front bumper on these models consists of a fascia. It covers the lower crossmember that functions as the actual bumper.

2   Raise the vehicle and support it securely on jackstands. On 2005 through 2007 models, remove the park-turn signal light from the fender, then remove the fascia mounting screws near the park lights.

3   Refer to Section 11 and remove the grille.

4   Remove the screws from the air dam that are visible from below.

5   On 2008 and later models, remove the inner fender splash shields (see Section 12).

6   Disconnect any electrical connectors, then remove the clips from the upper part of the fascia **(see illustration)**.

7   Remove the plastic rivets that secure the air dam to the splash shield **(see illustration)**.

8   Release the tabs below each headlight.

9   Release the clips located between the lights in the fascia **(see illustration 12.9)**

10   Disconnect the electrical connectors **(see illustration)** and remove the bumper fascia.

11   Installation is the reverse of the removal procedure.

Rear

Refer to illustrations 13.15, 13.16, 13.18, 13.19, 13.22, 13.23 and 13.24

12   The rear bumper consists of a fascia that covers a fascia support. The fascia support is riveted directly to the frame of the vehicle.

13   Loosen the rear wheel lug nuts. Raise the vehicle and support it securely on jackstands.

14   Remove the rear wheels.

15   Remove the five rivets that secure the rear section of each fender flare to the inner splash shields **(see illustration)**. **Note:** *Only remove the rear portion of the rear wheel flare moldings - do not remove the front door side flare moldings.*

16   Carefully pry to release the clips that secure the flares to the body and the fascia **(see illustration)**.

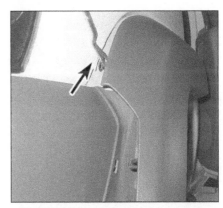

13.16 If you use a metal tool to release the flare clips, take care to protect the body and paint by putting tape under the tool

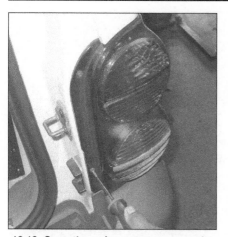

13.18 Open the swing gate to access the screws for the rear lights

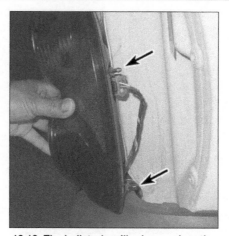

13.19 The ballstuds will release when the light assembly is pulled out

13.22 The rear fascia end bolts

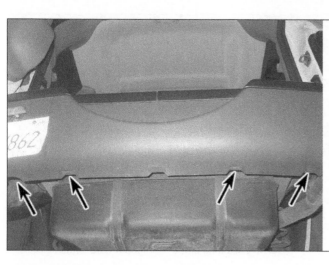

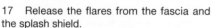

13.23 The bottom of the fascia is secured with pushpin-type plastic rivets. Push in on the center of the rivets to unlock them, then pull them out

24   Remove the three bolts along the top of the fascia **(see illustration)**.
25   Remove the fascia.
26   To reinstall the fascia, first install new ball-stud retainers in the quarter panel (if necessary).
27   The remainder of installation is the reverse of removal.

**14   Door trim panel -
        removal and installation**

Refer to illustrations 14.1, 14.2, 14.3, 14.5 and 14.6
Note: *The procedures for the front and rear trim panels are the same.*
1   On 2007 and earlier models, remove the screw cap from the inside door handle, then remove the screw **(see illustration)**. On 2008 and later models, carefully pry the handle trim from around the door handle using a trim removal tool.
2   Remove the screw from the pull handle **(see illustration)**.

2007 and earlier models

3   Insert a special door panel trim tool (available at auto parts stores) or a putty knife

17   Release the flares from the fascia and the splash shield.
18   With the swing gate open, remove the light housing screws **(see illustration)**.
19   Pull back on the outer side of the light assembly and then release the two ballstuds **(see illustration)**.
20   Disconnect the light assembly wiring.
21   Set the light assembly aside and discard

the plastic ballstud retainers from the quarter panel (it's recommended that they be replaced with new ones whenever the light assembly is removed, but if they're still a nice tight fit, they can be reused).
22   Remove the two bolts at each end of the fascia **(see illustration)**.
23   Remove the four pushpin rivets from the bottom of the fascia **(see illustration)**.

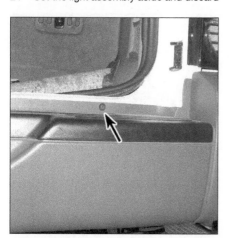

13.24 The upper fascia uses these screws (pushpins on later models)

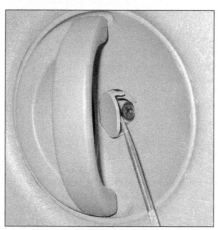

14.1 The screw for the inside door handle is under this cover

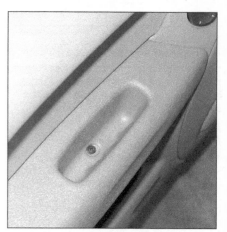

14.2 Remove this screw from the pull handle

14.3 Be careful of the painted surfaces as you pry off the door panel

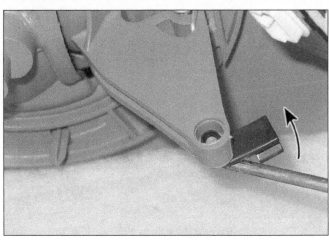

14.5 The inner door latch rod is held in place with this clip; swing the clip off the rod, then pull the rod out

between the trim panel and the door and disengage the retaining clips (see illustration). Work around the outer edge until the panel is free.
4   Once all of the clips are disengaged, detach the trim panel.

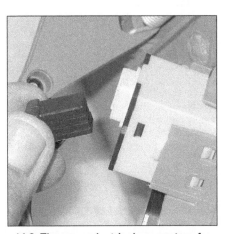

14.6 There are electrical connectors for the power windows (and the power mirrors on the driver's side panel) that will have to be disconnected if the vehicle is so equipped

15.1 Disconnect the wiring before proceeding with the other steps

5   Reach behind the panel and disconnect the rod from the handle mechanism (see illustration).
6   Disconnect the electrical connector(s) and remove the trim panel from the vehicle (see illustration).

2008 and later models
7   Carefully pry the trim cap from the bottom of the panel, then remove the trim panel mounting screw.
8   Carefully lift the door panel upwards to disengage the panel hooks from the door module.

All models
9   For access to the inner door, carefully peel back the plastic watershield.
10   Prior to installation of the door panel, be sure to reinstall any clips in the panel which may have come out during the removal procedure and remain in the door itself.
11   Connect the electrical connector(s), connect the handle rod and place the panel in position in the door. Press the trim panel into place until the clips are seated.
12   Install the screws in the pull handle and inside door handle.

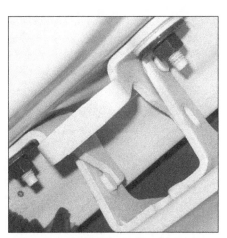

15.4 Remove the lower door hinge nuts and then the upper ones

15   Door - removal, installation and adjustment

Refer to illustrations 15.1 and 15.4

Note: *The procedures for the front and rear doors are the same.*
1   Disconnect the cable from the negative terminal of the battery (see Chapter 5). Disconnect the wiring connector at the front edge of the door (see illustration).
2   Place a jack under the door or have an assistant on hand to support it when the hinge nuts are removed. **Note:** *If a jack is used, place a cushion between it and the door to protect the door's painted surfaces.*
3   Remove the bolts securing the door check strap to the door jamb.
4   Remove the nuts and carefully lift off the door (see illustration).
5   Installation is the reverse of removal.
6   Following installation of the door, check the alignment and adjust it if necessary by moving the door lock striker.
7   The doors can also be adjusted vertically and horizontally by loosening the hinge-to-body fasteners. **Note:** *If the hinge-to-door connection must be adjusted, it will be necessary to remove the washers from the hinge. They are epoxied on at the factory and must be removed with a hammer and chisel. If possible, avoid making the adjustment at this point. Always adjust the rear door before adjusting the front door.*

16   Door latch, lock cylinder and handles - removal and installation

2007 and earlier models
Latch
Refer to illustrations 16.4, 16.5 and 16.6
1   Disconnect the cable from the negative terminal of the battery (see Chapter 5).
2   Refer to Section 14 and remove the interior door trim panel.

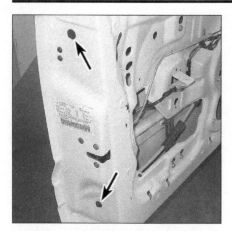

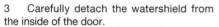

16.4 The bolts for the rear glass channel are accessible at the rear edge of the door

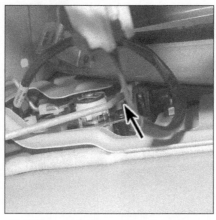

16.5 The rod for the outside door handle can be disconnected at the handle (shown here) or at the latch

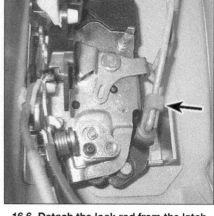

16.6 Detach the lock rod from the latch

3   Carefully detach the watershield from the inside of the door.
4   On front doors, remove the two bolts for the glass channel from the rear of the door **(see illustration)**. Move the rubber channel aside for clearance.
5   Disconnect the rod from the outside door handle **(see illustration)**.
6   Disconnect the rod for the lock button at the latch **(see illustration)**.
7   On front doors, disconnect the rod for the key lock at the latch.
8   Unbolt the latch assembly, pull it out and disconnect the wiring.
9   Installation is the reverse of the removal procedure.
10   Check the door to make sure it closes properly. Readjust the latch (by loosening the bolts and moving it) as necessary until the door closes smoothly (with the door handle flush with the door).

Lock cylinder

Refer to illustration 16.12
11   Remove the outside door handle (see below).
12   The lock cylinder is part of the outside handle assembly. Remove the screw to extract the lock cylinder from the handle **(see illustration)**.
13   Installation is the reverse of the removal procedure.

Outside handle

Refer to illustration 16.18
14   Refer to Section 17 and remove the window.
15   On vehicles with power locks, disconnect the switch wiring at the handle.
16   Disconnect the handle rod at the handle itself.
17   On front doors, disconnect the rod for the lock cylinder at the latch assembly
18   Remove the fasteners, then remove the handle **(see illustration)**.
19   Installation is the reverse of the removal procedure.

Inside handle

20   Pry the cap from the retaining screw. Note the position of the handle. Remove the screw and pull off the handle.
21   Installation is the reverse of removal.

2008 and later models

Latch
22   Remove the door module (see Section 18).
23   Disconnect the electrical connector to the latch.
24   Rotate the door module to the back side and slide the latch bracket to the left until the mounting clips are free.
25   Remove the latch assembly and bracket from the door module.
26   The latch assembly is connected to the bracket at three different attaching points. Disconnect the bracket from the latch at each point and remove the latch from the bracket.
27   Installation is the reverse of removal.

Lock cylinder
28   Remove the door module (see Section 18).
29   Remove the horse shoe type lock cylin-

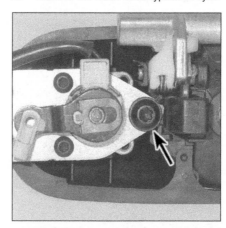

16.12 Remove this Torx-head screw to remove the lock cylinder

der retaining clip from the lock cylinder.
30   Remove the lock cylinder and gasket from the door.
31   Installation is the reverse of removal.

Outside handle
32   Remove the door trim panel (see Section 14).
33   Remove the access plug from the door.
34   Working through the access hole, remove the outside door handle mounting nut.
35   Remove the door module (see Section 18).
36   Remove the remaining outside door handle nut, then remove the handle from the outside of the door.
37   Installation is the reverse of removal.

Inside handle
38   Remove the door trim panel (see Section 14).
39   Unclip the interior lock rod from the handle.
40   Pull the push pin from the top of the handle.
41   Remove the spring and handle assembly from the door module.
42   Installation is the reverse of removal.

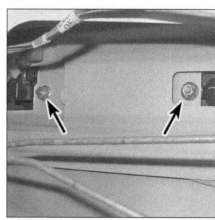

16.18 The outer handle is secured with these two nuts - later models also have one handle screw in the doorjamb

17.3 Be very careful of the paint (use protective tape) when removing the lower window weatherstrip

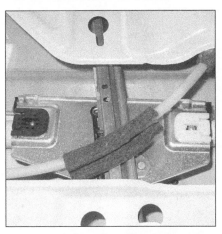

17.4 The window should be positioned so that the glass clips are accessible through the opening

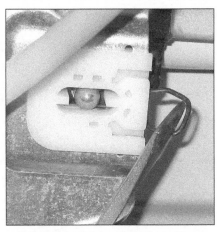

17.5 A screwdriver will work to release the glass clips. Be careful not to let the glass drop

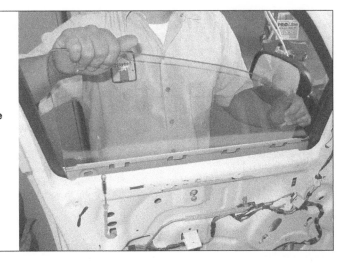

17.6 The rear of the glass is lifted up and forward

Rear door glass
Refer to illustrations 17.9 and 17.12
9   Position the glass so the clips can be accessed **(see illustration)**.
10   Release the clip by prying out **(see illustration 17.5)**. Pull the glass support from the lift plate.
11   Allow the glass to rest in the bottom of the door.
12   Remove the bolt from the vertical division bar in the door **(see illustration)**.
13   Turn the bar and release it from the door glass.
14   Lift the glass from the door.
15   Installation is the reverse of the removal procedure.

2008 and later models

Front door glass
16   Carefully pry up the weatherstrip channel from the base of the window opening.
17   Remove the tape covering the access plug in the door module.
18   Have an assistant push down on the window glass. Working through the access

---

17   Door window glass - removal and installation

1   Disconnect the cable from the negative terminal of the battery (see Chapter 5). Remove the door trim panel and watershield (see Section 14).

2007 and earlier models
2   Lower the glass.

Front door glass
Refer to illustrations 17.3, 17.4, 17.5 and 17.6
3   Pry off the outer piece of the lower window weatherstrip by pulling out on the lower part while pulling upward **(see illustration)**. Apply tape over the exposed metal edge to avoid scratching the glass.
4   Position the glass so that the clips are accessible through opening in the door **(see illustration)**.
5   Release the clips using an appropriate tool **(see illustration)**. Pull the glass support from the lift plate.
6   Pull the rear of the glass up and toward

the front to remove it from the door **(see illustration)**.
7   Put the clips back in place in the window regulator lift plate.
8   Installation is the reverse of the removal procedure.

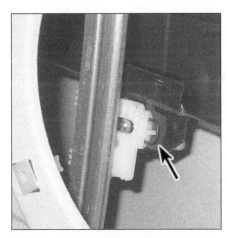

17.9 Position the glass so that the release clips are easy to reach

17.12 The division bar bolt must be removed to allow glass removal

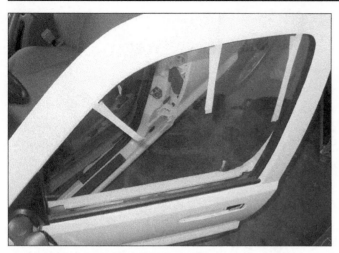

18.2 Use strong tape to hold the glass in the raised position temporarily

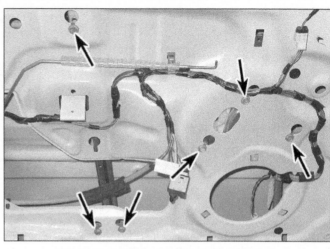

18.3 These bolts secure the window regulator

hole in the door module, use a flat-bladed screwdriver to carefully release the locking tab on the window regulator.

19   With the release tab held in this position lift the window glass up until it clears the locking tab, then remove the screwdriver.

20   Raise the window glass upwards, then rotate the front end downwards until the window can be completely removed.

21   Installation is the reverse of removal.

Rear door glass

22   Carefully pry up the weatherstrip channel from the base of the window opening.

23   Remove the stationary or quarter glass mounting bolt from the door.

24   Pull the inner widow seal out of the channel, then rotate the stationary glass counterclockwise until it can be completely removed with the seal.

25   Lower the window to the bottom of the door.

26   Use a flat-bladed screwdriver to carefully release the locking tab on the window regulator.

27   With the release tab held in this position, lift the window glass up until it clears the locking tab, then remove the screwdriver.

28   Remove the glass from the door.

29   Installation is the reverse of removal.

---

18   Door window regulator - removal and installation

---

1   Remove the door trim panel and watershield (see Section 14).

2007 and earlier models

Refer to illustrations 18.2, 18.3, 18.7a and 18.7b

2   Detach the glass from the regulator (see Section 17). **Note:** *The glass does not need to be completely removed from the door. The glass can be held out of the way in the up position by running lengths of tape across it and over the top of the door frame* **(see illustration).**

3   Loosen the window regulator-to-door attaching bolts **(see illustration).**

4   Disconnect the runout tube clip and then remove the regulator.

5   On power windows, disconnect the wiring.

6   Remove the regulator from the door.

7   Installation is the reverse of removal. Tighten all of the fasteners in the proper sequence **(see illustrations).**

2008 and later models

Note: *The window regulator is an integral component of the door module. If there is a problem with the regulator the door module must be replaced.*

8   Remove the access plug from the door.

9   Working through the access hole, remove the outside door handle mounting nut.

10   Remove the window (see Section 17).

11   Remove the three latch fasteners from the end of the door.

12   Remove the latch-to-door face mounting screw.

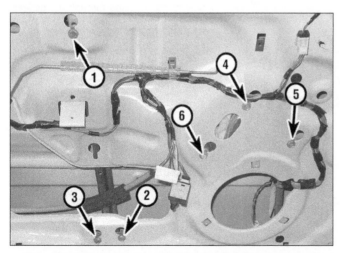

18.7a Power front window regulator bolt tightening sequence (manual windows follow the same order, eliminating number 6)

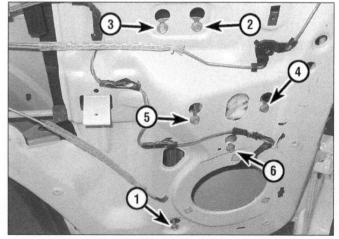

18.7b Power rear window regulator bolt tightening sequence (manual windows follow the same order, eliminating number 6)

13   Disconnect the main harness connector at the front edge of the door.

14   Remove the door module mounting screws from around the perimeter of the module.

15   Disconnect the outside mirror electrical connector.

16   Maneuver the module assembly out of the door body.

17   Installation is the reverse of removal. **Note:** *Make sure the door lock cylinder is aligned and engaged with the door module before installing the module mounting bolts.*

19   Center console - removal and installation

Refer to illustrations 19.1, 19.3, 19.4a, 19.4b, 19.5a and 19.5b

1   Carefully pry the shifter bezel from the console **(see illustration)**.

2   Pull the parking brake lever up, then pry out the parking brake handle boot with a plastic trim tool.

3   If equipped, pry the transfer case shifter

19.1 Use caution to avoid scratching the plastic when prying the bezel up

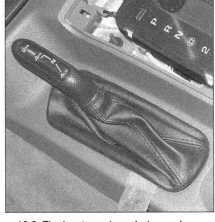

19.3 The boot can be pried up using a screwdriver or a putty knife

boot out of the console **(see illustration)**.

4   Remove the four mounting screws **(see illustrations)**.

5   Remove the window switch panels if so equipped **(see illustrations)**.

6   On 2008 and later models, pry the end

cap off of the center console using a trim tool, then disconnect the electrical connectors.

7   Raise the rear of the console first to disengage it, and then remove it.

8   Installation is the reverse of the removal procedure.

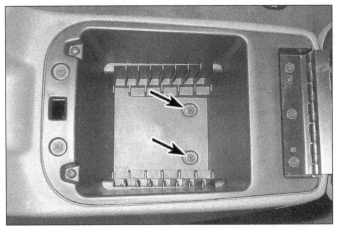

19.4a The two rear console screws at the bottom of the storage compartment

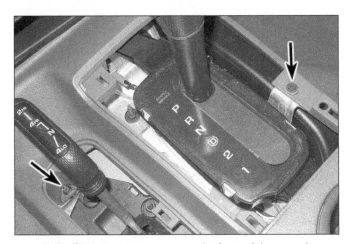

19.4b These two screws secure the front of the console

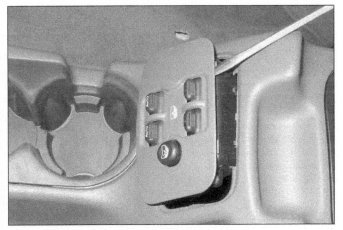

19.5a On vehicles with power windows, pry the window switch up and then disconnect the wiring

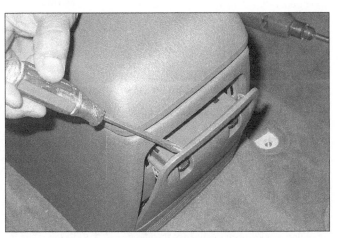

19.5b The switches for the rear windows are removed and disconnected just like the front switches

20.1 Carefully work the removal tool around the perimeter of the swing gate trim panel

21.2 Pull the weatherstrip from the rear opening

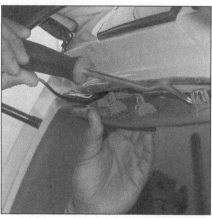

21.3 Detach the upper interior trim panel

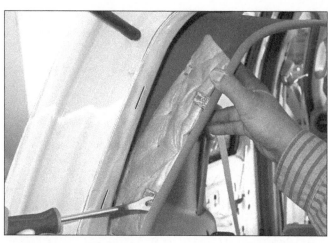

21.4 Pry the rear portion of the left quarter panel trim loose

21.5 Use a Torx bit to remove the left rear seat belt mounting bolt

20 Swing gate/liftgate trim panel - removal and installation

Note: *2007 and earlier models are equipped with a swing gate; 2008 and later models are equipped with a liftgate. Both styles use a flip-up type rear window glass.*

Swing gate

Refer to illustration 20.1
1   Using a screwdriver wrapped with tape, release the pushpin rivets securing the panel **(see illustration)**.
2   Lift the panel up so that it disengages from the upper clips.
3   For access to the swing gate inner panel, carefully peel back the plastic sound deadener.
4   Installation is the reverse of the removal procedure.

Liftgate

5   Open the liftgate, then remove the two mounting screws from the base of the liftgate panel.
6   Using a trim tool, release the trim panel

fasteners and remove the trim panel from the liftgate.
7   Installation is the reverse of removal.

21 Swing gate/liftgate - removal and installation

Note: *2007 and earlier models are equipped with a swing gate; 2008 and later models are equipped with a liftgate. Both styles use a flip-up type rear window glass.*

Swing gate

Refer to illustration 21.2, 21.3, 21.4, 21.5, 21.6, 21.7 and 21.9
Warning: *In addition to driver and passenger airbags, some later models may be equipped with pyrotechnic seat belt tensioners and/or side-impact airbags located in the seatbacks or along the inside edges of the headliner. Whenever working around these components, perform the procedure to disable the airbag system (see Chapter 12).*
1   Remove the spare tire.
2   Remove the weatherstrip **(see illustration)**.

3   Carefully pry off the upper interior header trim **(see illustration)**.
4   Pry loose the side trim pieces **(see illustration)**.
5   Remove the upper connection for the left rear seat belt **(see illustration)**.
6   Remove the bolts securing the check strap to the body **(see illustration)**.

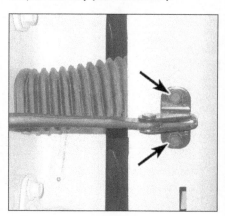

21.6 Remove the two bolts that hold the check strap to the body

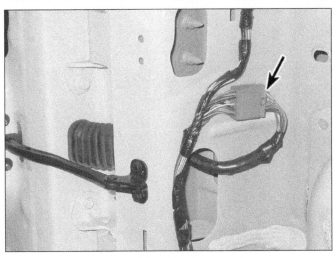

21.7 Disconnect the wiring for the swing gate

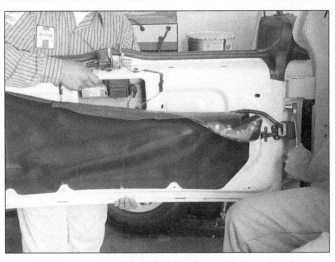

21.9 Have a helper support the gate while you remove the mounting bolts

7   Pull the quarter panel trim piece forward and then disconnect the wiring for the swing gate **(see illustration)**.
8   Have an assistant hold the weight of the swing gate or support it with a floor jack. If using a floor jack, pad the jack saddle with rags to prevent damage to the swing gate.
9   Mark the positions of the bolts, then remove the gate-to-hinge bolts from the swing gate hinges and remove the swing gate **(see illustration)**.
10   Installation is the reverse of the removal procedure. Be sure to line up the bolt heads with the marks made in Step 9.

Liftgate

11   Open the liftgate and carefully pry the headliner trim panel down from the opening.
12   Remove the liftgate panel (see Section 20).
13   Disconnect the liftgate harness and remove the grommet and harness from the liftgate.

14   Support the liftgate securely in the open position. **Caution:** *The liftgate is heavy and awkward; have an assistant help and make sure the liftgate is properly supported before removing the hinges.*
15   Remove the high-mount brake light (see Chapter 12).
16   Mark around the hinges so they can be installed in the same position.
17   With the help of an assistant, slowly remove the hinge nuts/bolts and remove the liftgate from the vehicle.
18   Installation is the reverse of removal.

22 Swing gate/liftgate glass support cylinders - replacement

Refer to illustration 22.2
Note: *2007 and earlier models are equipped with a swing gate; 2008 and later models are equipped with a liftgate. Both styles use a flip up type rear window glass.*

1   Open the rear flip-up glass. Support it securely with a piece of wood or a similar rod. Do not use a piece of metal that could scratch or possibly break the tempered glass if it were to slip.
2   Use a small flat-blade screwdriver to pull the clips partially out of each end of the cylinder **(see illustration)**. The clips will release the socket ends when they are pulled out a short distance. Do not try to remove them completely.
3   Remove the cylinder.
4   To install a replacement cylinder, first verify that the clips are pushed in at each end.
5   Snap the thin end of the cylinder to the glass and the thicker end to the body of the vehicle. The ends will snap solidly into place when installed correctly.

23 Dashboard trim panels - removal and installation

Warning: *The models covered by this manual are equipped with a Supplemental Restraint System (SRS), more commonly known as airbags. Always disarm the airbag system before working in the vicinity of any airbag system component to avoid the possibility of accidental deployment of the airbag, which could cause personal injury (see Chapter 12). Do not use a memory-saving device to preserve the PCM's memory when working on or near airbag system components.*

Dashboard top cover
Refer to illustrations 23.1, 23.3 and 23.4
1   Use a small screwdriver to pry the covers from the grab handle screws on the windshield pillar **(see illustration)**.
2   Remove the grab handle screws and remove the handle from the windshield pillar.

2007 and earlier models
3   Grasp the windshield pillar trim moldings and pull them from their retainers **(see illustration)**.

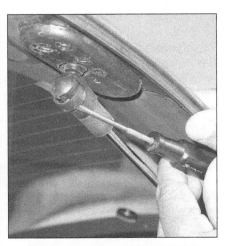

22.2 A small screwdriver is used to release the lift glass support struts

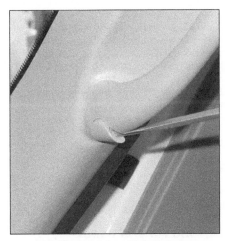

23.1 The grab handle screws are concealed beneath these covers; carefully pry them off

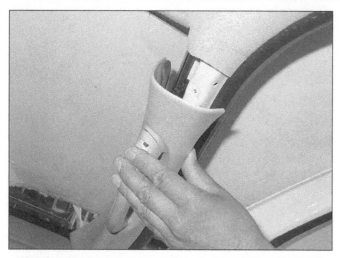

23.3 Starting from the top, pull the windshield pillar covers off

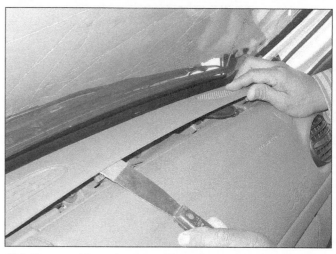

23.4 Try to avoid marring the soft plastic dash when prying off the dash top cover

23.12 The end caps can be removed by pulling them out from the bottom (2007 and earlier models shown, later models similar)

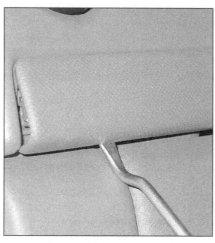

23.14 The dash bezels must be pried off with an appropriate tool

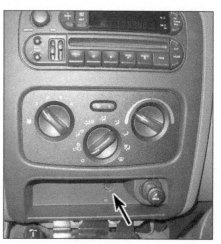

23.16 Remove this screw . . .

4   Use a screwdriver wrapped with tape to pry the dash top cover up **(see illustration)**.
5   Installation is the reverse of the removal procedure.

23.17 . . . then pull the bezel out of the dash

2008 and later models
6   Remove the A-pillar trim from each side **(see illustration 23.3)**.
7   Using a trim tool, carefully pry up the left side dashboard trim panel. Starting from the center of the instrument panel, working to the left, release the retaining clips along the front side edge.
8   Once all the clips are released, remove the trim panel.
9   Using a trim tool, carefully pry up the right side dashboard trim panel. Starting from the left end of the panel, working to the right, release the retaining clips along the front side edge.
10   Once all the clips are released, remove the trim panel.
11   Installation is the reverse of removal.

Dashboard end caps
Refer to illustration 23.12
12   With the door open, pull the end cap from the bottom **(see illustration)**. On some

models, there is a finger-indent in the end-cap to aid removal.
13   Installation is the reverse of the removal procedure.

Dashboard left-side bezels
Refer to illustration 23.14
14   The bezels are removed by carefully prying them off **(see illustration)**.
15   Install the bezels by pushing them firmly into place.

Dashboard center bezel
2007 and earlier models
Refer to illustrations 23.16 and 23.17
16   Remove the screw from the small storage compartment at the bottom of the bezel **(see illustration)**.
17   Pull the bezel back sharply to detach it **(see illustration)**.
18   With the bezel loose, disconnect the wiring and vacuum connections. Remove the bezel.
19   Installation is the reverse of the removal procedure.

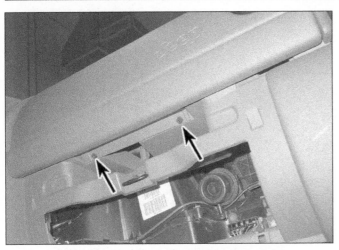

23.30 Remove these two screws to remove the right-side bezel

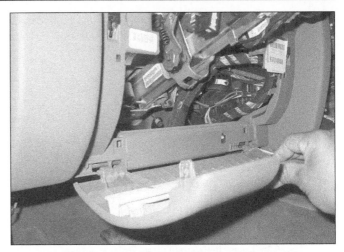

23.32 The lower dash trim panel is removed by hand from the top of the panel

2008 and later models

20   Set the parking brake, then place the shifter into the lowest gear (as far back as possible).

21   Remove the storage bin mat, then remove the storage bin mounting screws and carefully pry the storage bin up and out of the instrument panel using a plastic trim tool.

22   Carefully separate the upper section of the center bezel from the bin.

23   Using the trim tool, rotate the upper section of the bezel outwards to release the four clips (two parallel with the vents and two with the radio).

24   Lift the bezel upwards to unhook the two hooks at the base of the bezel.

25   Disconnect the electrical connectors and remove the bezel. **Note:** *The air conditioning/heating control head is mounted to the trim bezel.*

26   Installation is the reverse of removal.

Dashboard right-side bezel

Refer to illustration 23.30

Models with grab handles

27   Pry the covers from the grab handle ends.

28   Disconnect the electrical wiring.

29   Remove the grab handle screws.

All models

30   Remove the two screws, if so equipped. Pry the bezel off of the dash **(see illustration)**.

31   Installation is the reverse of the removal procedure.

Knee trim panel

Refer to illustration 23.32

32   Place your fingers in the opening at the top center of the trim panel. Pull the panel to the rear to unclip it at the top **(see illustration)**.

33   Lower the top of the panel.

34   Slide the panel to the left to release it from the lower hinges. **Note:** *On 2008 and later models, slide the panel to the left, then to the right to release it from the lower hinges.*

35   Installation is the reverse of the removal procedure.

Kick panels

36   Pry up the front of the door threshold molding, releasing the clips.

37   Pull the kick panels off of the body structure.

38   Installation is the reverse of the removal procedure.

---

24   Instrument cluster bezel - removal and installation

Refer to illustration 24.3

Warning: *The models covered by this manual are equipped with a Supplemental Restraint System (SRS), more commonly known as airbags. Always disarm the airbag system before working in the vicinity of any airbag system component to avoid the possibility of accidental deployment of the airbag, which could cause personal injury (see Chapter 12). Do not use a memory-saving device to preserve the PCM's memory when working on or near airbag system components.*

2007 and earlier models

1   Refer to Section 23 and remove the left-side trim bezels.

2   Refer to Section 23 and remove the dashboard top cover.

3   Remove the screws and detach the bezel from the instrument panel **(see illustration)**.

4   Installation is the reverse of the removal procedure.

2008 and later models

5   Remove the knee trim panel (see Section 23).

6   Remove the dashboard center bezel (see Section 23).

7   Remove the left and right top cover trim panels (see Section 23).

8   Remove the A-pillar trim **(see illustration 23.3)**.

9   Remove the instrument cluster bezel mounting screw from the top of the panel.

10   Lower the steering column as far as it can go.

11   Using a trim tool, release the bezel bottom mounting clips, then the upper mounting clips and carefully remove the bezel.

24.3 Remove all of the screws retaining the instrument cluster bezel. There are three screws at the top and four at the bottom

25.3a Lower instrument panel isolator screws

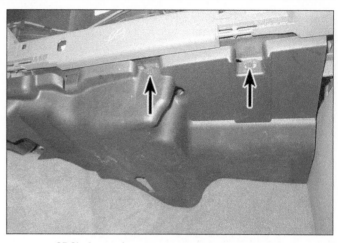

25.3b Lower instrument panel isolator screws

Note: *The instrument cluster bezel is quite large; it incorporates the left air conditioning vent, left speaker grille and a few inches past the middle of the instrument panel.*

12   Installation is the reverse of removal.

25   Instrument panel - removal and installation

Refer to illustrations 25.3a, 25.3b, 25.8, 25.9, 25.11, 25.14, 25.19, 25.20, 25.23, 25.26, 25.27, 25.29, 25.30 and 25.31

Warning: *The models covered by this manual are equipped with a Supplemental Restraint System (SRS), more commonly known as airbags. Always disarm the airbag system before working in the vicinity of any airbag system component to avoid the possibility of accidental deployment of the airbag, which could cause personal injury (see Chapter 12). Do not use a memory-saving device to preserve the PCM's memory when working on or near airbag system components.*

Note: *This procedure is lengthy and difficult,*
even for an experienced mechanic. Due to the number of electrical connections, fasteners used, and the various safety systems involved, we don't recommend instrument panel removal for the home mechanic.

1   Disconnect the cable from the negative terminal of the battery (see Chapter 5, Section 1).

2   Set the steering wheel to the straight-ahead position.

3   Refer to Section 23 and remove all of the instrument panel trim covers and bezels. Remove the two lower isolator panels **(see illustrations)**.

4   Refer to Section 26 and remove the steering column covers.

5   Remove the dash speakers.

6   Refer to Section 19 and remove the center console.

7   Refer to Chapter 12 and remove the radio.

8   Remove the lower center dash support **(see illustration)**.

9   Disconnect the airbag module wiring and the ground strap **(see illustration)**.

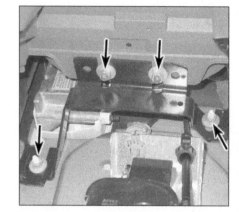

25.8 There are four nuts securing the center support

10   Turn the key to the On position.

11   If the vehicle is equipped with an automatic transmission, disconnect the shifter interlock cable at the steering column **(see illustration)**.

25.9 Unplug the electrical connector from the airbag module

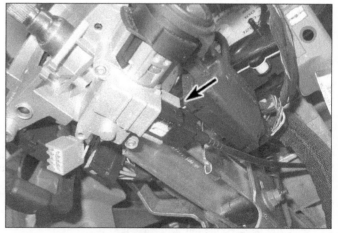

25.11 With the ignition key in the On position, depress the lock tab and pull the interlock cable out of the lock cylinder housing

25.14 There are four steering column mounting fasteners

25.19 The left kick panel has wiring that must be disconnected

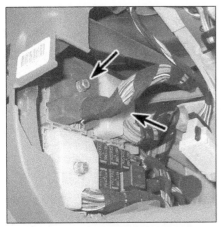

25.20 The wiring connectors shown here must be unbolted and disconnected

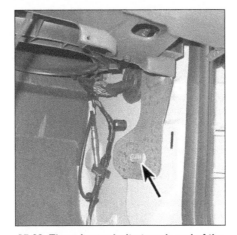

25.23 There is one bolt at each end of the instrument panel

the plastic retaining straps.

17   Remove the steering column from the vehicle.

18   Refer to Chapter 9 and disconnect the wiring from the brake light switch. Do not remove the switch.

19   Disconnect the wiring connector under the left kick panel **(see illustration)**.

20   Unbolt and disconnect the large light blue and green wiring connectors from the electrical junction block **(see illustration)**.

21   Disconnect the wiring connector at the pedal support.

22   Remove the four bolts from the front and bottom of the pedal support.

23   Remove the bolts at each end of the lower instrument panel **(see illustration)**.

24   Open the glove compartment door. Pull inward on the two tabs inside the compartment. Allow it to swing downward.

25   Slide the glove compartment door to the right to release it from the hinges.

26   Remove the two mounting bolts at the center of the heater unit **(see illustration)**.

27   Remove the heater unit bolts located just above the glove compartment latch and at the bottom outer corner of the glove compartment space **(see illustration)**.

12   Mark the relationship of the steering shaft to the coupling at the lower end of the shaft.

13   Refer to Chapter 10 and disconnect the steering shaft at the coupling. **Caution:** *Don't allow the steering column to turn more than 1/4 turn after the shaft has been detached; the airbag clockspring could be damaged.*

14   Remove the steering column mounting fasteners **(see illustration)**. Lower the column from the support.

15   Disconnect the wiring at the ignition switch and the combination switch. Remove the wiring harnesses from the column.

16   Remove the shifter interlock cable out of

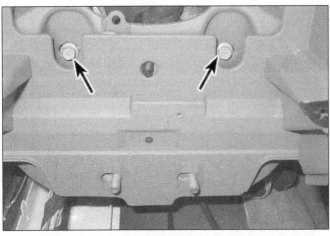

25.26 Remove these two bolts from the center of the heater unit

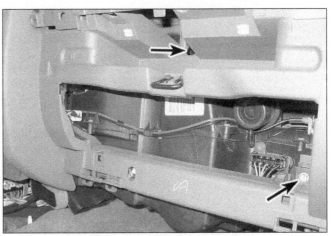

25.27 Remove the two heater unit bolts

25.29 Disconnect the wiring from the heater blower motor

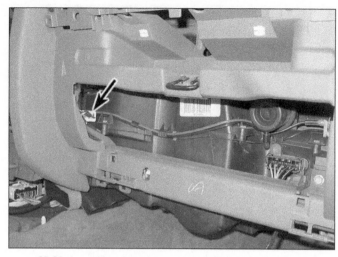

25.30 Location of the heater door electrical connector

28   Disconnect the wiring from the heater resistor and the check valve at the vacuum tank.
29   Disconnect the heater blower wiring connector **(see illustration)**.
30   Disconnect the heater door connector **(see illustration)**.
31   Remove the bolts from the top of the instrument panel **(see illustration)**.
32   Pull the upper part of the instrument panel to the rear.
33   Unfasten the wiring from the wiring guide in the back.
34   Disconnect the radio wiring harness clips.
35   Remove the instrument panel.
36   Installation is the reverse of the removal procedure. **Note 1:** *When placing the instrument panel in position, route the speaker wires through their openings.* **Note 2:** *When tightening the steering column support, first install the fasteners loosely, then tighten the lower nuts first, followed by the upper bolts.*

26   Steering column covers - removal and installation

Refer to illustration 26.2
Warning: *The models covered by this manual are equipped with a Supplemental Restraint System (SRS), more commonly known as airbags. Always disarm the airbag system before working in the vicinity of any airbag system component to avoid the possibility of accidental deployment of the airbag, which could cause personal injury (see Chapter 12). Do not use a memory-saving device to preserve the PCM's memory when working on or near airbag system components.*
1   Disconnect the negative cable from the battery.
2   Remove the screws in the lower column cover **(see illustration)**. **Note:** *On 2008 and later models, the driver's side knee panel must be removed to allow the lower steering column cover to be removed.*

3   Pinch the lower part of the upper column cover to release the locking tabs for the lower cover.
4   Carefully remove the lower and then the upper column covers.
5   Installation is the reverse of removal.

27   Rear view mirrors - removal and installation

Refer to illustration 27.7

Interior mirror
1   Disconnect the electrical connector, if so equipped.
2   Remove the setscrew, then slide the mirror up off the mount on the windshield.
3   Installation is the reverse of removal.
4   If the support button for the mirror has come off the windshield, it can be reattached with a special mirror adhesive kit available at auto parts stores. Clean the glass and the

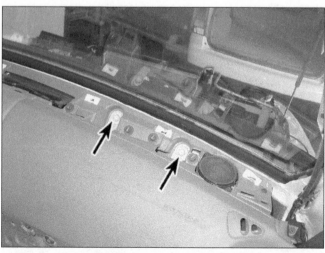

25.31 Remove the bolts from the top of the instrument panel (right side shown, left side similar)

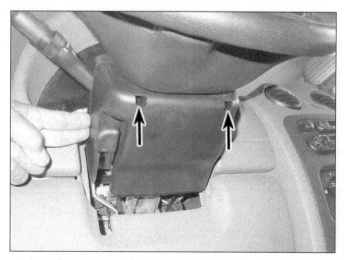

26.2 Remove the two screws from the lower column cover

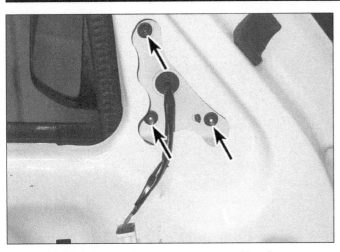

27.7 These three nuts secure the outside mirror

28.5 To remove the cowl retaining clips, pull the center of each clip out, then pry the clips out with a screwdriver or trim tool

mount thoroughly. Follow the directions on the adhesive package, allowing the base to bond in a warm area before attaching the mirror. Make certain to install the mount with the rounded end facing up and with the wider part away from the glass.

Exterior mirror

5   Remove the door trim panel (see Section 14).
6   Disconnect the wiring connector.
7   Remove the three nuts and remove the mirror **(see illustration)**.
8   Installation is the reverse of the removal procedure.

28   Cowl cover - removal and installation

Refer to illustration 28.5

1   Mark the position of the windshield wiper blades on the windshield with a wax marking pen or tape. Raise the wiper arms to the up position.
2   Pry the plastic covers from bases of the wiper arms.
3   Remove the wiper arm nuts.
4   Grasp the outer part of the lower wiper arm section and rock it firmly up and down to detach it from the pivot. **Note:** *If the arm is frozen to the pivot, a battery terminal puller may be used to remove it.*
5   Pry up and remove the four retaining clips **(see illustration)**.
6   Installation is the reverse of the removal procedure. Make sure to align the wiper blades with the marks made during removal.

29   Seats - removal and installation

Refer to illustrations 29.4, 29.6, 29.9 and 29.13

Warning 1: *The models covered by this manual are equipped with a Supplemental Restraint System (SRS), more commonly known as airbags. Some models are also equipped with seat belt pre-tensioners, which are also explosive devices. Always disarm the airbag system before working in the vicinity of any airbag system component to avoid the possibility of accidental deployment of the airbag, which could cause personal injury (see Chapter 12). Do not use a memory-saving device to preserve the PCM's memory when working on or near airbag system components.*

Warning 2: *In addition to driver and passenger airbags, some later models may be equipped with pyrotechnic seat belt tensioners and/or side-impact airbags located in the seatbacks or along the inside edges of the headliner. Whenever working around these components, perform the procedure to disable the airbag system (see Chapter 12).*

Front

1   Move the seat all the way forward.
2   Pry up the lower seat belt anchor cover at the outer side of the seat.

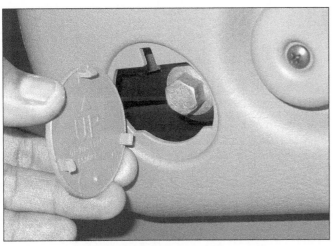

29.4 The outer seat belt bolts are concealed beneath these covers

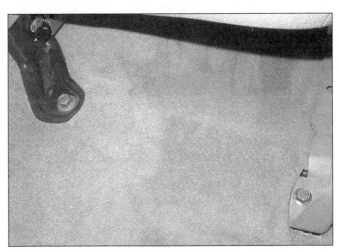

29.6 Front seat mounting bolts (two of four shown)

29.9 Outer seat belt anchor bolts

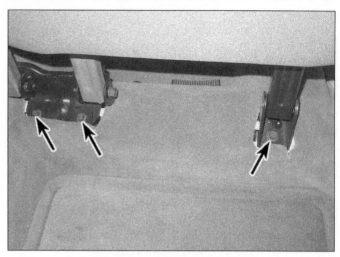

29.13 Rear seat frame bolts

3   Disconnect the seat belt tension sensor from the right seat, if so equipped.

4   Remove the bolt from the outer seat belt anchor **(see illustration)**.

5   Remove the rear bolt and nut from the seat mounting track.

6   Move the seat rearward and remove the bolts at the front. Disconnect any electrical connectors attached to the seat **(see illustration)**.

7   Lift the seat from the vehicle.

8   Installation is the reverse of removal.

Rear

9   Remove the bolts from the outer seat belt anchors **(see illustration)**.

10   Fold the rear seat forward. Remove the seat belt bolts at the floor.

11   Detach the rear cargo area carpet from the seat.

12   Remove the seat belt anchor bolt from the rear of the right rear seat.

13   Unbolt the seat frame from the floor **(see illustration)**.

14   Fold the seat backs forward and then remove the rear seat through the swing gate.

Chapter 12
Chassis electrical system

Contents

1   General information

The electrical system is a 12-volt, negative ground type. Power for the lights and all electrical accessories is supplied by a lead/acid-type battery that is charged by the alternator.

This Chapter covers repair and service procedures for the various electrical components not associated with the engine. Information on the battery, alternator, distributor and starter motor can be found in Chapter 5. **Warning:** *When working on the electrical system, disconnect the negative battery cable from the battery to prevent electrical shorts and/or fires* (see Chapter 5, Section 1).

2   Electrical troubleshooting - general information

Refer to illustrations 2.5a, 2.5b, 2.6, 2.9 and 2.15

A typical electrical circuit consists of an electrical component, any switches, relays, motors, fuses, fusible links or circuit breakers related to that component and the wiring and connectors that link the component to both the battery and the chassis. To help you pinpoint an electrical circuit problem, wiring diagrams are included at the end of this Chapter.

Before tackling any troublesome electrical circuit, first study the appropriate wiring diagrams to get a complete understanding of what makes up that individual circuit. Trouble spots, for instance, can often be narrowed down by noting if other components related to the circuit are operating properly. If several components or circuits fail at one time, chances are the problem is in a fuse or ground connection, because several circuits are often routed through the same fuse and ground connections.

Electrical problems usually stem from simple causes, such as loose or corroded connections, a blown fuse, a melted fusible link or a failed relay. Visually inspect the condition of all fuses, wires and connections in a problem circuit before troubleshooting the circuit.

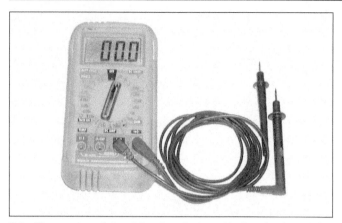

2.5a **The most useful tool for electrical troubleshooting is a digital multimeter that can check volts, amps, and test continuity**

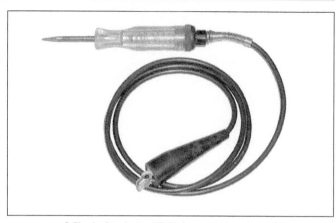

2.5b **A simple test light is a very handy tool used for testing voltage**

If test equipment and instruments are going to be utilized, use the diagrams to plan ahead of time where you will make the necessary connections in order to accurately pinpoint the trouble spot.

The basic tools needed for electrical troubleshooting include a circuit tester or voltmeter (a 12-volt bulb with a set of test leads can also be used), a continuity tester, which includes a bulb, battery and set of test leads, and a jumper wire, preferably with a circuit breaker incorporated, which can be used to bypass electrical components **(see illustrations)**. Before attempting to locate a problem with test instruments, use the wiring diagram(s) to decide where to make the connections.

Voltage checks

Voltage checks should be performed if a circuit is not functioning properly. Connect one lead of a circuit tester to either the negative battery terminal or a known good ground. Connect the other lead to a connector in the circuit being tested, preferably nearest to the battery or fuse **(see illustration)**. If the bulb

of the tester lights, voltage is present, which means that the part of the circuit between the connector and the battery is problem free. Continue checking the rest of the circuit in the same fashion. When you reach a point at which no voltage is present, the problem lies between that point and the last test point with voltage. Most of the time the problem can be traced to a loose connection. **Note:** *Keep in mind that some circuits receive voltage only when the ignition key is in the Accessory or Run position.*

Finding a short

One method of finding shorts in a circuit is to remove the fuse and connect a test light or voltmeter to the fuse terminals. There should be no voltage present in the circuit when it is turned off. Move the wiring harness from side-to-side while watching the test light. If the bulb goes on, there is a short to ground somewhere in that area, probably where the insulation has rubbed through. The same test can be performed on each component in the circuit, even a switch.

Ground check

Perform a ground test to check whether a component is properly grounded. Disconnect the battery and connect one lead of a continuity tester or multimeter (set to the ohms scale), to a known good ground. Connect the other lead to the wire or ground connection being tested. If the resistance is low (less than 5 ohms), the ground is good. If the bulb on a self-powered test light does not go on, the ground is not good.

Continuity check

A continuity check is done to determine if there are any breaks in a circuit - if it is passing electricity properly. With the circuit off (no power in the circuit), a self-powered continuity tester or multimeter can be used to check the circuit. Connect the test leads to both ends of the circuit (or to the "power" end and a good ground), and if the test light comes on the circuit is passing current properly **(see illustration)**. If the resistance is low (less than 5 ohms), there is continuity; if the reading is 10,000 ohms or higher, there is a

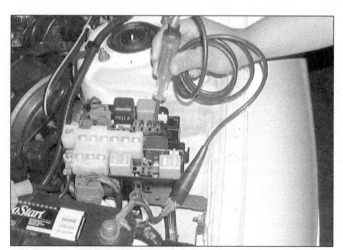

2.6 **In use, a basic test light's lead is clipped to a known good ground, then the pointed probe can test connectors, wires or electrical sockets - if the bulb lights, battery voltage is present at the test point**

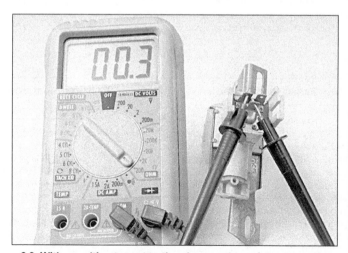

2.9 **With a multimeter set to the ohms scale, resistance can be checked across two terminals - when checking for continuity, a low reading indicates continuity, a very high or infinite reading indicates lack of continuity**

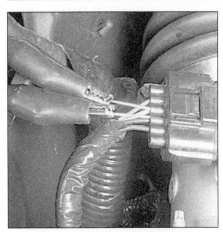

2.15 To backprobe a connector, insert a small, sharp probe (such as a straight-pin) into the back of the connector alongside the desired wire until it contacts the metal terminal inside; connect your meter leads to the probes - this allows you to test a functioning circuit

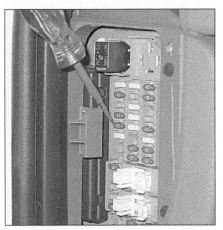

3.1a The interior fuse block is located in the left end of the dash, and is accessed by pulling of the trim panel from the bottom - these fuses can be tested on top with a test light without removing the fuse

break somewhere in the circuit. The same procedure can be used to test a switch, by connecting the continuity tester to the switch terminals. With the switch turned On, the test light should come on (or low resistance should be indicated on a meter).

Finding an open circuit

When diagnosing for possible open circuits, it is often difficult to locate them by sight because the connectors hide oxidation or terminal misalignment. Merely wiggling a connector on a sensor or in the wiring harness may correct the open circuit condition. Remember this when an open circuit is indicated when troubleshooting a circuit. Intermittent problems may also be caused by oxidized or loose connections.

Electrical troubleshooting is simple if you keep in mind that all electrical circuits are basically electricity running from the battery, through the wires, switches, relays, fuses and fusible links to each electrical component (light bulb, motor, etc.) and to ground, from which it is passed back to the battery. Any

electrical problem is an interruption in the flow of electricity to and from the battery.

Connectors

Most electrical connections on these vehicles are made with multi-wire plastic connectors. The mating halves of many connectors are secured with locking clips molded into the plastic connector shells. The mating halves of large connectors, such as some of those under the instrument panel, are held together by a bolt through the center of the connector.

To separate a connector with locking clips, use a small screwdriver to pry the clips apart carefully, then separate the connector halves. Pull only on the shell, never pull on the wiring harness as you may damage the individual wires and terminals inside the connectors. Look at the connector closely before trying to separate the halves. Often the locking clips are engaged in a way that is not immediately clear. Additionally, many connectors have more than one set of clips.

Each pair of connector terminals has a male half and a female half. When you look at the end view of a connector in a diagram, be

sure to understand whether the view shows the harness side or the component side of the connector. Connector halves are mirror images of each other, and a terminal shown on the right side end-view of one half will be on the left side end view of the other half.

It is often necessary to take circuit voltage measurements with a connector connected. Whenever possible, carefully insert a small straight pin (not your meter probe) into the rear of the connector shell to contact the terminal inside, then clip your meter lead to the pin. This kind of connection is called "backprobing" **(see illustration)**. When inserting a test probe into a male terminal, be careful not to distort the terminal opening. Doing so can lead to a poor connection and corrosion at that terminal later. Using the small straight pin instead of a meter probe results in less chance of deforming the terminal connector.

3 Fuses - general information

Refer to illustrations 3.1a, 3.1b and 3.3

The electrical circuits of the vehicle are protected by a combination of fuses and circuit breakers. The main fuse panel is located at the left end of the instrument panel **(see illustration)**. There is a fuse and relay block, called the power distribution center, located on the left side of the engine compartment **(see illustration)**.

Each of the fuses is designed to protect a specific circuit, and the various circuits are identified on the fuse panel itself.

Different sizes of fuses are employed in the fuse blocks. There are "mini" and "maxi" sizes, with the larger located in the power distribution center. The maxi fuses can be removed with your fingers, but the mini fuses require the use of pliers or the small plastic fuse-puller tool found in most fuse boxes. If an electrical component fails, always check the fuse first. The best way to check the fuses is with a test light. Check for power at the exposed terminal tips of each fuse. If power is present at one side of the fuse but not the other, the fuse is blown. A blown fuse can also be identified by visually inspecting it **(see illustration)**.

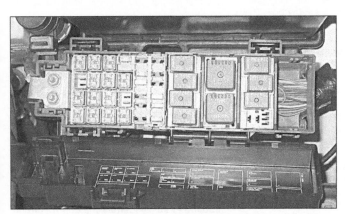

3.1b The power distribution center is located at the left side of the engine compartment

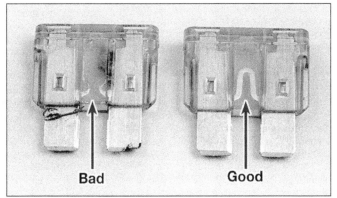

3.3 When a fuse blows, the element between the terminals melts - the fuse on the left is blown, the one on the right is good

Be sure to replace blown fuses with the correct type. Fuses of different ratings are physically interchangeable, but only fuses of the proper rating should be used. Replacing a fuse with one of a higher or lower value than specified is not recommended. Each electrical circuit needs a specific amount of protection. The amperage rating of each fuse is molded into the fuse body.

If the replacement fuse immediately fails, don't replace it again until the cause of the problem is isolated and corrected. In most cases, the cause will be a short circuit in the wiring caused by a broken or deteriorated wire.

4   Fusible links - general information

The wiring between the power distribution center and the alternator is protected by a fusible link. This link functions like a fuse, in that it melts when the circuit is overloaded, but resembles a large-gauge wire. To replace a fusible link, first disconnect the negative cable from the battery. Disconnect the burned-out link and replace it with a new one (available from your dealer or auto parts store). Always determine the cause for the overload that melted the fusible link before installing a new one.

5   Circuit breakers - general information

Circuit breakers protect certain heavy-load circuits. Depending on the vehicle's accessories, there may be one to three circuit breakers located in the main fuse panel and also in the power distribution center.

Because the circuit breakers reset automatically, an electrical overload in a circuit-breaker-protected system will cause the circuit to fail momentarily, then come back on. If the circuit does not come back on, check it immediately.

For a basic check, pull the circuit breaker up out of its socket on the fuse panel, but just far enough to probe with a voltmeter. The breaker should still contact the sockets.

With the voltmeter negative lead on a good chassis ground, touch each end prong of the circuit breaker with the positive meter probe. There should be battery voltage at each end. If there is battery voltage only at one end, the circuit breaker must be replaced.

6   Relays - general information

1   Many electrical accessories in the vehicle utilize relays to transmit current to the component. If the relay is defective, the component won't operate properly.
2   Some relays are located in the power distribution center, which is a fuse and relay box located in the engine compartment (see illustration 3.1b).
3   Other relays are located in the fuse panel

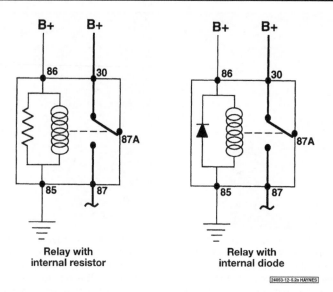

Relay with internal resistor   **Relay with internal diode**

24053-12-6.2a HAYNES

6.5a Typical ISO relay designs, terminal numbering and circuit connections

at the left end of the instrument panel **(see illustration 3.1a)**.
4   If a faulty relay is suspected, it can be removed and tested using the procedure below or by a dealer service department or a repair shop. Defective relays must be replaced as a unit.

Testing

Refer to illustrations 6.5a and 6.5b
5   Most of the relays used in these vehicles are of a type often called "ISO" relays, which refers to the International Standards Organization. The terminals of ISO relays are numbered to indicate their usual circuit connections and functions. There are two basic layouts of terminals on the relays used in the vehicles covered by this manual **(see illustrations)**.
6   Refer to the wiring diagram for the circuit to determine the proper connections for the relay you're testing. If you can't determine the correct connection from the wiring diagrams, however, you may be able to determine the test connections from the information that follows.
7   Two of the terminals are the relay control circuit and connect to the relay coil. The other relay terminals are the power circuit. When the relay is energized, the coil creates a magnetic field that closes the larger contacts of the power circuit to provide power to the circuit loads.
8   Terminals 85 and 86 are normally the control circuit. If the relay contains a diode, terminal 86 must be connected to battery positive (B+) voltage and terminal 85 to ground. If the relay contains a resistor, terminals 85 and 86 can be connected in either direction with respect to B+ and ground.
9   Terminal 30 is normally connected to the battery voltage (B+) source for the circuit loads. Terminal 87 is connected to the ground side of the circuit, either directly or through a load. If the relay has several alternate termi-

nals for load or ground connections, they usually are numbered 87A, 87B, 87C, and so on.
10   Use an ohmmeter to check continuity through the relay control coil.
   a) *Connect the meter according to the polarity shown in the illustration for one check; then reverse the ohmmeter leads and check continuity in the other direction.*
   b) *If the relay contains a resistor, resistance will be indicated on the meter, and should be the same value with the ohmmeter in either direction.*
   c) *If the relay contains a diode, resistance should be higher with the ohmmeter in the forward polarity direction than with the meter leads reversed.*
   d) *If the ohmmeter shows infinite resistance in both directions, replace the relay.*

11   Remove the relay from the vehicle and use the ohmmeter to check for continuity between the relay power circuit terminals. There should be no continuity between terminal 30 and 87 with the relay de-energized.
12   Connect a fused jumper wire to terminal 86 and the positive battery terminal. Connect another jumper wire between terminal 85 and

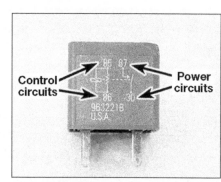

6.5b Most relays are marked on the outside to easily identify the control circuit and power circuits - this one is of the four-terminal type

7.6 The combination switch is pulled upwards to remove it

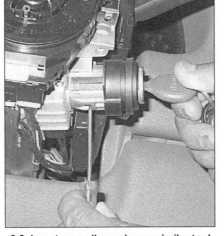

8.6 Insert a small punch or a similar tool into the bottom of the switch to release it when the key is pulled

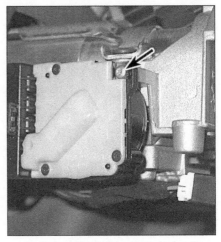

8.15 Location of the tamper-proof Torx screw

ground. When the connections are made, the relay should click.

13   With the jumper wires connected, check for continuity between the power circuit terminals. Now there should be continuity between terminals 30 and 87.

14   If the relay fails any of the above tests, replace it.

7   Multi-function switch - replacement

Refer to illustration 7.6

Warning: *The models covered by this manual are equipped with a Supplemental Restraint System (SRS), more commonly known as airbags. Always disarm the airbag system before working in the vicinity of any airbag system component to avoid the possibility of accidental deployment of the airbag, which could cause personal injury (see Section 25). Do not use a memory-saving device to preserve the PCM's memory when working on or near airbag system components.*

1   The multi-function switch consists of a main body that fits over the top of the steering column and two stalks. One stalk is on the left of the column and controls the lighting functions such as the headlights and interior lights; the other is on the right of the column and controls the front and rear wiper and washer systems.

2   Disconnect the cable from the negative terminal of the battery (see Chapter 5).

2007 and earlier models

3   On vehicles with tilt steering columns, lower the column as much as possible and allow the tilt lever to remain in the released position.

4   Remove the steering column covers (see Chapter 11).

5   Disconnect both electrical connectors from the switch.

6   Grasp the switch and pull it upward while wiggling it gently to release it from the locating tabs **(see illustration)**.

7   Installation is the reverse of the removal procedure.

2008 and later models

8   The multi-function switch on these models consists of three parts: the left side is the headlight/turn signal switch; the center is the clockspring; and the right side is the wiper/washer switch. All three sections can be replaced separately and the procedure for both switches is the same. See Chapter 10 for clockspring replacement.

9   Remove the steering column covers (see Chapter 11).

10   Disconnect the electrical connector to the rear of the switch.

11   Remove the retaining screw.

12   Slide the switch away from the center section/clockspring until the slide tabs on the switch housing can be disengaged from the channels.

13   Disconnect the electrical connector on the inboard end of the switch.

14   Remove the switch.

15   Installation is the reverse of removal.

8   Ignition switch and key lock cylinder - replacement

Warning: *The models covered by this manual are equipped with a Supplemental Restraint System (SRS), more commonly known as airbags. Always disarm the airbag system before working in the vicinity of any airbag system component to avoid the possibility of accidental deployment of the airbag, which could cause personal injury (see Section 25). Do not use a memory-saving when working on or near airbag system components.*

Note: *If the key is difficult to turn, the problem may not be in the switch or lock cylinder.*

On automatic transmission vehicles with floor console-mounted shifters, the Park/Lock cable may be out of adjustment. It can be adjusted; refer to Chapter 7B. If it is defective inside the steering column, the steering column must be replaced. On manual transmission vehicles, the small button behind the key cylinder may be defective. If it is defective, the steering column must be replaced.

Key lock cylinder

Refer to illustration 8.6

1   Put the key into the key cylinder.

2   On vehicles equipped with an automatic transmission, place the shifter in Park.

3   Refer to Chapter 11 and remove the steering column covers.

4   Remove the keyless entry module and the ring around the lock cylinder.

5   Turn the key to the On position.

6   Push a small punch or a similar tool into the hole in the bottom of the switch **(see illustration)**.

7   Pull out on the key while pushing in on the punch. The cylinder will slide out with the key.

8   To install a new cylinder, first put the key into it. Turn it to the On position.

9   Line the cylinder up in the correct "On" orientation with the housing and slide it in. It will snap into place.

10   Turn the key to the Lock position.

11   The remainder of installation is the reverse of removal.

Ignition switch

Refer to illustrations 8,15, 8.16 and 8.17

12   Remove the ignition key lock cylinder (see Steps 1 through 7).

13   Refer to Section 7 and remove the multi-function switch.

14   Disconnect the wiring from the switch.

15   Use a tamper-proof Torx screwdriver to remove the screw securing the switch **(see illustration)**. These special screwdrivers are available at tool dealers and some auto supply stores.

8.16 The switch is pulled straight out to release it

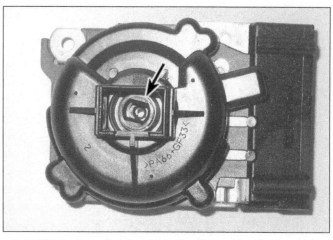

8.17 This is the switch turned to the ON position (it must be in this position before it can be installed)

16   Release the switch from the lock tabs by pulling it straight out **(see illustration)**.
17   To install the switch, first turn it to the On position **(see illustration)**.
18   The remainder of installation is the reverse of removal.

---

9   Rear window defogger switch and switch pod – replacement

Warning: *The models covered by this manual are equipped with a Supplemental Restraint System (SRS), more commonly known as airbags. Always disarm the airbag system before working in the vicinity of any airbag system component to avoid the possibility of accidental deployment of the airbag, which could cause personal injury (see Section 25). Do not use a memory-saving device to preserve the PCM's memory when working on or near airbag system components.*

Rear window defogger switch

Note: *On 2008 and later models, the defogger switch is integrated into the air conditioning/heating temperature control module and*

can't be replaced separately. If there is a problem with the switch, the control module must be replaced (see Chapter 3).
1   The rear window defogger switch is a part of the heater control assembly and it is not serviceable separately. If it is defective, the entire heater control unit must be replaced. Refer to Chapter 3 for information on the replacement of the heater control assembly
2   When depressed, the switch connects to ground. The switch can be tested for correct grounding after removal of the heater control unit from the instrument panel.
3   The defogger circuit includes the Body Control Module (BCM), a computer module that controls a number of body system communication tasks. If the fuse, switch and relay are OK but the defogger still doesn't work, take the vehicle to a dealer service department where they can use a scan tool to check for a Diagnostic

Switch pod
4   The switch pod housing has multiple configurations; all configurations will have the hazard warning switch but may include trailer brake control switches, Electronic Stability

Control (ESC) OFF switch and tow/haul mode switch or heated seats switch.
Note: *The individual switches in the instrument panel switch pod cannot be repaired and are not serviced individually. If any component within the switch pod is defective, the entire switch pod must be replaced.*
5   Disconnect the cable from the negative battery terminal (see Chapter 5).
6   Remove the center instrument trim bezel (see Chapter 11).
7   Disconnect the electrical connectors to the pod.
8   Remove the pod four mounting screws and separate the pod from the bezel.
9   Installation is the reverse of removal. rouble Code (DTC) in the BCM.

---

10   Rear window defogger - repair

1   The rear window defogger consists of a number of horizontal heating elements baked onto the inside surface of the glass. Power is supplied through a relay and fuse from the interior fuse/relay box. The heater is controlled by the instrument panel switch. The defogger is programmed to operate for about ten minutes and then shut off. The second time the switch is pressed, it will operate for about five minutes.
2   Small breaks in the element can be repaired without removing the rear window.

Check
Refer to illustrations 10.5, 10.6 and 10.8
3   Turn the ignition and defogger switches to the ON position.
4   Using a voltmeter, place the positive probe against the defogger grid positive side and the negative probe against the ground side. If battery voltage is not indicated, check that the ignition switch is On and that the feed and ground wires are properly connected. Check the two fuses, defogger switch, defogger relay and related wiring. The dealer can scan the body control module if necessary. If voltage is indicated, but all or part of the defogger doesn't heat, proceed with the following tests.

10.5 When measuring voltage at the rear window defogger grid, wrap a piece of aluminum foil around the positive probe of the voltmeter and press the foil against the wire with your finger

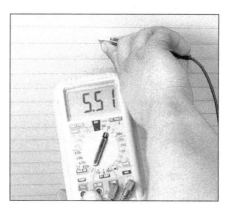

10.6 To determine if a heating element has broken, check the voltage at the center of each element - if the voltage is 6-volts, the element is unbroken

10.8 To find the break, place the voltmeter negative lead against the defogger ground terminal, place the voltmeter positive lead with the foil strip against the heat wire at the positive terminal end and slide it toward the negative terminal end - the point at which the voltmeter deflects from several volts to zero volts is the point at which the wire is broken

5   When measuring voltage during the next two tests, wrap a piece of aluminum foil around the tip of the voltmeter positive probe and press the foil against the heating element with your finger **(see illustration)**. Place the negative probe on the defogger grid ground terminal.

6   Check the voltage at the center of each heating element **(see illustration)**. If the voltage is 5 to 6 volts, the element is okay (there is no break). If the voltage is 0 volts, the element is broken between the center of the element and the positive end. If the voltage is 10 to 12 volts, the element is broken between the center of the element and the ground side. Check each heating element.

7   If none of the elements are broken, connect the negative probe to a good chassis ground. The voltage reading should stay the same, if it doesn't the ground connection is bad.

8   To find the break, place the voltmeter negative probe against the defogger ground terminal. Place the voltmeter positive probe with the foil strip against the heating element at the positive side and slide it toward the negative side. The point at which the voltme-

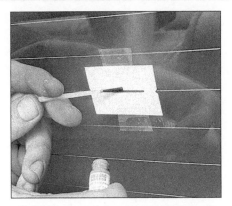

10.14 To use a defogger repair kit, apply masking to the inside of the window at the damaged area, then brush on the special conductive coating

ter deflects from several volts to zero is the point where the heating element is broken **(see illustration)**.

Repair
Refer to illustration 10.14

9   Repair the break in the element using a repair kit specifically for this purpose, such as Dupont paste No. 4817 (or equivalent). The kit includes conductive plastic epoxy.

10   Before repairing a break, turn off the system and allow it to cool for a few minutes.

11   Lightly buff the element area with fine steel wool; then clean it thoroughly with rubbing alcohol.

12   Use masking tape to mask off the area being repaired.

13   Thoroughly mix the epoxy, following the kit instructions.

14   Apply the epoxy material to the slit in the masking tape, overlapping the undamaged area about 3/4-inch on either end **(see illustration)**.

15   Allow the repair to cure for 24 hours before removing the tape and using the system.

---

11 Headlight bulb - replacement

Refer to illustration 11.2

Warning: *Halogen bulbs are gas-filled and*

under pressure and may shatter if the surface is scratched or the bulb is dropped. Wear eye protection and handle the bulbs carefully, grasping only the base whenever possible. Don't touch the surface of the bulb with your fingers because the oil from your skin could cause it to overheat and fail prematurely. If you do touch the bulb surface, clean it with rubbing alcohol.

1   Reach behind the headlight assembly. **Note:** *If you're replacing the driver's-side bulb, release the power distribution center from its clips for access.*

2   Rotate the lockring counterclockwise and then remove the bulb from the headlight housing **(see illustration)**.

3   Disconnect the wiring from the socket.

4   Remove the bulb along with its socket from the lock ring.

5   Remove the old bulb and install a new bulb into the socket.

6   Insert the new bulb and socket into the lockring.

7   Connect the wiring.

8   Seat the bulb socket securely into the headlight housing.

9   Engage the lockring with the rear of the headlight housing and turn it clockwise.

10   Connect the battery terminal.

---

12 Headlights - adjustment

Warning: *The headlights must be aimed correctly. If adjusted incorrectly, they could temporarily blind the driver of an oncoming vehicle and cause an accident or seriously reduce your ability to see the road. The headlights should be checked for proper aim every 12 months and any time a new headlight is installed or front-end bodywork is performed. The following procedure is only an interim step to provide temporary adjustment until the headlights can be adjusted by a properly equipped shop.*

Headlights
Refer to illustrations 12.1 and 12.2

1   Each headlight has one adjusting screw for up-and-down movement only **(see illustration)**.

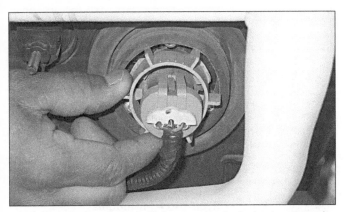

11.2 Turn the headlamp lockring counterclockwise to release it

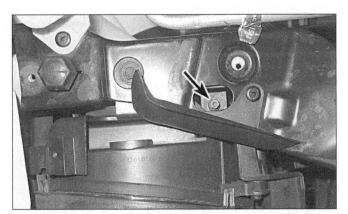

12.1 The headlamp vertical adjustment screw is visible when looking straight down at the headlamp

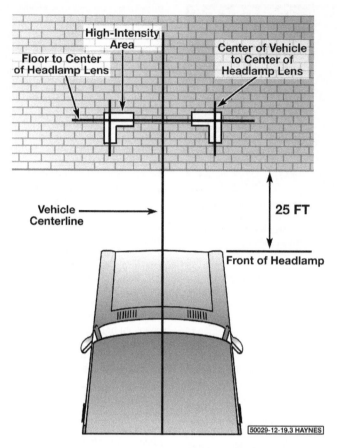

12.2 Headlight adjustment screen details

13.2 There are two mounting screws at the inboard side of each headlamp

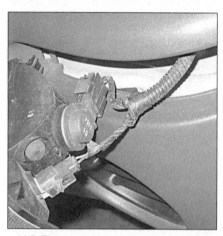

14.2 The outer side of the lens must be pulled out first; the inner side will then release

2   There are several methods of adjusting the headlights. The simplest method requires an open area with a blank wall and a level floor **(see illustration)**.
3   Position masking tape vertically on the wall in reference to the vehicle centerline and the centerlines of both headlights.
4   Position a horizontal tape line in reference to the centerline of the headlights. **Note:** *It may be easier to position the tape on the wall with the vehicle parked only a few inches away.*
5   Adjustment should be made with the vehicle parked 25 feet from the wall, sitting level, the gas tank full and no unusually heavy load in the vehicle.
6   The high intensity zone should be vertically centered with the exact center, about three inches below the horizontal line.
7   Have the headlights adjusted by a dealer service department at the earliest opportunity.

Fog lights
8   Fog lights are optional in these vehicles.
9   Park the vehicle 25 feet from the wall.
10   Tape a horizontal line on the wall that represents the height of the fog lights and tape another line four inches below that line.
11   Using the adjusting screw on each fog light, adjust the pattern on the wall so that the top of the fog light beam meets the lower line on the wall.

Light bar
12   The roof-mounted light bar is optional equipment.
13   Park the vehicle 25 feet from the wall.
14   Attach a horizontal piece of tape to the wall at seven feet, nine inches from the ground.
15   Cover three of the four lights and aim the fourth using the adjustment screw in the rear. The center of its high intensity zone should be on the tape line.
16   Repeat the procedure for the other three lights.

13   Headlight housing - removal and installation

Refer to illustration 13.2
1   Refer to Chapter 11 and remove the grille.
2   On 2007 and earlier models, remove the two screws securing the headlight housing tabs on the fender side of the housing **(see illustration)**.
3   On 2007 and earlier models, push the headlight housing toward the fender until the tab on the inboard side of the housing comes free of the radiator support panel.
4   On 2008 and later models, remove the four headlight housing screws.
5   Pull the headlight unit out from the body

far enough to disconnect the wiring connectors.
6   Remove the headlight unit from the vehicle.
7   Installation is the reverse of the removal procedure. **Note:** *Make certain that the two locating tabs on the inner side of the housing align with their pins.*

14   Bulb replacement

Front turn signal, parking and side marker lights
2007 and earlier models
Refer to illustration 14.2
1   Remove the lens screw.
2   Pull out on the outer part of the lens to release the inner tab **(see illustration)**. Pull the light housing out from the fascia.
3   Turn the bulb socket counterclockwise and then remove it from the housing.
4   Pull the bulb straight out from the socket.
5   Installation is the reverse of removal.

2008 and later models
6   Turn the steering wheel all the way to the right to access the left front turn signal or

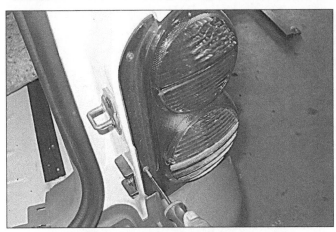

14.12 The screws are visible with the swing gate open

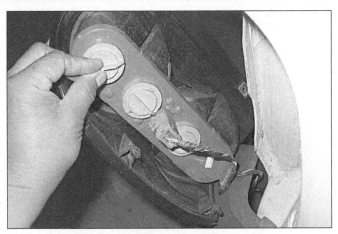

14.14 First remove the bulb socket, then remove the bulb from the socket

side marker and the opposite way for the other side.

7   Remove the push-pins from the inner fender splash shield.

8   Carefully pull the splash shield forward until the bulb can be accessed.

9   Rotate the bulb socket counterclockwise, and remove the socket assembly from the housing.

10   Pull the bulb straight out from the socket.

11   Installation is the reverse of removal.

Rear tail light/brake light/turn signal

Refer to illustrations 14.12 and 14.14

12   With the tailgate/liftgate open, remove the mounting screws **(see illustration)**.

13   Pull back on the outer part of the lens to release the two ballstuds that secure it to the body. **Note:** *If the ballstuds are not a tight fit in their plastic retainers, replace them with new ones.*

14   On 2007 and earlier models, turn the bulb socket counterclockwise and remove it from the housing **(see illustration)**, then pull the bulb straight out from the socket.

15   On 2008 and later models, remove the back plate mounting screws and separate the back plate from the tail light housing. Pull the bulb straight out from the back plate.

16   Installation is the reverse of removal.

High-mounted brake light

Refer to illustration 14.17

Note: *On 2008 and later models, the high-mounted brake light LEDs are an integral part of the housing, and are not separately serviceable.*

17   Remove the screws from the light housing **(see illustration)**.

18   Pull the light housing out for access to the bulb holders.

19   Turn the bulb socket counterclockwise and then remove it from the housing.

20   Pull the bulb straight out from the socket.

21   Installation is the reverse of removal.

Instrument cluster lights

Note: *On 2008 and later models, the instrument cluster LEDs are an integral part of the cluster circuit board, and are not separately serviceable.*

22   Refer to Section 18 and remove the instrument cluster.

23   Turn the bulb socket counterclockwise, then remove it from the housing.

24   Pull the bulb straight out from the socket.

25   Installation is the reverse of removal.

License plate light

2007 and earlier models

Refer to illustration 14.26

26   Remove the screws and pull the light through the hole in the fascia **(see illustration)**.

27   Pull straight out to remove the bulb socket from the housing.

28   Pull the bulb straight out from the socket.

29   Installation is the reverse of removal.

2008 and later models

30   Working under the liftgate handle, insert a small screwdriver into the slot in the lens to depress the release tab, then remove the lens.

31   Pull the bulb straight out from the socket.

32   Installation is the reverse of removal.

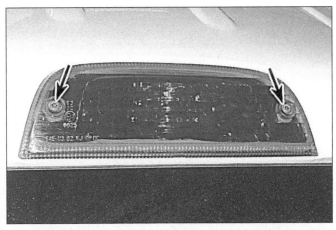

14.17 Remove these screws and then pull the housing out; the bulb can then be removed from the housing

14.26 The license plate light can be removed through this opening in the fascia after removing the two screws

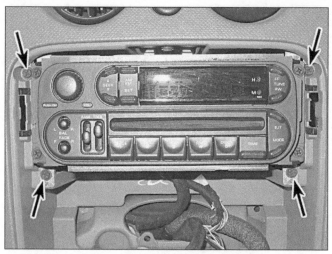

15.3 There are four radio mounting screws (2007 and earlier models shown, later models similar)

15.5 The antenna cable has a lock sleeve that secures it; don't try to remove it without pulling out the lock sleeve first

Fog lights

Warning: *Halogen bulbs are gas-filled and under pressure and may shatter if the surface is scratched or the bulb is dropped. Wear eye protection and handle the bulbs carefully, grasping only the base whenever possible. Don't touch the surface of the bulb with your fingers because the oil from your skin could cause it to overheat and fail prematurely. If you do touch the bulb surface, clean it with rubbing alcohol.*

33   Turn the wheels as far as possible toward the light that is being replaced.

34   Working inside the wheel housing, remove the cover over the rear of the fog light assembly.

35   Turn the bulb socket counterclockwise and then remove it from the housing.

36   Pull the bulb straight out from the socket.

37   Installation is the reverse of removal.

Interior lights

Warning: *The models covered by this manual are equipped with a Supplemental Restraint System (SRS), more commonly known as airbags. Always disarm the airbag system before*

working in the vicinity of any airbag system component to avoid the possibility of accidental deployment of the airbag, which could cause personal injury (see Section 25). Do not use a memory-saving device to preserve the PCM's memory when working on or near airbag system components.

38   These vehicles have many optional and standard interior lights available for various functions. All are replaceable using the same general procedures outlined above.

39   Light lenses such as the cargo light lens, reading light lens, vanity light lens and the courtesy light lens are pried from their housings using a small screwdriver wrapped with tape. The bulbs are of several different styles, but most are pulled straight out to remove them. Round-shaped bulb holders are turned to release them.

15   Radio and speakers - removal and installation

Warning: *The models covered by this manual are equipped with a Supplemental Restraint System (SRS), more commonly known as air-*

bags. Always disarm the airbag system before working in the vicinity of any airbag system component to avoid the possibility of accidental deployment of the airbag, which could cause personal injury (see Section 25). Do not use a memory-saving device to preserve the PCM's memory when working on or near airbag system components.

Note: *On AM/FM Stereo with a 6-disc CD/DVD changer (sales code "REQ"), the radio only must be put into "Transportation Mode" before removing. Turn the ignition switch to the "RUN" or "ACCESSORY" position then, press SET and DISK buttons at the same time until the radio displays "TRANSPORTATION" on the display. If the display is inoperable hold the two buttons down for 5 seconds before disconnecting the battery.*

Radio

Refer to illustrations 15.3 and 15.5

1   Disconnect the cable from the negative terminal of the battery (see Chapter 5).

2   Refer to Chapter 11 and remove the dashboard center trim panel.

3   Remove the screws **(see illustration)** and pull the radio from the dash.

4   Disconnect the electrical wiring.

5   Disconnect the antenna conductor by pulling out on the lock sleeve **(see illustration)**. Do not pull on the cable itself - it could be damaged.

6   Installation is the reverse of the removal procedure.

Speakers (front and rear doors)

7   Refer to Chapter 11 and remove the front door trim panel.

8   Remove the speaker mounting screws and remove the speaker.

9   Installation is the reverse of the removal procedure.

Instrument panel

10   Refer to Chapter 11 for removal and installation of the top cover of the instrument panel and the speakers.

16.3 Antenna fender connection

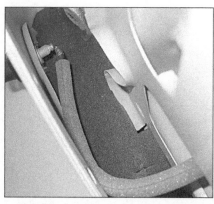

16.9 The front fender can be carefully pulled out to get access to the inside of the antenna mount

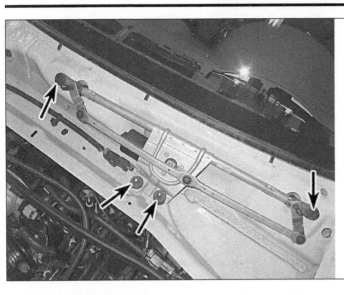

17.6 There are two outer bolts and two inner nuts securing the wiper assembly

16 Antenna - removal and installation

Refer to illustrations 16.3 and 16.9

Note 1: *Some later models are equipped with a grid-type antenna that is bonded to the right-rear quarter glass. This antenna can only be replaced with the glass, which is fixed glass that should only be replaced by a glass shop.*

Note 2: *Some later models are equipped with optional satellite radio, which uses a small antenna located on the rear of the roof exterior.*

1　Disconnect the cable from the negative terminal of the battery (see Chapter 5).
2　Unscrew the antenna mast from the fitting on the right fender.
3　Remove the cover, nut and bezel fitting **(see illustration)**.
4　Refer to Chapter 11 and remove the right kick panel.
5　Disconnect the antenna cable splice inside the kick panel area.
6　Tie a string or length of wire several feet long to the end of the antenna cable.
7　Refer to Chapter 11 and remove the top fender bolts.
8　Refer to Chapter 11 and remove the rear fender bolts in the area of the top door hinge.

9　Carefully pull out on the upper rear corner of the fender just enough to gain access to the antenna **(see illustration)**.
10　Cut the string and remove the antenna cable from the vehicle.
11　Tie the string to the replacement cable end.
12　Pull the cable into the interior of the vehicle with the string.
13　The remainder of installation is the reverse of removal.
14　The antenna cable that runs from the kick panel to the radio can be replaced by disconnecting it at each end and then pulling it out with a string attached to it. Refer to Section 15 for disconnecting it at the radio end and Chapter 11 for removing the right kick panel.
15　On 2008 and later models and models equipped with satellite radio, the antenna can be replaced by pulling back the headliner at the rear of the interior. This may be a job best left to an experienced shop, as the headliner is fragile. Disconnect the electrical cable at the antenna base, then use a small screwdriver to move the two locating tabs so the antenna comes free of the roof. Installation is the reverse of removal.

17 Wiper motor - replacement

Front wiper motor assembly

Refer to illustration 17.6

1　Disconnect the cable from the negative terminal of the battery (see Chapter 5).
2　Use tape to mark the wiper blade positions on the windshield, then remove the covers and nuts.
3　Raise the wiper arm. Grasp the lower wiper arm section and firmly wiggle the end up and down to loosen it from the pivot. Detach the wiper arms. **Note:** *If the wiper arm is stuck tight, don't use a prying tool under it. Use a battery cable terminal puller to release it without damage.*
4　Remove the cowl cover (see Chapter 11).
5　Disconnect the wiring from the motor assembly.
6　Remove the two end mounting bolts and two center nuts **(see illustration)**.
7　Pull the assembly off of the mounting studs and remove it from the vehicle.
8　Installation is the reverse of the removal procedure.

Rear wiper motor

Refer to illustrations 17.11 and 17.15

9　Disconnect the cable from the negative terminal of the battery (see Chapter 5).
10　Flip up the base cover from the wiper arm. Remove the nut and then remove the arm. Refer to Steps 2 and 3.
11　Carefully pry the cover off of the shaft nut **(see illustration)**. **Caution:** *To avoid damage to the paint, apply several layers of tape in the area before removing the cover.*
12　Remove the wiper nut, the bezel beneath it and the gasket.
13　Refer to Chapter 11 and remove the tailgate/liftgate inner trim panel.
14　Disconnect the wiring connectors.
15　Loosen but do not remove the two screws at the mounting bracket **(see illustration)**.
16　Slide the wiper motor forward and out of the tailgate/liftgate.
17　To install a replacement motor, first loosely install the assembly in its mounting holes.

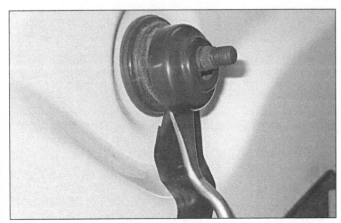

17.11 It is necessary to protect the paint when prying off the shaft cover

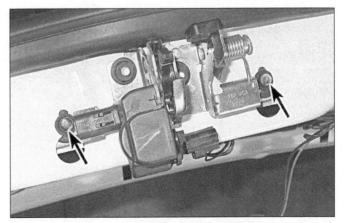

17.15 Loosen the two mounting screws, but don't remove them completely; the motor can now be slid out

18.3 There are two screws at the sides and two at the top center of the instrument cluster (which aren't visible in this photo) - typical 2007 and earlier models, later models similar

19.2 Horn mounting details

1	*Electrical connectors*	2	*Mounting bolt*

18 Place the pivot shaft in the exact center of the tailgate/liftgate hole and then install the gasket and bezel.

19 Install the pivot shaft mounting nut. Tighten the motor mounting fasteners.

20 The remainder of installation is the reverse of removal.

18 Instrument cluster - removal and installation

Refer to illustration 18.3

Warning: *The models covered by this manual are equipped with a Supplemental Restraint System (SRS), more commonly known as airbags. Always disarm the airbag system before working in the vicinity of any airbag system component to avoid the possibility of accidental deployment of the airbag, which could cause personal injury (see Section 25). Do not use a memory-saving device to preserve the PCM's memory when working on or near airbag system components.*

1 Disconnect the cable from the negative terminal of the battery (see Chapter 5).

2 Refer to Chapter 11 and remove the instrument panel cluster bezel.

3 Remove the mounting screws **(see illustration)**.

4 Gently pull the cluster to the rear far enough to reach the wiring connections.

5 Disconnect the wiring connectors from the cluster. Remove the panel.

6 Installation is the reverse of the removal procedure.

19 Horn - replacement

Refer to illustration 19.2

1 On 2007 and earlier models, the horns are located under the power distribution center in the engine compartment. On 2008 and later models, the horns are located behind the headlight housing.

2 Disconnect the wiring connectors **(see illustration)**.

3 Remove the mounting bolt and remove the horns. Each horn can be removed from the mounting bracket by unscrewing the nut on the horn post.

4 Installation is the reverse of removal.

20 Cruise control system - description

2007 and earlier models

Refer to illustrations 20.1 and 20.3

1 The cruise control system maintains vehicle speed with a servo motor connected to the throttle linkage by a cable. The system consists of the servo motor, brake switch, clutch switch, control switches and associated wiring and vacuum hoses **(see illustration)**. Some features of the system require the use of a scan tool and diagnostic procedures which are beyond the scope of the

home mechanic. Listed below are some general procedures that may be used to locate common problems.

2 Locate and check the fuse (see Section 3).

3 Visually inspect the vacuum hose connected to the servo and check the control linkage between the cruise control servo and the throttle linkage and replace as necessary **(see illustration)**.

4 If the speedometer is inoperative along with the cruise control system, refer to Chapter 6 and check the vehicle speed sensor.

5 Test-drive the vehicle to determine if the cruise control is now working. If it isn't, take it to a dealer service department or an automotive electrical specialist for further diagnosis and repair.

2008 and later models

6 The cruise control system maintains vehicle speed with the Powertrain Control Module (PCM), throttle actuator control motor, brake switch, ElectroMechanical instrument

20.1 The cruise control servo motor is located along the right side of the engine compartment

20.3 The cruise control cable at the throttle body

cluster, control switches and associated wiring. There is no mechanical connection, such as a vacuum servo or cable. Some features of the system require special testers and diagnostic procedures that are beyond the scope of the home mechanic. Listed below are some general procedures that may be used to locate common problems:

a) Check the fuses.

b) The Brake Pedal Position (BPP) switch (or brake light switch) deactivates the cruise control system. Have an assistant press the brake pedal while you check the brake light operation. If the brake lights do not operate properly, correct the problem and retest the cruise control.

c) Check the wiring between the PCM and throttle actuator motor for opens or shorts and repair as necessary.

d) The cruise control system uses information from the PCM, including the Vehicle Speed Sensor (VSS), which is located in the transaxle.

e) Test drive the vehicle to determine if the cruise control is now working. If it isn't, take it to a dealer service department or other qualified repair shop for further diagnosis.

21 Power window system - description

1 The power window system operates the electric motors mounted in the doors which lower and raise the windows. The system consists of the control switches, the motors (regulators), glass mechanisms and associated wiring. When the ignition is On, power flows from the fuse in the power distribution center to the circuit breaker in the fuse block, then to the master switch.

2 Power windows are wired so they can be lowered and raised from the master control switch by the driver or by remote switches located at the individual windows. Each window has a separate motor that is reversible. The position of the control switch determines the polarity and therefore the direction of operation. Some systems are equipped with relays that control current to the motors.

3 Some vehicles are equipped with a separate circuit breaker for each motor in addition to the fuse or circuit breaker protecting the whole circuit. This prevents one stuck window from disabling the whole system.

4 The power window system will only operate when the ignition switch is ON. In addition, many models have a window lockout switch at the master control switch which, when activated, disables the switches at the rear windows and, sometimes, the switch at the passenger's window also. Always check these items before troubleshooting a window problem.

5 These procedures are general in nature, so if you can't find the problem using them, take the vehicle to a dealer service department.

6 If the power windows don't work at all, check the fuse or circuit breaker.

7 If only the rear windows are inoperative, or if the windows only operate from the master control switch, check the rear window lockout switch for continuity in the unlocked position. Replace it if it doesn't have continuity.

8 Check the wiring between the switches and fuse panel for continuity. Repair the wiring, if necessary.

9 If only one window is inoperative from the master control switch, try the other control switch at the window. **Note:** This doesn't apply to the driver's door window.

10 If the same window works from one switch, but not the other, check the switch for continuity.

11 If the switch tests OK, check for a short or open in the wiring between the affected switch and the window motor.

12 If one window is inoperative from both switches, remove the trim panel from the affected door and check for voltage at the switch and at the motor while the switch is operated. First check for voltage at the connectors. With the ignition key On and connectors in place on the switch, backprobe at the designated wire (see the wiring diagrams at the end of this Chapter) with a grounded test light. Pushing the driver's window switch Down, there should be voltage at one terminal. Pushing the same switch Up, there should be voltage at another terminal. If these voltage are OK, disconnect the electrical connector at the driver's motor, and check for voltage there when the switch is operated.

13 If voltage is reaching the motor and the switch is OK, disconnect the glass from the regulator (see Chapter 11). Move the window up and down by hand while checking for binding and damage. Also check for binding and damage to the regulator. If the regulator is not damaged and the window moves up and down smoothly, replace the motor. If there's binding or damage, lubricate, repair or replace parts, as necessary.

14 If voltage isn't reaching the motor, check the wiring in the circuit for continuity between the switches and motors. You'll need to consult the wiring diagram for the vehicle. Some power window circuits are equipped with relays. If equipped, check that the relays are grounded properly and receiving voltage from the switches. Also check that each relay sends voltage to the motor when the switch is turned on. If it doesn't, replace the relay.

15 Test the windows after you are done to confirm proper repairs. If the main power window switch is to be replaced, pry the switch unit out of the door panel, then disconnect the two connectors.

22 Power door lock and keyless entry system - description

1 The power door lock system operates the door lock actuators mounted in each door. The system consists of the switches, actuators, relays and associated wiring. Diagnosis can usually be limited to simple checks of the wiring connections and actuators for minor faults that can be easily repaired.

2 Power door lock systems are operated by bi-directional solenoids located in the doors. The lock switches have two operating positions: Lock and Unlock. These switches activate a relay, which in turn connects voltage to the door lock solenoids. Depending on which way the relay is activated, it reverses polarity, allowing the two sides of the circuit to be used alternately as the feed (positive) and ground side.

3 Some vehicles may have anti-theft systems incorporated into the power locks. If you are unable to locate the trouble using the following general steps, consult your dealer service department.

4 Always check the circuit protection (fuses and relays) first.

5 Operate the door lock switches in both directions (Lock and Unlock) with the engine off. Listen for the faint click of the relay operating.

6 If there's no click, check for voltage at the switches. If no voltage is present, check the wiring between the fuse block and the switches for shorts and opens.

7 If voltage is present but no click is heard, test the switch for continuity. Replace it if there's no continuity in either switch position.

8 If the switch has continuity but the relay doesn't click, check the wiring between the switch and relay for continuity. Repair the wiring if there's not continuity.

9 If the relay is receiving voltage from the switch but is not sending voltage to the solenoids, check for a bad ground at the relay case. If the relay case is grounding properly, replace the relay.

10 If all but one lock solenoids operate, remove the trim panel from the affected door (see Chapter 11) and check for voltage at the solenoid while the lock switch is operated. One of the wires should have voltage in the Lock position; the other should have voltage in the unlock position.

11 If the inoperative solenoid is receiving voltage, replace the solenoid.

12 If the inoperative solenoid isn't receiving voltage, check for an open or short in the wire between the lock solenoid and the relay.

Note: It's common for wires to break in the portion of the harness between the body and door (opening and closing the door fatigues and eventually breaks the wires).

Keyless entry system
Refer to illustrations 22.15 and 22.16

13 The keyless entry system consists of a remote control transmitter that sends a coded infrared signal to the remote keyless entry module attached to the body control module (BCM). This operates the door lock system.

14 Replace the transmitter batteries when the red LED light on the side of the case doesn't light when the button is pushed.

22.15 A coin works well to open the keyless entry system unit to replace batteries

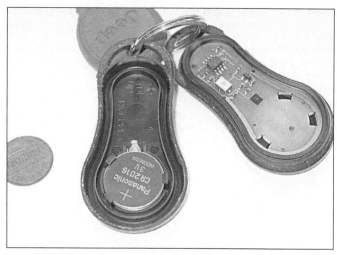

22.16 This unit uses two #2016 batteries; some models use one #2032 battery

15   Use a small screwdriver or a coin to carefully separate the case halves **(see illustration)**.

16   Replace the batteries **(see illustration)**. If equipped with two batteries, use 2016 batteries. If equipped with one battery, use a 2032.

17   Snap the case halves together.

---

23   Power seats - description

Warning: *The models covered by this manual are equipped with a Supplemental Restraint System (SRS), more commonly known as airbags. Some later models have side-curtain airbags along the sides of the headliner. Additionally, some models are equipped with seat belt pre-tensioners, which are explosive devices. Always disarm the airbag/restraint system before working in the vicinity of any airbag/restraint system component to avoid the possibility of accidental deployment of the airbag/seat belt pre-tensioners, which could cause personal injury (see Section 25). Do not use a memory-saving device to preserve the PCM's memory when working on or near airbag system components.*

1   These models feature a six-way seat that goes forward and backward, up and down and tilts forward and backward. The seats are powered by three reversible motors mounted in one housing that are controlled by switches on the side of the seat. Each switch changes the direction of seat travel by reversing polarity to the drive motor.

2   Diagnosis is usually a simple matter, using the following procedures.

3   Look under the seat for any object which may be preventing the seat from moving.

4   If the seat won't work at all, check the circuit breaker and the fuse in the fuse block.

5   With the engine off to reduce the noise level, operate the seat controls in all directions and listen for sound coming from the seat motors.

6   If the motor doesn't work or make noise, check for voltage at the motor while an assistant operates the switch.

7   If the motor is getting voltage but doesn't run, test it off the vehicle with jumper wires. If it still doesn't work, replace it. On most power seat systems, the individual components are not available separately. The whole power-seat track must be purchased as an assembly.

8   If the motor isn't getting voltage, remove the seat side panel to access the switch and check for voltage. If there's no voltage at the switch, check the wiring between the fuse block and the switch. If there's battery voltage at the switch, check the circuit breaker. Check the black wire and replace the switch if there's no continuity. If the switch is OK, check for a short or open in the wiring between the switch and motor.

9   Test the completed repairs.

---

24   Electric rear view mirrors - description

Side mirror

1   The electric rear view mirrors use two motors to move the glass; one for up and down adjustments and one for left-right adjustments. Some models also have a heating element for defogging.

2   The control switch has a selector portion which sends voltage to the left or right side mirror. With the ignition switch in the ACC position, roll down the windows and operate the mirror control switch through all functions (left-right and up-down) for both the left and right side mirrors.

3   Listen carefully for the sound of the electric motors running in the mirrors.

4   If the motors can be heard but the mirror glass doesn't move, the problem is usually with the drive mechanism inside the mirror, which will necessitate replacement of the mirror.

5   If the mirrors don't operate and no sound comes from the mirrors, check the fuse in the fuse block (see Section 3).

6   If the fuse is OK, refer to Chapter 11 and remove the door panel for access to the back of the mirror control switch, without disconnecting the wires attached to it. Turn the ignition ON and check for voltage at the switch. There should be voltage at one terminal. If there's no voltage at the switch, check for an open in the wiring between the fuse panel and the switch.

7   If there's voltage at the switch, disconnect it. Check the switch for continuity in all its operating positions. If the switch does not have continuity, replace it.

8   Re-connect the switch. Locate the wire going from the switch to ground. Leaving the switch connected, connect a jumper wire between this wire and ground. If the mirror works normally with this wire in place, repair the faulty ground connection.

9   If the mirror still doesn't work, remove the mirror and check the wires at the mirror for voltage. Check with the ignition ON and the mirror selector switch on the appropriate side. Operate the mirror switch in all its positions. There should be voltage at one of the switch-to-mirror wires in each switch position (except the neutral "off" position).

10   If voltage is not present in each switch position, check the wiring between the mirror and control switch for opens and shorts.

11   If there's voltage, remove the mirror and test it off the vehicle with jumper wires. Replace the mirror if it fails this test.

Automatic/day night mirror

12   To reduce glare, the automatic day/night mirror adjusts the amount of light reflected according to conditions. This is achieved with two photocell sensors, one facing forward and one facing rearward, that darken or lighten the thin layer of electrochromic material incorporated into the mirror glass.

13   With the ignition switch On, place the

mirror switch on non-sunroof models, in the High or Low position or on sunroof equipped models, in the On position.

14 Cover the forward facing sensor, shine a light into the rear facing sensor, make sure the glass darkens, then shift the transmission into reverse and make sure it lightens. If it doesn't, disconnect the electrical connector. With the ignition On, cavity 1 should have battery voltage, cavity 2 should have continuity to ground and cavity 3 should have continuity with the transmission in reverse.

15 If it fails any of these tests, replace the mirror (see Chapter 11).

25 Airbag - general information

Refer to illustration 25.5

These models are equipped with a Supplemental Restraint System (SRS), more commonly called an airbag system. There are two airbags, one for the driver and one for the front seat passenger. The SRS system is designed to protect the driver and passenger from serious injury in the event of a head-on or frontal collision. The airbag control module is located on the transmission tunnel between the front seats. Side curtain airbags are optional on these models. Vehicles with this option can be identified by the airbag logo printed on the headliner above the front of the rear windows.

Airbag module

The airbag module contains a housing incorporating the cushion (airbag) and inflator unit. The inflator assembly is mounted on the back of the housing over a hole through which gas is expelled, inflating the bag almost instantaneously when an electrical signal is sent from the system. On the driver's airbag, the specially wound wire that carries this signal to the module is called a clockspring. The clockspring is a flat, ribbon-like electrically conductive tape that winds and unwinds as the steering wheel is turned so it can transmit an electrical signal regardless of wheel position.

A second air bag assembly for the front seat passenger is located in the top of the dashboard, above the glove compartment. Optional side curtain airbags run nearly the length of the vehicle above the side windows.

Crash sensors

The front crash sensors are basically pressure sensitive switches that complete an electrical circuit during an impact of sufficient G force; they're mounted on the backside of the radiator support. Side curtain sensors are

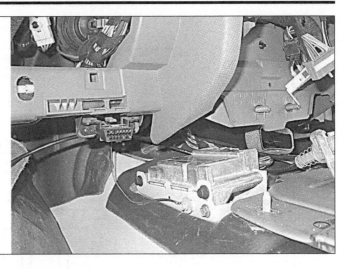

25.5 The airbag control module is located near the front of the console beneath the dash

at the base of the pillar separating the front and rear doors. The electrical signal from the crash sensors is sent to the safing sensor in the airbag control module, which then completes the circuit and inflates the airbag(s).

Airbag control module

The ASDM contains the safing sensor and an on-board microprocessor that monitors the operation of the system **(see illustration)**. It checks this system every time the vehicle is started, causing the AIRBAG light to go on, then off, if the system is operating properly. If there is a fault in the system, the light will go on and stay on and the airbag control module will store fault codes indicating the nature of the fault. If the AIRBAG light doesn't come on, or goes on and stays on, or the light blinks at all times, the vehicle should be taken to your dealer immediately for service.

Servicing components near the SRS system

There are times when you may need to remove the steering wheel, radio or other components on or near the dashboard. At these times you'll be working around components and wire harnesses for the SRS system. Do not use electrical test equipment on airbag system wires; it could cause the airbag(s) to deploy. **ALWAYS DISABLE THE SRS SYSTEM BEFORE WORKING NEAR THE SRS SYSTEM COMPONENTS OR RELATED WIRING.**

Disabling the system

Whenever working in the vicinity of the steering wheel, steering column, floor console or near other components of the airbag system, the system should be disarmed. To do this perform the following steps:

a) *Turn the ignition switch to the Off position.*
b) *Disconnect the cable from the negative terminal of the battery (see Chapter 5).*
c) *Wait for at least two minutes for the backup power supply to be depleted before beginning work.*

Enabling the system

To enable the airbag system, perform the following steps:

a) *Turn the ignition switch to the Off position.*
b) *Connect the negative battery cable to the battery.*
c) *Without putting your body in front of either airbag, turn the ignition switch to the On position. Observe that the airbag warning light glows for six to eight seconds and then goes out, indicating that the system is functioning properly.*

26 Wiring diagrams - general information

Since it isn't possible to include all wiring diagrams for every year covered by this manual, the following diagrams are those that are typical and most commonly needed.

Prior to troubleshooting any circuits, check the fuse and circuit breakers (if equipped) to make sure they are in good condition. Make sure the battery is properly charged and has clean, tight cable connections (see Chapter 1).

When checking the wiring system, make sure that all electrical connectors are clean, with no broken or loose pins. When disconnecting an electrical connector, do not pull on the wires, only on the connector housings.

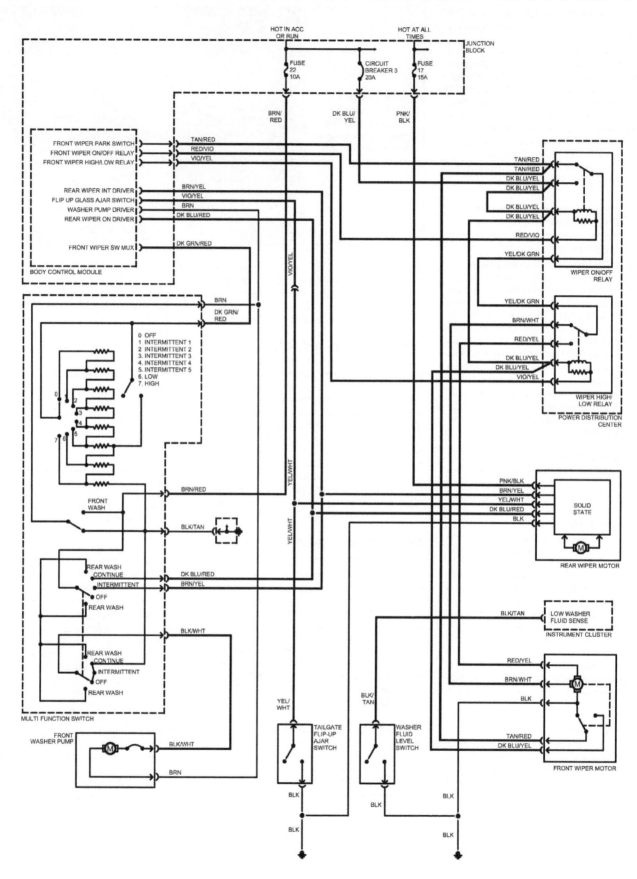

Windshield wiper/washer system

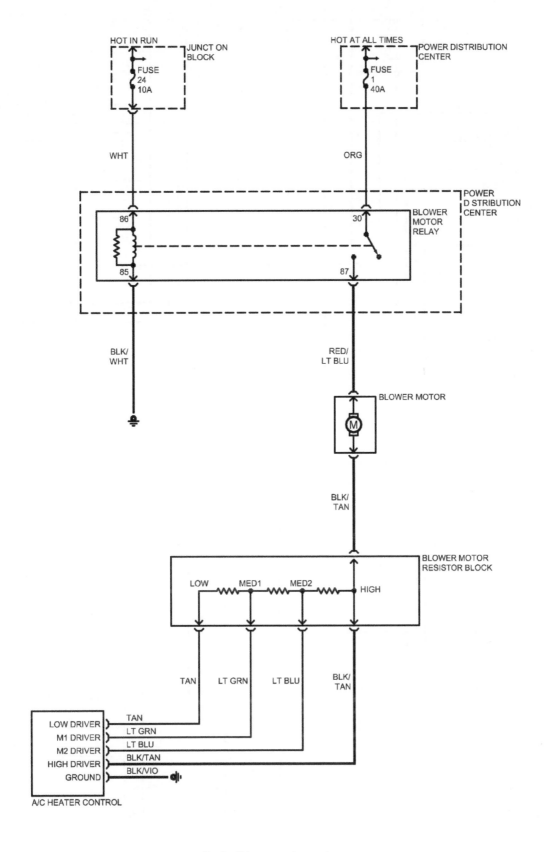

Heater/blower motor system

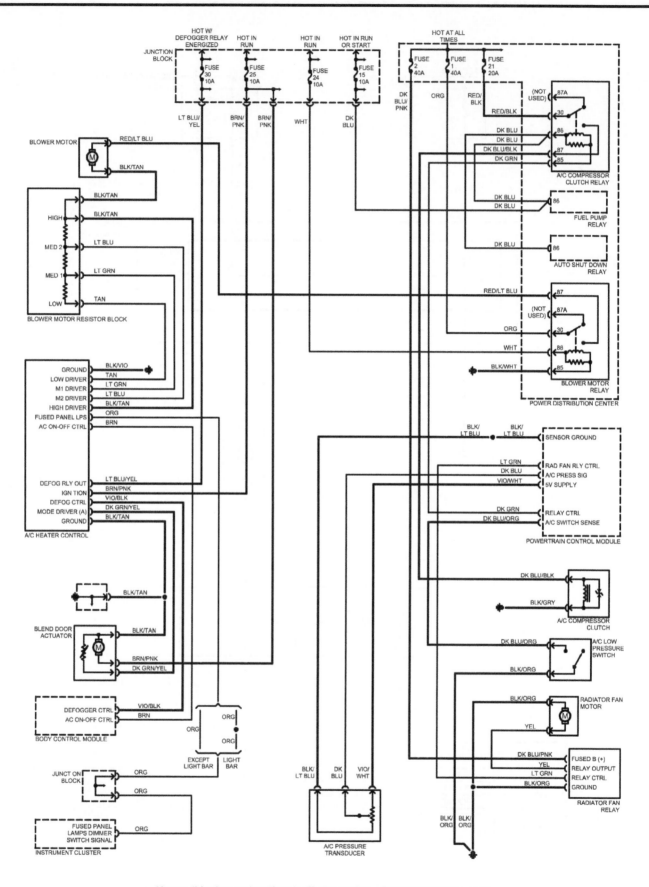

Air conditioning and engine cooling systems - 2002 and 2003 models

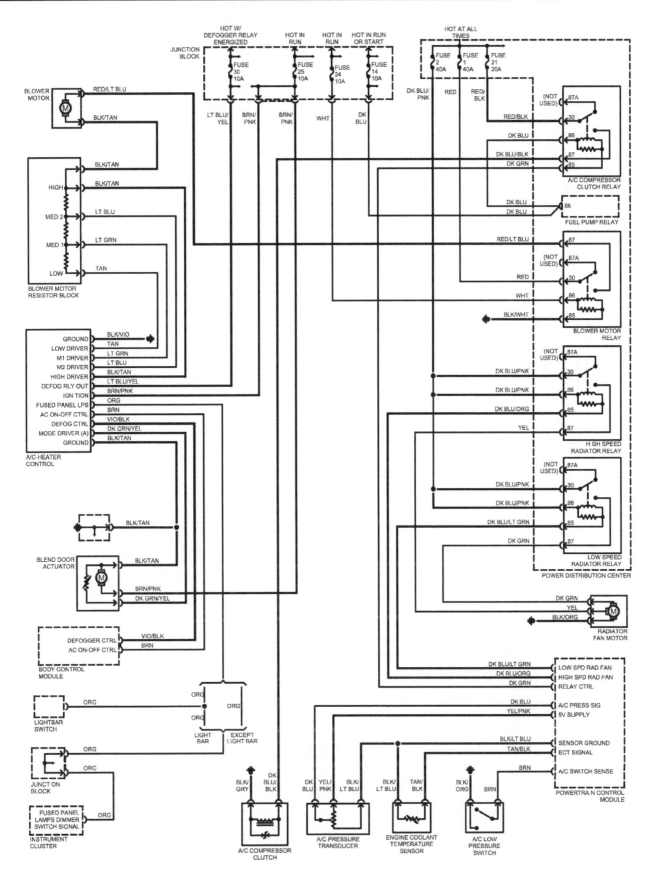

Air conditioning and engine cooling systems - 2004 models

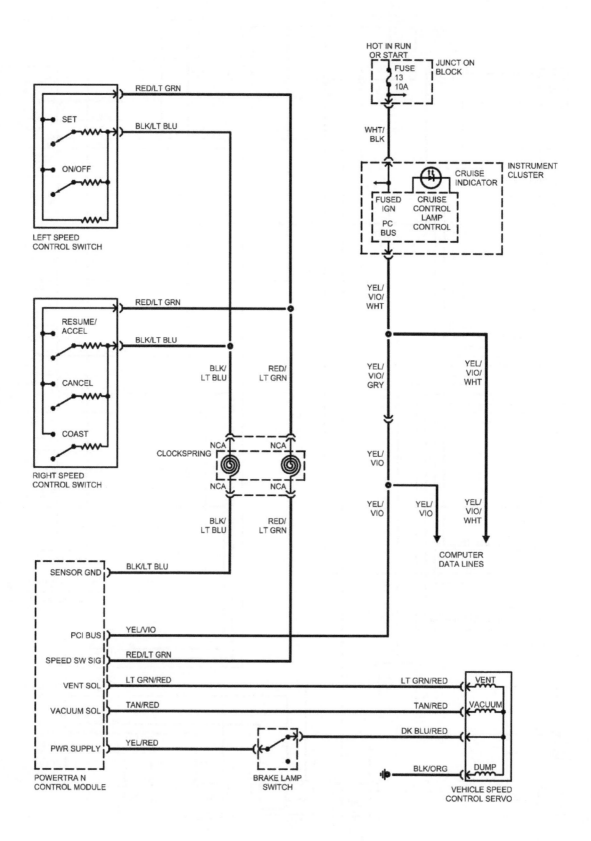

Cruise control system - 2002 and 2003 models

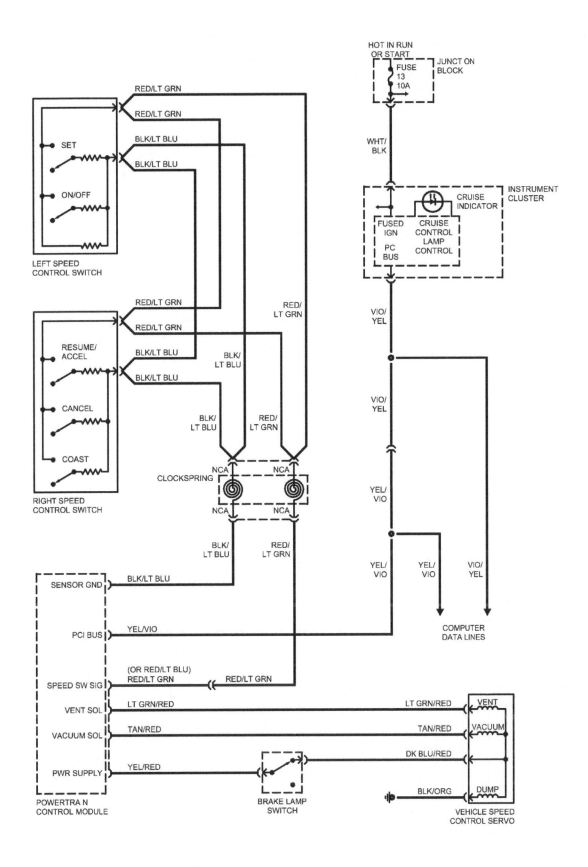

Cruise control system - 2004 models

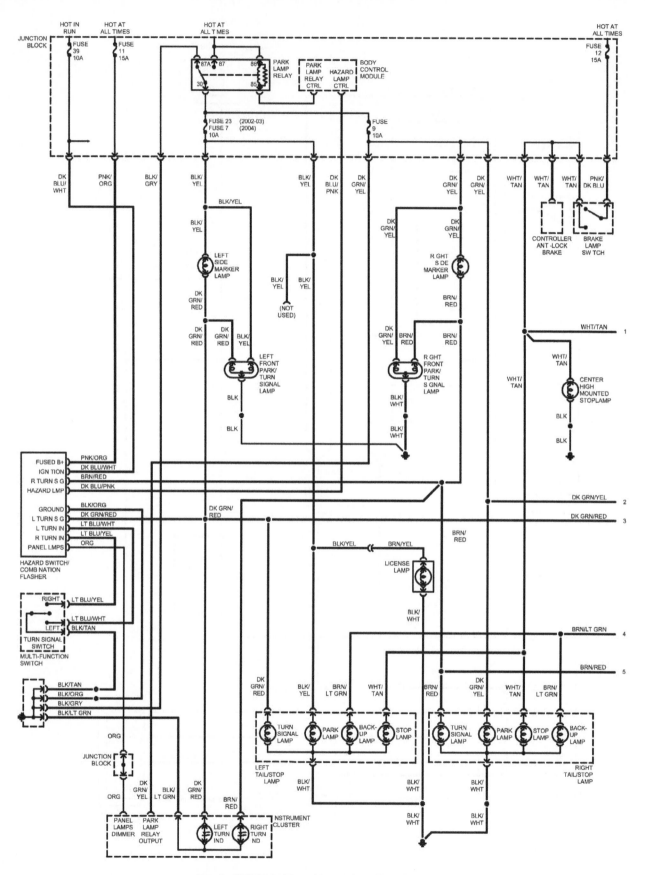

Exterior lighting system (except headlights) - 1 of 2

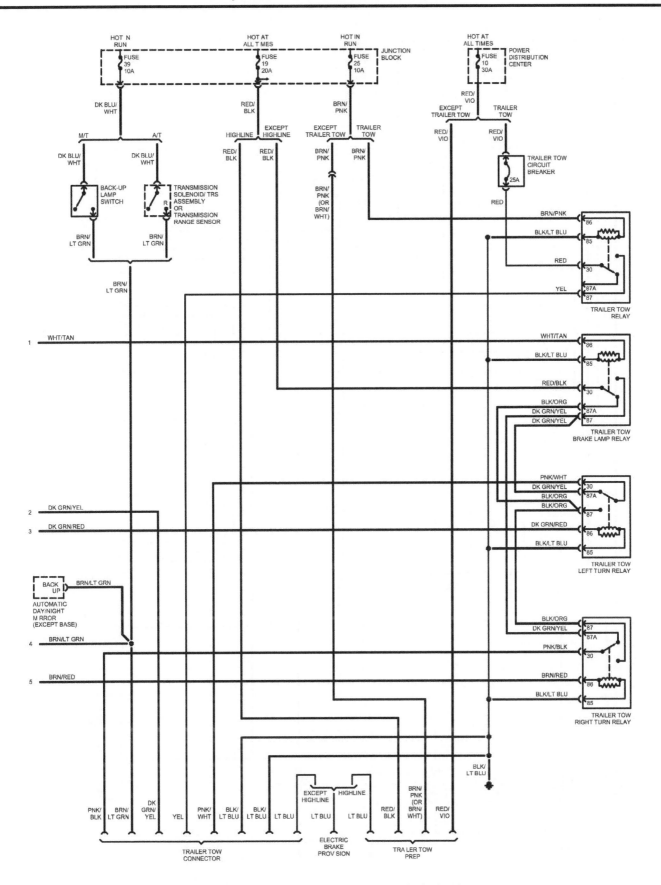

Exterior lighting system (except headlights) - 2 of 2

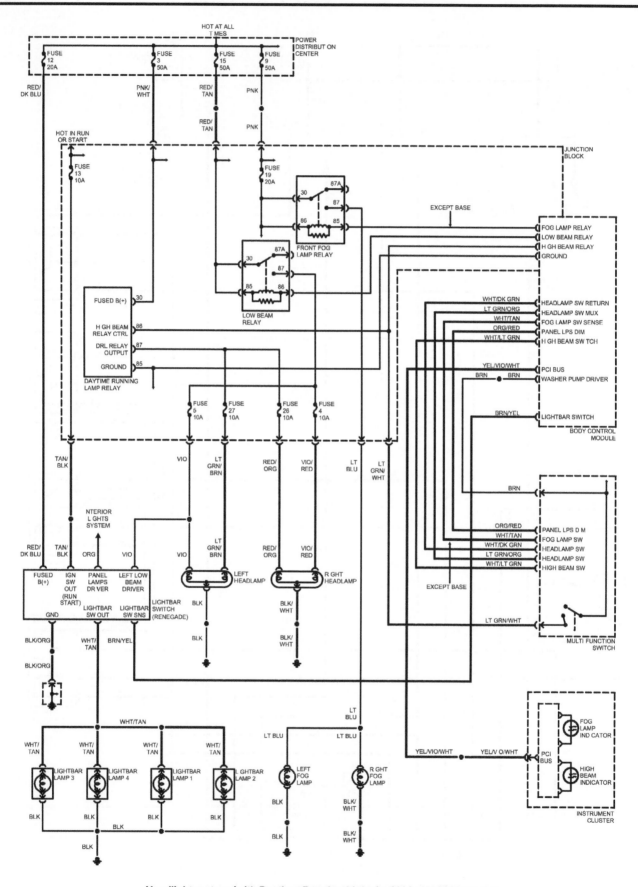

Headlight system (with Daytime Running Lights) - 2002 and 2003 models

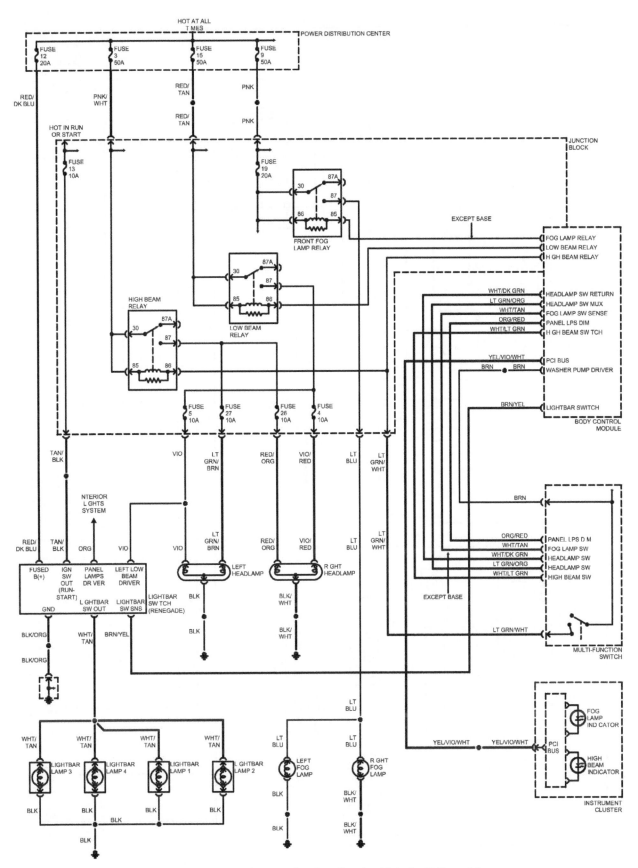

Headlight system (without Daytime Running Lights) - 2004 models

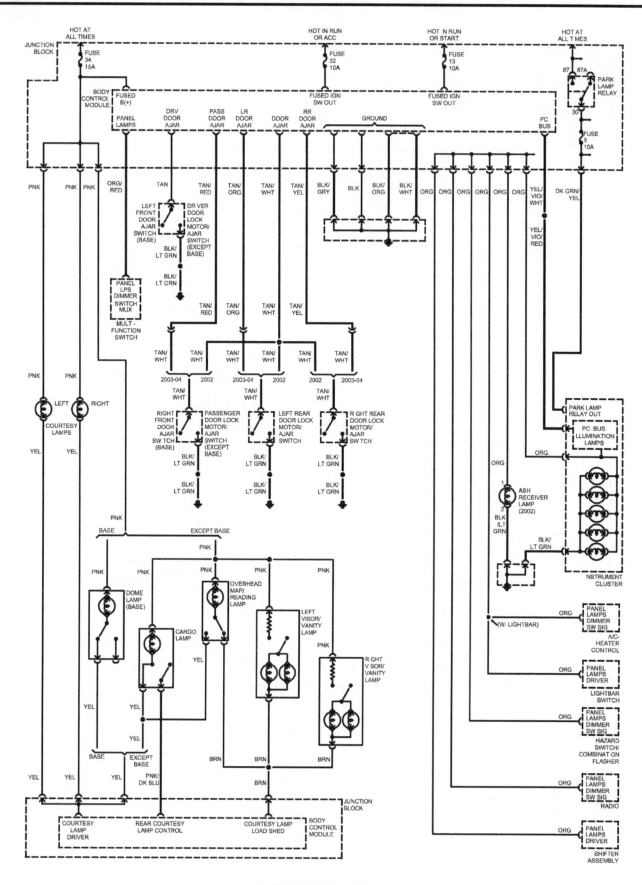

Interior lighting systems

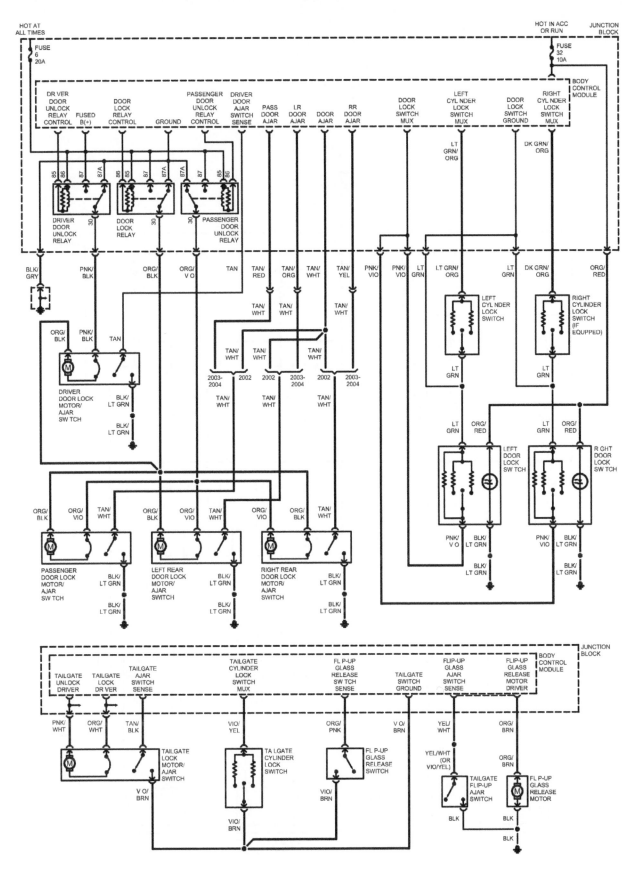

Power door lock system

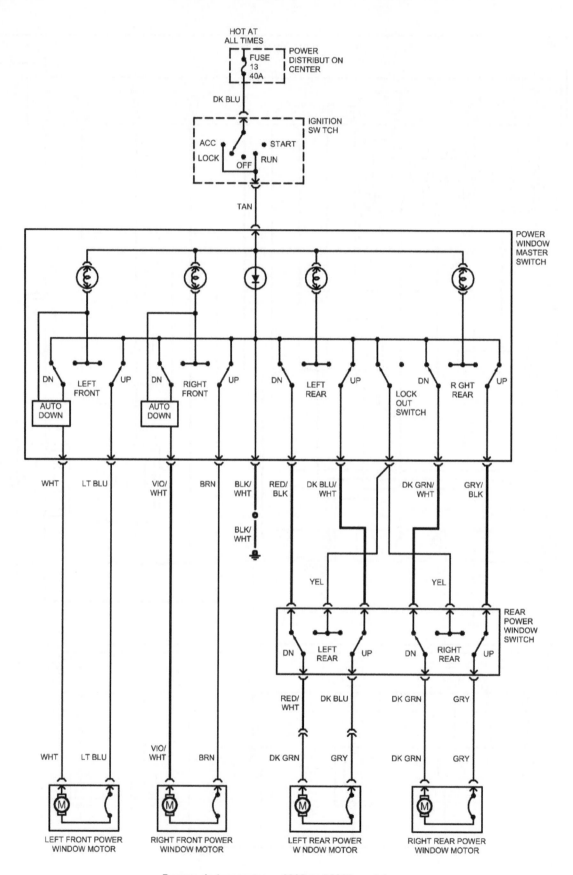

Power window system - 2002 and 2003 models

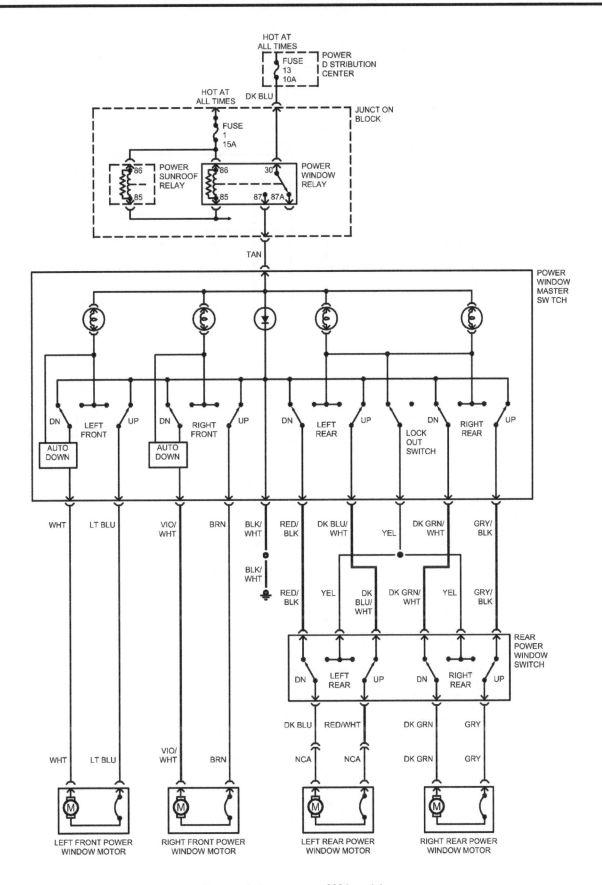

Power window system - 2004 models

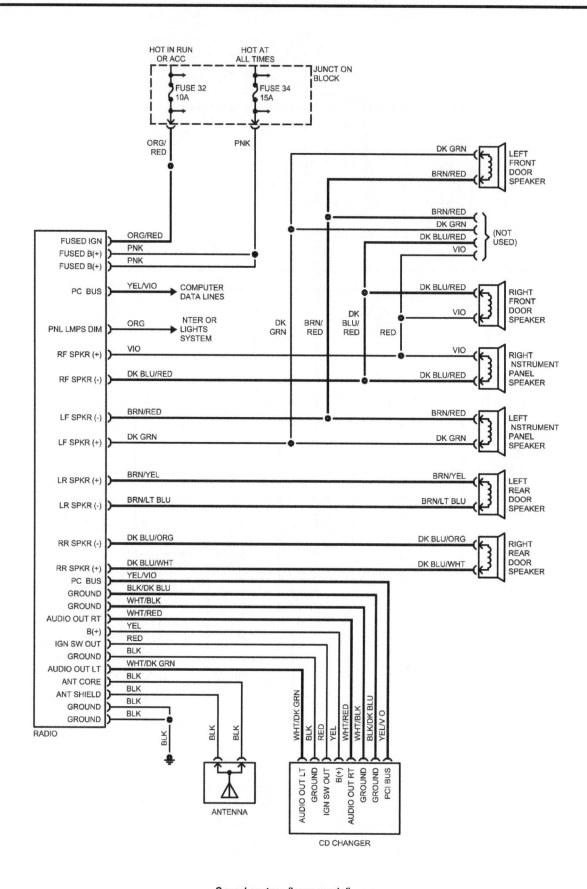

Sound system (base model)

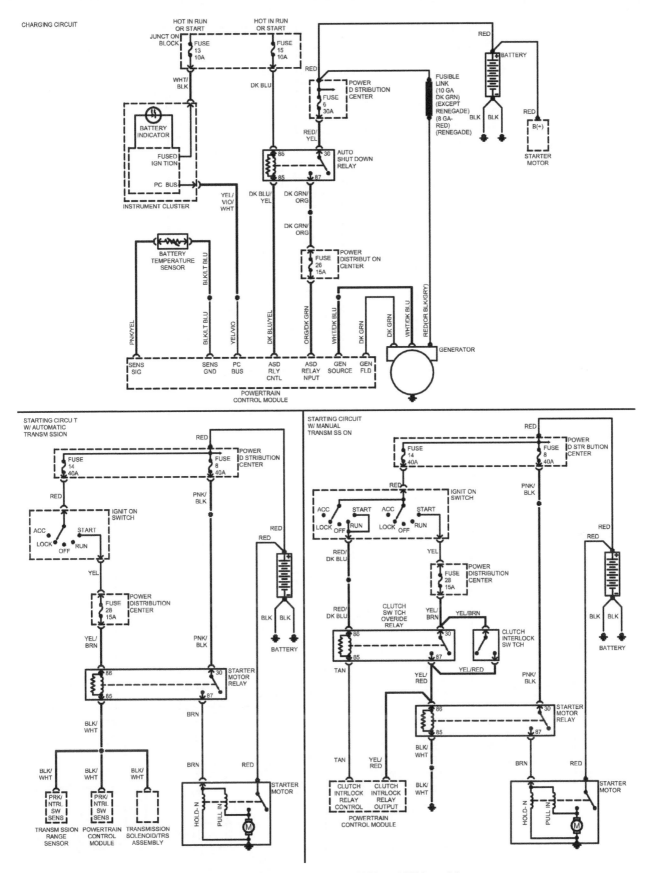

Starting and charging systems - 2002 and 2003 models

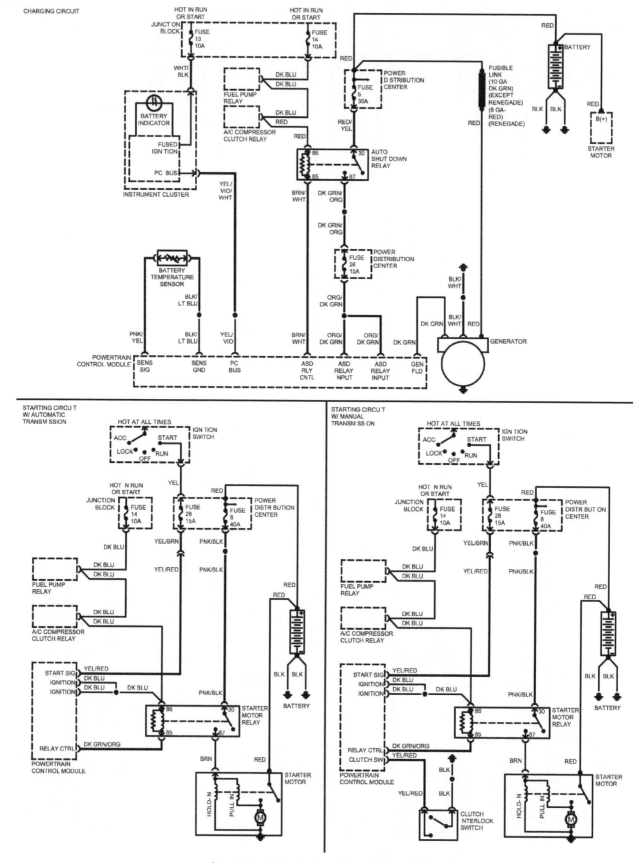

Starting and charging systems - 2004 models

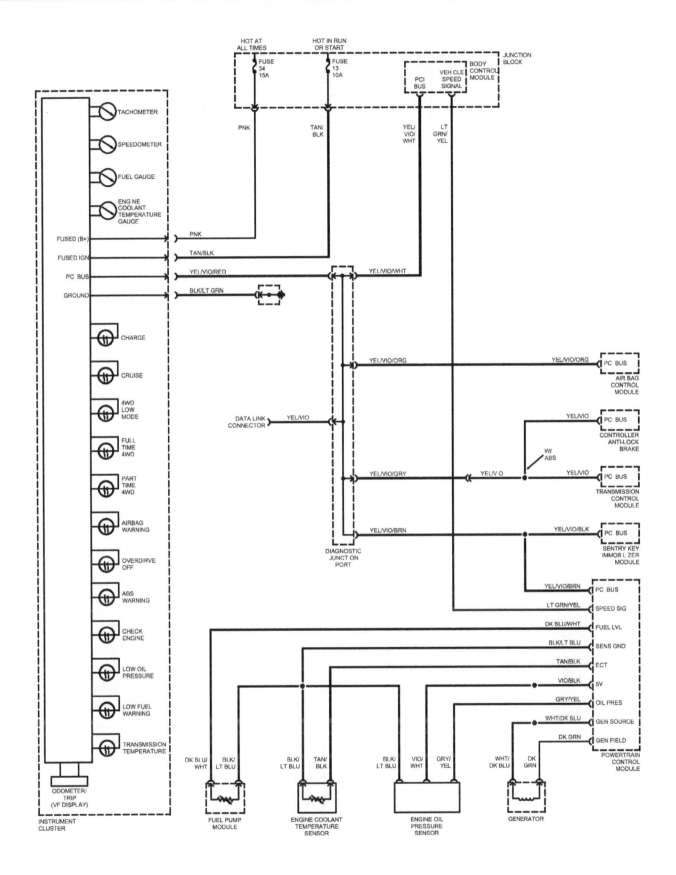

Warning systems - 2002 models

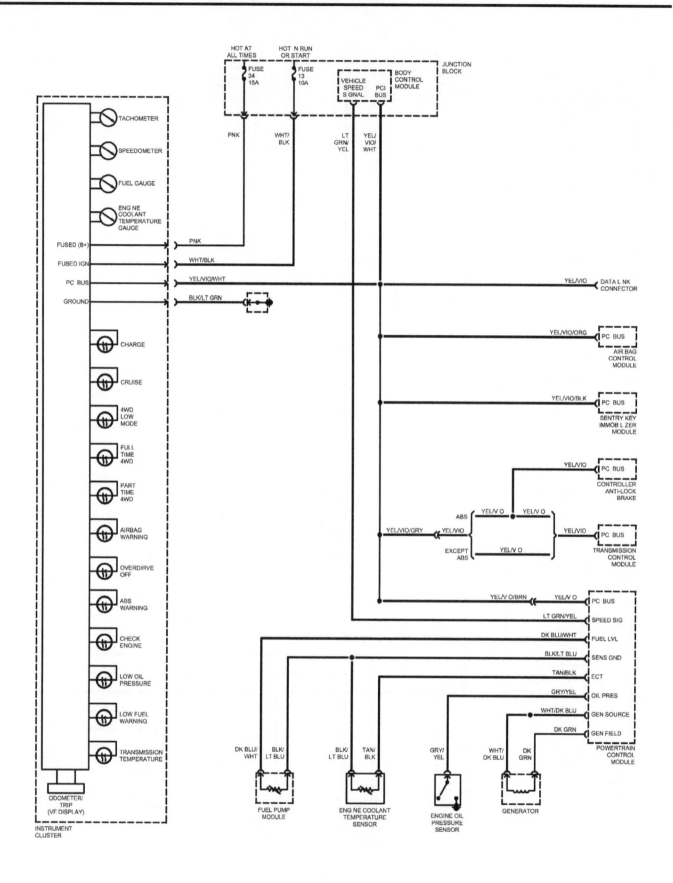

Warning systems - 2003 models

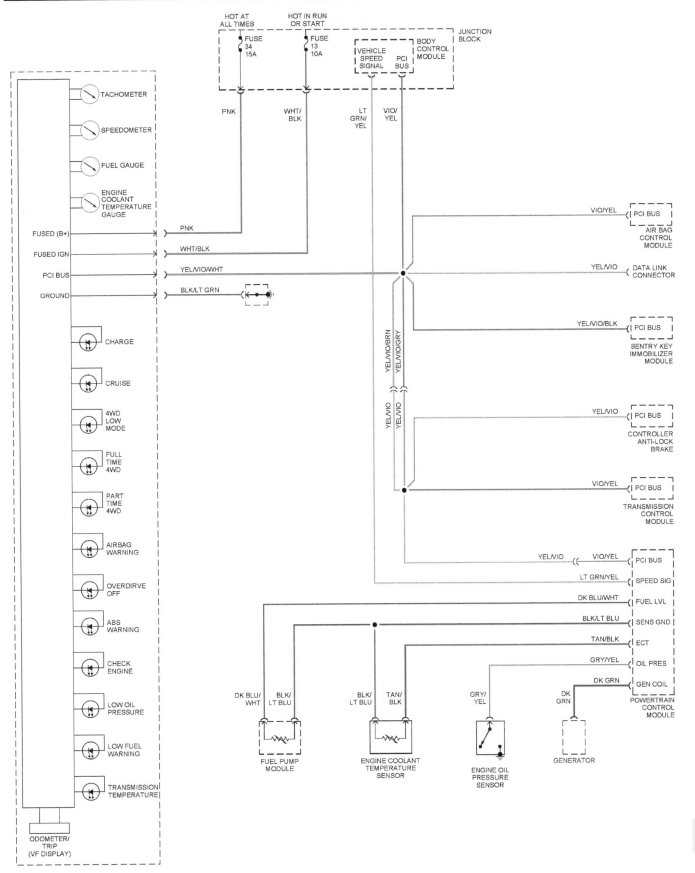

Warning systems - 2004 models

Notes

Index

A

Haynes Automotive Manuals

NOTE: If you do not see a listing for your vehicle, please visit **haynes.com** for the latest product information and check out our **Online Manuals!**

ACURA
12020 Integra '86 thru '89 & **Legend** '86 thru '90
12021 Integra '90 thru '93 & **Legend** '91 thru '95
Integra '94 thru '00 - see HONDA Civic (42025)
MDX '01 thru '07 - see HONDA Pilot (42037)
12050 Acura TL all models '99 thru '08

AMC
14020 Mid-size models '70 thru '83
14025 (Renault) Alliance & Encore '83 thru '87

AUDI
15020 4000 all models '80 thru '87
15025 5000 all models '77 thru '83
15026 5000 all models '84 thru '88
Audi A4 '96 thru '01 - see VW Passat (96023)
15030 Audi A4 '02 thru '08

AUSTIN-HEALEY
Sprite - see MG Midget (66015)

BMW
18020 3/5 Series '82 thru '92
18021 3-Series incl. Z3 models '92 thru '98
18022 3-Series incl. Z4 models '99 thru '05
18023 3-Series '06 thru '14
18025 320i all 4-cylinder models '75 thru '83
18050 1500 thru 2002 except Turbo '59 thru '77

BUICK
19010 Buick Century '97 thru '05
Century (front-wheel drive) - see GM (38005)
19020 Buick, Oldsmobile & Pontiac Full-size (Front-wheel drive) '85 thru '05
Buick Electra, LeSabre and Park Avenue; Oldsmobile Delta 88 Royale, Ninety Eight and Regency; Pontiac Bonneville
19025 Buick, Oldsmobile & Pontiac Full-size (Rear wheel drive) '70 thru '90
Buick Estate, Electra, LeSabre, Limited, Oldsmobile Custom Cruiser, Delta 88, Ninety-eight, Pontiac Bonneville, Catalina, Grandville, Parisienne
19027 Buick LaCrosse '05 thru '13
Enclave - see GENERAL MOTORS (38001)
Rainier - see CHEVROLET (24072)
Regal - see GENERAL MOTORS (38010)
Riviera - see GENERAL MOTORS (38030, 38031)
Roadmaster - see CHEVROLET (24046)
Skyhawk - see GENERAL MOTORS (38015)
Skylark - see GENERAL MOTORS (38020, 38025)
Somerset - see GENERAL MOTORS (38025)

CADILLAC
21015 CTS & CTS-V '03 thru '14
21030 Cadillac Rear Wheel Drive '70 thru '93
Cimarron - see GENERAL MOTORS (38015)
DeVille - see GENERAL MOTORS (38031 & 38032)
Eldorado - see GENERAL MOTORS (38030)
Fleetwood - see GENERAL MOTORS (38031)
Seville - see GM (38030, 38031 & 38032)

CHEVROLET
10305 Chevrolet Engine Overhaul Manual
24010 Astro & GMC Safari Mini-vans '85 thru '05
24013 Aveo '04 thru '11
24015 Camaro V8 all models '70 thru '81
24016 Camaro all models '82 thru '92
24017 Camaro & Firebird '93 thru '02
Cavalier - see GENERAL MOTORS (38016)
Celebrity - see GENERAL MOTORS (38005)
24018 Camaro '10 thru '15
24020 Chevelle, Malibu & El Camino '69 thru '87
Cobalt - see GENERAL MOTORS (38017)
24024 Chevette & Pontiac T1000 '76 thru '87
Citation - see GENERAL MOTORS (38020)
24027 Colorado & GMC Canyon '04 thru '12
24032 Corsica & Beretta all models '87 thru '96
24040 Corvette all V8 models '68 thru '82
24041 Corvette all models '84 thru '96
24042 Corvette all models '97 thru '13
24044 Cruze '11 thru '19
24045 Full-size Sedans Caprice, Impala, Biscayne, Bel Air & Wagons '69 thru '90
24046 Impala SS & Caprice and Buick Roadmaster '91 thru '96
Impala '00 thru '05 - see LUMINA (24048)
24047 Impala & Monte Carlo all models '06 thru '11
Lumina '90 thru '94 - see GM (38010)
24048 Lumina & Monte Carlo '95 thru '05
Lumina APV - see GM (38035)
24050 Luv Pick-up all 2WD & 4WD '72 thru '82
24051 Malibu '13 thru '19
24055 Monte Carlo all models '70 thru '88
Monte Carlo '95 thru '01 - see LUMINA (24048)
24059 Nova all V8 models '69 thru '79

24060 Nova and Geo Prizm '85 thru '92
24064 Pick-ups '67 thru '87 - Chevrolet & GMC
24065 Pick-ups '88 thru '98 - Chevrolet & GMC
24066 Pick-ups '99 thru '06 - Chevrolet & GMC
24067 Chevrolet Silverado & GMC Sierra '07 thru '14
24068 Chevrolet Silverado & GMC Sierra '14 thru '19
24070 S-10 & S-15 Pick-ups '82 thru '93, Blazer & Jimmy '83 thru '94,
24071 S-10 & Sonoma Pick-ups '94 thru '04, including Blazer, Jimmy & Hombre
24072 Chevrolet TrailBlazer, GMC Envoy & Oldsmobile Bravada '02 thru '09
24075 Sprint '85 thru '88 & Geo Metro '89 thru '01
24080 Vans - Chevrolet & GMC '68 thru '96
24081 Chevrolet Express & GMC Savana Full-size Vans '96 thru '19

CHRYSLER
10310 Chrysler Engine Overhaul Manual
25015 Chrysler Cirrus, Dodge Stratus, Plymouth Breeze '95 thru '00
25020 Full-size Front-Wheel Drive '88 thru '93
K-Cars - see DODGE Aries (30008)
Laser - see DODGE Daytona (30030)
25025 Chrysler LHS, Concorde, New Yorker, Dodge Intrepid, Eagle Vision, '93 thru '97
25026 Chrysler LHS, Concorde, 300M, Dodge Intrepid, '98 thru '04
25027 Chrysler 300 '05 thru '18, Dodge Charger '06 thru '18, Magnum '05 thru '08 & Challenger '08 thru '18
25030 Chrysler & Plymouth Mid-size front wheel drive '82 thru '95
Rear-wheel Drive - see Dodge (30050)
25035 PT Cruiser all models '01 thru '10
25040 Chrysler Sebring '95 thru '06, Dodge Stratus '01 thru '06 & Dodge Avenger '95 thru '00
25041 Chrysler Sebring '07 thru '10, 200 '11 thru '17 Dodge Avenger '08 thru '14

DATSUN
28005 200SX all models '80 thru '83
28012 240Z, 260Z & 280Z Coupe '70 thru '78
28014 280ZX Coupe & 2+2 '79 thru '83
300ZX - see NISSAN (72010)
28018 510 & PL521 Pick-up '68 thru '73
28020 510 all models '78 thru '81
28022 620 Series Pick-up all models '73 thru '79
720 Series Pick-up - see NISSAN (72030)

DODGE
400 & 600 - see CHRYSLER (25030)
30008 Aries & Plymouth Reliant '81 thru '89
30010 Caravan & Plymouth Voyager '84 thru '95
30011 Caravan & Plymouth Voyager '96 thru '02
30012 Challenger & Plymouth Sapporo '78 thru '83
30013 Caravan, Chrysler Voyager & Town & Country '03 thru '07
30014 Grand Caravan & Chrysler Town & Country '08 thru '18
30016 Colt & Plymouth Champ '78 thru '87
30020 Dakota Pick-ups all models '87 thru '96
30021 Durango '98 & '99 & Dakota '97 thru '99
30022 Durango '00 thru '03 & Dakota '00 thru '04
30023 Durango '04 thru '09 & Dakota '05 thru '11
30025 Dart, Demon, Plymouth Barracuda, Duster & Valiant 6-cylinder models '67 thru '76
30030 Daytona & Chrysler Laser '84 thru '89
Intrepid - see CHRYSLER (25025, 25026)
30034 Neon all models '95 thru '99
30035 Omni & Plymouth Horizon '78 thru '90
30036 Dodge & Plymouth Neon '00 thru '05
30040 Pick-ups full-size models '74 thru '93
30042 Pick-ups full-size models '94 thru '08
30043 Pick-ups full-size models '09 thru '18
30045 Ram 50/D50 Pick-ups & Raider and Plymouth Arrow Pick-ups '79 thru '93
30050 Dodge/Plymouth/Chrysler RWD '71 thru '89
30055 Shadow & Plymouth Sundance '87 thru '94
30060 Spirit & Plymouth Acclaim '89 thru '95
30065 Vans - Dodge & Plymouth '71 thru '03

EAGLE
Talon - see MITSUBISHI (68030, 68031)
Vision - see CHRYSLER (25025)

FIAT
34010 124 Sport Coupe & Spider '68 thru '78
34025 X1/9 all models '74 thru '80

FORD
10320 Ford Engine Overhaul Manual
10355 Ford Automatic Transmission Overhaul
11500 Mustang '64-1/2 thru '70 Restoration Guide
36004 Aerostar Mini-vans all models '86 thru '97
36006 Contour & Mercury Mystique '95 thru '00
36008 Courier Pick-up all models '72 thru '82

36012 Crown Victoria & Mercury Grand Marquis '88 thru '11
36014 Edge '07 thru '19 & Lincoln MKX '07 thru '18
36016 Escort & Mercury Lynx all models '81 thru '90
36020 Escort & Mercury Tracer '91 thru '02
36022 Escape '01 thru '17, Mazda Tribute '01 thru '11, & Mercury Mariner '05 thru '11
36024 Explorer & Mazda Navajo '91 thru '01
36025 Explorer & Mercury Mountaineer '02 thru '10
36026 Explorer '11 thru '17
36028 Fairmont & Mercury Zephyr '78 thru '83
36030 Festiva & Aspire '88 thru '97
36032 Fiesta all models '77 thru '80
36034 Focus all models '00 thru '11
36035 Focus '12 thru '14
36045 Fusion '06 thru '14 & Mercury Milan '06 thru '11
36048 Mustang V8 all models '64-1/2 thru '73
36049 Mustang II 4-cylinder, V6 & V8 models '74 thru '78
36050 Mustang & Mercury Capri '79 thru '93
36051 Mustang all models '94 thru '04
36052 Mustang '05 thru '14
36054 Pick-ups & Bronco '73 thru '79
36058 Pick-ups & Bronco '80 thru '96
36059 F-150 '97 thru '03, Expedition '97 thru '17, F-250 '97 thru '99, F-150 Heritage '04 & Lincoln Navigator '98 thru '17
36060 Super Duty Pick-ups & Excursion '99 thru '10
36061 F-150 full-size '04 thru '14
36062 Pinto & Mercury Bobcat '75 thru '80
36063 F-150 full-size '15 thru '17
36064 Super Duty Pick-ups '11 thru '16
36066 Probe all models '89 thru '92
Probe '93 thru '97 - see MAZDA 626 (61042)
36070 Ranger & Bronco II gas models '83 thru '92
36071 Ranger '93 thru '11 & Mazda Pick-ups '94 thru '09
36074 Taurus & Mercury Sable '86 thru '95
36075 Taurus & Mercury Sable '96 thru '07
36076 Taurus '08 thru '14, Five Hundred '05 thru '07, Mercury Montego '05 thru '07 & Sable '08 thru '09
36078 Tempo & Mercury Topaz '84 thru '94
36082 Thunderbird & Mercury Cougar '83 thru '88
36086 Thunderbird & Mercury Cougar '89 thru '97
36090 Vans all V8 Econoline models '69 thru '91
36094 Vans full size '92 thru '14
36097 Windstar '95 thru '03, Freestar & Mercury Monterey Mini-van '04 thru '07

GENERAL MOTORS
10360 GM Automatic Transmission Overhaul
38001 GMC Acadia '07 thru '16, Buick Enclave '08 thru '17, Saturn Outlook '07 thru '10 & Chevrolet Traverse '09 thru '17
38005 Buick Century, Chevrolet Celebrity, Oldsmobile Cutlass Ciera & Pontiac 6000 all models '82 thru '96
38010 Buick Regal '88 thru '04, Chevrolet Lumina '88 thru '04, Oldsmobile Cutlass Supreme '88 thru '97 & Pontiac Grand Prix '88 thru '07
38015 Buick Skyhawk, Cadillac Cimarron, Chevrolet Cavalier, Oldsmobile Firenza, Pontiac J-2000 & Sunbird '82 thru '94
38016 Chevrolet Cavalier & Pontiac Sunfire '95 thru '05
38017 Chevrolet Cobalt '05 thru '10, HHR '06 thru '11, Pontiac G5 '07 thru '09, Pursuit '05 thru '06 & Saturn ION '03 thru '07
38020 Buick Skylark, Chevrolet Citation, Oldsmobile Omega, Pontiac Phoenix '80 thru '85
38025 Buick Skylark '86 thru '98, Somerset '85 thru '87, Oldsmobile Achieva '92 thru '98 & Calais '85 thru '91, & Pontiac Grand Am all models '85 thru '98
38026 Chevrolet Malibu '97 thru '03, Classic '04 thru '05, Oldsmobile Alero '99 thru '03, Cutlass '97 thru '00, & Pontiac Grand Am '99 thru '03
38027 Chevrolet Malibu '04 thru '12, Pontiac G6 '05 thru '10 & Saturn Aura '07 thru '10
38030 Cadillac Eldorado, Seville, Oldsmobile Toronado & Buick Riviera '71 thru '85
38031 Cadillac Eldorado, Seville, DeVille, Fleetwood, Oldsmobile Toronado & Buick Riviera '86 thru '93
38032 Cadillac DeVille '94 thru '05, Seville '92 thru '04 & Cadillac DTS '06 thru '10
38035 Chevrolet Lumina APV, Oldsmobile Silhouette & Pontiac Trans Sport all models '90 thru '96
38036 Chevrolet Venture '97 thru '05, Oldsmobile Silhouette '97 thru '04, Pontiac Trans Sport '97 thru '98 & Montana '99 thru '05
38040 Chevrolet Equinox '05 thru '17, GMC Terrain '10 thru '17 & Pontiac Torrent '06 thru '09

GEO
Metro - see CHEVROLET Sprint (24075)
Prizm - '85 thru '92 see CHEVY (24060), '93 thru '02 see TOYOTA Corolla (92036)
40030 Storm all models '90 thru '93
Tracker - see SUZUKI Samurai (90010)

(Continued on other side)

Haynes North America, Inc. • (805) 498-6703 • www.haynes.com

Haynes Automotive Manuals (continued)

NOTE: If you do not see a listing for your vehicle, please visit **haynes.com** for the latest product information and check out our **Online Manuals!**

GMC
Acadia - see GENERAL MOTORS (38001)
Pick-ups - see CHEVROLET (24027, 24068)
Vans - see CHEVROLET (24081)

HONDA
42010 **Accord CVCC** all models '76 thru '83
42011 **Accord** all models '84 thru '89
42012 **Accord** all models '90 thru '93
42013 **Accord** all models '94 thru '97
42014 **Accord** all models '98 thru '02
42015 **Accord** '03 thru '12 & **Crosstour** '10 thru '14
42016 **Accord** '13 thru '17
42020 **Civic 1200** all models '73 thru '79
42021 **Civic 1300 & 1500 CVCC** '80 thru '83
42022 **Civic 1500 CVCC** all models '75 thru '79
42023 **Civic** all models '84 thru '91
42024 **Civic & del Sol** '92 thru '95
42025 **Civic** '96 thru '00, **CR-V** '97 thru '01 & **Acura Integra** '94 thru '00
42026 **Civic** '01 thru '11 & **CR-V** '02 thru '11
42027 **Civic** '12 thru '15 & **CR-V** '12 thru '16
42030 **Fit** '07 thru '13
42035 **Odyssey** all models '99 thru '10
Passport - see ISUZU Rodeo (47017)
42037 **Honda Pilot** '03 thru '08, **Ridgeline** '06 thru '14 & **Acura MDX** '01 thru '07
42040 **Prelude CVCC** all models '79 thru '89

HYUNDAI
43010 **Elantra** all models '96 thru '19
43015 **Excel & Accent** all models '86 thru '13
43050 **Santa Fe** all models '01 thru '12
43055 **Sonata** all models '99 thru '14

INFINITI
G35 '03 thru '08 - see NISSAN 350Z (72011)

ISUZU
Hombre - see CHEVROLET S-10 (24071)
47017 **Rodeo** '91 thru '02, **Amigo** '89 thru '94 & '98 thru '02 & **Honda Passport** '95 thru '02
47020 **Trooper** '84 thru '91 & **Pick-up** '81 thru '93

JAGUAR
49010 **XJ6** all 6-cylinder models '68 thru '86
49011 **XJ6** all models '88 thru '94
49015 **XJ12 & XJS** all 12-cylinder models '72 thru '85

JEEP
50010 **Cherokee, Comanche & Wagoneer Limited** all models '84 thru '01
50011 **Cherokee** '14 thru '19
50020 **CJ** all models '49 thru '86
50025 **Grand Cherokee** all models '93 thru '04
50026 **Grand Cherokee** '05 thru '19 & **Dodge Durango** '11 thru '19
50029 **Grand Wagoneer & Pick-up** '72 thru '91 Grand Wagoneer '84 thru '91, Cherokee & Wagoneer '72 thru '83, Pick-up '72 thru '88
50030 **Wrangler** all models '87 thru '17
50035 **Liberty** '02 thru '12 & **Dodge Nitro** '07 thru '11
50050 **Patriot & Compass** '07 thru '17

KIA
54050 **Optima** '01 thru '10
54060 **Sedona** '02 thru '14
54070 **Sephia** '94 thru '01, **Spectra** '00 thru '09, **Sportage** '05 thru '20
54077 **Sorento** '03 thru '13

LEXUS
ES 300/330 - see TOYOTA Camry (92007, 92008)
ES 350 - see TOYOTA Camry (92009)
RX 300/330/350 - see TOYOTA Highlander (92095)

LINCOLN
MKX - see FORD (36014)
Navigator - see FORD Pick-up (36059)
59010 **Rear-Wheel Drive Continental** '70 thru '87, **Mark Series** '70 thru '92 & **Town Car** '81 thru '10

MAZDA
61010 **GLC** (rear-wheel drive) '77 thru '83
61011 **GLC** (front-wheel drive) '81 thru '85
61012 **Mazda3** '04 thru '11
61015 **323 & Protegé** '90 thru '03
61016 **MX-5 Miata** '90 thru '14
61020 **MPV** all models '89 thru '98
Navajo - see Ford Explorer (36024)
61030 **Pick-ups** '72 thru '93
Pick-ups '94 thru '09 - see Ford Ranger (36071)
1035 **RX-7** all models '79 thru '85
36 **RX-7** all models '86 thru '91
626 (rear-wheel drive) all models '79 thru '82
6 **& MX-6** (front-wheel drive) '83 thru '92
'93 thru '01 & **MX-6/Ford Probe** '93 thru '02
8 '03 thru '13

MERCEDES-BENZ
63012 **123 Series Diesel** '76 thru '85
63015 **190 Series** 4-cylinder gas models '84 thru '88
63020 **230/250/280** 6-cylinder SOHC models '68 thru '72
63025 **280 123 Series** gas models '77 thru '81
63030 **350 & 450** all models '71 thru '80
63040 **C-Class:** C230/C240/C280/C320/C350 '01 thru '07

MERCURY
64200 **Villager & Nissan Quest** '93 thru '01
All other titles, see FORD Listing.

MG
66010 **MGB** Roadster & GT Coupe '62 thru '80
66015 **MG Midget, Austin Healey Sprite** '58 thru '80

MINI
67020 **Mini** '02 thru '13

MITSUBISHI
68020 **Cordia, Tredia, Galant, Precis & Mirage** '83 thru '93
68030 **Eclipse, Eagle Talon & Plymouth Laser** '90 thru '94
68031 **Eclipse** '95 thru '05 & **Eagle Talon** '95 thru '98
68035 **Galant** '94 thru '12
68040 **Pick-up** '83 thru '96 & **Montero** '83 thru '93

NISSAN
72010 **300ZX** all models including Turbo '84 thru '89
72011 **350Z & Infiniti G35** all models '03 thru '08
72015 **Altima** all models '93 thru '06
72016 **Altima** '07 thru '12
72020 **Maxima** all models '85 thru '92
72021 **Maxima** all models '93 thru '08
72025 **Murano** '03 thru '14
72030 **Pick-ups** '80 thru '97 & **Pathfinder** '87 thru '95
72031 **Frontier** '98 thru '04, **Xterra** '00 thru '04, & **Pathfinder** '96 thru '04
72032 **Frontier & Xterra** '05 thru '14
72037 **Pathfinder** '05 thru '14
72040 **Pulsar** all models '83 thru '86
72042 **Roque** all models '08 thru '20
72050 **Sentra** all models '82 thru '94
72051 **Sentra & 200SX** all models '95 thru '06
72060 **Stanza** all models '82 thru '90
72070 **Titan pick-ups** '04 thru '10, **Armada** '05 thru '10 & **Pathfinder Armada** '04
72080 **Versa** all models '07 thru '19

OLDSMOBILE
73015 **Cutlass** V6 & V8 gas models '74 thru '88
For other OLDSMOBILE titles, see BUICK, CHEVROLET or GENERAL MOTORS listings.

PLYMOUTH
For PLYMOUTH titles, see DODGE listing.

PONTIAC
79008 **Fiero** all models '84 thru '88
79018 **Firebird** V8 models except Turbo '70 thru '81
79019 **Firebird** all models '82 thru '92
79025 **G6** all models '05 thru '09
79040 **Mid-size Rear-wheel Drive** '70 thru '87
Vibe '03 thru '10 - see TOYOTA Corolla (92037)
For other PONTIAC titles, see BUICK, CHEVROLET or GENERAL MOTORS listings.

PORSCHE
80020 **911** Coupe & Targa models '65 thru '89
80025 **914** all 4-cylinder models '69 thru '76
80030 **924** all models including Turbo '76 thru '82
80035 **944** all models including Turbo '83 thru '89

RENAULT
Alliance & Encore - see AMC (14025)

SAAB
84010 **900** all models including Turbo '79 thru '88

SATURN
87010 **Saturn** all S-series models '91 thru '02
Saturn Ion '03 thru '07- see GM (38017)
Saturn Outlook - see GM (38001)
87020 **Saturn L-series** all models '00 thru '04
87040 **Saturn VUE** '02 thru '09

SUBARU
89002 **1100, 1300, 1400 & 1600** '71 thru '79
89003 **1600 & 1800** 2WD & 4WD '80 thru '94
89080 **Impreza** '02 thru '11, **WRX** '02 thru '14, & **WRX STI** '04 thru '14
89100 **Legacy** all models '90 thru '99
89101 **Legacy & Forester** '00 thru '09
89102 **Legacy** '10 thru '16 & **Forester** '12 thru '16

SUZUKI
90010 **Samurai/Sidekick & Geo Tracker** '86 thru '01

TOYOTA
92005 **Camry** all models '83 thru '91
92006 **Camry** '92 thru '96 & **Avalon** '95 thru '96
92007 **Camry, Avalon, Solara, Lexus ES 300** '97 thru '01

92008 **Camry, Avalon, Lexus ES 300/330** '02 thru '06 & **Solara** '02 thru '08
92009 **Camry, Avalon & Lexus ES 350** '07 thru '17
92015 **Celica Rear-wheel Drive** '71 thru '85
92020 **Celica Front-wheel Drive** '86 thru '99
92025 **Celica Supra** all models '79 thru '92
92030 **Corolla** all models '75 thru '79
92032 **Corolla** all rear-wheel drive models '80 thru '87
92035 **Corolla** all front-wheel drive models '84 thru '92
92036 **Corolla & Geo/Chevrolet Prizm** '93 thru '02
92037 **Corolla** '03 thru '19, **Matrix** '03 thru '14, & **Pontiac Vibe** '03 thru '10
92040 **Corolla Tercel** all models '80 thru '82
92045 **Corona** all models '74 thru '82
92050 **Cressida** all models '78 thru '82
92055 **Land Cruiser FJ40, 43, 45, 55** '68 thru '82
92056 **Land Cruiser FJ60, 62, 80, FZJ80** '80 thru '96
92060 **Matrix** '03 thru '11 & **Pontiac Vibe** '03 thru '10
92065 **MR2** all models '85 thru '87
92070 **Pick-up** all models '69 thru '78
92075 **Pick-up** all models '79 thru '95
92076 **Tacoma** '95 thru '04, **4Runner** '96 thru '02 & **T100** '93 thru '08
92077 **Tacoma** all models '05 thru '18
92078 **Tundra** '00 thru '06 & **Sequoia** '01 thru '07
92079 **4Runner** all models '03 thru '09
92080 **Previa** all models '91 thru '95
92081 **Prius** all models '01 thru '12
92082 **RAV4** all models '96 thru '12
92085 **Tercel** all models '87 thru '94
92090 **Sienna** all models '98 thru '10
92095 **Highlander** '01 thru '19 & **Lexus RX330/330/350** '99 thru '19
92179 **Tundra** '07 thru '19 & **Sequoia** '08 thru '19

TRIUMPH
94007 **Spitfire** all models '62 thru '81
94010 **TR7** all models '75 thru '81

VW
96008 **Beetle & Karmann Ghia** '54 thru '79
96009 **New Beetle** '98 thru '10
96016 **Rabbit, Jetta, Scirocco & Pick-up** gas models '75 thru '92 & Convertible '80 thru '92
96017 **Golf, GTI & Jetta** '93 thru '98, **Cabrio** '95 thru '02
96018 **Golf, GTI, Jetta** '99 thru '05
96019 **Jetta, Rabbit, GLI, GTI & Golf** '05 thru '11
96020 **Rabbit, Jetta & Pick-up** diesel '77 thru '84
96021 **Jetta** '11 thru '18 & **Golf** '15 thru '19
96023 **Passat** '98 thru '05 & **Audi A4** '96 thru '01
96030 **Transporter 1600** all models '68 thru '79
96035 **Transporter 1700, 1800 & 2000** '72 thru '79
96040 **Type 3 1500 & 1600** all models '63 thru '73
96045 **Vanagon Air-Cooled** all models '80 thru '83

VOLVO
97010 **120, 130 Series & 1800 Sports** '61 thru '73
97015 **140 Series** all models '66 thru '74
97020 **240 Series** all models '76 thru '93
97040 **740 & 760 Series** all models '82 thru '88
97050 **850 Series** all models '93 thru '97

TECHBOOK MANUALS
10205 **Automotive Computer Codes**
10206 **OBD-II & Electronic Engine Management**
10210 **Automotive Emissions Control Manual**
10215 **Fuel Injection Manual** '78 thru '85
10225 **Holley Carburetor Manual**
10230 **Rochester Carburetor Manual**
10305 **Chevrolet Engine Overhaul Manual**
10320 **Ford Engine Overhaul Manual**
10330 **GM and Ford Diesel Engine Repair Manual**
10331 **Duramax Diesel Engines** '01 thru '19
10332 **Cummins Diesel Engine Performance Manual**
10333 **GM, Ford & Chrysler Engine Performance Manual**
10334 **GM Engine Performance Manual**
10340 **Small Engine Repair Manual,** 5 HP & Less
10341 **Small Engine Repair Manual,** 5.5 thru 20 HP
10345 **Suspension, Steering & Driveline Manual**
10355 **Ford Automatic Transmission Overhaul**
10360 **GM Automatic Transmission Overhaul**
10405 **Automotive Body Repair & Painting**
10410 **Automotive Brake Manual**
10411 **Automotive Anti-lock Brake (ABS) Systems**
10420 **Automotive Electrical Manual**
10425 **Automotive Heating & Air Conditioning**
10435 **Automotive Tools Manual**
10445 **Welding Manual**
10450 **ATV Basics**

 Over a 100 Haynes motorcycle manuals also available

10/22

Haynes North America, Inc. • (805) 498-6703 • www.haynes.com